APPLIED CALCULUS

Sixth Edition

APPLIED CALCULUS

Sixth Edition

Produced by the Calculus Consortium and initially funded by a National Science Foundation Grant.

Deborah Hughes-Hallett
University of Arizona

Patti Frazer Lock
St. Lawrence University

Daniel E. Flath
Macalester College

Andrew M. Gleason
Harvard University

Guadalupe I. Lozano
University of Arizona

Karen Rhea
University of Michigan

Eric Connally
Harvard University Extension

William G. McCallum
University of Arizona

Ayşe Şahin
Wright State University

Selin Kalaycıoğlu
New York University

Brad G. Osgood
Stanford University

Adam H. Spiegler
Loyola University Chicago

Brigitte Lahme
Sonoma State University

Cody L. Patterson
University of Texas at San Antonio

Jeff Tecosky-Feldman
Haverford College

David O. Lomen
University of Arizona

Douglas Quinney
University of Keele

Thomas W. Tucker
Colgate University

David Lovelock
University of Arizona

Aaron D. Wootton
University of Portland

with the assistance of
Otto K. Bretscher
Colby College

Coordinated by
Elliot J. Marks

John Wiley & Sons, Inc.

VICE PRESIDENT AND DIRECTOR	Laurie Rosatone
ACQUISITIONS EDITOR	Shannon Corliss
DEVELOPMENTAL EDITOR	Adria Giattino
SENIOR PRODUCT DESIGNER	David Dietz
FREELANCE DEVELOPMENTAL EDITOR	Anne Scanlan-Rohrer / Two Ravens Editorial
FREELANCE EDITOR	Teresa Ward
MARKETING MANAGER	John LaVacca
SENIOR PRODUCTION EDITOR	Laura Abrams
SENIOR EDITORIAL ASSISTANT	Giana Milazzo
DEVELOPMENTAL EDITORIAL ASSISTANT	Kimberly Eskin
COVER DESIGNER	Maureen Eide
COVER AND CHAPTER OPENING PHOTO	©Patrick Zephyr/Patrick Zephyr Nature Photography

Problems from *Calculus: The Analysis of Functions*, by Peter D. Taylor (Toronto: Wall & Emerson, Inc., 1992). Reprinted with permission of the publisher.

This book was set in Times Roman by the Consortium using TeX, Mathematica, and the package AsTeX, which was written by Alex Kasman. It was printed and bound by Quad Graphics / Versailles. The cover was printed by Quad Graphics / Versailles.

This book is printed on acid-free paper.

Founded in 1807, John Wiley & Sons, Inc. has been a valued source of knowledge and understanding for more than 200 years, helping people around the world meet their needs and fulfill their aspirations. Our company is built on a foundation of principles that include responsibility to the communities we serve and where we live and work. In 2008, we launched a Corporate Citizenship Initiative, a global effort to address the environmental, social, economic, and ethical challenges we face in our business. Among the issues we are addressing are carbon impact, paper specifications and procurement, ethical conduct within our business and among our vendors, and community and charitable support. For more information, please visit our website: www.wiley.com/go/citizenship.

This material is based upon work supported by the National Science Foundation under Grant No. DUE-9352905. Opinions expressed are those of the authors and not necessarily those of the Foundation.

ISBN-13: 978-1-119-39935-3

The inside back cover will contain printing identification and country of origin if omitted from this page. In addition, if the ISBN on the back cover differs from the ISBN on this page, the one on the back cover is correct.

Printed in the United States of America

V10015140_103119

Hughes-Hallett • Gleason • Lock • Flath • et al.

APPLIED CALCULUS

SIXTH EDITION

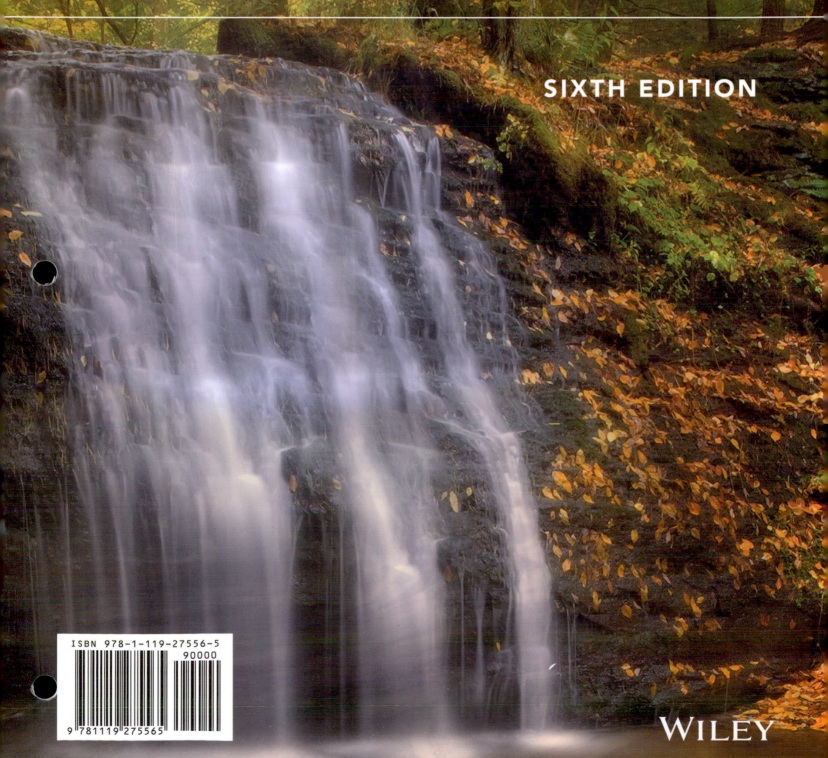

ISBN 978-1-119-27556-5
90000

9 781119 275565

WILEY

We dedicate this book to Andrew M. Gleason.

His brilliance and the extraordinary kindness and dignity with which he treated others made an enormous difference to us, and to many, many people. Andy brought out the best in everyone.

Deb Hughes Hallett
for the Calculus Consortium

PREFACE

Calculus is one of the greatest achievements of the human intellect. Inspired by problems in astronomy, Newton and Leibniz developed the ideas of calculus 300 years ago. Since then, each century has demonstrated the power of calculus to illuminate questions in mathematics, the physical sciences, engineering, business, and the social and biological sciences.

Calculus has been so successful because of its extraordinary power to reduce complicated problems to simple rules and procedures. Therein lies the danger in teaching calculus: it is possible to teach the subject as nothing but the rules and procedures—thereby losing sight of both the mathematics and of its practical value. This edition of *Applied Calculus* continues our effort to promote courses in which understanding reinforces computation.

Embracing e-Learning

Paper books are playing a smaller role in courses than in the past and are being replaced by electronic materials. This Sixth Edition provides opportunities for students to experience the concepts of calculus in ways that would not be possible in a traditional textbook. The enhanced e-text of *Applied Calculus*, powered by VitalSource, provides embedded videos and the complete solutions from the Student Solutions Manual. The enhanced e-text also contains additional content not found in the print edition:

- Worked example **videos**, which provide students the opportunity to see and hear over one hundred of the book's examples being explained and worked out in detail, have been created to accompany the sixth edition.
- Strengthen Your Understanding true/false problems that focus on conceptual understanding.
- Appendices that extend ideas in the course.
- Chapter 10, on Geometric Series.
- Chapter summaries, giving a concise overview of each chapter.

WileyPLUS

In addtion to the enhanced e-text, Students and instructors can access a wide variety of rescources through WileyPLUS with ORION, Wiley's digital learning environment. ORION Learning provides an adaptive, personalized learning experience that delivers easy-to-use analytics so instructors and students can see exactly where they're excelling and where they need help. WileyPLUS with ORION features the following resources:

- Homework management tools, which enable the instructor to assign questions easily and grade them automatically, using a rich set of options and controls.
- QuickStart pre-designed reading and homework assignments. Use them as-is or customize them to fit the needs of your classroom.
- Intelligent Tutoring questions, in which students are prompted for responses as they step through a problem solution and receive targeted feedback based on those responses.
- Algebra Refresher material, delivered through ORION, provides students with an opportunity to brush up on material necessary to master Calculus, as well as to determine areas that require further review.
- Graphing Calculator Manual, to help students get the most out of their graphing calculator, and to show how they can apply the numerical and graphing functions of their calculators to their study of calculus.

Flexible Balance: Concepts and Modeling

The first goal of a calculus course is to acquire a clear intuitive picture of the central ideas. After this foundation has been laid, there is a choice of direction. All students benefit from both mathematical concepts and modeling, but the balance may differ for different groups of students. For instructors wishing to emphasize the connection between calculus and other fields, the text includes:

- A variety of problems and examples from the **biological sciences**, **economics**, and **business**.
- Models from the **health sciences** and of **population growth**.
- Problems on **sustainability**.
- Case studies on **medicine** by David E. Sloane, MD.

Active Learning: Good Problems

As instructors ourselves, we know that interactive classrooms and well-crafted problems promote student learning. Since its inception, the hallmark of our work has been its innovative and engaging problems. These problems probe student understanding in ways often taken for granted. Praised for their creativity and variety, the influence of these problems has extended far beyond the users of our textbook.

The Sixth Edition continues this tradition. Under our approach, which we called the "Rule of Four," ideas are presented graphically, numerically, symbolically, and verbally, thereby encouraging students with a variety of learning styles to deepen their understanding. This edition continues to provide a wide variety of problem types:

- End of **Section Problems** reinforce the ideas of that section and make connections with earlier sections.
- **ConcepTests** promote active learning in the classroom. These can be used with or without any polling software, and have been shown to dramatically improve student learning. ConcepTests are particularly useful to instructors teaching in a flipped classroom. ConcepTests are available online for instructors in PPT or PDF format in WileyPLUS or on the Instructor Book Companion site at www.wiley.com/college/hughes-hallett.
- **Chapter Review Problems**, reserved for instructor-only use in WileyPLUS, provide opportunities to review ideas from the whole chapter.
- **Projects** for each chapter provide opportunities for a sustained investigation, often using skills from different parts of the course. These include business applications, issues in sustainability, and medical case studies based on clinical practice.
- True-False **Strengthen Your Understanding** questions, available online for every chapter, enable students to check their progress.
- **Spreadsheet Projects** in the online Appendix provide the opportunity for students to develop their spreadsheet skills while deepening their understanding of functions and calculus.
- **Focus on Practice** exercises at the end of Chapter 3 and 6 (Derivatives and Antiderivatives) build student skill and confidence.

Origin of the Text: A Community of Instructors

This text, like others we write, draws on the experience of a diverse group of authors and users. We have benefitted enormously from input from a broad spectrum of instructors—at research universities, four-year colleges, community colleges, and secondary schools. For *Applied Calculus*, the contributions of colleagues in biology, economics, medicine, business, and other life and social sciences have been equally central to the development of the text. It is the collective wisdom of this community of mathematicians, teachers, and natural and social scientists that forms the basis for the new edition.

What Student Background is Expected?

This book is intended for students in business, the social sciences, and the life sciences. A background in trigonometry is *not* required; the sections involving trigonometry are optional.

We have found the material to be thought-provoking for well-prepared students while still accessible to students with limited algebra backgrounds. Providing numerical and graphical approaches as well as the algebraic gives students several ways of mastering the material. This approach encourages students to persist, thereby lowering failure rates. A pre-test over background material is available at the student book companion site: www.wiley.com/college/hughes-hallett. An ORION algebra refresher is available in WileyPLUS.

Mathematical Skills: A Balance Between Symbolic Manipulation and Technology

To use calculus effectively, students need familiarity with both symbolic manipulation and the use of technology. The balance between them may vary, depending on the needs of the students and the wishes of the instructor. The book is adaptable to many different combinations.

The book does not require any specific software or technology. Students may use whatever is readily available—a graphing calculator or online tools.

The Sixth Edition

Because different users often choose very different topics to cover in a one-semester applied calculus course, we have designed this book for either a one-semester course (with much flexibility in choosing topics) or a two-semester course. Sample syllabi are provided in the Instructor's Manual.

The sixth edition has the same vision as previous editions. In preparing this edition, we solicited comments from a large number of mathematics instructors who had used the text. We continued to discuss with our colleagues in client disciplines the mathematical needs of their students. We were offered many valuable suggestions, which we have tried to incorporate, while maintaining our original commitment to a focused treatment of a limited number of topics. The changes we have made include:

- About 400 **additional problems** have been added to the WileyPLUS course, expanding instructors' options for online homework assignments.
- **Worked example videos** have been added for every section in the text.
- **Updated data** and **fresh applications** throughout the book, including
  - Problems on **sustainability**.
  - Case studies on **medicine** by David E. Sloane, MD.
- Many **new problems** have been added, designed to build student confidence with basic concepts and to reinforce skills.
- As in the previous edition, a **Pre-test** is included for students whose skills may need a refresher prior to taking the course. It is available online at www.wiley.com/college/hughes-hallett.

Content

This content represents our vision of how applied calculus can be taught. It is flexible enough to accommodate individual course needs and requirements. Topics can easily be added or deleted, or the order changed.

Chapter 1: Functions and Change

Chapter 1 introduces the concept of a function and the idea of change, including the distinction between total change, rate of change, and relative change. All elementary functions are introduced here. Although the functions are probably familiar, the graphical, numerical, verbal, and modeling approach to them is likely to be new. We introduce exponential functions early, since they are fundamental to the understanding of real-world processes. The trigonometric functions are optional.

Chapter 2: Rate of Change: The Derivative

Chapter 2 presents the key concept of the derivative according to the Rule of Four. The purpose of this chapter is to give the student a practical understanding of the meaning of the derivative and its interpretation as an instantaneous rate of change. Students will learn how the derivative can be used to represent relative rates of change. After finishing this chapter, a student will be able to approximate derivatives numerically by taking difference quotients, visualize derivatives graphically as the slope of the graph, and interpret the meaning of first and second derivatives in various applications. The student will also understand the concept of marginality and recognize the derivative as a function in its own right.

Focus on Theory: This section discusses limits and continuity and presents the symbolic definition of the derivative.

Chapter 3: Short-Cuts to Differentiation

The derivatives of all the functions in Chapter 1 are introduced, as well as the rules for differentiating products, quotients, and composite functions. Students learn how to find relative rates of change using logarithms.

Focus on Theory: This section uses the definition of the derivative to obtain the differentiation rules.

Focus on Practice: This section provides a collection of differentiation problems for skill-building.

Chapter 4: Using the Derivative

The aim of this chapter is to enable the student to use the derivative in solving problems, including optimization and graphing. It is not necessary to cover all the sections.

Chapter 5: Accumulated Change: The Definite Integral

Chapter 5 presents the key concept of the definite integral, in the same spirit as Chapter 2. The purpose of this chapter is to give the student a practical understanding of the definite integral as a limit of Riemann sums, and to bring out the connection between the derivative and the definite integral in the Fundamental Theorem of Calculus. We use the same method as in Chapter 2, introducing the fundamental concept in depth without going into technique. The student will finish the chapter with the ability to approximate a definite integral numerically and interpret it graphically. The chapter includes applications of definite integrals in a variety of contexts, including the average value of a function.

Chapter 5 can be covered immediately after Chapter 2 without difficulty.

Focus on Theory: This section presents the Second Fundamental Theorem of Calculus and the properties of the definite integral.

Chapter 6: Antiderivatives and Applications

This chapter covers antiderivatives from a graphical, numerical, and algebraic point of view. The Fundamental Theorem of Calculus is used to evaluate definite integrals.

Sections 6.4–6.7 are optional. Application sections are included on consumer and producer surplus and on present and future value; the integrals in these sections can be evaluated numerically or using the Fundamental Theorem. The chapter concludes with sctions on integration by substitution and integration by parts.

Focus on Practice: This section provides a collection of integration problems for skill-building.

Chapter 7: Probability

This chapter covers probability density functions, cumulative distribution functions, the median and the mean.

Chapter 8: Functions of Several Variables

This chapter introduces functions of two variables from several points of view, using contour diagrams, formulas, and tables. It gives students the skills to read contour diagrams and think graphically, to read tables and think numerically, and to apply these skills, along with their algebraic skills, to modeling. The idea of the partial derivative is introduced from graphical, numerical, and symbolic viewpoints. Partial derivatives are then applied to optimization problems, ending with a discussion of constrained optimization using Lagrange multipliers.

Focus on Theory: This section uses optimization to derive the formula for the regression line.

Chapter 9: Mathematical Modeling Using Differential Equations

This chapter introduces differential equations. The emphasis is on modeling, qualitative solutions, and interpretation. This chapter includes applications of systems of differential equations to population models, the spread of disease, and predator-prey interactions.

Focus on Theory: This section explains the technique of separation of variables.

Chapter 10: Geometric Series (Available online and in the e-text)

This chapter covers geometric series and their applications to business, economics, and the life sciences.

Appendices (Available online and in the e-text)

Appendix A introduces the student to fitting formulas to data; Appendix B provides further discussion of compound interest and the definition of the number e. Appendix C contains selection of spreadsheet projects.

Supplementary Materials

Supplements for the instructor can be obtained online at the book companion site or by contacting your Wiley representative. The following supplementary materials are available for this edition:

- **Instructor's Manual** containing teaching tips, sample syllabi, calculator programs, and overhead transparency masters, also available as PowerPoint slides.

- **Instructor's Solution Manual** with complete solutions to all problems.

- **Student's Solution Manual** with complete solutions to half the odd-numbered problems.

- **Additional Material for Instructors**, elaborating specially marked points in the text, lecture notes, and course notes, as well as password protected electronic versions of the instructor ancillaries, can be found on the web at the book companion site: www.wiley.com/college/hughes-hallett.

- **Additional Material for Students**, including an algebra refresher, is available via WileyPLUS.

ConcepTests

ConcepTests, or clicker questions, modeled on the pioneering work of Harvard physicist Eric Mazur, are questions designed to promote active learning during class, particularly (but not exclusively) in large lectures. Evaluation data showed that students taught with ConcepTests outperformed students taught by traditional lecture methods 73% versus 17% on conceptual questions, and 63% versus 54% on computational problems.[1] A supplement to *Applied Calculus*, 6th edn, containing ConcepTests by section, is available through WileyPLUS and at the book companion site, www.wiley.com/college/hughes-hallett.

WileyPLUS Studio

The WileyPLUS studio is an online community that brings WileyPLUS users together in an engaging, virtual environment. It's a space where you can share insights, identify best practices, provide product feedback, learn from peers and get rewarded for your efforts.

Acknowledgements

First and foremost, we want to express our appreciation to the National Science Foundation for their faith in our ability to produce a revitalized calculus curriculum and, in particular, to Louise Raphael, John Kenelly, John Bradley, Bill Haver, and James Lightbourne. We also want to thank the members of our Advisory Board, Benita Albert, Lida Barrett, Bob Davis, Lovenia DeConge-Watson, John Dossey, Ron Douglas, Don Lewis, Seymour Parter, John Prados, and Steve Rodi for their ongoing guidance and advice.

---

[1]"Peer Instruction in Physics and Mathematics" by Scott Pilzer in *Primus*, Vol XI, No 2, June 2001. At the start of Calculus II, students earned 73% on conceptual questions and 63% on computational questions if they were taught with ConcepTests in Calculus I; 17% and 54% otherwise.

In addition, we want to thank all the people across the country who encouraged us to write this book and who offered so many helpful comments. We would like to thank the following people, for all that they have done to help our project succeed: Enrique Acosta, Ruth Baruth, Graeme Bird, Jeanne Bowman, Lucille Buonocore, Scott Clark, Jeff Edmunds, Sunny Fawcett, Lynn Garner, Sheldon P. Gordon, Ole Hald, Jenny Harrison, Adrian Iovita, Thomas Judson, Christopher Kennedy, Hannah Knight, Donna Krawczyk, Suzanne Lenhart, Madelyn Lesure, Kevin Martin, Rosalind Horn Martin, Georgia Kamvosoulis Mederer, Nolan Miller, David Muñoz Ramírez, Andrew Pasquale, Richard D. Porter, Laurie Rosatone, Kenneth Santor, Anne Scanlan-Rohrer, Alfred Schipke, Virginia Stallings, Ralph Teixeira, Joe B. Thrash, J. Jerry Uhl, Rachel Deyette Werkema, Hannah Winkler, and Hung-Hsi Wu.

Reports from the following reviewers were most helpful in shaping the sixth edition:

Jill E Guerra, Nicole Williams, Ben Wehrung, Lauren Fern, Dipa Sarkar-Dey, Alicia Frost, Pamela D Nemeth, Alice Deanin, Pam Crawford, Christopher Dona, Victor Roeske, Steven Leonhardi, and Christopher Goodrich.

Reports from the following reviewers were most helpful in shaping the fifth edition:

Anthony Barcellos, Catherine Benincasa, Bill Blubagh, Carol Demas, Darlene Diaz, Lauren Fern, Wesley Griffith, Juill Guerra, Molly Martin, Rebecca McKay, Barry Peratt, Karl Schaffer, Randy Scott, Paul Vicknair, Tracy Whelan, P. Jay Zeltner.

Deborah Hughes-Hallett	David O. Lomen	Karen Rhea
Patti Frazer Lock	David Lovelock	Ayşe Şahin
Daniel E. Flath	Guadalupe I. Lozano	Adam Spiegler
Andrew M. Gleason	William G. McCallum	Jeff Tecosky-Feldman
Eric Connally	Brad G. Osgood	Thomas W. Tucker
Selin Kalaycıoğlu	Cody L. Patterson	Aaron D. Wooton
Brigitte Lahme	Douglas Quinney	

APPLICATIONS INDEX

Life Sciences and Ecology

To Students: How to Learn from this Book

- This book may be different from other math textbooks that you have used, so it may be helpful to know about some of the differences in advance. At every stage, this book emphasizes the *meaning* (in practical, graphical or numerical terms) of the symbols you are using. There is much less emphasis on "plug-and-chug" and using formulas, and much more emphasis on the interpretation of these formulas than you may expect. You will often be asked to explain your ideas in words or to explain an answer using graphs.

- The book contains the main ideas of calculus in plain English. Success in using this book will depend on reading, questioning, and thinking hard about the ideas presented. It will be helpful to read the text in detail, not just the worked examples.

- There are few examples in the text that are exactly like the homework problems, so homework problems can't be done by searching for similar–looking "worked out" examples. Success with the homework will come by grappling with the ideas of calculus.

- For many problems in the book, there is more than one correct approach and more than one correct solution. Sometimes, solving a problem relies on common sense ideas that are not stated in the problem explicitly but which you know from everyday life.

- Some problems in this book assume that you have access to a graphing calculator or computer. There are many situations where you may not be able to find an exact solution to a problem, but you can use a calculator or computer to get a reasonable approximation.

- This book attempts to give equal weight to four methods for describing functions: graphical (a picture), numerical (a table of values), algebraic (a formula), and verbal (words). Sometimes it's easier to translate a problem given in one form into another. For example, you might replace the graph of a parabola with its equation, or plot a table of values to see its behavior. It is important to be flexible about your approach: if one way of looking at a problem doesn't work, try another.

- Students using this book have found discussing these problems in small groups helpful. There are a great many problems which are not cut-and-dried; it can help to attack them with the other perspectives your colleagues can provide. If group work is not feasible, see if your instructor can organize a discussion session in which additional problems can be worked on.

- You are probably wondering what you'll get from the book. The answer is, if you put in a solid effort, you will get a real understanding of one of the crowning achievements of human creativity—calculus—as well as a real sense of the power of mathematics in the age of technology.

CONTENTS

6 ANTIDERIVATIVES AND APPLICATIONS   281

7 PROBABILITY   321

8 FUNCTIONS OF SEVERAL VARIABLES   339

9 MATHEMATICAL MODELING USING DIFFERENTIAL EQUATIONS   391

10 GEOMETRIC SERIES   *Digital*

APPENDICES   *Digital*

Chapter 1

FUNCTIONS AND CHANGE

CONTENTS

1.1   WHAT IS A FUNCTION?

In mathematics, a *function* is used to represent the dependence of one quantity upon another.

Let's look at an example. In 2015, Boston, Massachusetts, had the highest annual snowfall, 110.6 inches, since recording started in 1872. Table 1.1 shows one 14-day period in which the city broke another record with a total of 64.4 inches.[1]

Table 1.1   *Daily snowfall in inches for Boston, January 27 to February 9, 2015*

Day	1	2	3	4	5	6	7	8	9	10	11	12	13	14
Snowfall	22.1	0.2	0	0.7	1.3	0	16.2	0	0	0.8	0	0.9	7.4	14.8

You may not have thought of something so unpredictable as daily snowfall as being a function, but it *is* a function of day, because each day gives rise to one snowfall total. There is no formula for the daily snowfall (otherwise we would not need a weather bureau), but nevertheless the daily snowfall in Boston does satisfy the definition of a function: Each day, t, has a unique snowfall, S, associated with it.

We define a function as follows:

> A **function** is a rule that takes certain numbers as inputs and assigns to each a definite output number. The set of all input numbers is called the **domain** of the function and the set of resulting output numbers is called the **range** of the function.

The input is called the *independent variable* and the output is called the *dependent variable*. In the snowfall example, the domain is the set of days $\{1, 2, 3, 4, 5, 6, 7, 8, 9, 10, 11, 12, 13, 14\}$ and the range is the set of daily snowfalls $\{0, 0.2, 0.7, 0.8, 0.9, 1.3, 7.4, 14.8, 16.2, 22.1\}$. We call the function f and write $S = f(t)$. Notice that a function may have identical outputs for different inputs (Days 8 and 9, for example).

Some quantities, such as a day or date, are *discrete*, meaning they take only certain isolated values (days must be integers). Other quantities, such as time, are *continuous* as they can be any number. For a continuous variable, domains and ranges are often written using interval notation:

The set of numbers t such that $a \leq t \leq b$ is called a *closed interval* and written $[a, b]$.

The set of numbers t such that $a < t < b$ is called an *open interval* and written (a, b).

The Rule of Four: Tables, Graphs, Formulas, and Words

Functions can be represented by tables, graphs, formulas, and descriptions in words. For example, the function giving the daily snowfall in Boston can be represented by the graph in Figure 1.1, as well as by Table 1.1.

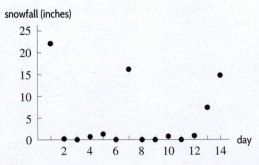

Figure 1.1: Boston snowfall, starting January 27, 2015

[1] http://w2.weather.gov/climate/xmacis.php?wfo=box. Accessed June 2015.

Other functions arise naturally as graphs. Figure 1.2 contains electrocardiogram (EKG) pictures showing the heartbeat patterns of two patients, one normal and one not. Although it is possible to construct a formula to approximate an EKG function, this is seldom done. The pattern of repetitions is what a doctor needs to know, and these are more easily seen from a graph than from a formula. However, each EKG gives electrical activity as a function of time.

Functions, such as the EKG, whose graphs can be drawn without lifting pencil from paper are called *continuous functions*. See the Focus on Theory section on page 130 for details.

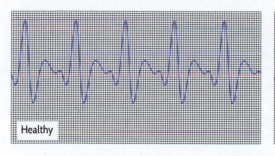
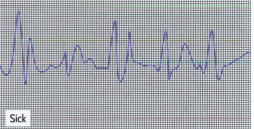

Figure 1.2: EKG readings on two patients

As another example of a function, consider the snowy tree cricket. Surprisingly enough, all such crickets chirp at essentially the same rate if they are at the same temperature. That means that the chirp rate is a function of temperature. In other words, if we know the temperature, we can determine the chirp rate. Even more surprisingly, the chirp rate, C, in chirps per minute, increases steadily with the temperature, T, in degrees Fahrenheit, and can be computed, to a fair degree of accuracy, using the formula

$$C = f(T) = 4T - 160.$$

Mathematical Modeling

A *mathematical model* is a mathematical description of a real situation. In this book we consider models that are functions, such as $C = f(T) = 4T - 160$.

Modeling almost always involves some simplification of reality. We choose which variables to include and which to ignore—for example, we consider the dependence of chirp rate on temperature, but not on other variables. The choice of variables is based on knowledge of the context (the biology of crickets, for example), not on mathematics. To test the model, we compare its predictions with observations.

In this book, we often model a situation that has a discrete domain with a continuous function whose domain is an interval of numbers. For example, the annual US gross domestic product (GDP) has a value for each year, $t = 0, 1, 2, 3, \ldots$. We may model it by a function of the form $G = f(t)$, with values for t in a continuous interval. In doing this, we expect that the values of $f(t)$ match the values of the GDP at the points $t = 0, 1, 2, 3, \ldots$, and that information obtained from $f(t)$ closely matches observed values.

Used judiciously, a mathematical model captures trends in the data to enable us to analyze and make predictions. A common way of finding a model is described in Appendix A.

Function Notation and Intercepts

We write $y = f(t)$ to express the fact that y is a function of t. The independent variable is t, the dependent variable is y, and f is the name of the function. The graph of a function has an *intercept* where it crosses the horizontal or vertical axis. Horizontal intercepts are also called the *zeros* of the function.

Example 1   (a) Graph the cricket chirp rate function, $C = f(T) = 4T - 160$.
(b) Solve $f(T) = 0$ and interpret the result.

Solution   (a) The graph is in Figure 1.3.
(b) Solving $f(T) = 0$ gives the horizontal intercept:

$$4T - 160 = 0$$
$$T = \frac{160}{4} = 40.$$

Thus at a temperature of 40°F, the chirp rate is zero.
    For temperatures below 40°F, the model would predict negative values of C, so we conclude that the model does not apply for such temperature values.

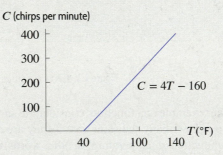

Figure 1.3: Cricket chirp rate as a function of temperature

Example 2   The value of a car, V, is a function of the age of the car, a, so $V = g(a)$, where g is the name we are giving to this function.

(a) Interpret the statement $g(5) = 9.97$ in terms of the value of a car if V is in thousands of dollars and a is in years.
(b) In the same units, the value of a Honda[2] is approximated by $g(a) = 18.97 - 1.8a$. Find and interpret the vertical and horizontal intercepts of the graph of this depreciation function g.

Solution   (a) Since $V = g(a)$, the statement $g(5) = 9.97$ means $V = 9.97$ when $a = 5$. This tells us that the car is worth $9970 when it is 5 years old.
(b) Since $V = g(a)$, a graph of the function g has the value of the car on the vertical axis and the age of the car on the horizontal axis. The vertical intercept is the value of V when $a = 0$. It is $V = g(0) = 18.97$, so the Honda was valued at $18,970 when new. The horizontal intercept is the value of a such that $g(a) = 0$, so

$$18.97 - 1.8a = 0$$
$$a = \frac{18.97}{1.8} = 10.54.$$

At age 10 years, the Honda has no value.

Increasing and Decreasing Functions

In the previous examples, the chirp rate increases with temperature, while the value of the Honda decreases with age. We express these facts saying that f is an increasing function, while g is decreasing. See Figure 1.4. In general:

---

[2]Data obtained from the Kelley Blue Book, based on a 2015 Honda Accord sedan. www.kbb.com, accessed September 14, 2016.

A function f is **increasing** if the values of $f(x)$ increase as x increases.
A function f is **decreasing** if the values of $f(x)$ decrease as x increases.

The graph of an *increasing* function *climbs* as we move from left to right.
The graph of a *decreasing* function *descends* as we move from left to right.

Figure 1.4: Increasing and decreasing functions

Problems for Section 1.1

1. Which graph in Figure 1.5 best matches each of the following stories?[3] Write a story for the remaining graph.

   (a) I had just left home when I realized I had forgotten my books, so I went back to pick them up.
   (b) Things went fine until I had a flat tire.
   (c) I started out calmly but sped up when I realized I was going to be late.

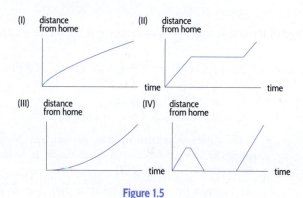

Figure 1.5

■ In Problems 2–5, use the description of the function to sketch a possible graph. Put a label on each axis and state whether the function is increasing or decreasing.

2. The height of a sand dune is a function of time, and the wind erodes away the sand dune over time.

3. The amount of carbon dioxide in the atmosphere is a function of time, and is going up over time.

4. The number of air conditioning units sold is a function of temperature, and goes up as the temperature goes up.

5. The noise level, in decibels, is a function of distance from the source of the noise, and the noise level goes down as the distance increases.

6. The population of Washington DC grew from 1900 to 1950, stayed approximately constant during the 1950s, and decreased from about 1960 to 2005. Graph the population as a function of years since 1900.

7. Let $W = f(t)$ represent wheat production in Argentina,[4] in millions of metric tons, where t is years since 2010. Interpret the statement $f(5) = 49.2$ in terms of wheat production.

8. The chirp rate, C, in chirps per minute, of the snowy tree cricket is given by $C = f(T) = 4T - 160$ where T is degrees Fahrenheit.

   (a) Find an appropriate domain of f in the context of the model assuming a maximum temperature of $134°F$, the highest recorded at a weather station.[5]
   (b) Find the range of f on this domain.

9. The concentration of carbon dioxide, $C = f(t)$, in the atmosphere, in parts per million (ppm), is a function of years, t, since 2000.

   (a) Interpret $f(15) = 400$ in terms of carbon dioxide.[6]
   (b) What is the meaning of $f(20)$?

[3] Adapted from Jan Terwel, "Real Math in Cooperative Groups in Secondary Education." *Cooperative Learning in Mathematics*, ed. Neal Davidson, p. 234 (Reading: Addison Wesley, 1990).
[4] www.world-grain.com, accessed September 14, 2016.
[5] http://www.guinnessworldrecords.com. Accessed January 2017.
[6] en.wikipedia.org, accessed September 14, 2016.

10. (a) The graph of $r = f(p)$ is in Figure 1.6. What is the value of r when p is 0? When p is 3?

(b) What is $f(2)$?

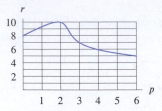

Figure 1.6

■ For the functions in Problems **11–15**, find $f(5)$.

11. $f(x) = 2x + 3$ **12.** $f(x) = 10x - x^2$

13.

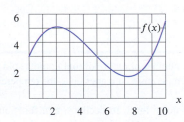

14.

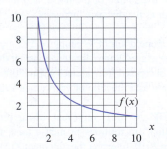

15.

x	1	2	3	4	5	6	7	8
$f(x)$	2.3	2.8	3.2	3.7	4.1	5.0	5.6	6.2

16. Let $y = f(x) = x^2 + 2$.

(a) Find the value of y when x is zero.

(b) What is $f(3)$?

(c) What values of x give y a value of 11?

(d) Are there any values of x that give y a value of 1?

■ In Problems **17–20** the function $S = f(t)$ gives the average annual sea level, S, in meters, in Aberdeen, Scotland,[7] as a function of t, the number of years before 2012. Write a mathematical expression that represents the given statement.

17. In 2000 the average annual sea level in Aberdeen was 7.049 meters.

18. The average annual sea level in Aberdeen in 2012.

19. The average annual sea level in Aberdeen was the same in 1949 and 2000.

[7]www.gov.uk, accessed January 7, 2015.

20. The average annual sea level in Aberdeen decreased by 8 millimeters from 2011 to 2012.

21. (a) A potato is put in an oven to bake at time $t = 0$. Which of the graphs in Figure 1.7 could represent the potato's temperature as a function of time?

(b) What does the vertical intercept represent in terms of the potato's temperature?

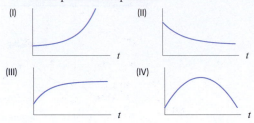

Figure 1.7

22. An object is put outside on a cold day at time $t = 0$ minutes. Its temperature, $H = f(t)$, in °C, is graphed in Figure 1.8.

(a) What does the statement $f(30) = 10$ mean in terms of temperature? Include units for 30 and for 10 in your answer.

(b) Explain what the vertical intercept, a, and the horizontal intercept, b, represent in terms of temperature of the object and time outside.

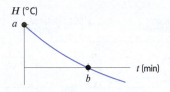

Figure 1.8

23. In the Andes mountains in Peru, the number, N, of species of bats is a function of the elevation, h, in feet above sea level, so $N = f(h)$.

(a) Interpret the statement $f(500) = 100$ in terms of bat species.

(b) What are the meanings of the vertical intercept, k, and horizontal intercept, c, in Figure 1.9?

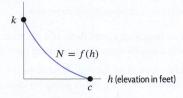

Figure 1.9

24. In tide pools on the New England coast, snails eat algae. Describe what Figure 1.10 tells you about the effect of snails on the diversity of algae.[8] Does the graph support the statement that diversity peaks at an intermediate number of snails per square meter?

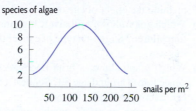

Figure 1.10

25. Figure 1.11 shows the amount of nicotine, $N = f(t)$, in mg, in a person's bloodstream as a function of the time, t, in hours, since the person finished smoking a cigarette.

(a) Estimate $f(3)$ and interpret it in terms of nicotine.
(b) About how many hours have passed before the nicotine level is down to 0.1 mg?
(c) What is the vertical intercept? What does it represent in terms of nicotine?
(d) If this function had a horizontal intercept, what would it represent?

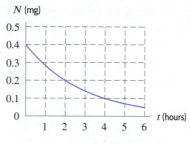

Figure 1.11

26. A deposit is made into an interest-bearing account. Figure 1.12 shows the balance, B, in the account t years later.

(a) What was the original deposit?
(b) Estimate $f(10)$ and interpret it.
(c) When does the balance reach $5000?

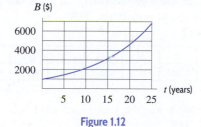

Figure 1.12

27. The use of CFCs (chlorofluorocarbons) has declined since the 1987 Montreal Protocol came into force to reduce the use of substances that deplete the ozone layer. World annual CFC consumption, $C = f(t)$, in million tons, is a function of time, t, in years since 1987. (CFCs are measured by the weight of ozone that they could destroy.)

(a) Interpret $f(10) = 0.2$ in terms of CFCs.[9]
(b) Interpret the vertical intercept of the graph of this function in terms of CFCs.
(c) Interpret the horizontal intercept of the graph of this function in terms of CFCs.

28. When the US Federal government spends more money than it takes in, the amount of the difference is called the deficit. In Figure 1.13, the function $D = f(t)$ gives a good approximation of the deficit from 2011 to 2015, where D is the deficit in billions of dollars and t is the number of years since 2010.

(a) Is the deficit increasing or decreasing over this time period?
(b) Interpret $f(4) = 485$ in terms of the deficit.
(c) In Figure 1.13, label the axes and put a dot at the point identified in part (b).
(d) If we extended the domain of this function to include a vertical intercept, what would the vertical intercept represent?
(e) If we extended the domain of this function to include a horizontal intercept, what would the horizontal intercept represent?

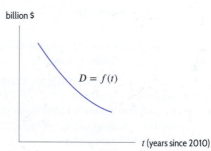

Figure 1.13

29. When a patient with a rapid heart rate takes a drug, the heart rate plunges dramatically and then slowly rises again as the drug wears off. Sketch the heart rate against time from the moment the drug is administered.

30. The gas mileage of a car (in miles per gallon) is highest when the car is going about 45 miles per hour and is lower when the car is going faster or slower than 45 mph. Graph gas mileage as a function of speed of the car.

[8]Rosenzweig, M.L., *Species Diversity in Space and Time*, p. 343 (Cambridge: Cambridge University Press, 1995).
[9]The Worldwatch Institute, *Vital Signs 2007-2008* (New York: W.W. Norton & Company, 2007), p. 47.

31. After an injection, the concentration of a drug in a patient's body increases rapidly to a peak and then slowly decreases. Graph the concentration of the drug in the body as a function of the time since the injection was given. Assume that the patient has none of the drug in the body before the injection. Label the peak concentration and the time it takes to reach that concentration.

32. Financial investors know that, in general, the higher the expected rate of return on an investment, the higher the corresponding risk.

   (a) Graph this relationship, showing expected return as a function of risk.
   (b) On the figure from part (a), mark a point with high expected return and low risk. (Investors hope to find such opportunities.)

33. The number of sales per month, S, is a function of the amount, a (in dollars), spent on advertising that month, so $S = f(a)$.

   (a) Interpret the statement $f(1000) = 3500$.
   (b) Which of the graphs in Figure 1.14 is more likely to represent this function?
   (c) What does the vertical intercept of the graph of this function represent, in terms of sales and advertising?

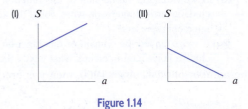

Figure 1.14

34. Figure 1.15 shows fifty years of fertilizer use in the US, India, and the former Soviet Union.[10]

   (a) Estimate fertilizer use in 1970 in the US, India, and the former Soviet Union.
   (b) Write a sentence for each of the three graphs describing how fertilizer use has changed in each region over this 50-year period.

Figure 1.15

35. The six graphs in Figure 1.16 show frequently observed patterns of age-specific cancer incidence rates, in number of cases per 1000 people, as a function of age.[11] The scales on the vertical axes are equal.

   (a) For each of the six graphs, write a sentence explaining the effect of age on the cancer rate.
   (b) Which graph shows a relatively high incidence rate for children? Suggest a type of cancer that behaves this way.
   (c) Which graph shows a brief decrease in the incidence rate at around age 50? Suggest a type of cancer that might behave this way.
   (d) Which graph or graphs might represent a cancer that is caused by toxins which build up in the body over time? (For example, lung cancer.) Explain.

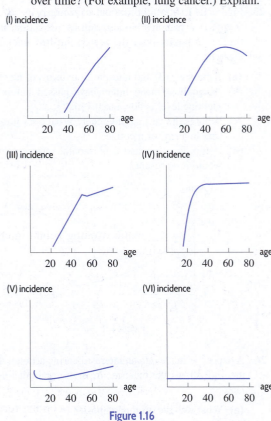

Figure 1.16

36. Table 1.2 shows the average annual sea level, S, in meters, in Aberdeen, Scotland,[12] as a function of time, t, measured in years before 2008.

Table 1.2

t	0	25	50	75	100	125
S	7.094	7.019	6.992	6.965	6.938	6.957

   (a) What was the average sea level in Aberdeen in 2008?

---

[10]The Worldwatch Institute, *Vital Signs 2001,* p. 32 (New York: W.W. Norton, 2001).
[11]Abraham M. Lilienfeld, *Foundations of Epidemiology,* p. 155 (New York: Oxford University Press, 1976).
[12]www.decc.gov.uk, accessed June 2011.

(b) In what year was the average sea level 7.019 meters? 6.957 meters?

(c) Table 1.3 gives the average sea level, S, in Aberdeen as a function of the year, x. Complete the missing values.

Table 1.3

x	1883	?	1933	1958	1983	2008
S	?	6.938	?	6.992	?	?

■ Problems **37–40** ask you to plot graphs based on the following story: "As I drove down the highway this morning, at first traffic was fast and uncongested, then it crept nearly bumper-to-bumper until we passed an accident, after which traffic flow went back to normal until I exited."

37. Driving speed against time on the highway

38. Distance driven against time on the highway

39. Distance from my exit vs time on the highway

40. Distance between cars vs distance driven on the highway

1.2 LINEAR FUNCTIONS

Probably the most commonly used functions are the *linear functions*, whose graphs are straight lines. The chirp-rate and the Honda-depreciation functions in the previous section are both linear. We now look at more examples of linear functions.

Olympic and World Records

During the early years of the Olympics, the height of the men's winning pole vault increased approximately 20 centimeters every four years. Table 1.4 shows that the height started at 330 cm in 1900, and increased by the equivalent of 5 cm a year between 1900 and 1912. So the height was a linear function of time.

Table 1.4 *Winning height (approximate) for men's Olympic pole vault*

Year	1900	1904	1908	1912
Height (cm)	330	350	370	390

If y is the winning height in cm and t is the number of years since 1900, then y is predicted approximately by

$$y = f(t) = 330 + 5t.$$

Since $y = f(t)$ increases with t, we see that f is an increasing function. The coefficient 5 tells us the rate, in cm per year, at which the height increases. This rate is the *slope* of the line in Figure 1.17. The slope is given by the ratio

$$\text{Slope} = \frac{\text{Rise}}{\text{Run}} = \frac{370 - 350}{8 - 4} = \frac{20}{4} = 5 \text{ cm/year.}$$

Calculating the slope (rise/run) using any other two points on the line gives the same value.

What about the constant 330? This represents the initial height in 1900, when $t = 0$. Geometrically, 330 is the intercept on the vertical axis.

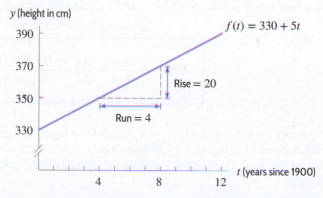

Figure 1.17: Olympic pole vault records

You may wonder whether the linear trend continues beyond 1912. Not surprisingly, it does not give a very good prediction. The formula $y = 330 + 5t$ predicts that the height in the Rio 2016 Olympics would be 910 cm, which is considerably higher than the actual value of 603 cm.[13] There is clearly a danger in *extrapolating* too far from the given data. You should also observe that the data in Table 1.4 is discrete, because it is given only at specific points (every four years). However, we have treated the variable t as though it were continuous, because the function $y = 330 + 5t$ makes sense for all values of t.

The graph in Figure 1.17 is of the continuous function because it is a solid line, rather than four separate points representing the years in which the Olympics were held.

Example 1   If y is the world record time to run the mile, in seconds, and t is the number of years since 1900, then records show that, approximately,

$$y = g(t) = 260 - 0.4t.$$

Explain the meaning of the intercept, 260, and the slope, -0.4, in terms of the world record time to run the mile and sketch the graph.

Solution   The intercept, 260, tells us that the world record was 260 seconds in 1900 (at $t = 0$). The slope, -0.4, tells us that the world record decreased at a rate of about 0.4 seconds per year. See Figure 1.18.

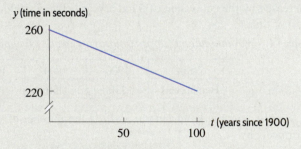

Figure 1.18: World record time to run the mile

Slope and Rate of Change

We use the symbol Δ (the Greek letter capital delta) to mean "change in," so Δx means change in x and Δy means change in y. The slope of a linear function $y = f(x)$ can be calculated from values of the function at two points, given by x_1 and x_2, using the formula

$$\text{Slope} = \frac{\text{Rise}}{\text{Run}} = \frac{\Delta y}{\Delta x} = \frac{f(x_2) - f(x_1)}{x_2 - x_1}.$$

The quantity $(f(x_2) - f(x_1))/(x_2 - x_1)$ is called a *difference quotient* because it is the quotient of two differences. (See Figure 1.19.) Since slope $= \Delta y/\Delta x$, the slope represents the *rate of change* of y with respect to x. The units of the slope are y-units over x-units.

[13] www.rio2016.com/en/athletics-standings-at-mens-pole-vault, accessed September 14, 2016.

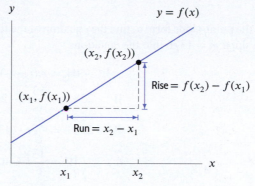

Figure 1.19: Difference quotient $= \dfrac{f(x_2) - f(x_1)}{x_2 - x_1}$

Linear Functions in General

A **linear function** can be put in the *slope-intercept form*

$$y = f(x) = b + mx.$$

Its graph is a line such that

- m is the **slope**, or rate of change of y with respect to x.
- b is the **vertical intercept** or value of y when x is zero.

If the slope, m, is positive, then f is an increasing function. If m is negative, then f is decreasing.

Notice that if the slope, m, is zero, we have $y = b$, a horizontal line. For a line of slope m through the point (x_0, y_0), we have

$$\text{Slope} = m = \frac{y - y_0}{x - x_0}.$$

Therefore we can write the equation of the line in the *point-slope form*:

The equation of a line of slope m through the point (x_0, y_0) is

$$y - y_0 = m(x - x_0).$$

Example 2   The solid waste generated each year in the cities of the US is increasing. The solid waste generated,[14] in millions of tons, was 238.3 in 2000 and 251.3 in 2006.

(a) Assuming that the amount of solid waste generated by US cities is a linear function of time, find a formula for this function by finding the equation of the line through these two points.

(b) Use this formula to predict the amount of solid waste generated in the year 2020.

Solution   (a) We think of the amount of solid waste, in millions of tons W, as a function of year, t, and the two points are $(2000, 238.3)$ and $(2006, 251.3)$. The slope of the line is

$$m = \frac{\Delta W}{\Delta t} = \frac{251.3 - 238.3}{2006 - 2000} = \frac{13}{6} = 2.167 \text{ million tons/year.}$$

[14] *Statistical Abstracts of the US*, 2009, Table 361, accessed January 8, 2017.

We use the point-slope form to find the equation of the line. We substitute the point $(2000, 238.3)$ and the slope $m = 13/6$ into the equation:

$$W - W_0 = m(t - t_0)$$

$$W - 238.3 = \frac{13}{6}(t - 2000)$$

$$W - 238.3 = \frac{13}{6}t - 4333.333$$

$$W = \frac{13}{6}t - 4095.033.$$

The equation of the line is $W = (13/6)t - 4095.033$. Alternatively, we could use the slope-intercept form of a line to find the vertical intercept.

(b) To calculate solid waste predicted for the year 2020, we substitute $t = 2020$ into the equation of the line, $W = -4095.033 + (13/6)t$, and calculate W:

$$W = -4095.033 + \frac{13}{6}(2020) = 281.63.$$

The formula predicts that in the year 2020, there will be 281.63 million tons of solid waste.

> **Recognizing Data from a Linear Function:** Values of x and y in a table could come from a linear function $y = b + mx$ if differences in y-values are constant for equal differences in x.

Example 3   Which of the following tables of values could represent a linear function?

x	0	1	2	3
$f(x)$	25	30	35	40

x	0	2	4	6
$g(x)$	10	16	26	40

t	20	30	40	50
$h(t)$	2.4	2.2	2.0	1.8

Solution   Since $f(x)$ increases by 5 for every increase of 1 in x, the values of $f(x)$ could be from a linear function with slope $= 5/1 = 5$.

Between $x = 0$ and $x = 2$, the value of $g(x)$ increases by 6 as x increases by 2. Between $x = 2$ and $x = 4$, the value of y increases by 10 as x increases by 2. Since the differences in y-values are not constant for equal differences in x, $g(x)$ could not be a linear function.

Since $h(t)$ decreases by 0.2 for every increase of 10 in t, the values of $h(t)$ could be from a linear function with slope $= -0.2/10 = -0.02$.

Example 4   The data in the following table lie on a line. Find formulas for each of the following functions, and give units for the slope in each case:

(a) q as a function of p

(b) p as a function of q

p (dollars)	5	10	15	20
q (tons)	100	90	80	70

Solution   (a) If we think of q as a linear function of p, then q is the dependent variable and p is the independent variable. We can use any two points to find the slope. The first two points give

$$\text{Slope} = m = \frac{\Delta q}{\Delta p} = \frac{90 - 100}{10 - 5} = \frac{-10}{5} = -2.$$

The units are the units of q over the units of p, or tons per dollar.

To write q as a linear function of p, we use the equation $q = b + mp$. We know that $m = -2$, and we can use any of the points in the table to find b. Substituting $p = 10$, $q = 90$ gives

$$q = b + mp$$
$$90 = b + (-2)(10)$$
$$90 = b - 20$$
$$110 = b.$$

Thus, the equation of the line is

$$q = 110 - 2p.$$

(b) If we now consider p as a linear function of q, then p is the dependent variable and q is the independent variable. We have

$$\text{Slope} = m = \frac{\Delta p}{\Delta q} = \frac{10 - 5}{90 - 100} = \frac{5}{-10} = -0.5.$$

The units of the slope are dollars per ton.

Since p is a linear function of q, we have $p = b + mq$ and $m = -0.5$. To find b, we substitute any point from the table, such as $p = 10$, $q = 90$, into this equation:

$$p = b + mq$$
$$10 = b + (-0.5)(90)$$
$$10 = b - 45$$
$$55 = b.$$

Thus, the equation of the line is

$$p = 55 - 0.5q.$$

Alternatively, we could take our answer to part (a), that is $q = 110 - 2p$, and solve for p.

Appendix A shows how to fit a linear function to data that is not exactly linear.

Families of Linear Functions

Formulas such as $f(x) = b + mx$, in which the constants m and b can take on various values, represent a *family of functions*. All the functions in a family share certain properties—in this case, the graphs are lines. The constants m and b are called *parameters*. Figures 1.20 and 1.21 show graphs with several values of m and b. Notice the greater the magnitude of m, the steeper the line.

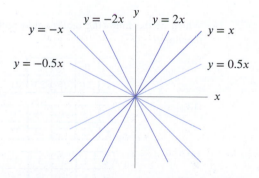

Figure 1.20: The family $y = mx$ (with $b = 0$)

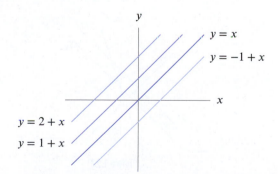

Figure 1.21: The family $y = b + x$ (with $m = 1$)

Problems for Section 1.2

■ For Problems 1–4, find an equation for the line that passes through the given points.

1. $(0, 2)$ and $(2, 3)$

2. $(0, 0)$ and $(1, 1)$

3. $(-2, 1)$ and $(2, 3)$

4. $(4, 5)$ and $(2, -1)$

■ For Problems 5–8, determine the slope and the y-intercept of the line whose equation is given.

5. $7y + 12x - 2 = 0$

6. $3x + 2y = 8$

7. $12x = 6y + 4$

s **8.** $-4y + 2x + 8 = 0$

9. Figure 1.22 shows four lines given by equation $y = b + mx$. Match the lines to the conditions on the parameters m and b.

(a) $m > 0, b > 0$

(b) $m < 0, b > 0$

(c) $m > 0, b < 0$

(d) $m < 0, b < 0$

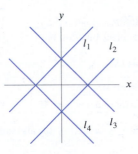

Figure 1.22

10. (a) Which two lines in Figure 1.23 have the same slope? Of these two lines, which has the larger y-intercept?

(b) Which two lines have the same y-intercept? Of these two lines, which has the larger slope?

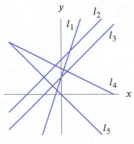

Figure 1.23

11. A city's population was 28,600 in the year 2015 and is growing by 750 people a year.

(a) Give a formula for the city's population, P, as a function of the number of years, t, since 2015.

(b) What is the population predicted to be in 2020?

(c) When is the population expected to reach 35,000?

12. A car rental company charges a daily fee of $35 plus $0.20 per mile driven. Find a formula for the daily charge, C, in dollars, as a function of the number of miles, m, driven that day.

13. A company rents cars at $40 a day and 15 cents a mile. Its competitor's cars are $50 a day and 10 cents a mile.

(a) For each company, give a formula for the cost of renting a car for a day as a function of the distance traveled.

(b) On the same axes, graph both functions.

(c) How should you decide which company is cheaper?

14. A phone company charges $14.99 a month and 10 cents for every minute above 120 minutes. Write an expression for the monthly phone charge, P, in dollars, as a function of the number of minutes, m, used that month.

15. Figure 1.24 shows the distance from home, in miles, of a person on a 5-hour trip.

(a) Estimate the vertical intercept. Give units and interpret it in terms of distance from home.

(b) Estimate the slope of this linear function. Give units, and interpret it in terms of distance from home.

(c) Give a formula for distance, D, from home as a function of time, t in hours.

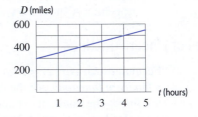

Figure 1.24

16. Which of the following tables could represent linear functions?

(a)

x	0	1	2	3
y	27	25	23	21

(b)

t	15	20	25	30
s	62	72	82	92

(c)

u	1	2	3	4
w	5	10	18	28

17. For each table in Problem 16 that could represent a linear function, find a formula for that function.

18. Annual revenue R from McDonald's restaurants worldwide between 2013 and 2015 can be estimated by $R = 28.1 - 1.35t$, where R is in billion dollars and t is in years since January 1, 2013.[15]

    (a) What is the slope of this function? Include units. Interpret the slope in terms of McDonald's revenue.
    (b) What is the vertical intercept of this function? Include units. Interpret the vertical intercept in terms of McDonald's revenue.
    (c) What annual revenue does the function predict for 2018?
    (d) If revenue continues to decline at the same rate, when is annual revenue predicted to fall to 22 billion dollars?

■ Problems **19–21** use Figure 1.25 showing how the quantity, Q, of grass (kg/hectare) in different parts of Namibia depended on the average annual rainfall, r, (mm), in two different years.[16]

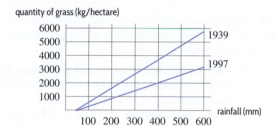

Figure 1.25

19. (a) For 1939, find the slope of the line, including units.
    (b) Interpret the slope in this context.
    (c) Find the equation of the line.

20. (a) For 1997, find the slope of the line, including units.
    (b) Interpret the slope in this context.
    (c) Find the equation of the line.

21. Which of the two functions in Figure 1.25 has the larger difference quotient $\Delta Q / \Delta r$? What does this tell us about grass in Namibia?

22. US imports of crude oil and petroleum have been increasing.[17] There have been many ups and downs, but the general trend is shown by the line in Figure 1.26.

    (a) Find the slope of the line. Include its units of measurement.

(b) Write an equation for the line. Define your variables, including their units.
(c) Assuming the trend continues, when does the linear model predict imports will reach 18 million barrels per day? Do you think this is a reliable prediction? Give reasons.

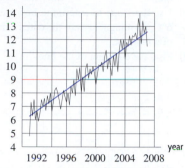

Figure 1.26

23. A company's pricing schedule in Table 1.5 is designed to encourage large orders. (A gross is 12 dozen.) Find a formula for:

    (a) q as a linear function of p.
    (b) p as a linear function of q.

Table 1.5

q (order size, gross)	3	4	5	6
p (price/dozen)	15	12	9	6

24. World milk production rose at an approximately constant rate between 2000 and 2012.[18] See Figure 1.27.

    (a) Estimate the vertical intercept and interpret it in terms of milk production.
    (b) Estimate the slope and interpret it in terms of milk production.
    (c) Give an approximate formula for milk production, M, as a function of t.

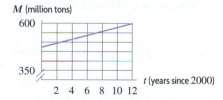

Figure 1.27

---

[15]Based on McDonald's Annual Report 2015, corporate.mcdonalds.com/mcd/investors/financial-information/annual-report.html, accessed September 15, 2016.

[16]David Ward and Ben T. Ngairorue, "Are Namibia's Grasslands Desertifying?", *Journal of Range Management* 53, 2000, 138–144.

[17]http://www.theoildrum.com/node/2767. Accessed May 2015.

[18]www.dairyco.org.uk/library/market-information/datum/world-milk-production.aspx, accessed September 15, 2016.

25. Marmots are large squirrels that hibernate in the winter and come out in the spring. Figure 1.28 shows the date (days after Jan 1) that they are first sighted each year in Colorado as a function of the average minimum daily temperature for that year.[19]

   (a) Find the slope of the line, including units.

   (b) What does the sign of the slope tell you about marmots?

   (c) Use the slope to determine how much difference 6°C warming makes to the date of first appearance of a marmot.

   (d) Find the equation of the line.

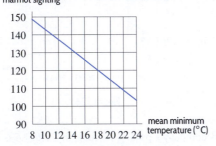

Figure 1.28

26. In Colorado spring has arrived when the bluebell first flowers. Figure 1.29 shows the date (days after Jan 1) that the first flower is sighted in one location as a function of the first date (days after Jan 1) of bare (snow-free) ground.[20]

   (a) If the first date of bare ground is 140, how many days later is the first bluebell flower sighted?

   (b) Find the slope of the line, including units.

   (c) What does the sign of the slope tell you about bluebells?

   (d) Find the equation of the line.

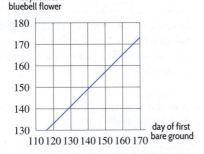

Figure 1.29

27. On March 5, 2015, Capracotta, Italy, received 256 cm (100.787 inches) of snow in 18 hours.[21]

   (a) Assuming the snow fell at a constant rate and there were already 100 cm of snow on the ground, find a formula for $f(t)$, in cm, for the depth of snow as a function of t hours since the snowfall began on March 5.

   (b) What are the domain and range of f?

28. The approximate percentage of people, P, below the poverty level in the US[22] is given in Table 1.6.

   (a) Find a formula for the percentage in poverty as a linear function of time in years since 2007.

   (b) Use the formula to predict the percentage in poverty in 2013.

   (c) What is the difference between the prediction and the actual percentage, 14.5%?

Table 1.6

Year (since 2007)	0	1	2	3
P (percentage)	12.5	13.4	14.3	15.2

29. World grain production was 1241 million tons in 1975 and 2048 million tons in 2005, and has been increasing at an approximately constant rate.[23]

   (a) Find a linear function for world grain production, P, in million tons, as a function of t, the number of years since 1975.

   (b) Using units, interpret the slope in terms of grain production.

   (c) Using units, interpret the vertical intercept in terms of grain production.

   (d) According to the linear model, what is the predicted world grain production in 2015?

   (e) According to the linear model, when is grain production predicted to reach 2500 million tons?

30. Annual sales of music compact discs (CDs) have declined since 2000. Sales were 942.5 million in 2000 and 384.7 million in 2008.[24]

   (a) Find a formula for annual sales, S, in millions of music CDs, as a linear function of the number of years, t, since 2000.

   (b) Give units for and interpret the slope and the vertical intercept of this function.

   (c) Use the formula to predict music CD sales in 2012.

[19]David W. Inouye, Billy Barr, Kenneth B. Armitage, and Brian D. Inouye, "Climate Change is Affecting Altitudinal Migrants and Hibernating Species", *PNAS* 97, 2000, 1630–1633.

[20]David W. Inouye, Billy Barr, Kenneth B. Armitage, and Brian D. Inouye, "Climate Change is Affecting Altitudinal Migrants and Hibernating Species", *PNAS* 97, 2000, 1630–1633.

[21]http://iceagenow.info/2015/03/official-italy-captures-world-one-day-snowfall-record/

[22]Data based on: www.huffingtonpost.com/2014/09/16/poverty-household-income_n_5828974.html, accessed September 15, 2016.

[23]The Worldwatch Institute, *Vital Signs 2007-2008* (New York: W.W. Norton & Company, 2007), p. 21.

[24]*The World Almanac and Book of Facts 2008* (New York).

31. A \$25,000 vehicle depreciates \$2000 a year as it ages. Repair costs are \$1500 per year.

    (a) Write formulas for each of the two linear functions at time t, value, $V(t)$, and repair costs to date, $C(t)$. Graph them.

    (b) One strategy is to replace a vehicle when the total cost of repairs is equal to the current value. Find this time.

    (c) Another strategy is to replace the vehicle when the value of the vehicle is some percent of the original value. Find the time when the value is 6%.

32. Search and rescue teams work to find lost hikers. Members of the search team separate and walk parallel to one another through the area to be searched. Table 1.7 shows the percent, P, of lost individuals found for various separation distances, d, of the searchers.[25]

    (a) Explain how you know that the percent found, P, could be a linear function of separation distance, d.

    (b) Find P as a linear function of d.

    (c) What is the slope of the function? Give units and interpret the answer.

    (d) What are the vertical and horizontal intercepts of the function? Give units and interpret the answers.

Table 1.7

Separation distance d (ft)	20	40	60	80	100
Approximate percent found, P	90	80	70	60	50

33. In a Washington town, the charge for commercial waste collection is \$694.55 for 5 tons and \$1098.32 for 8 tons of waste.

    (a) Find a linear formula for the cost, C, of waste collection as a function of the weight, w, in tons.

    (b) What is the slope of the line found in part (a)? Give units and interpret your answer in terms of the cost of waste collection.

    (c) What is the vertical intercept of the line found in part (a)? Give units and interpret your answer in terms of the cost of waste collection.

34. The number of species of coastal dune plants in Australia decreases as the latitude, in °S, increases. There are 34 species at 11°S and 26 species at 44°S.[26]

    (a) Find a formula for the number, N, of species of coastal dune plants in Australia as a linear function of the latitude, l, in °S.

    (b) Give units for and interpret the slope and the vertical intercept of this function.

    (c) Graph this function between $l = 11$°S and $l = 44$°S. (Australia lies entirely within these latitudes.)

35. Table 1.8 gives the required standard weight, w, in kilograms, of American soldiers, aged between 21 and 27, for height, h, in centimeters.[27]

    (a) How do you know that the data in this table could represent a linear function?

    (b) Find weight, w, as a linear function of height, h. What is the slope of the line? What are the units for the slope?

    (c) Find height, h, as a linear function of weight, w. What is the slope of the line? What are the units for the slope?

Table 1.8

h (cm)	172	176	180	184	188	192	196
w (kg)	79.7	82.4	85.1	87.8	90.5	93.2	95.9

■ Problems **36–41** concern the maximum heart rate (MHR), which is the maximum number of times a person's heart can safely beat in one minute. If MHR is in beats per minute and a is age in years, the formulas used to estimate MHR are

$$\text{For females: MHR} = 226 - a,$$

$$\text{For males: MHR} = 220 - a.$$

36. Which of the following is the correct statement?

    (a) As you age, your maximum heart rate decreases by one beat per year.

    (b) As you age, your maximum heart rate decreases by one beat per minute.

    (c) As you age, your maximum heart rate decreases by one beat per minute per year.

37. Which of the following is the correct statement for a male and female of the same age?

    (a) Their maximum heart rates are the same.

    (b) The male's maximum heart rate exceeds the female's.

    (c) The female's maximum heart rate exceeds the male's.

38. What can be said about the ages of a male and a female with the same maximum heart rate?

---

[25] J. Wartes, *An Experimental Analysis of Grid Sweep Searching* (Explorer Search and Rescue, Western Region, 1974).

[26] M. L. Rosenzweig, *Species Diversity in Space and Time* (Cambridge: Cambridge University Press, 1995), p. 292.

[27] Adapted from usmilitary.about.com, accessed March 29, 2015.

39. Recently[28] it has been suggested that a more accurate predictor of MHR for both males and females is given by

$$\text{MHR} = 208 - 0.7a.$$

(a) At what age do the old and new formulas give the same MHR for females? For males?

(b) Which of the following is true?

  (i) The new formula predicts a higher MHR for young people and a lower MHR for older people than the old formula.

  (ii) The new formula predicts a lower MHR for young people and a higher MHR for older people than the old formula.

(c) When testing for heart disease, doctors ask patients to walk on a treadmill while the speed and incline are gradually increased until their heart rates reach 85 percent of the MHR. For a 65-year-old male, what is the difference in beats per minute between the heart rate reached if the old formula is used and the heart rate reached if the new formula is used?

40. Experiments[29] suggest that the female MHR decreases by 12 beats per minute by age 21, and by 19 beats per minute by age 33. Is this consistent with MHR being approximately linear with age?

41. Experiments[30] suggest that the male MHR decreases by 9 beats per minute by age 21, and by 26 beats per minute by age 33. Is this consistent with MHR being approximately linear with age?

42. Let y be the percent increase in annual US national production during a year when the unemployment rate changes by u percent. (For example, $u = 2$ if unemployment increases from 4% to 6%.) Okun's law states that

$$y = 3.5 - 2u.$$

(a) What is the meaning of the number 3.5 in Okun's law?

(b) What is the effect on national production of a year when unemployment rises from 5% to 8%?

(c) What change in the unemployment rate corresponds to a year when production is the same as the year before?

(d) What is the meaning of the coefficient -2 in Okun's law?

43. An Australian[31] study found that, if other factors are constant (education, experience, etc.), taller people receive higher wages for the same work. The study reported a "height premium" for men of 3% of the hourly wage for a 10 cm increase in height; for women the height premium reported was 2%. We assume that hourly wages are a linear function of height. The slope is given by the height premium at the average hourly wage for that gender, AU\$29.40 for men and AU\$24.78 for women.

(a) The average hourly wage[32] for a 178 cm Australian man is AU\$29.40. Express the average hourly wage of an Australian man as a function of his height, x cm.

(b) The average hourly wage for a 164 cm Australian woman is AU\$24.78. Express the average hourly wage of an Australian woman as a function of her height, y cm.

(c) What is the difference in average hourly wages between men and women of height 178 cm?

(d) Is there a height for which men and women are predicted to have the same wage? If so, what is it?

1.3   AVERAGE RATE OF CHANGE AND RELATIVE CHANGE

Average Rate of Change

In the previous section, we saw that the height of the winning Olympic pole vault increased at an approximately constant rate of 5 cm/year between 1900 and 1912. Similarly, the world record for the mile decreased at an approximately constant rate of 0.4 seconds/year. We now see how to calculate rates of change when they are not constant.

Example 1   Table 1.9 shows the height of the winning pole vault at the Olympics[33] during the 1960s and more recently. Find the rate of change of the winning height between 1960 and 1968, and between 2008 and 2016. In which of these two periods did the height increase faster than during the period 1900–1912?

Table 1.9   *Winning height in men's Olympic pole vault (approximate)*

Year	1960	1964	1968	...	2008	2012	2016
Height (cm)	470	510	540	...	596	597	603

---

[28]www.physsportsmed.com/issues/2001/07_01/jul01news.htm, accessed January 4, 2005.

[29]www.css.edu/users/tboone2/asep/May2002JEPonline.html, accessed January 4, 2005.

[30]www.css.edu/users/tboone2/asep/May2002JEPonline.html, accessed January 4, 2005.

[31]"Study Finds Tall People at Top of Wages Ladder", Yahoo News, May 17, 2009.

[32]Australian Fair Pay Commission, August 2007.

[33]*The World Almanac and Book of Facts, 2005*, p. 866 (New York).

Solution   From 1900 to 1912, the height increased by 5 cm/year. To compare the 1960s and the more recent period, we calculate

$$\text{Average rate of change of height} \atop \text{1960 to 1968} = \frac{\text{Change in height}}{\text{Change in time}} = \frac{540 - 470}{1968 - 1960} = 8.75 \text{ cm/year.}$$

$$\text{Average rate of change of height} \atop \text{2008 to 2016} = \frac{\text{Change in height}}{\text{Change in time}} = \frac{603 - 596}{2016 - 2008} = 0.88 \text{ cm/year.}$$

Thus, on average, the height was increasing more quickly during the 1960s than from 1900 to 1912, and was increasing more slowly from 2008 to 2016 than from 1900 to 1912.

In Example 1, the function does not have a constant rate of change (it is not linear). However, we can compute an *average rate of change* over any interval. The word average is used because the rate of change may vary within the interval. We have the following general formula.

> If y is a function of t, so $y = f(t)$, then
>
> $$\textbf{Average rate of change} \text{ of } y \atop \text{between } t = a \text{ and } t = b = \frac{\Delta y}{\Delta t} = \frac{f(b) - f(a)}{b - a}.$$
>
> The units of average rate of change of a function are units of y per unit of t.

The average rate of change of a linear function is the slope, and a function is linear if the average rate of change is the same on all intervals.

Example 2   Using Figure 1.30, estimate the average rate of change of the number of farms[34] in the US between 1950 and 1970.

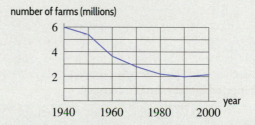

number of farms (millions)

Figure 1.30: Number of farms in the US (in millions)

Solution   Figure 1.30 shows that the number, N, of farms in the US was approximately 5.4 million in 1950 and approximately 2.8 million in 1970. If time, t, is in years, we have

$$\text{Average rate of change} = \frac{\Delta N}{\Delta t} = \frac{2.8 - 5.4}{1970 - 1950} = -0.13 \text{ million farms per year.}$$

The average rate of change is negative because the number of farms is decreasing. During this period, the number of farms decreased at an average rate of 0.13 million, or 130,000, per year.

---

[34]*The World Almanac and Book of Facts, 2005*, p. 136 (New York).

We have looked at how an Olympic record and the number of farms change over time. In the next example, we look at average rate of change with respect to a quantity other than time.

Example 3   Polychlorinated biphenyl (PCB) is an industrial pollutant thought dangerous to wildlife. Table 1.10 shows data on PCB levels and the thickness of pelican egg shells.[35] (Thin shells break more easily.)

Find the average rate of change in the thickness of the shell as the PCB concentration changes from 87 ppm to 452 ppm. Give units and explain what the sign of the answer tells us.

Table 1.10   *Thickness of pelican eggshells and PCB concentration in the eggshells*

Concentration, c, in parts per million (ppm)	87	147	204	289	356	452
Thickness, h, in millimeters (mm)	0.44	0.39	0.28	0.23	0.22	0.14

Solution   Since we are looking for the average rate of change of thickness with respect to change in PCB concentration, we have

$$\text{Average rate of change of thickness} = \frac{\text{Change in the thickness}}{\text{Change in the PCB level}} = \frac{\Delta h}{\Delta c} = \frac{0.14 - 0.44}{452 - 87}$$

$$= -0.00082 \frac{\text{mm}}{\text{ppm}}.$$

The units are thickness units (mm) over PCB concentration units (ppm), or millimeters over parts per million. The fact that the average rate of change is negative tells us that, for this data, the thickness of the eggshell decreases as the PCB concentration increases. The thickness of pelican eggs decreases by an average of 0.00082 mm for every additional part per million of PCB in the eggshell.

Visualizing Rate of Change

For a function $y = f(x)$, the change in the value of the function between $x = a$ and $x = c$ is $\Delta y = f(c) - f(a)$. Since Δy is a difference of two y-values, it is represented by the vertical distance in Figure 1.31. The average rate of change of f between $x = a$ and $x = c$ is represented by the slope of the line joining the points A and C in Figure 1.32. This line is called the *secant line* between $x = a$ and $x = c$.

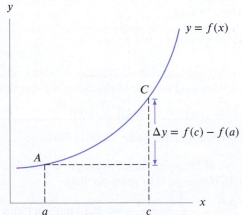

Figure 1.31: The change in a function is represented by a vertical distance

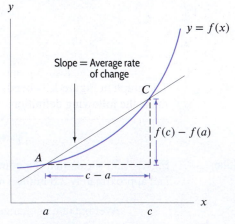

Figure 1.32: The average rate of change is represented by the slope of the line

---

[35]R. W. Risebrough, "Effects of Environmental Pollutants upon Animals Other than Man." *Proceedings of the 6th Berkeley Symposium on Mathematics and Statistics, VI*, p. 443–463 (Berkeley: University of California Press, 1972). PCBs were banned in the US in 1979.

Example 4
    (a) Find the average rate of change of $y = f(x) = \sqrt{x}$ between $x = 1$ and $x = 4$.
    (b) Graph $f(x)$ and represent this average rate of change as the slope of a line.
    (c) Which is larger, the average rate of change of the function between $x = 1$ and $x = 4$ or the average rate of change between $x = 4$ and $x = 5$? What does this tell us about the graph of the function?

Solution
    (a) Since $f(1) = \sqrt{1} = 1$ and $f(4) = \sqrt{4} = 2$, between $x = 1$ and $x = 4$, we have

$$\text{Average rate of change} = \frac{\Delta y}{\Delta x} = \frac{f(4) - f(1)}{4 - 1} = \frac{2 - 1}{3} = \frac{1}{3}.$$

    (b) A graph of $f(x) = \sqrt{x}$ is given in Figure 1.33. The average rate of change of f between 1 and 4 is the slope of the secant line between $x = 1$ and $x = 4$.

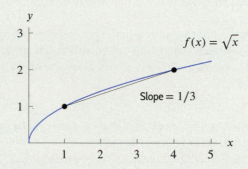

Figure 1.33: Average rate of change = Slope of secant line

    (c) Since the secant line between $x = 1$ and $x = 4$ is steeper than the secant line between $x = 4$ and $x = 5$, the average rate of change between $x = 1$ and $x = 4$ is larger than it is between $x = 4$ and $x = 5$. The rate of change is decreasing. This tells us that the graph of this function is bending downward.

Concavity

We now look at the graphs of functions whose rates of change are increasing throughout an interval or decreasing throughout an interval.

Figure 1.33 shows a graph that is bending downward because the rate of change is decreasing. The graph in Figure 1.31 bends upward because the rate of change of the function is increasing. We make the following definitions.

> The graph of a function is **concave up** if it bends upward as we move left to right; the graph is **concave down** if it bends downward. (See Figure 1.34.) A line is neither concave up nor concave down.

Figure 1.34: Concavity of a graph

Example 5   Using Figure 1.35, estimate the intervals over which:

(a)   The function is increasing; decreasing.   (b)   The graph is concave up; concave down.

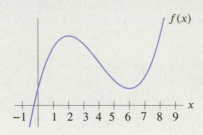

Figure 1.35

Solution   (a) The graph suggests that the function is increasing for $x < 2$ and for $x > 6$. It appears to be decreasing for $2 < x < 6$.

(b) The graph is concave down on the left and concave up on the right. It is difficult to tell exactly where the graph changes concavity, although it appears to be about $x = 4$. Approximately, the graph is concave down for $x < 4$ and concave up for $x > 4$.

Example 6   From the following values of $f(t)$, does f appear to be increasing or decreasing? Do you think its graph is concave up or concave down?

t	0	5	10	15	20	25	30
$f(t)$	12.6	13.1	14.1	16.2	20.0	29.6	42.7

Solution   Since the given values of $f(t)$ increase as t increases, f appears to be increasing. As we read from left to right, the change in $f(t)$ starts small and gets larger (for constant change in t), so the graph is climbing faster. Thus, the graph appears to be concave up. Alternatively, plot the points and notice that a curve through these points bends up.

Distance, Velocity, and Speed

A grapefruit is thrown up in the air. The height of the grapefruit above the ground first increases and then decreases. See Table 1.11.

Table 1.11   *Height, y, of the grapefruit above the ground t seconds after it is thrown*

t (sec)	0	1	2	3	4	5	6
y (feet)	6	90	142	162	150	106	30

Example 7   Find the change and average rate of change of the height of the grapefruit during the first 3 seconds. Give units and interpret your answers.

Solution   The change in height during the first 3 seconds is $\Delta y = 162 - 6 = 156$ ft. This means that the grapefruit goes up a total of 156 feet during the first 3 seconds. The average rate of change during this 3 second interval is $156/3 = 52$ ft/sec. During the first 3 seconds, the grapefruit is rising at an average rate of 52 ft/sec.

The average rate of change of height with respect to time is *velocity*. You may recognize the units (feet per second) as units of velocity.

$$\text{Average velocity} \quad = \frac{\text{Change in distance}}{\text{Change in time}} = \quad \begin{array}{c} \text{Average rate of change of distance} \\ \text{with respect to time} \end{array}$$

There is a distinction between *velocity* and *speed*. Suppose an object moves along a line. If we pick one direction to be positive, the velocity is positive if the object is moving in that direction and negative if it is moving in the opposite direction. For the grapefruit, upward is positive and downward is negative. Speed is the magnitude of velocity, so it is always positive or zero.

Example 8   Find the average velocity of the grapefruit over the interval $t = 4$ to $t = 6$. Explain the sign of your answer.

Solution   Since the height is $y = 150$ feet at $t = 4$ and $y = 30$ feet at $t = 6$, we have

$$\text{Average velocity} = \frac{\text{Change in distance}}{\text{Change in time}} = \frac{\Delta y}{\Delta t} = \frac{30 - 150}{6 - 4} = -60 \text{ ft/sec.}$$

The negative sign means the height is decreasing and the grapefruit is moving downward.

Example 9   A car travels away from home on a straight road. Its distance from home at time t is shown in Figure 1.36. Is the car's average velocity greater during the first hour or during the second hour?

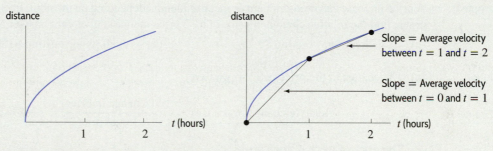

Figure 1.36: Distance of car from home   **Figure 1.37**: Average velocities of the car

Solution   Average velocity is represented by the slope of a secant line. Figure 1.37 shows that the secant line between $t = 0$ and $t = 1$ is steeper than the secant line between $t = 1$ and $t = 2$. Thus, the average velocity is greater during the first hour.

Relative Change

Is a population increase of 1000 a significant change? It depends on the original size of the community. If the town of Coyote, NM, population 128, increases by 1000 people, the townspeople would definitely notice.[36] On the other hand, if New York City, population 8.5 million,[37] increases by 1000 people, almost no one will notice. To visualize the impact of the increase on the two different communities, we look at the change, 1000, as a fraction, or percentage, of the initial population. This percent change is called the *relative change*.

Example 10   If the population increases by 1000 people, find the relative change in the population for

(a) Coyote, NM (population 128)
(b) New York City (population 8,500,000)

---

[36] en.wikipedia.org/wiki/Coyote,_Rio_Arriba_County,_New_Mexico, accessed February 3, 2017.
[37] en.wikipedia.org/wiki/Demographics_of_New_York_City, accessed February 3, 2017.

Solution   (a) The population increases by 1000 from 128 so

$$\text{Relative change } = \frac{\text{Change in population}}{\text{Initial population}} = \frac{1000}{128} = 7.813.$$

The population has increased by 781.3%, a significant increase.

(b) The population increases by 1000 from 8,500,000 so

$$\text{Relative change } = \frac{\text{Change in population}}{\text{Initial population}} = \frac{1000}{8,500,000} = 0.00012.$$

The population has increased by 0.012%, or less than one-tenth of one percent.

In general, when a quantity P changes from P_0 to P_1, we define

$$\text{Relative change in } P = \frac{\text{Change in } P}{P_0} = \frac{P_1 - P_0}{P_0}.$$

The relative change is a number, without units. It is often expressed as a percentage.

Example 11   A price increase can be significant or inconsequential depending on the item. In each of the following cases, find the relative change in price of a $2 price increase; give your answer as a percent.

(a) A gallon of gas costing $2.25   (b) A cell phone costing $180

Solution   (a) The change in the price is $2 so we have

$$\text{Relative change in price of gas } = \frac{\text{Change in price}}{\text{Initial price}} = \frac{2}{2.25} = 0.889.$$

The price of gas has gone up 88.9%.

(b) We have

$$\text{Relative change in price of cell phone } = \frac{\text{Change in price}}{\text{Initial price}} = \frac{2}{180} = 0.011.$$

The price of the cell phone has gone up only 1.1%.

Relative change can be positive or negative, as we see in the following example.

Example 12   Find the relative change in the price of a $75.99 pair of jeans if the sale price is $52.99.

Solution   The price has dropped from $75.99 to $52.99. We have

$$\text{Relative change } = \frac{52.99 - 75.99}{75.99} = \frac{-23}{75.99} = -0.303.$$

The price has been reduced by 30.3% for the sale.

Example 13   The number of sales per week of a $75.99 pair of jeans is 25 pairs. The number of weekly sales goes up to 45 pairs when the price is reduced by 30.3%. Find the relative change in the weekly sales.

Solution   The quantity sold goes up from 25 pairs of jeans to 45 pairs. We have

$$\text{Relative change } = \frac{45 - 25}{25} = \frac{20}{25} = 0.8.$$

The number of sales per week grows by 80%.

Ratio of Relative Changes: Elasticity

In Examples 12 and 13 we see that a 30.3% reduction in the price of a $75.99 pair of jeans increased the weekly sales by 80%. If we are interested in the percent change in quantity as a result of a 1% change in price, we look at the ratio:

$$\left| \frac{\text{Relative change in quantity}}{\text{Relative change in price}} \right| = \left| \frac{0.8}{-0.303} \right| = 2.64.$$

So the number of pair of jeans sold increases by 2.64% when the price drops by 1%. This ratio is called the elasticity. It gives information about how change in price of a product effects the change in quantity. Elasticity will be discussed in further detail in Section 4.6.

Problems for Section 1.3

■ In Problems 1–4, decide whether the graph is concave up, concave down, or neither.

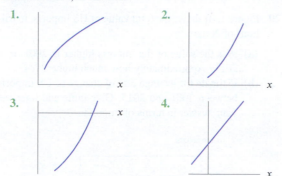

1.   2.

3.   4.

5. Table 1.12 gives values of a function $w = f(t)$. Does this function appear to be increasing or decreasing? Does its graph appear to be concave up or concave down?

Table 1.12

t	0	4	8	12	16	20	24
w	100	58	32	24	20	18	17

6. Graph a function $f(x)$ which is increasing everywhere and concave up for negative x and concave down for positive x.

7. For which pairs of consecutive points in Figure 1.38 is the function graphed:

(a) Increasing and concave up?

(b) Increasing and concave down?

(c) Decreasing and concave up?

(d) Decreasing and concave down?

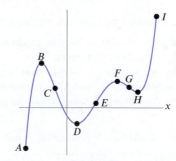

Figure 1.38

8. Find the average rate of change of $f(x) = 2x^2$ between $x = 1$ and $x = 3$.

9. Find the average rate of change of $f(x) = 3x^2 + 4$ between $x = -2$ and $x = 1$. Illustrate your answer graphically.

10. When a deposit of $1000 is made into an account paying 2% interest, compounded annually, the balance, B, in the account after t years is given by $B = 1000(1.02)^t$. Find the average rate of change in the balance over the interval $t = 0$ to $t = 5$. Give units and interpret your answer in terms of the balance in the account.

■ In Problems **11–15**, interpret the expression in terms of Arctic Sea ice extent, the area of sea covered by ice.[38] Let $E(x)$ and $F(t)$ be the Arctic Sea ice extent, both in millions of square kilometers, as a function of time, x, in years, since February 1979, and time, t, in days, since December 31, 2014.

11. $E(29) = 15$

12. $E(4) = 16$

13. $\dfrac{E(29) - E(4)}{29 - 4} = -0.04$

14. $F(32) = 13.97$

15. $\dfrac{F(59) - F(32)}{59 - 32} = 0.0159$

16. Table 1.13 gives the net sales of The Gap, a clothing retailer.[39]

    (a) Find the change in net sales between 2013 and 2015.
    (b) Find the average rate of change in net sales between 2011 and 2015. Give units and interpret your answer.
    (c) From 2011 to 2015, were there any one-year intervals during which the average rate of change was negative? If so, when?

Table 1.13   *Gap net sales, in billions of dollars*

Year	2011	2012	2013	2014	2015
Sales	14.55	15.65	16.15	16.44	15.80

17. Table 1.14 shows world bicycle production.[40]

    (a) Find the change in bicycle production between 1950 and 2000. Give units.
    (b) Find the average rate of change in bicycle production between 1950 and 2000. Give units and interpret your answer in terms of bicycle production.

Table 1.14   *World bicycle production, in millions*

Year	1950	1960	1970	1980	1990	2000
Bicycles	11	20	36	62	91	95

18. Table 1.15 shows attendance at NFL football games.[41]

    (a) Find the average rate of change in the attendance from 2011 to 2015. Give units.
    (b) Find the annual increase in the average attendance for each year from 2011 to 2015. (Your answer should be four numbers.)

(c) Show that the average rate of change found in part (a) is the average of the four yearly changes found in part (b).

Table 1.15   *Attendance at NFL football games, in millions of fans*

Year	2011	2012	2013	2014	2015
Attendance (millions)	17.12	17.18	17.30	17.36	17.26

19. An observer tracks the distance a plane has rolled along the runway after touching down. Figure 1.39 shows this distance, x, in thousands of feet, as a function of time, t, in seconds since touchdown. Find the average velocity of the plane over the following time intervals:

    (a) Between 0 and 20 seconds.
    (b) Between 20 and 40 seconds.

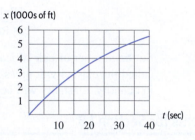

Figure 1.39

20. Figure 1.40 shows the total value of US imports, in trillions of dollars.[42]

    (a) Was the value of the imports higher in 2001 or in 2015? Approximately how much higher?
    (b) Estimate the average rate of change of US imports between 2001 and 2015. Give units and interpret your answer in terms of imports.

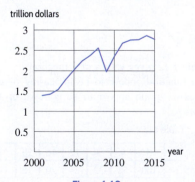

Figure 1.40

---

[38] Sea ice extent definition and data values from nsidc.org/cryosphere/seaice/data/terminology.html. Accessed March 2015.

[39] www.last10k.com, 10-K March 2016, accessed December 30, 2016.

[40] www.earth-policy.org, accessed September 15, 2016.

[41] http://www.statista.com/statistics/193420/regular-season-attendance-in-the-nfl-since-2006/, accessed Sept 15, 2016.

[42] data.worldbank.org/indicator/, accessed September 16, 2016.

21. Figure 1.41 shows a particle's distance from a point as a function of time, t. What is the particle's average velocity from $t = 0$ to $t = 3$?

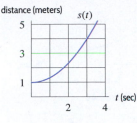

Figure 1.41

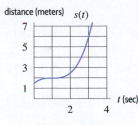

Figure 1.42

22. Figure 1.42 shows a particle's distance from a point as a function of time, t. What is the particle's average velocity from $t = 1$ to $t = 3$?

23. At time t in seconds, a particle's distance $s(t)$, in cm, from a point is given in the table. What is the average velocity of the particle from $t = 3$ to $t = 10$?

t	0	3	6	10	13
$s(t)$	0	72	92	144	180

24. Table 1.16 shows world population, P, in billions of people, world passenger automobile production, A, in millions of cars, and world cell phone subscribers, C, in millions of subscribers.[43]

(a) Find the average rate of change, with units, for each of P, A, and C between 1995 and 2005.

(b) Between 1995 and 2005, which increased faster:

   (i) Population or the number of automobiles?

   (ii) Population or the number of cell phone subscribers?

Table 1.16

Year	1995	2000	2005
P (billions)	5.68	6.07	6.45
A (millions)	36.1	41.3	45.9
C (millions)	91	740	2168

25. Table 1.17 gives sales of Pepsico, which operates two major businesses: beverages (including Pepsi) and snack foods.[44]

(a) Find the change in sales between 2003 and 2015.

(b) Find the average rate of change in sales between 2003 and 2015. Give units and interpret your answer.

Table 1.17 *Pepsico sales, in billions of dollars*

Year	2003	2004	2005	2006	2007	2008	2009
Sales	26.97	29.26	32.56	35.14	39.47	43.25	43.23
Year	2010	2011	2012	2013	2014	2015	
Sales	57.84	66.50	65.49	66.42	66.68	63.06	

26. Do you expect the average rate of change (in units per year) of each of the following to be positive or negative? Explain your reasoning.

(a) Number of acres of rain forest in the world.

(b) Population of the world.

(c) Number of polio cases each year in the US, since 1950.

(d) Height of a sand dune that is being eroded.

(e) Cost of living in the US.

27. Table 1.18 shows the production of tobacco in the US.[45]

(a) What is the average rate of change in tobacco production between 2003 and 2015? Give units and interpret your answer in terms of tobacco production.

(b) During this twelve-year period, are there any one-year intervals during which the average rate of change was positive? If so, when?

Table 1.18 *Tobacco production, in millions of pounds*

Year	2003	2004	2005	2006	2007	2008	2009
Production	803	882	645	728	788	801	823
Year	2010	2011	2012	2013	2014	2015	
Production	718	598	763	724	876	711	

28. Figure 1.43 shows the length, L, in cm, of a sturgeon (a type of fish) as a function of the time, t, in years.[46]

(a) Is the function increasing or decreasing? Is the graph concave up or concave down?

(b) Estimate the average rate of growth of the sturgeon between $t = 5$ and $t = 15$. Give units and interpret your answer in terms of the sturgeon.

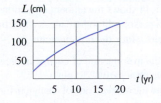

Figure 1.43

---

[43]The Worldwatch Institute, *Vital Signs 2007-2008* (New York: W.W. Norton & Company, 2007), pp. 51 and 67.
[44]www.pepsico.com, accessed August, 2012 and December 30, 2016.
[45]www.statista.com, accessed September 18, 2016.
[46]Data from L. von Bertalanffy, *General System Theory*, p. 177 (New York: Braziller, 1968).

29. Table 1.19 shows the total US labor force, L. Find the average rate of change between 1940 and 2000; between 1940 and 1960; between 1980 and 2000. Give units and interpret your answers in terms of the labor force.[47]

Table 1.19 *US labor force, in thousands of workers*

Year	1940	1960	1980	2000
L	47,520	65,778	99,303	136,891

30. The total world marine catch[48] of fish, in metric tons, was 17 million in 1950 and 99 million in 2001. What was the average rate of change in the marine catch during this period? Give units and interpret your answer.

31. Table 1.20 gives the revenues, R, of General Motors, formerly the world's largest auto manufacturer.[49]

(a) Find the change in revenues between 2003 and 2015.

(b) Find the average rate of change in revenues between 2003 and 2015. Give units and interpret your answer.

(c) From 2003 to 2015, were there any one-year intervals during which the average rate of change was negative? If so, which?

Table 1.20 *GM revenues, billions of dollars*

Year	2003	2004	2005	2006	2007	2008	2009
R	184.0	192.9	193.1	205.6	181.1	149.0	142.0

Year	2010	2011	2012	2013	2014	2015	
R	135.0	150.2	152.2	155.4	155.9	152.2	

32. The number of US households with cable television[50] was 12,168,450 in 1977 and 73,365,880 in 2003. Estimate the average rate of change in the number of US households with cable television during this 26-year period. Give units and interpret your answer.

33. Figure 1.44 shows the amount of nicotine $N = f(t)$, in mg, in a person's bloodstream as a function of the time, t, in hours, since the last cigarette.

(a) Is the average rate of change in nicotine level positive or negative? Explain.

(b) Find the average rate of change in the nicotine level between $t = 0$ and $t = 3$. Give units and interpret your answer in terms of nicotine.

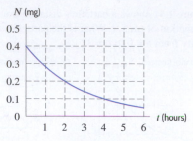

Figure 1.44

34. Table 1.21 shows the concentration, c, of creatinine in the bloodstream of a dog.[51]

(a) Including units, find the average rate at which the concentration is changing between the

(i) 6th and 8th minutes. (ii) 8th and 10th minutes.

(b) Explain the sign and relative magnitudes of your results in terms of creatinine.

Table 1.21

t (minutes)	2	4	6	8	10
c (mg/ml)	0.439	0.383	0.336	0.298	0.266

35. The population of the world reached 1 billion in 1804, 2 billion in 1927, 3 billion in 1960, 4 billion in 1974, 5 billion in 1987 and 6 billion in 1999. Find the average rate of change of the population of the world, in people per minute, during each of these intervals (that is, from 1804 to 1927, 1927 to 1960, etc.).

■ Problems **36–37** refer to Figure 1.45, which shows the contraction velocity of a muscle as a function of the load it pulls against.

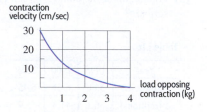

Figure 1.45

36. In terms of the muscle, interpret the

(a) Vertical intercept (b) Horizontal intercept

37. (a) Find the change in muscle contraction velocity when the load changes from 1 kg to 3 kg. Give units.

(b) Find the average rate of change in the contraction velocity between 1 kg and 3 kg. Give units.

[47] *The World Almanac and Book of Facts 2005*, p. 144 (New York).
[48] *The World Almanac and Book of Facts 2005*, p. 143 (New York).
[49] www.gm.com, accessed May 23, 2009 and September 18, 2016.
[50] *The World Almanac and Book of Facts 2005*, p. 310 (New York).
[51] From M. R. Cullen, *Linear Models in Biology* (Chichester: Ellis Horwood, 1985).

38. Table 1.22 gives the sales, S, of Intel Corporation, a leading manufacturer of integrated circuits.[52]

(a) Find the change in sales between 2000 and 2015.
(b) Find the average rate of change in sales between 2000 and 2015. Give units and interpret your answer.

Table 1.22 *Intel sales, in billions of dollars*

Year	2000	2005	2010	2015
S	33.7	38.8	43.6	55.36

39. Let $f(t)$ be the number of US billionaires in year t.

(a) Express the following statements[53] in terms of f.
  (i) In 2001 there were 272 US billionaires.
  (ii) In 2014 there were 525 US billionaires.
(b) Find the average yearly increase in the number of US billionaires from 2001 to 2014. Express this using f.
(c) Assuming the yearly increase remains constant, find a formula predicting the number of US billionaires in year t.

40. Figure 1.46 shows the position of an object at time t.

(a) Draw a line on the graph whose slope represents the average velocity between $t = 2$ and $t = 8$.
(b) Is average velocity greater between $t = 0$ and $t = 3$ or between $t = 3$ and $t = 6$?
(c) Is average velocity positive or negative between $t = 6$ and $t = 9$?

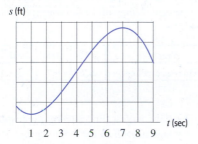

s (ft)

t (sec)

Figure 1.46

41. In an experiment, a lizard is encouraged to run as fast as possible. Figure 1.47 shows the distance run in meters as a function of the time in seconds.[54]

(a) If the lizard were running faster and faster, what would be the concavity of the graph? Does this match what you see?
(b) Estimate the average velocity of the lizard during this 0.8 second experiment.

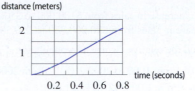

distance (meters)

time (seconds)

Figure 1.47

42. Values of $F(t)$, $G(t)$, and $H(t)$ are in Table 1.23. Which graph is concave up and which is concave down? Which function is linear?

Table 1.23

t	$F(t)$	$G(t)$	$H(t)$
10	15	15	15
20	22	18	17
30	28	21	20
40	33	24	24
50	37	27	29
60	40	30	35

43. Experiments suggest that the male maximum heart rate (the most times a male's heart can safely beat in a minute) decreases by 9 beats per minute during the first 21 years of his life, and by 26 beats per minute during the first 33 years.[55] If you model the maximum heart rate as a function of age, should you use a function that is increasing or decreasing? Concave up or concave down?

44. A car starts slowly and then speeds up. Eventually the car slows down and stops. Graph the distance that the car has traveled against time.

■ In Problems 45–48, find the relative, or percent, change.

45. B changes from 12,000 to 15,000

46. S changes from 400 to 450

47. W changes from 0.3 to 0.05

48. R changes from 50 to 47

■ In Problems 49–52, which relative change is bigger in magnitude? Justify your answer.

49. The change in the Dow Jones average from 164.6 to 77.9 in 1931; the change in the Dow Jones average from 13,261.8 to 8776.4 in 2008.

50. The change in the US population from 5.2 million to 7.2 million from 1800 to 1810; the change in the US population from 151.3 to 179.3 from 1950 to 1960.

51. An increase in class size from 5 to 10; an increase in class size from 30 to 50.

[52]www.statista.com/statistics/, accessed September 18, 2016.
[53]www.statista.com, accessed March 18, 2015.
[54]Data from R. B. Huey and P. E. Hertz, "Effects of Body Size and Slope on the Acceleration of a Lizard," *J. Exp. Biol.*, Volume 110, 1984, p. 113-123.
[55]www.css.edu/users/tboone2/asep/May2002JEPonline.html, accessed January 4, 2005.

52. An increase in sales from $100,000 to $500,000; an increase in sales from $20,000,000 to $20,500,000.

53. Find the relative change of a population if it changes

   (a) From 1000 to 2000   **(b)** From 2000 to 1000
   (c) From 1,000,000 to 1,001,000

54. On Black Monday, October 28, 1929, the stock market on Wall Street crashed. The Dow Jones average dropped from 298.94 to 260.64 in one day. What was the relative change in the index?

55. On January 22, 2017, the cost to mail a letter in the US[56] was raised from 47 cents to 49 cents. Find the relative change in the cost.

56. The US Consumer Price Index (CPI) is a measure of the cost of living. The inflation rate is the annual relative rate of change of the CPI. Use the January data[57] in Table 1.24 to estimate the inflation rate for each of the years 2007–2012.

Table 1.24

Year	2007	2008	2009	2010	2011	2012
CPI	202.416	211.08	211.143	216.687	220.223	226.655

57. During 2008 the US economy stopped growing and began to shrink. Table 1.25 gives quarterly data on the US Gross Domestic Product (GDP), which measures the size of the economy.[58]

   (a) Estimate the relative growth rate (percent per year) at the first four times in the table.
   (b) Economists often say an economy is in recession if the GDP decreases for two quarters in a row. Was the US in recession in 2008?

Table 1.25

t (years since 2008)	0	0.25	0.5	0.75	1.0
GDP (trillion dollars)	14.67	14.81	14.84	14.55	14.38

58. An alternative to petroleum-based diesel fuel, biodiesel, is derived from renewable resources such as food crops, algae, and animal oils. The table shows the recent annual percent growth in US biodiesel exports.[59]

   (a) Find the largest time interval over which the percentage growth in the US exports of biodiesel was an increasing function of time. Interpret what increasing means, practically speaking, in this case.
   (b) Find the largest time interval over which the actual US exports of biodiesel was an increasing function of time. Interpret what increasing means, practically speaking, in this case.

Year	2010	2011	2012	2013	2014
% growth over previous yr	−60.5	−30.5	69.9	53.0	−57.8

59. Hydroelectric power is electric power generated by the force of moving water. Figure 1.48 shows[60] the annual percent growth in hydroelectric power consumption by the US industrial sector between 2006 and 2014.

   (a) Find the largest time interval over which the percentage growth in the US consumption of hydroelectric power was an increasing function of time. Interpret what increasing means, practically speaking, in this case.
   (b) Find the largest time interval over which the actual US consumption of hydroelectric power was a decreasing function of time. Interpret what decreasing means, practically speaking, in this case.

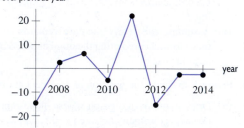

Figure 1.48

60. Solar panels are arrays of photovoltaic cells that convert solar radiation into electricity. The table shows the annual percent change in the US price per watt of a solar panel.[61]

Year	2005	2006	2007	2008	2009	2010
% growth over previous yr	6.7	9.7	−3.7	3.6	−20.1	−29.7

   (a) Find the largest time interval over which the percentage growth in the US price per watt of a solar panel was a decreasing function of time. Interpret what decreasing means, practically speaking, in this case.
   (b) Find the largest time interval over which the actual price per watt of a solar panel was a decreasing function of time. Interpret what decreasing means, practically speaking, in this case.

---

[56]www.wikipedia.com, accessed January 8, 2017.
[57]www.inflationdata.com, accessed December 30, 2016.
[58]www.bea.gov, accessed December 30, 2016.
[59]www.eia.doe.gov, accessed March 29, 2015.
[60]Yearly values have been joined with line segments to highlight trends in the data; however, values in between years should not be inferred from the segments. From www.eia.doe.gov, accessed March 29, 2015.
[61]We use the official price per peak watt, which uses the maximum number of watts a solar panel can produce under ideal conditions. From www.eia.doe.gov, accessed March 29, 2015.

61. School organizations raise money by selling candy door-to-door. When the price is \$1 a school organization sells 2765 candies and when the price goes up to \$1.25 the quantity of candy sold drops down to 2440.

    (a) Find the relative change in the price of candy.
    (b) Find the relative change in the quantity of candy

sold.
    (c) Find and interpret the ratio

$$\left| \frac{\text{Relative change in quantity}}{\text{Relative change in price}} \right|$$

and interpret your answer.

1.4 APPLICATIONS OF FUNCTIONS TO ECONOMICS

In this section, we look at some of the functions of interest to decision-makers in a firm or industry.

The Cost Function

> The **cost function**, $C(q)$, gives the total cost of producing a quantity q of some good.

What sort of function do you expect C to be? The more goods that are made, the higher the total cost, so C is an increasing function. Costs of production can be separated into two parts: the *fixed costs,* which are incurred even if nothing is produced, and the *variable costs,* which depend on how many units are produced.

An Example: Manufacturing Costs

Let's consider a company that makes radios. The factory and machinery needed to begin production are fixed costs, which are incurred even if no radios are made. The costs of labor and raw materials are variable costs since these quantities depend on how many radios are made. The fixed costs for this company are \$24,000 and the variable costs are \$7 per radio. Then,

$$\text{Total costs for the company} = \text{Fixed costs} + \text{Variable costs}$$
$$= 24{,}000 + 7 \cdot \text{Number of radios},$$

so, if q is the number of radios produced,

$$C(q) = 24{,}000 + 7q.$$

This is the equation of a line with slope 7 and vertical intercept 24,000.

    The variable cost for one additional unit is called the *marginal cost.* For a linear cost function, the marginal cost is the rate of change, or slope, of the cost function.

Example 1   Graph the cost function $C(q) = 24{,}000 + 7q$. Label the fixed costs and marginal cost.

Solution   The graph of $C(q)$ is the line in Figure 1.49. The fixed costs are represented by the vertical intercept of 24,000. The marginal cost is represented by the slope of 7, which is the change in cost corresponding to a unit change in production.

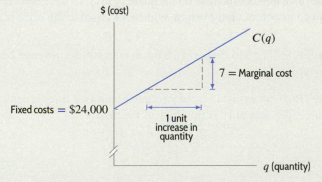

Figure 1.49: Cost function for the radio manufacturer

If $C(q)$ is a linear cost function,

- Fixed costs are represented by the vertical intercept.
- Marginal cost is represented by the slope.

Example 2   In each case, draw a graph of a linear cost function satisfying the given conditions:

    (a) Fixed costs are large but marginal cost is small.
    (b) There are no fixed costs but marginal cost is high.

Solution   (a) The graph is a line with a large vertical intercept and a small slope. See Figure 1.50.
    (b) The graph is a line with a vertical intercept of zero (so the line goes through the origin) and a large positive slope. See Figure 1.51. Figures 1.50 and 1.51 have the same scales.

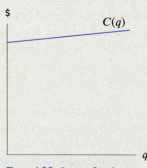

Figure 1.50: Large fixed costs, small marginal cost

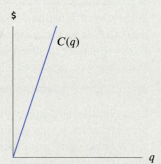

Figure 1.51: No fixed costs, high marginal cost

The Revenue Function

The **revenue function**, $R(q)$, gives the total revenue received by a firm from selling a quantity, q, of some good.

If the good sells for a price of p per unit, and the quantity sold is q, then

$$\text{Revenue} = \text{Price} \cdot \text{Quantity}, \quad \text{so} \quad R = pq.$$

If the price does not depend on the quantity sold, so p is a constant, the graph of revenue as a function of q is a line through the origin, with slope equal to the price p.

Example 3   If radios sell for \$15 each, sketch the manufacturer's revenue function. Show the price of a radio on the graph.

Solution   Since $R(q) = pq = 15q$, the revenue graph is a line through the origin with a slope of 15. See Figure 1.52. The price is the slope of the line.

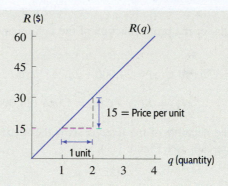

Figure 1.52: Revenue function for the radio manufacturer

Example 4   Graph the cost function $C(q) = 24{,}000 + 7q$ and the revenue function $R(q) = 15q$ on the same axes. For what values of q does the company make money?

Solution   The company makes money whenever revenues are greater than costs, so we find the values of q for which the graph of $R(q)$ lies above the graph of $C(q)$. See Figure 1.53.

We find the point at which the graphs of $R(q)$ and $C(q)$ cross:

$$\text{Revenue} = \text{Cost}$$
$$15q = 24{,}000 + 7q$$
$$8q = 24{,}000$$
$$q = 3000.$$

The company makes a profit if it produces and sells more than 3000 radios. The company loses money if it produces and sells fewer than 3000 radios.

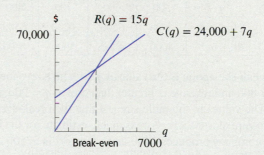

Figure 1.53: Cost and revenue functions for the radio manufacturer: What values of q generate a profit?

The Profit Function

Decisions are often made by considering the profit, usually written[62] as π to distinguish it from the price, p. We have

$$\boxed{\text{Profit} = \text{Revenue} - \text{Cost} \quad \text{so} \quad \pi = R - C.}$$

The *break-even point* for a company is the point where the profit is zero and revenue equals cost. See Figure 1.53.

---

[62]This π has nothing to do with the area of a circle, and merely stands for the Greek equivalent of the letter "p."

Example 5   Find a formula for the profit function of the radio manufacturer. Graph it, marking the break-even point.

Solution   Since $R(q) = 15q$ and $C(q) = 24{,}000 + 7q$, we have

$$\pi(q) = R(q) - C(q) = 15q - (24{,}000 + 7q) = -24{,}000 + 8q.$$

Notice that the negative of the fixed costs is the vertical intercept and the break-even point is the horizontal intercept. See Figure 1.54.

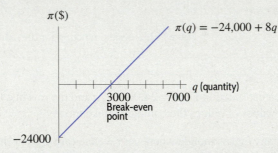

Figure 1.54: Profit for radio manufacturer

Example 6   (a) Using Table 1.26, estimate the break-even point for this company.
(b) Find the company's profit if 1000 units are produced.
(c) What price do you think the company is charging for its product?

Table 1.26   *Company's estimates of cost and revenue for a product*

q	500	600	700	800	900	1000	1100
$C(q)$, in $	5000	5500	6000	6500	7000	7500	8000
$R(q)$, in $	4000	4800	5600	6400	7200	8000	8800

Solution   (a) The break-even point is the value of q for which revenue equals cost. Since revenue is below cost at $q = 800$ and revenue is greater than cost at $q = 900$, the break-even point is between 800 and 900. The values in the table suggest that the break-even point is closer to 800, as the cost and revenue are closer there. A reasonable estimate for the break-even point is $q = 830$.
(b) If the company produces 1000 units, the cost is $7500 and the revenue is $8000, so the profit is $8000 - 7500 = 500$ dollars.
(c) From the table, it appears that $R(q) = 8q$. This indicates the company is selling the product for $8 each.

The Marginal Cost, Marginal Revenue, and Marginal Profit

Just as we used the term marginal cost to mean the rate of change, or slope, of a linear cost function, we use the terms *marginal revenue* and *marginal profit* to mean the rate of change, or slope, of linear revenue and profit functions, respectively. The term *marginal* is used because we are looking at how the cost, revenue, or profit change "at the margin," that is, by the addition of one more unit. For example, for the radio manufacturer, the marginal cost is 7 dollars/item (the additional cost of producing one more item is $7), the marginal revenue is 15 dollars/item (the additional revenue from selling one more item is $15), and the marginal profit is 8 dollars/item (the additional profit from selling one more item is $8).

The Depreciation Function

Suppose that the radio manufacturer has a machine that costs \$20,000 and is sold ten years later for \$3000. We say the value of the machine *depreciates* from \$20,000 today to a resale value of \$3000 in ten years. The depreciation formula gives the value, $V(t)$, in dollars, of the machine as a function of the number of years, t, since the machine was purchased. We assume that the value of the machine depreciates linearly.

The value of the machine when it is new ($t = 0$) is \$20,000, so $V(0) = 20,000$. The resale value at time $t = 10$ is \$3000, so $V(10) = 3000$. We have

$$\text{Slope} = m = \frac{3000 - 20,000}{10 - 0} = \frac{-17,000}{10} = -1700 \text{ dollars per year.}$$

This slope tells us that the value of the machine is decreasing at a rate of \$1700 per year. Since $V(0) = 20,000$, the vertical intercept is 20,000, so

$$V(t) = 20,000 - 1700t \text{ dollars.}$$

Supply and Demand Curves

The quantity, q, of an item that is manufactured and sold depends on its price, p. As the price increases, manufacturers are usually willing to supply more of the product, whereas the quantity demanded by consumers falls.

> The **supply curve**, for a given item, relates the quantity, q, of the item that manufacturers are willing to make per unit time to the price, p, for which the item can be sold.
> The **demand curve** relates the quantity, q, of an item demanded by consumers per unit time to the price, p, of the item.

Economists often think of the quantities supplied and demanded as functions of price. However, for historical reasons, the economists put price (the independent variable) on the vertical axis and quantity (the dependent variable) on the horizontal axis. (The reason for this state of affairs is that economists originally took price to be the dependent variable and put it on the vertical axis. Later, when the point of view changed, the axes did not.) Thus, typical supply and demand curves look like those shown in Figure 1.55.

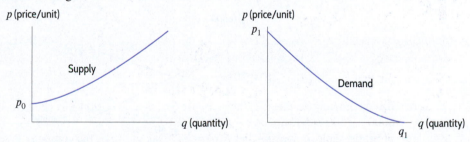

Figure 1.55: Supply and demand curves

Example 7   What is the economic meaning of the prices p_0 and p_1 and the quantity q_1 in Figure 1.55?

Solution   The vertical axis corresponds to a quantity of zero. Since the price p_0 is the vertical intercept on the supply curve, p_0 is the price at which the quantity supplied is zero. In other words, for prices below p_0, the suppliers will not produce anything. The price p_1 is the vertical intercept on the demand curve, so it corresponds to the price at which the quantity demanded is zero. In other words, for prices above p_1, consumers buy none of the product.

The horizontal axis corresponds to a price of zero, so the quantity q_1 on the demand curve is the quantity demanded if the price were zero—the quantity that could be given away free.

Equilibrium Price and Quantity

If we plot the supply and demand curves on the same axes, as in Figure 1.56, the graphs cross at the *equilibrium point*. The values p^* and q^* at this point are called the *equilibrium price* and *equilibrium quantity*, respectively. It is assumed that the market naturally settles to this equilibrium point. (See Problem 28.)

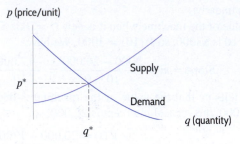

Figure 1.56: The equilibrium price and quantity

Example 8   Find the equilibrium price and quantity if

$$\text{Quantity supplied} = 3p - 50 \quad \text{and} \quad \text{Quantity demanded} = 100 - 2p.$$

Solution   To find the equilibrium price and quantity, we find the point at which

$$\text{Supply} = \text{Demand}$$
$$3p - 50 = 100 - 2p$$
$$5p = 150$$
$$p = 30.$$

The equilibrium price is \$30. To find the equilibrium quantity, we use either the demand curve or the supply curve. At a price of \$30, the quantity produced is $100 - 2 \cdot 30 = 40$ items. The equilibrium quantity is 40 items. In Figure 1.57, the demand and supply curves intersect at $p^* = 30$ and $q^* = 40$.

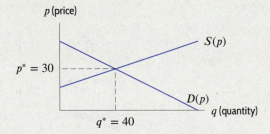

Figure 1.57: Equilibrium: $p^* = 30$, $q^* = 40$

The Effect of Taxes on Equilibrium

What effect do taxes have on the equilibrium price and quantity for a product? We distinguish between two types of taxes.[63] A *specific tax* is a fixed amount per unit of a product sold regardless of the selling price. This is the case with such items as gasoline, alcohol, and cigarettes. A specific tax is usually imposed on the producer. A *sales tax* is a fixed percentage of the selling price. Many cities and states collect sales tax on a wide variety of items. A sales tax is usually imposed on the consumer. We consider a specific tax now; a sales tax is considered in Problems 44 and 45.

---

[63] Adapted from Barry Bressler, *A Unified Approach to Mathematical Economics*, p. 81–88 (New York: Harper & Row, 1975).

Example 9   A specific tax of $5 per unit is now imposed upon suppliers in Example 8. What are the new equilibrium price and quantity?

Solution   The consumers pay p dollars per unit, but the suppliers receive only $p - 5$ dollars per unit because $5 goes to the government as taxes. Since

$$\text{Quantity supplied} = 3(\text{Amount per unit received by suppliers}) - 50,$$

the new supply equation is

$$\text{Quantity supplied} = 3(p - 5) - 50 = 3p - 65;$$

the demand equation is unchanged:

$$\text{Quantity demanded} = 100 - 2p.$$

At the equilibrium price, we have

$$\text{Demand} = \text{Supply}$$
$$100 - 2p = 3p - 65$$
$$165 = 5p$$
$$p = 33.$$

The equilibrium price is $33. The equilibrium quantity is 34 units, since the quantity demanded is $q = 100 - 2 \cdot 33 = 34$.

In Example 8, the equilibrium price was $30; with the imposition of a $5 tax in Example 9, the equilibrium price is $33. Thus the equilibrium price increases by $3 as a result of the tax. Notice that this is less than the amount of the tax. The consumer ends up paying $3 more than if the tax did not exist. However the government receives $5 per item. The producer pays the other $2 of the tax, retaining $28 of the price paid per item. Although the tax was imposed on the producer, some of the tax is passed on to the consumer in terms of higher prices. The tax has increased the price and reduced the number of items sold. See Figure 1.58. Notice that the taxes have the effect of moving the supply curve up by $5 because suppliers have to be paid $5 more to produce the same quantity.

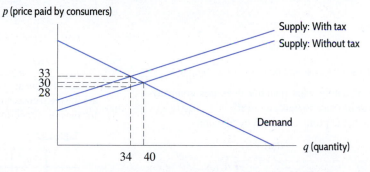

Figure 1.58: Specific tax shifts the supply curve, altering the equilibrium price and quantity

A Budget Constraint

An ongoing debate in the federal government concerns the allocation of money between defense and social programs. In general, the more that is spent on defense, the less that is available for social programs, and vice versa. Let's simplify the example to guns and butter. Assuming a constant budget, we show that the relationship between the number of guns and the quantity of butter is linear. Suppose that there is $12,000 to be spent and that it is to be divided between guns, costing $400 each, and butter, costing $2000 a ton. Suppose the number of guns bought is g, and the number of tons of butter is b. Then the amount of money spent on guns is $\$400g$, and the amount spent on butter is $\$2000b$. Assuming all the money is spent,

$$\text{Amount spent on guns} + \text{Amount spent on butter} = \$12{,}000$$

or

$$400g + 2000b = 12{,}000.$$

Thus, dividing both sides by 400,

$$g + 5b = 30.$$

This equation is the budget constraint. Since the budget constraint can be written as

$$g = 30 - 5b,$$

the graph of the budget constraint is a line. See Figure 1.59.

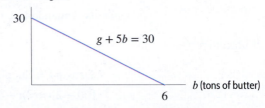

Figure 1.59: Budget constraint

Problems for Section 1.4

1. In Figure 1.60, which shows the cost and revenue functions for a product, label each of the following:

   (a) Fixed costs   **(b)** Break-even quantity

   (c) Quantities at which the company:

      (i) Makes a profit   **(ii)** Loses money

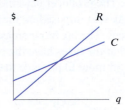

Figure 1.60

2. Figure 1.61 shows cost and revenue for a company.

   (a) Approximately what quantity does this company have to produce to make a profit?

   (b) Estimate the profit generated by 600 units.

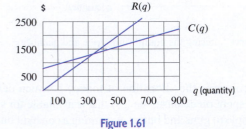

Figure 1.61

3. (a) Estimate the fixed costs and the marginal cost for the cost function in Figure 1.62.

   (b) Estimate $C(10)$ and interpret it in terms of cost.

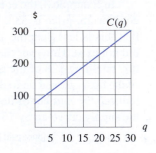

Figure 1.62

4. Values of a linear cost function are in Table 1.27. What are the fixed costs and the marginal cost? Find a formula for the cost function.

Table 1.27

q	0	5	10	15	20
$C(q)$	5000	5020	5040	5060	5080

5. The cost C, in millions of dollars, of producing q items is given by $C = 5.7 + 0.002q$. Interpret the 5.7 and the 0.002 in terms of production. Give units.

6. (a) Give an example of a possible company where the fixed costs are zero (or very small).
  (b) Give an example of a possible company where the marginal cost is zero (or very small).

7. Suppose that $q = f(p)$ is the demand curve for a product, where p is the selling price in dollars and q is the quantity sold at that price.
  (a) What does the statement $f(12) = 60$ tell you about demand for this product?
  (b) Do you expect this function to be increasing or decreasing? Why?

8. A company has cost and revenue functions, in dollars, given by $C(q) = 6000 + 10q$ and $R(q) = 12q$.
  (a) Find the cost and revenue if the company produces 500 units. Does the company make a profit? What about 5000 units?
  (b) Find the break-even point and illustrate it graphically.

9. The demand curve for a quantity q of a product is $q = 5500 - 100p$ where p is price in dollars. Interpret the 5500 and the 100 in terms of demand. Give units.

10. A demand curve is given by $75p + 50q = 300$, where p is the price of the product, in dollars, and q is the quantity demanded at that price. Find the p- and q-intercepts and interpret them in terms of consumer demand.

■ In Problems **11–14**, give the cost, revenue, and profit functions.

11. An online seller of T-shirts pays $500 to start up the website and $6 per T-shirt, then sells the T-shirts for $12 each.

12. A car wash operator pays $35,000 for a franchise, then spends $10 per car wash, which costs the consumer $15.

13. A couple running a house-cleaning business invests $5000 in equipment, and they spend $15 in supplies to clean a house, for which they charge $60.

14. A lemonade stand operator sets up the stand for free in front of the neighbor's house, makes 5 quarts of lemonade for $4, then sells each 8-oz cup for 25 cents.

15. A company that makes Adirondack chairs has fixed costs of $5000 and variable costs of $30 per chair. The company sells the chairs for $50 each.
  (a) Find formulas for the cost and revenue functions.
  (b) Find the marginal cost and marginal revenue.
  (c) Graph the cost and the revenue functions on the same axes.
  (d) Find the break-even point.

16. An amusement park charges an admission fee of $21 per person as well as an additional $4.50 for each ride.
  (a) For one visitor, find the park's total revenue $R(n)$ as a function of the number of rides, n, taken.
  (b) Find $R(2)$ and $R(8)$ and interpret your answers in terms of amusement park fees.

17. A photocopying company has two different price lists. The first price list is $100 plus 3 cents per copy; the second price list is $200 plus 2 cents per copy.
  (a) For each price list, find the total cost as a function of the number of copies needed.
  (b) Determine which price list is cheaper for 5000 copies.
  (c) For what number of copies do both price lists charge the same amount?

18. A company has cost function $C(q) = 4000 + 2q$ dollars and revenue function $R(q) = 10q$ dollars.
  (a) What are the fixed costs for the company?
  (b) What is the marginal cost?
  (c) What price is the company charging for its product?
  (d) Graph $C(q)$ and $R(q)$ on the same axes and label the break-even point, q_0. Explain how you know the company makes a profit if the quantity produced is greater than q_0.
  (e) Find the break-even point q_0.

19. A movie theater has fixed costs of $5000 per day and variable costs averaging $6 per customer. The theater charges $11 per ticket.
  (a) How many customers per day does the theater need in order to make a profit?
  (b) Find the cost and revenue functions and graph them on the same axes. Mark the break-even point.

20. A company producing jigsaw puzzles has fixed costs of $6000 and variable costs of $2 per puzzle. The company sells the puzzles for $5 each.
  (a) Find formulas for the cost function, the revenue function, and the profit function.
  (b) Sketch a graph of $R(q)$ and $C(q)$ on the same axes. What is the break-even point, q_0, for the company?

21. Production costs for manufacturing running shoes consist of a fixed overhead of $650,000 plus variable costs of $20 per pair of shoes. Each pair of shoes sells for $70.
  (a) Find the total cost, $C(q)$, the total revenue, $R(q)$, and the total profit, $\pi(q)$, as a function of the number of pairs of shoes produced, q.
  (b) Find the marginal cost, marginal revenue, and marginal profit.
  (c) How many pairs of shoes must be produced and sold for the company to make a profit?

22. A $15,000 robot depreciates linearly to zero in 10 years.
  (a) Find a formula for its value as a function of time.
  (b) How much is the robot worth three years after it is purchased?

23. A $50,000 tractor has a resale value of $10,000 twenty years after it was purchased. Assume that the value of the tractor depreciates linearly from the time of purchase.

(a) Find a formula for the value of the tractor as a function of the time since it was purchased.
(b) Graph the value of the tractor against time.
(c) Find the horizontal and vertical intercepts, give units, and interpret them.

24. A new bus worth $100,000 in 2015 depreciates linearly to $25,000 in 2030.

(a) Find a formula for the value of the bus, V, as a function of time, t, in years since 2015.
(b) What is the value of the bus in 2020?
(c) Find and interpret the vertical and horizontal intercepts of the graph of the function.
(d) Assuming depreciation continues at the same linear rate, what is the domain of the function?

25. A corporate office provides the demand curve in Figure 1.63 to its ice cream shop franchises. At a price of $1.00 per scoop, 240 scoops per day can be sold.

(a) Estimate how many scoops could be sold per day at a price of 50¢ per scoop. Explain.
(b) Estimate how many scoops per day could be sold at a price of $1.50 per scoop. Explain.

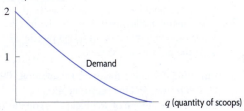

p (price per scoop in dollars)

Demand

q (quantity of scoops)

Figure 1.63

26. The table shows the cost of manufacturing various quantities of an item and the revenue obtained from their sale.

Quantity	0	10	20	30	40	50	60	70	80
Cost ($)	120	400	600	780	1000	1320	1800	2500	3400
Revenue ($)	0	300	600	900	1200	1500	1800	2100	2400

(a) What range of production levels appears to be profitable?
(b) Calculate the profit or loss for each of the quantities shown. Estimate the most profitable production level.

27. One of Tables 1.28 and 1.29 represents a supply curve; the other represents a demand curve.

(a) Which table represents which curve? Why?
(b) At a price of $155, approximately how many items would consumers purchase?
(c) At a price of $155, approximately how many items would manufacturers supply?
(d) Will the market push prices higher or lower than $155?
(e) What would the price have to be if you wanted consumers to buy at least 20 items?
(f) What would the price have to be if you wanted manufacturers to supply at least 20 items?

Table 1.28

p ($/unit)	182	167	153	143	133	125	118
q (quantity)	5	10	15	20	25	30	35

Table 1.29

p ($/unit)	6	35	66	110	166	235	316
q (quantity)	5	10	15	20	25	30	35

28. Figure 1.64 shows supply and demand for a product.

(a) What is the equilibrium price for this product? At this price, what quantity is produced?
(b) Choose a price above the equilibrium price—for example, $p = 12$. At this price, how many items are suppliers willing to produce? How many items do consumers want to buy? Use your answers to these questions to explain why, if prices are above the equilibrium price, the market tends to push prices lower (toward the equilibrium).
(c) Now choose a price below the equilibrium price—for example, $p = 8$. At this price, how many items are suppliers willing to produce? How many items do consumers want to buy? Use your answers to these questions to explain why, if prices are below the equilibrium price, the market tends to push prices higher (toward the equilibrium).

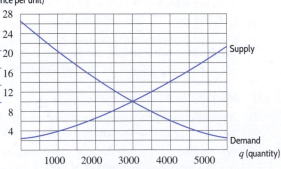

p (price per unit)

Supply

Demand

q (quantity)

Figure 1.64

29. A company produces and sells shirts. The fixed costs are $7000 and the variable costs are $5 per shirt.

    (a) Shirts are sold for $12 each. Find cost and revenue as functions of the quantity of shirts, q.
    (b) The company is considering changing the selling price of the shirts. Demand is $q = 2000 - 40p$, where p is price in dollars and q is the number of shirts. What quantity is sold at the current price of $12? What profit is realized at this price?
    (c) Use the demand equation to write cost and revenue as functions of the price, p. Then write profit as a function of price.
    (d) Graph profit against price. Find the price that maximizes profits. What is this profit?

30. When the price, p, charged for a boat tour was $25, the average number of passengers per week, N, was 500. When the price was reduced to $20, the average number of passengers per week increased to 650. Find a formula for the demand curve, assuming that it is linear.

31. Table 1.30 gives data for the linear demand curve for a product, where p is the price of the product and q is the quantity sold every month at that price. Find formulas for the following functions. Interpret their slopes in terms of demand.

    (a) q as a function of p.   (b) p as a function of q.

    Table 1.30

p (dollars)	16	18	20	22	24
q (tons)	500	460	420	380	340

32. The demand curve for a product is given by $q = 120{,}000 - 500p$ and the supply curve is given by $q = 1000p$ for $0 \le q \le 120{,}000$, where price is in dollars.

    (a) At a price of $100, what quantity are consumers willing to buy and what quantity are producers willing to supply? Will the market push prices up or down?
    (b) Find the equilibrium price and quantity. Does your answer to part (a) support the observation that market forces tend to push prices closer to the equilibrium price?

33. World production, Q, of zinc in thousands of metric tons and the value, P, in dollars per metric ton are given[64] in Table 1.31. Plot the value as a function of production. Sketch a possible supply curve.

Table 1.31 *World zinc production*

Year	2003	2004	2005	2006	2007	2008	2009	2010
Q	9520	9600	10,000	10,300	11,000	11,700	11,400	12,000
P	896	1160	1480	3500	3400	1960	1720	2250

34. A taxi company has an annual budget of $720,000 to spend on drivers and car replacement. Drivers cost the company $30,000 each and car replacements cost $20,000 each.

    (a) What is the company's budget constraint equation? Let d be the number of drivers paid and c be the number of cars replaced.
    (b) Find and interpret both intercepts of the graph of the equation.

35. A company has a total budget of $500,000 and spends this budget on raw materials and personnel. The company uses m units of raw materials, at a cost of $100 per unit, and hires r employees, at a cost of $25,000 each.

    (a) What is the equation of the company's budget constraint?
    (b) Solve for m as a function of r.
    (c) Solve for r as a function of m.

36. You have a budget of $2000 for the year to cover your books and social outings. Books cost (on average) $80 each and social outings cost (on average) $20 each. Let b denote the number of books purchased per year and s denote the number of social outings in a year.

    (a) What is the equation of your budget constraint?
    (b) Graph the budget constraint. (It does not matter which variable you put on which axis.)
    (c) Find the vertical and horizontal intercepts, and give a financial interpretation for each.

37. A bakery owner knows that customers buy a total of q cakes when the price, p, is no more than $p = d(q) = 20 - q/20$ dollars. She is willing to make and supply as many as q cakes at a price of $p = s(q) = 11 + q/40$ dollars each. The graphs of $d(q)$ and $s(q)$ are in Figure 1.65.

    (a) Why, in terms of the context, is the slope of $d(q)$ negative and the slope of $s(q)$ positive?
    (b) Is each of the ordered pairs (q, p) a solution to the inequality $p \le 20 - q/20$? Interpret your answers in terms of the context.

    (60, 18)   (120, 12)

    (c) Graph in the qp-plane the solution set of the system of inequalities $p \le 20 - q/20$, $p \ge 11 + q/40$. What does this solution set represent in terms of the context?
    (d) What is the rightmost point of the solution set you graphed in part (c)? Interpret your answer in terms of the context.

[64]http://minerals.usgs.gov/ds/2005/140/, accessed September, 2012.

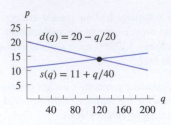

Figure 1.65

38. Linear supply and demand curves are shown in Figure 1.66, with price on the vertical axis.

    (a) Label the equilibrium price p_0 and the equilibrium quantity q_0 on the axes.

    (b) Explain the effect on equilibrium price and quantity if the slope, $\Delta p/\Delta q$, of the supply curve increases. Illustrate your answer graphically.

    (c) Explain the effect on equilibrium price and quantity if the slope, $\Delta p/\Delta q$, of the demand curve becomes more negative. Illustrate your answer graphically.

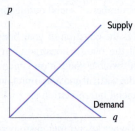

Figure 1.66

39. The demand for a product is given by $p = 90 - 10q$. Find the ratio

$$\left| \frac{\text{Relative change in demand}}{\text{Relative change in price}} \right|$$

if the price changes from $p = 50$ to $p = 51$. Interpret this ratio.

40. A demand curve has equation $q = 100 - 5p$, where p is price in dollars. A \$2 tax is imposed on consumers. Find the equation of the new demand curve. Sketch both curves.

41. A supply curve has equation $q = 4p - 20$, where p is price in dollars. A \$2 tax is imposed on suppliers. Find the equation of the new supply curve. Sketch both curves.

42. A tax of \$8 per unit is imposed on the supplier of an item. The original supply curve is $q = 0.5p - 25$ and the demand curve is $q = 165 - 0.5p$, where p is price in dollars. Find the equilibrium price and quantity before and after the tax is imposed.

43. The demand and supply curves for a product are given in terms of price, p, by

$$q = 2500 - 20p \quad \text{and} \quad q = 10p - 500.$$

    (a) Find the equilibrium price and quantity. Represent your answers on a graph.

    (b) A specific tax of \$6 per unit is imposed on suppliers. Find the new equilibrium price and quantity. Represent your answers on the graph.

    (c) How much of the \$6 tax is paid by consumers and how much by producers?

    (d) What is the total tax revenue received by the government?

44. The demand and supply curves are given by $q = 100 - 2p$ and $q = 3p - 50$, respectively; the equilibrium price is \$30 and the equilibrium quantity is 40 units. A sales tax of 5% is imposed on the consumer.

    (a) Find the equation of the new demand and supply curves.

    (b) Find the new equilibrium price and quantity.

    (c) How much is paid in taxes on each unit? How much of this is paid by the consumer and how much by the producer?

    (d) How much tax does the government collect?

45. Answer the questions in Problem 44, assuming that the 5% sales tax is imposed on the supplier instead of the consumer.

1.5   EXPONENTIAL FUNCTIONS

The function $f(x) = 2^x$, where the power is variable, is an *exponential function*. The number 2 is called the base. Exponential functions of the form $f(x) = k \cdot a^x$, where a is a positive constant, are used to represent many phenomena in the natural and social sciences.

Population Growth

The population of Burkina Faso, a sub-Saharan African country,[65] from 2007 to 2013 is given in Table 1.32. To see how the population is growing, we look at the increase in population in the third column. If the population had been growing linearly, all the numbers in the third column would be the same.

[65] data.worldbank.org, accessed September 20, 2016.

Table 1.32   *Population of Burkina Faso (estimated), 2007–2013*

Year	Population (millions)	Change in population (millions)
2007	14.235	
		0.425
2008	14.660	
		0.435
2009	15.095	
		0.445
2010	15.540	
		0.455
2011	15.995	
		0.465
2012	16.460	
		0.474
2013	16.934	

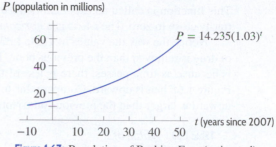

Figure 1.67: Population of Burkina Faso (estimated): Exponential growth

Suppose we divide each year's population by the previous year's population. For example,

$$\frac{\text{Population in 2008}}{\text{Population in 2007}} = \frac{14.660 \text{ million}}{14.235 \text{ million}} = 1.03$$

$$\frac{\text{Population in 2009}}{\text{Population in 2008}} = \frac{15.095 \text{ million}}{14.660 \text{ million}} = 1.03.$$

The fact that both calculations give 1.03 shows the population grew by about 3% between 2007 and 2008 *and* between 2008 and 2009. Similar calculations for other years show that the population grew by a factor of about 1.03, or 3%, every year. Whenever we have a constant percent increase (here 3%), we have *exponential growth*. If t is the number of years since 2007 then projected population in millions is,

When $t = 0$,   projected population $= 14.235 = 14.235(1.03)^0$.

When $t = 1$,   projected population $= 14.662 = 14.235(1.03)^1$.

When $t = 2$,   projected population $= 15.102 = 14.662(1.03) = 14.235(1.03)^2$.

When $t = 3$,   projected population $= 15.555 = 15.102(1.03) = 14.235(1.03)^3$.

So P, the population in millions t years after 2007, is given approximately by

$$P = 14.235(1.03)^t \text{ million}.$$

Since the variable t is in the exponent, this is an exponential function. The base, 1.03, represents the factor by which the population grows each year and is called the *growth factor*. Assuming that the formula holds for 50 years, the population graph has the shape in Figure 1.67. The population is growing, so the function is increasing. Since the population grows faster as time passes, the graph is concave up. This behavior is typical of an exponential function. Even exponential functions that climb slowly at first, such as this one, eventually climb extremely quickly.

Elimination of a Drug from the Body

Now we look at a quantity that is decreasing instead of increasing. When a patient is given medication, the drug enters the bloodstream. The rate at which the drug is metabolized and eliminated depends on the particular drug. For the antibiotic ampicillin, approximately 40% of the drug is eliminated every hour. A typical dose of ampicillin is 250 mg. Suppose $Q = f(t)$, where Q is the quantity of ampicillin, in mg, in the bloodstream at time t hours since the drug was given. At $t = 0$, we have $Q = 250$. Since the quantity remaining at the end of each hour is 60% of the quantity remaining the hour before, we have

$$f(0) = 250$$
$$f(1) = 250(0.6)$$
$$f(2) = 250(0.6)(0.6) = 250(0.6)^2$$
$$f(3) = 250(0.6)^2(0.6) = 250(0.6)^3.$$

So, after t hours,

$$Q = f(t) = 250(0.6)^t.$$

This function is called an *exponential decay* function. As t increases, the function values get arbitrarily close to zero. The t-axis is a *horizontal asymptote* for this function.

Notice the way the values in Table 1.33 are decreasing. Each additional hour a smaller quantity of drug is removed than the previous hour (100 mg the first hour, 60 mg the second, and so on). This is because as time passes, there is less of the drug in the body to be removed. Thus, the graph in Figure 1.68 bends upward. Compare this to the exponential growth in Figure 1.67, where each step upward is larger than the previous one. Notice that both graphs are concave up.

Table 1.33   *Value of decay function*

t (hours)	Q (mg)
0	250
1	150
2	90
3	54
4	32.4
5	19.4

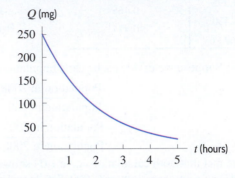

Figure 1.68: Drug elimination: Exponential decay

The General Exponential Function

Exponential growth is often described in terms of percent growth rates. The population of Burkina Faso is growing at 3% per year, so it increases by a factor of $a = 1 + 0.03 = 1.03$ every year. Similarly, 40% of the ampicillin is removed every hour, so the quantity remaining decays by a factor of $a = 1 - 0.40 = 0.6$ each hour. We have the following general formulas.

> We say that P is an **exponential function** of t with base a if
>
> $$P = P_0 a^t,$$
>
> where P_0 is the initial quantity (when $t = 0$) and a is the factor by which P changes when t increases by 1. If $a > 1$, we have **exponential growth**; if $0 < a < 1$, we have **exponential decay**. The factor a is given by
>
> $$a = 1 + r$$
>
> where r is the decimal representation of the percent rate of change; r may be positive (for growth) or negative (for decay).

The largest possible domain for the exponential function is all real numbers,[66] provided $a > 0$.

Comparison Between Linear and Exponential Functions

Every exponential function changes at a constant percent, or *relative*, rate. For example, the population of Burkina Faso increased approximately 3% per year. Every linear function changes at a constant absolute rate. For example, the Olympic pole vault record increased by 5 cm per year.

---

[66]The reason we do not want $a \leq 0$ is that, for example, we cannot define $a^{1/2}$ if $a < 0$. Also, we do not usually have $a = 1$, since $P = P_0 a^t = P_0 1^t = P_0$ is then a constant function.

A **linear** function has a constant rate of change.
An **exponential** function has a constant percent, or relative, rate of change.

Example 1 The amount of adrenaline in the body can change rapidly. Suppose the initial amount is 15 mg. Find a formula for A, the amount in mg, at a time t minutes later if A is:

(a) Increasing by 0.4 mg per minute.
(b) Decreasing by 0.4 mg per minute.
(c) Increasing by 3% per minute.
(d) Decreasing by 3% per minute.

Solution (a) This is a linear function with initial quantity 15 and slope 0.4, so

$$A = 15 + 0.4t.$$

(b) This is a linear function with initial quantity 15 and slope -0.4, so

$$A = 15 - 0.4t.$$

(c) This is an exponential function with initial quantity 15 and base $1 + 0.03 = 1.03$, so

$$A = 15(1.03)^t.$$

(d) This is an exponential function with initial quantity 15 and base $1 - 0.03 = 0.97$, so

$$A = 15(0.97)^t.$$

Example 2 Wolves were once common in the western US. By the 1990s, the wolf population in Wyoming had been wiped out by hunters and wolves were put on the endangered species list. In 1995, wolves were reintroduced to Wyoming from Canada. Starting with 14 wolves, their number increased to 207 wolves in 2012.[67] Assuming the Wyoming wolf population was growing exponentially, find a function of the form $P = P_0 a^t$, where P is the population t years after 1995. What is the annual percent growth rate?

Solution We know that $P_0 = 14$ when $t = 0$. In 2012, when $t = 17$, we have $P = 207$. Substituting in $P = P_0 a^t$ gives an equation we can solve for a:

$$207 = 14a^{17}.$$

Dividing both sides by 14 gives

$$\frac{207}{14} = a^{17}.$$

Taking the 17th root of both sides, we get

$$a = (207/14)^{1/17} = 1.172.$$

Since $a = 1.172$, the Wyoming wolf population as a function of the number of years since 1995 is given by

$$P = 14(1.172)^t.$$

During this period, the population increased by about 17% per year.

---

[67]Based on "Wyoming Drops Federal Protection of Gray Wolves", BBC, August 31, 2012, and *Wyoming Wolf Program Status Report*, US Fish and Wildlife Services, September, 2012.

Example 3   Suppose that $Q = f(t)$ is an exponential function of t. If $f(20) = 88.2$ and $f(23) = 91.4$:

(a)   Find the base.   (b)   Find the percent growth rate.

(c)   Evaluate $f(25)$.

Solution   (a) Let $Q = Q_0 a^t$. Substituting $t = 20, Q = 88.2$ and $t = 23, Q = 91.4$ gives two equations for Q_0 and a:

$$88.2 = Q_0 a^{20} \quad \text{and} \quad 91.4 = Q_0 a^{23}.$$

Dividing the two equations enables us to eliminate Q_0:

$$\frac{91.4}{88.2} = \frac{Q_0 a^{23}}{Q_0 a^{20}} = a^3.$$

Solving for the base, a, gives

$$a = \left(\frac{91.4}{88.2}\right)^{1/3} = 1.012.$$

(b) Since $a = 1.012$, the percent growth rate is 1.2%.

(c) We want to evaluate $f(25) = Q_0 a^{25} = Q_0(1.012)^{25}$. First we find Q_0 from the equation

$$88.2 = Q_0(1.012)^{20}.$$

Solving gives $Q_0 = 69.5$. Thus,

$$f(25) = 69.5(1.012)^{25} = 93.6.$$

The Family of Exponential Functions and the Number e

The formula $P = P_0 a^t$ gives a family of exponential functions with parameters P_0 (the initial quantity) and a (the base). The base tells us whether the function is increasing ($a > 1$) or decreasing ($0 < a < 1$). Since a is the factor by which P changes when t is increased by 1, large values of a mean fast growth; values of a near 0 mean fast decay. (See Figures 1.69 and 1.70.) All members of the family $P = P_0 a^t$ are concave up if $P_0 > 0$.

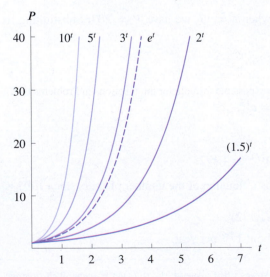

Figure 1.69: Exponential growth: $P = a^t$, for $a > 1$

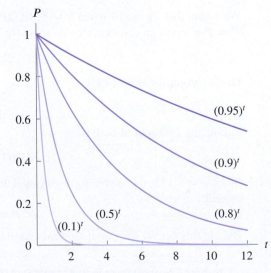

Figure 1.70: Exponential decay: $P = a^t$, for $0 < a < 1$

In practice the most commonly used base is the number $e = 2.71828\ldots$. The fact that most calculators have an e^x button is an indication of how important e is. Since e is between 2 and 3, the graph of $y = e^t$ in Figure 1.69 is between the graphs of $y = 2^t$ and $y = 3^t$.

The base e is used so often that it is called the natural base. At first glance, this is somewhat mysterious: What could be natural about using 2.71828... as a base? The full answer to this question must wait until Chapter 3, where you will see that many calculus formulas come out more neatly when e is used as the base. (See Appendix B for the relation to compound interest.)

Problems for Section 1.5

1. The following functions give the populations of four towns with time t in years.

   (i) $P = 600(1.12)^t$ (ii) $P = 1{,}000(1.03)^t$
   (iii) $P = 200(1.08)^t$ (iv) $P = 900(0.90)^t$

   (a) Which town has the largest percent growth rate? What is the percent growth rate?
   (b) Which town has the largest initial population? What is that initial population?
   (c) Are any of the towns decreasing in size? If so, which one(s)?

2. Each of the following functions gives the amount of a substance present at time t. In each case, give the amount present initially (at $t = 0$), state whether the function represents exponential growth or decay, and give the percent growth or decay rate.

   (a) $A = 100(1.07)^t$ (b) $A = 5.3(1.054)^t$
   (c) $A = 3500(0.93)^t$ (d) $A = 12(0.88)^t$

3. Figure 1.71 shows $Q = 50(1.2)^t$, $Q = 50(0.6)^t$, $Q = 50(0.8)^t$, and $Q = 50(1.4)^t$. Match each formula to a graph.

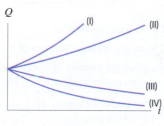

Figure 1.71

4. Figure 1.72 shows graphs of several cities' populations against time. Match each of the following descriptions to a graph and write a description to match each of the remaining graphs.

   (a) The population increased at 5% per year.
   (b) The population increased at 8% per year.
   (c) The population increased by 5000 people per year.
   (d) The population was stable.

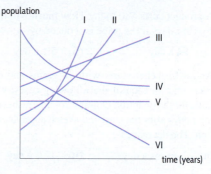

Figure 1.72

5. (a) The exponential functions in Figure 1.73 have b, d, q positive. Which of the constants a, c, and p must be positive?
   (b) Which of the constants a, b, c, d, p, and q must be between 0 and 1?
   (c) Which two of the constants a, b, c, d, p, and q must be equal?
   (d) What information about the constants a and b does the point $(1, 1)$ provide?

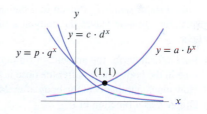

Figure 1.73

■ Give a possible formula for the functions in Problems 6–9.

6.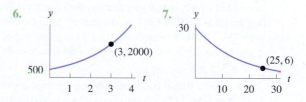

7.

8.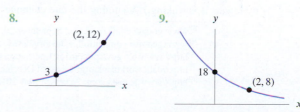

9.

10. The gross domestic product, G, of Switzerland[68] was 685.4 billion dollars in 2013. Give a formula for G (in billions of dollars) t years after 2013 if G increases by
    (a) 2% per year   (b) 7 billion dollars per year

11. A town has a population of 1000 people at time $t = 0$. In each of the following cases, write a formula for the population, P, of the town as a function of year t.

    (a) The population increases by 50 people a year.
    (b) The population increases by 5% a year.

12. A product costs $80 today. How much will the product cost in t days if the price is reduced by

    (a) $4 a day   (b) 5% a day

13. An air-freshener starts with 30 grams and evaporates over time. In each of the following cases, write a formula for the quantity, Q grams, of air-freshener remaining t days after the start and sketch a graph of the function. The decrease is:

    (a) 2 grams a day   (b) 12% a day

14. World population[69] is approximately $P = 7.32(1.0109)^t$, with P in billions and t in years since 2015.

    (a) What is the yearly percent rate of growth of the world population?
    (b) What was the world population in 2015? What does this model predict for the world population in 2020?
    (c) Use part (b) to find the projected average rate of change of the world population between 2015 and 2020.

15. A 50 mg dose of quinine is given to a patient to prevent malaria. Quinine leaves the body at a rate of 6% per hour.

    (a) Find a formula for the amount, A (in mg), of quinine in the body t hours after the dose is given.
    (b) How much quinine is in the body after 24 hours?
    (c) Graph A as a function of t.
    (d) Use the graph to estimate when 5 mg of quinine remains.

16. The consumer price index (CPI) for a given year is the amount of money in that year that has the same purchasing power as $100 in 1983. At the start of 2015, the CPI was 234.[70] Write a formula for the CPI as a function of t years after 2015, assuming that the CPI increases by 1.8% every year.

17. In the 1990s, Brazil had a deforestation rate of about 0.48% per year.[71] (This is the rate at which land covered by forests is shrinking.) Assuming the rate continues, what percent of the land in Brazil covered by forests in 1990 will be forested in 2030?

18. Graph $y = 100e^{-0.4x}$. Describe what you see.

19. (a) Make a table of values for $y = e^x$ using $x = 0, 1, 2, 3$.
    (b) Plot the points found in part (a). Does the graph look like an exponential growth or decay function?
    (c) Make a table of values for $y = e^{-x}$ using $x = 0, 1, 2, 3$.
    (d) Plot the points found in part (c). Does the graph look like an exponential growth or decay function?

■ For Problems 20–21, find a possible formula for the function represented by the data.

20.

x	0	1	2	3
$f(x)$	4.30	6.02	8.43	11.80

21.

t	0	1	2	3
$g(t)$	5.50	4.40	3.52	2.82

■ In Problems 22–23, find all the tables that have the given characteristic.

(A)

x	0	40	80	160
y	2.2	2.2	2.2	2.2

(B)

x	−8	−4	0	8
y	51	62	73	95

(C)

x	−4	−3	4	6
y	18	0	4.5	−2.25

(D)

x	3	4	5	6
y	18	9	4.5	2.25

22. y could be a linear function of x.

23. y could be an exponential function of x.

24. The table gives the number of North American houses (millions) with analog cable TV.[72]

    (a) Plot the number of houses, H, in millions, with cable TV versus year, Y.
    (b) Could H be a linear function of Y? Why or why not?
    (c) Could H be an exponential function of Y? Why or why not?

Year	2010	2011	2012	2013	2014	2015
Houses	18.3	13	7.8	3.9	1	0.5

[68] www.google.co.uk, accessed September 20, 2016.
[69] www.geohive.com/earth/his_history3.aspx, accessed September 20, 2016.
[70] http://www.bls.gov/cpi/cpid1501.pdf, accessed September 20, 2016.
[71] http://data.worldbank.org, accessed December 15, 2016. The rate has slowed since the 1990s.
[72] http://www.statista.com. Accessed May 2015.

■ In Problems **25–28**, a quantity P is an exponential function of time t. Use the given information about the function $P = P_0 a^t$ to:

(a) Find values for the parameters a and P_0.

(b) State the initial quantity and the percent rate of growth or decay.

25. $P_0 a^3 = 75$ and $P_0 a^2 = 50$

26. $P_0 a^4 = 18$ and $P_0 a^3 = 20$

27. $P = 320$ when $t = 5$ and $P = 500$ when $t = 3$

28. $P = 1600$ when $t = 3$ and $P = 1000$ when $t = 1$

29. If the world's population increased exponentially from 5.937 billion in 1998 to 7.238 billion in 2014 and continued to increase at the same percentage rate in 2014 and 2015, calculate what the world's population would have been in 2015. How does this compare to the Population Reference Bureau estimate of 7.34 billion in July 2015? [73]

30. The number of passengers using a railway fell from 190,205 to 174,989 during a 5-year period. Find the annual percentage decrease over this period.

31. The company that produces Cliffs Notes (abridged versions of classic literature) was started in 1958 with $4000 and sold in 1998 for $14,000,000. Find the annual percent increase in the value of this company over the 40 years.

32. Find a formula for the number of zebra mussels in a bay as a function of the number of years since 2014, given that there were 2700 at the start of 2014 and 3186 at the start of 2015.

(a) Assume that the number of zebra mussels is growing linearly. Give units for the slope of the line and interpret it in terms of zebra mussels.

(b) Assume that the number of zebra mussels is growing exponentially. What is the annual percent growth rate of the zebra mussel population?

33. Worldwide, wind energy[74] generating capacity, W, was 236,733 megawatts at the end of 2011 and 371,374 megawatts at the end of 2014. (Generating capacity is the maximum rate at which energy can be produced, measured in this problem in megawatts.)

(a) Use the values given to write W, in megawatts, as a linear function of t, the number of years since 2011.

(b) Use the values given to write W as an exponential function of t.

(c) Graph the functions found in parts (a) and (b) on the same axes. Label the given values.

(d) Use the functions found in parts (a) and (b) to predict the wind energy generating capacity in 2015. Generating capacity at the end of 2015 was

434,944 megawatts. Which estimate is closer to the real value?

34. Which of the following tables of values could correspond to an exponential function, a linear function, or neither? For those which could correspond to an exponential or linear function, find a formula for the function.

(a)

x	$f(x)$
0	16
1	24
2	36
3	54
4	81

(b)

x	$g(x)$
0	14
1	20
2	24
3	29
4	35

(c)

x	$h(x)$
0	5.3
1	6.5
2	7.7
3	8.9
4	10.1

35. (a) Which (if any) of the functions in the following table could be linear? Find formulas for those functions.

(b) Which (if any) of these functions could be exponential? Find formulas for those functions.

x	$f(x)$	$g(x)$	$h(x)$
-2	12	16	37
-1	17	24	34
0	20	36	31
1	21	54	28
2	18	81	25

36. Determine whether each of the following tables of values could correspond to a linear function, an exponential function, or neither. For each table of values that could correspond to a linear or an exponential function, find a formula for the function.

(a)

x	$f(x)$
0	10.5
1	12.7
2	18.9
3	36.7

(b)

t	$s(t)$
-1	50.2
0	30.12
1	18.072
2	10.8432

(c)

u	$g(u)$
0	27
2	24
4	21
6	18

37. The 2004 US presidential debates questioned whether the minimum wage has kept pace with inflation. Decide the question using the following information:[75] In 1938, the minimum wage was 25¢; in 2004, it was $5.15. During the same period, inflation averaged 4.3%.

38. (a) Niki invested $10,000 in the stock market. The investment was a loser, declining in value 10% per year each year for 10 years. How much was the investment worth after 10 years?

(b) After 10 years, the stock began to gain value at 10% per year. After how long will the investment regain its initial value ($10,000)?

---

[73]www.prb.org/pdf15/2015-world-population-data-sheet_eng.pdf, accessed September 20, 2016.

[74]www.wwindea.org, 2015 and 2016 half-year reports, accessed December 12, 2016.

[75]http://www.dol.gov/esa/minwage/chart.htm#5.

39. Whooping cough was thought to have been almost wiped out by vaccinations. It is now known that the vaccination wears off, leading to an increase[76] in the number of cases, w, from 1248 in 1981 to 25,827 in 2004.

   (a) With t in years since 1980, find an exponential function that fits this data.

   (b) What does your answer to part (a) give as the average annual percent growth rate of the number of cases?

   (c) On May 4, 2005, the *Arizona Daily Star* newspaper reported (correctly) that the number of cases had more than doubled between 2000 and 2004. Does your model confirm this report? Explain.

40. Aircraft require longer takeoff distances, called takeoff rolls, at high-altitude airports because of diminished air density. The table shows how the takeoff roll for a certain light airplane depends on the airport elevation. (Takeoff rolls are also strongly influenced by air temperature; the data shown assume a temperature of 0° C.) Determine a formula for this particular aircraft that gives the takeoff roll as an exponential function of airport elevation.

Elevation (ft)	Sea level	1000	2000	3000	4000
Takeoff roll (ft)	670	734	805	882	967

■ Problems **41–42** concern biodiesel, a fuel derived from renewable resources such as food crops, algae, and animal oils. The table shows the percent growth over the previous year in US biodiesel consumption.[77]

Year	2008	2009	2010	2011	2012	2013	2014
% growth	−14.1	5.9	−19.3	240.8	1.0	56.9	1.1

41. (a) According to the US Department of Energy, the US consumed 322 million gallons of biodiesel in 2009. Approximately how much biodiesel (in millions of gallons) did the US consume in 2010? In 2011?

   (b) Graph the points showing the annual US consumption of biodiesel, in millions of gallons of biodiesel, for the years 2009 to 2014. Label the scales on the horizontal and vertical axes.

42. (a) True or false: The annual US consumption of biodiesel grew exponentially from 2008 to 2010. Justify your answer without doing any calculations.

   (b) According to this data, during what single year(s), if any, did the US consumption of biodiesel at least triple?

43. Hydroelectric power is electric power generated by the force of moving water. The table shows the annual percent change in hydroelectric power consumption by the US industrial sector.[78]

Year	2009	2010	2011	2012	2013	2014
% growth over previous yr	6.3	−4.9	22.4	−15.5	−3.0	−3.4

(a) According to the US Department of Energy, the US industrial sector consumed about 2.65 quadrillion[79] BTUs of hydroelectric power in 2009. Approximately how much hydroelectric power (in quadrillion BTUs) did the US consume in 2010? In 2011?

(b) Graph the points showing the annual US consumption of hydroelectric power, in quadrillion BTUs, for the years 2009 to 2014. Label the scales on the horizontal and vertical axes.

(c) According to this data, when did the largest yearly decrease, in quadrillion BTUs, in the US consumption of hydroelectric power occur? What was this decrease?

■ Problems **44–45** concern wind power, which has been used for centuries to propel ships and mill grain. Modern wind power is obtained from windmills that convert wind energy into electricity. Figure 1.74 shows the annual percent growth in US wind power consumption[80] between 2009 and 2014.

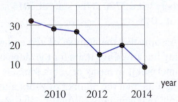

Figure 1.74

44. (a) According to the US Department of Energy, the US consumption of wind power was 721 trillion BTUs in 2009. How much wind power did the US consume in 2010? In 2011?

   (b) Graph the points showing the annual US consumption of wind power, in trillion BTUs, for the years 2009 to 2014. Label the scales on the horizontal and vertical axes.

   (c) Based on this data, in what year did the largest yearly increase, in trillion BTUs, in the US consumption of wind power occur? What was this increase?

---

[76] www.cdc.gov/pertussis/surv-reporting/cases-by-year.html, accessed Jan 22, 2017.

[77] www.eia.doe.gov, accessed March 29, 2015.

[78] From www.eia.doe.gov, accessed March 31, 2015.

[79] 1 quadrillion BTU=10^{15} BTU.

[80] Yearly values have been joined with line segments to highlight trends in the data. Actual values in between years should not be inferred from the segments. From www.eia.doe.gov, accessed April 1, 2015.

45. (a) According to Figure 1.74, during what single year(s), if any, did the US consumption of wind power energy increase by at least 25%? Decrease by at least 25%?
    (b) True or false: The US consumption of wind power energy doubled from 2008 to 2011?

1.6 THE NATURAL LOGARITHM

If t is in years since 2000, the population of Nevada (in millions) can be modeled by the function

$$P = f(t) = 2.020(1.036)^t,$$

How do we find when the population is projected to reach 4 million? We want to find the value of t for which

$$4 = f(t) = 2.020(1.036)^t.$$

We use logarithms to solve for a variable in an exponent.

Definition and Properties of the Natural Logarithm

We define the natural logarithm of x, written $\ln x$, as follows:

> The **natural logarithm** of x, written $\ln x$, is the power of e needed to get x. In other words,
>
> $$\ln x = c \quad \text{means} \quad e^c = x.$$
>
> The natural logarithm, sometimes written $\log_e x$, is called the *inverse function* of e^x.

For example, $\ln e^3 = 3$ since 3 is the power of e needed to give e^3. Similarly, $\ln(1/e) = \ln e^{-1} = -1$. A calculator gives $\ln 5 = 1.6094$, because $e^{1.6094} = 5$. However if we try to find $\ln(-7)$ on a calculator, we get an error message because e to any power is never negative or 0. In general,

> $\ln x$ is not defined if x is negative or 0.

To work with logarithms, we use the following properties:

> **Properties of the Natural Logarithm**
> 1. $\ln(AB) = \ln A + \ln B$
> 2. $\ln\left(\dfrac{A}{B}\right) = \ln A - \ln B$
> 3. $\ln(A^p) = p \ln A$
> 4. $\ln e^x = x$
> 5. $e^{\ln x} = x$
> In addition, $\ln 1 = 0$ because $e^0 = 1$, and $\ln e = 1$ because $e^1 = e$.

Using the $\boxed{\text{LN}}$ button on a calculator, we get the graph of $f(x) = \ln x$ in Figure 1.75. Observe that, for large x, the graph of $y = \ln x$ climbs very slowly as x increases. The x-intercept is $x = 1$, since $\ln 1 = 0$. For $x > 1$, the value of $\ln x$ is positive; for $0 < x < 1$, the value of $\ln x$ is negative.

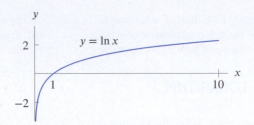

Figure 1.75: The natural logarithm function climbs very slowly

Solving Equations Using Logarithms

Natural logs can be used to solve for unknown exponents.

Example 1   Find t such that $3^t = 10$.

Solution   First, notice that we expect t to be between 2 and 3, because $3^2 = 9$ and $3^3 = 27$. To find t exactly, we take the natural logarithm of both sides and solve for t:

$$\ln(3^t) = \ln 10.$$

The third property of logarithms tells us that $\ln(3^t) = t \ln 3$, so we have

$$t \ln 3 = \ln 10$$
$$t = \frac{\ln 10}{\ln 3}.$$

Using a calculator to find the natural logs gives

$$t = 2.096.$$

Example 2   We return to the question of when the population of Nevada reaches 4 million. To get an answer, we solve $4 = 2.020(1.036)^t$ for t, using logs.

Solution   Dividing both sides of the equation by 2.020, we get

$$\frac{4}{2.020} = (1.036)^t.$$

Now take natural logs of both sides:

$$\ln\left(\frac{4}{2.020}\right) = \ln(1.036^t).$$

Using the fact that $\ln(1.036^t) = t \ln 1.036$, we get

$$\ln\left(\frac{4}{2.020}\right) = t \ln(1.036).$$

Solving this equation using a calculator to find the logs, we get

$$t = \frac{\ln(4/2.020)}{\ln(1.036)} = 19.317 \text{ years.}$$

Since $t = 0$ in 2000, this value of t corresponds to the year 2019.

Example 3   Find t such that $12 = 5e^{3t}$.

Solution   It is easiest to begin by isolating the exponential, so we divide both sides of the equation by 5:

$$2.4 = e^{3t}.$$

Now take the natural logarithm of both sides:

$$\ln 2.4 = \ln(e^{3t}).$$

Since $\ln(e^x) = x$, we have

$$\ln 2.4 = 3t,$$

so, using a calculator, we get

$$t = \frac{\ln 2.4}{3} = 0.2918.$$

Exponential Functions with Base e

An exponential function with base a has formula

$$P = P_0 a^t.$$

For any positive number a, we can write $a = e^k$ where $k = \ln a$. Thus, the exponential function can be rewritten as

$$P = P_0 a^t = P_0 (e^k)^t = P_0 e^{kt}.$$

If $a > 1$, then k is positive, and if $0 < a < 1$, then k is negative. We conclude:

> Writing $a = e^k$, so $k = \ln a$, any exponential function can be written in two forms
>
> $$P = P_0 a^t \quad \text{or} \quad P = P_0 e^{kt}.$$
>
> - If $a > 1$, we have exponential growth; if $0 < a < 1$, we have exponential decay.
> - If $k > 0$, we have exponential growth; if $k < 0$, we have exponential decay.
> - k is called the *continuous* growth or decay rate.

The word continuous in continuous growth rate is used in the same way to describe continuous compounding of interest earned on money. (See Appendix A.)

Example 4   (a) Convert the function $P = 1000e^{0.05t}$ to the form $P = P_0 a^t$.
(b) Convert the function $P = 500(1.06)^t$ to the form $P = P_0 e^{kt}$.

Solution   (a) Since $P = 1000e^{0.05t}$, we have $P_0 = 1000$. We want to find a so that

$$1000a^t = 1000e^{0.05t} = 1000(e^{0.05})^t.$$

We take $a = e^{0.05} = 1.0513$, so the following two functions give the same values:

$$P = 1000e^{0.05t} \quad \text{and} \quad P = 1000(1.0513)^t.$$

So a continuous growth rate of 5% is equivalent to a growth rate of 5.13% per unit time.

(b) We have $P_0 = 500$ and we want to find k with

$$500(1.06)^t = 500(e^k)^t,$$

so we take

$$1.06 = e^k$$
$$k = \ln(1.06) = 0.0583.$$

The following two functions give the same values:

$$P = 500(1.06)^t \quad \text{and} \quad P = 500e^{0.0583t}.$$

So a growth rate of 6% per unit time is equivalent to a continuous growth rate of 5.83%.

Example 5   Sketch graphs of $P = e^{0.5t}$, a continuous growth rate of 50%, and $Q = 5e^{-0.2t}$, a continuous decay rate of 20%.

Solution   The graph of $P = e^{0.5t}$ is in Figure 1.76. Notice that the graph is the same shape as the previous exponential growth curves: increasing and concave up. The graph of $Q = 5e^{-0.2t}$ is in Figure 1.77; it has the same shape as other exponential decay functions.

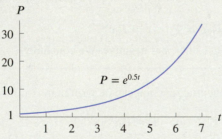

Figure 1.76: Continuous exponential growth function

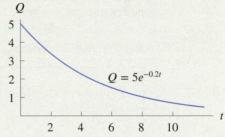

Figure 1.77: Continuous exponential decay function

Problems for Section 1.6

■ For Problems **1–20**, solve for t using natural logarithms.

1. $10 = 2^t$

2. $5^t = 7$

3. $2 = (1.02)^t$

4. $130 = 10^t$

5. $10 = e^t$

6. $100 = 25(1.5)^t$

7. $50 = 10 \cdot 3^t$

8. $5 = 2e^t$

9. $e^{3t} = 100$

10. $10 = 6e^{0.5t}$

11. $40 = 100e^{-0.03t}$

12. $a = b^t$

13. $B = Pe^{rt}$

14. $2P = Pe^{0.3t}$

15. $7 \cdot 3^t = 5 \cdot 2^t$

16. $5e^{3t} = 8e^{2t}$

17. $Ae^{2t} = Be^t$

18. $2e^t - 5 = 0$

19. $0 = 7 - 3e^t$

20. $Pe^{4t} - Qe^{-t} = 0$

■ The functions in Problems **21–24** represent exponential growth or decay. What is the initial quantity? What is the growth rate? State if the growth rate is continuous.

21. $P = 5(1.07)^t$

22. $P = 7.7(0.92)^t$

23. $P = 15e^{-0.06t}$

24. $P = 3.2e^{0.03t}$

■ Write the functions in Problems 25–28 in the form $P = P_0 a^t$. Which represent exponential growth and which represent exponential decay?

25. $P = 15e^{0.25t}$

26. $P = 2e^{-0.5t}$

27. $P = P_0 e^{0.2t}$

28. $P = 7e^{-\pi t}$

■ In Problems 29–32, put the functions in the form $P = P_0 e^{kt}$.

29. $P = 15(1.5)^t$

30. $P = 10(1.7)^t$

31. $P = 174(0.9)^t$

32. $P = 4(0.55)^t$

■ In Problems 33–34, a quantity P is an exponential function of time t. Use the given information about the function $P = P_0 e^{kt}$ to:

(a) Find values for the parameters k and P_0.

(b) State the initial quantity and the continuous percent rate of growth or decay.

33. $P = 140$ when $t = 3$ and $P = 100$ when $t = 1$

34. $P = 40$ when $t = 4$ and $P = 50$ when $t = 3$

35. **(a)** What is the continuous percent growth rate for $P = 100e^{0.06t}$, with time, t, in years?

   (b) Write this function in the form $P = P_0 a^t$. What is the annual percent growth rate?

36. **(a)** What is the annual percent decay rate for $P = 25(0.88)^t$, with time, t, in years?

   (b) Write this function in the form $P = P_0 e^{kt}$. What is the continuous percent decay rate?

37. What annual percent growth rate is equivalent to a continuous percent growth rate of 8%?

38. What continuous percent growth rate is equivalent to an annual percent growth rate of 10%?

39. The following formulas give the populations of four different towns, A, B, C, and D, with t in years from now.

$$P_A = 600e^{0.08t} \qquad P_B = 1000e^{-0.02t}$$
$$P_C = 1200e^{0.03t} \qquad P_D = 900e^{0.12t}$$

   (a) Which town is growing fastest (that is, has the largest percentage growth rate)?

   (b) Which town is the largest now?

   (c) Are any of the towns decreasing in size? If so, which one(s)?

40. A 2008 study of 300 oil fields producing a total of 84 million barrels per day reported that daily production was decaying at a continuous rate of 9.1% per year.[81] Find the estimated production in these fields in 2025 if the decay continues at the same rate.

41. A city's population is 1000 and growing at 5% a year.

   (a) Find a formula for the population at time t years from now assuming that the 5% per year is an:

   (i)   Annual rate   (ii)   Continuous annual rate

   (b) In each case in part (a), estimate the population of the city in 10 years.

42. The population, P, in millions, of Nicaragua[82] was 5.97 million in 2016 and growing at an annual rate of 0.99%. Let t be time in years since 2016.

   (a) Express P as a function in the form $P = P_0 a^t$.

   (b) Express P as an exponential function using base e.

   (c) Compare the annual and continuous growth rates.

43. The gross world product is $W = 74.31(1.029)^t$, where W is in trillions of dollars and t is years since 2013.[83] Find a formula for gross world product using a continuous growth rate.

44. The population of the world[84] can be represented by $P = 7.32(1.0109)^t$, where P is in billions of people and t is years since 2015. Find a formula for the population of the world using a continuous growth rate.

45. A fishery stocks a pond with 1000 young trout. The number of trout t years later is given by $P(t) = 1000e^{-0.5t}$.

   (a) How many trout are left after six months? After 1 year?

   (b) Find $P(3)$ and interpret it in terms of trout.

   (c) At what time are there 100 trout left?

   (d) Graph the number of trout against time, and describe how the population is changing. What might be causing this?

46. The Hershey Company is the largest US producer of chocolate. In 2011, annual net sales were 6.1 billion dollars and were increasing at a continuous rate of 7% per year.[85]

   (a) Write a formula for annual net sales, S, as a function of time, t, in years since 2011.

   (b) Estimate annual net sales in 2015.

   (c) Use a graph to estimate the year in which annual net sales are expected to pass 10 billion dollars and check your estimate using logarithms.

47. During a recession a firm's revenue declines continuously so that the revenue, R (measured in millions of dollars), in t years' time is given by $R = 5e^{-0.15t}$.

   (a) Calculate the current revenue and the revenue in two years' time.

   (b) After how many years will the revenue decline to $2.7 million?

---

[81] International Energy Agency, *World Energy Outlook*, 2008.
[82] www.cia.gov/library/publications/the-world-factbook, accessed December 30, 2016.
[83] www.indexmundi.com/world, accessed September 22, 2016.
[84] www.geohive.com/earth/his_history3.aspx, accessed December 12, 2016
[85] 2011 Annual Report to Stockholders, accessed at www.thehersheycompany.com, accessed December 12, 2016.

48. The population of a city is 50,000 in 2014 and is growing at a continuous rate of 4.5% per year.

    (a) Give the population of the city as a function of the number of years since 2014. Sketch a graph of the population against time.

    (b) What will be the city's population in the year 2020?

    (c) Calculate the time for the population of the city to reach 100,000. This is called the doubling time of the population.

49. For children and adults with diseases such as asthma, the number of respiratory deaths per year increases by 0.33% when pollution particles increase by a microgram per cubic meter of air.[86]

    (a) Write a formula for the number of respiratory deaths per year as a function of quantity of pollution in the air. (Let Q_0 be the number of deaths per year with no pollution.)

    (b) What quantity of air pollution results in twice as many respiratory deaths per year as there would be without pollution?

50. The concentration of the car exhaust fume nitrous oxide, NO_2, in the air near a busy road is a function of distance from the road. The concentration decays exponentially at a continuous rate of 2.54% per meter.[87] At what distance from the road is the concentration of NO_2 half what it is on the road?

51. With time, t, in years since the start of 1980, textbook prices have increased at 6.7% per year while inflation has been 3.3% per year.[88] Assume both rates are continuous growth rates.

    (a) Find a formula for $B(t)$, the price of a textbook in year t if it cost $\$B_0$ in 1980.

    (b) Find a formula for $P(t)$, the price of an item in year t if it cost $\$P_0$ in 1980 and its price rose according to inflation.

    (c) A textbook cost $50 in 1980. When is its price predicted to be double the price that would have resulted from inflation alone?

52. In 2014, the populations of China and India were approximately 1.355 and 1.255 billion people,[89] respectively. However, due to central control the annual population growth rate of China was 0.44% while the population of India was growing by 1.25% each year. If these growth rates remain constant, when will the population of India exceed that of China?

53. In 2010, there were about 246 million vehicles (cars and trucks) and about 308.7 million people in the US.[90] The number of vehicles grew 15.5% over the previous decade, while the population has been growing at 9.7% per decade. If the growth rates remain constant, when will there be, on average, one vehicle per person?

1.7   EXPONENTIAL GROWTH AND DECAY

Many quantities in nature change according to an exponential growth or decay function of the form $P = P_0 e^{kt}$, where P_0 is the initial quantity and k is the continuous growth or decay rate.

Example 1   The Environmental Protection Agency (EPA) investigated a spill of radioactive iodine. The radiation level at the site was about 2.4 millirems/hour (four times the maximum acceptable limit of 0.6 millirems/hour), so the EPA ordered an evacuation of the surrounding area. The level of radiation from an iodine source decays at a continuous hourly rate of $k = -0.004$.

    (a) What was the level of radiation 24 hours later?

    (b) Find the number of hours until the level of radiation reached the maximum acceptable limit, and the inhabitants could return.

Solution   **(a)** The level of radiation, R, in millirems/hour, at time t, in hours since the initial measurement, is given by

$$R = 2.4e^{-0.004t},$$

so the level of radiation 24 hours later was

$$R = 2.4e^{(-0.004)(24)} = 2.18 \text{ millirems per hour.}$$

---

[86]R. D. Brook, B. Franklin, W. Cascio, Y. Hong, G. Howard., M. Lipsett, R. Luepker, M. Mittleman, J. Samet, and S. C. Smith, "Air Pollution and Cardiovascular Disease." *Circulation,* 2004;109:2655–2671.

[87]P. Rickwood, and D. Knight, "The Health Impacts of Local Traffic Pollution on Primary School Age Children", *State of Australian Cities 2009 Conference Proceedings*, www.be.unsw.edu.au, accessed March 24, 2016.

[88]Data from "Textbooks Headed for Ash Heap of History", http://educationtechnews.com, Vol 5, 2010.

[89]www.indexmundi.com, accessed April 11, 2015.

[90]http://www.autoblog.com/2010/01/04/report-number-of-cars-in-the-u-s-dropped-by-four-million-in-20/ and http://2010.census.gov/news/releases/operations/cb10-cn93.html. Accessed February 2012.

(b) A graph of $R = 2.4e^{-0.004t}$ is in Figure 1.78. The maximum acceptable value of R is 0.6 millirems per hour, which occurs at approximately $t = 350$. Using logarithms, we have

$$0.6 = 2.4e^{-0.004t}$$
$$0.25 = e^{-0.004t}$$
$$\ln 0.25 = -0.004t$$
$$t = \frac{\ln 0.25}{-0.004} = 346.57.$$

The inhabitants will not be able to return for 346.57 hours, or about 15 days.

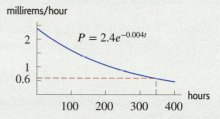

Figure 1.78: The level of radiation from radioactive iodine

Example 2   The population of Kenya[91] was 18.9 million in 1984 and 46.1 million in 2015. Assuming the population increases exponentially, find a formula for the population of Kenya as a function of time.

Solution   We measure the population, P, in millions and time, t, in years since 1984. We can express P in terms of t using the continuous growth rate k by

$$P = P_0e^{kt} = 18.9e^{kt},$$

where $P_0 = 18.9$ is the initial value of P. We find k using the fact that $P = 46.1$ when $t = 31$:

$$46.1 = 18.9e^{k\cdot31}.$$

Divide both sides by 18.9, giving

$$\frac{46.1}{18.9} = e^{31k}.$$

Take natural logs of both sides:

$$\ln\left(\frac{46.1}{18.9}\right) = \ln(e^{31k}).$$

Since $\ln(e^{31k}) = 31k$, this becomes

$$\ln\left(\frac{46.1}{18.9}\right) = 31k.$$

We get

$$k = \frac{1}{31}\ln\left(\frac{46.1}{18.9}\right) = 0.029,$$

and therefore

$$P = 18.9e^{0.029t}.$$

Since $k = 0.029 = 2.9\%$, the population of Kenya was growing at a continuous rate of 2.9% per year.

---

[91] www.worldometers.info/world-population/kenya-population/, accessed September 23, 2016.

Doubling Time and Half-Life

Every exponential growth function has a constant doubling time and every exponential decay function has a constant half-life.

> The **doubling time** of an exponentially increasing quantity is the time required for the quantity to double.
> The **half-life** of an exponentially decaying quantity is the time required for the quantity to be reduced by a factor of one half.

Example 3   Show algebraically that every exponentially growing function has a fixed doubling time.

Solution   Consider the exponential function $P = P_0 a^t$. For any base a with $a > 1$, there is a positive number d such that $a^d = 2$. We show that d is the doubling time. If the population is P at time t, then at time $t + d$, the population is

$$P_0 a^{t+d} = P_0 a^t a^d = (P_0 a^t)(2) = 2P.$$

So, no matter what the initial quantity and no matter what the initial time, the size of the population is doubled d time units later.

Example 4   The release of chlorofluorocarbons used in air conditioners and household sprays (hair spray, shaving cream, etc.) destroys the ozone in the upper atmosphere. The quantity of ozone, Q, is decaying exponentially at a continuous rate of 0.25% per year. What is the half-life of ozone? In other words, at this rate, how long will it take for half the ozone to disappear?

Solution   If Q_0 is the initial quantity of ozone and t is in years, then

$$Q = Q_0 e^{-0.0025t}.$$

We want to find the value of t making $Q = Q_0/2$, so

$$\frac{Q_0}{2} = Q_0 e^{-0.0025t}.$$

Dividing both sides by Q_0 and taking natural logs gives

$$\ln\left(\frac{1}{2}\right) = -0.0025t,$$

so

$$t = \frac{\ln(1/2)}{-0.0025} = 277 \text{ years.}$$

Half the present atmospheric ozone will be gone in 277 years.

Financial Applications: Compound Interest

We deposit $100 in a bank paying interest at a rate of 2% per year. How much is in the account at the end of the year? This depends on how often the interest is compounded. If the interest is paid into the account *annually*, that is, only at the end of the year, then the balance in the account after one year is $102. However, if the interest is paid twice a year, then 1% is paid at the end of the first six months and 1% at the end of the year. Slightly more money is earned this way, since the interest paid early in the year will earn interest during the rest of the year. This effect is called *compounding*.

In general, the more often interest is compounded, the more money is earned (although the increase may not be large). What happens if interest is compounded more frequently, such as every minute or every second? The benefit of increasing the frequency of compounding becomes negligible beyond a certain point. When that point is reached, we find the balance using the number e and we say that the interest per year is *compounded continuously*. If we have deposited $100 in an account paying 2% interest per year compounded continuously, the balance after one year is $100e^{0.02} = \$102.02$. Compounding is discussed further in Appendix B. In general:

> An amount P_0 is deposited in an account paying interest at a rate of r per year. Let P be the balance in the account after t years.
> - If interest is compounded annually, then $P = P_0(1 + r)^t$.
> - If interest is compounded continuously, then $P = P_0e^{rt}$, where $e = 2.71828....$

We write P_0 for the initial deposit because it is the value of P when $t = 0$. Note that for a 1.5% interest rate, $r = 0.015$. If a rate is continuous, we will say so explicitly.

Example 5   A bank advertises an interest rate of 2% per year. If you deposit $10,000, how much is in the account 4 years later if the interest is compounded (a)   Annually?   (b)   Continuously?

Solution   (a) For annual compounding, $P = P_0(1 + r)^t = 10{,}000(1.02)^4 = \$10{,}824.32$.
(b) For continuous compounding, $P = P_0e^{rt} = 10{,}000e^{0.02 \cdot 4} = \$10{,}832.87$. As expected, the amount in the account 4 years later is larger if the interest is compounded continuously ($10,832.87) than if the interest is compounded annually ($10,824.32).

Example 6   If $10,000 is deposited in an account paying interest at a rate of 1.5% per year, compounded continuously, how long does it take for the balance in the account to reach $12,000?

Solution   Since interest is compounded continuously, we use $P = P_0e^{rt}$ with $r = 0.015$ and $P_0 = 10{,}000$. We want to find the value of t for which $P = 12{,}000$. The equation is

$$12{,}000 = 10{,}000e^{0.015t}.$$

Now divide both sides by 10,000, then take logarithms and solve for t:

$$1.2 = e^{0.015t}$$
$$\ln(1.2) = \ln(e^{0.015t})$$
$$\ln(1.2) = 0.015t$$
$$t = \frac{\ln(1.2)}{0.015} = 12.1548.$$

It takes about 12.2 years for the balance in the account to reach $12,000.

Example 7   (a) Calculate the doubling time, D, for interest rates of 1%, 2%, 3%, and 4% per year, compounded annually.
(b) Use your answers to part (a) to check that an interest rate of i% gives a doubling time approximated for small values of i by

$$D \approx \frac{70}{i} \text{ years.}$$

This is the "Rule of 70" used by bankers: To compute the approximate doubling time of an investment, divide 70 by the percent annual interest rate.

Solution   (a) We find the doubling time for an interest rate of 1% per year using the formula $P = P_0(1.01)^t$ with t in years. To find the value of t for which $P = 2P_0$, we solve

$$2P_0 = P_0(1.01)^t$$
$$2 = (1.02)^t$$
$$\ln 2 = \ln(1.01)^t$$
$$\ln 2 = t \ln(1.01) \qquad \text{(using the third property of logarithms)}$$
$$t = \frac{\ln 2}{\ln 1.01} = 69.661 \text{ years.}$$

With an annual interest rate of 1%, it takes about 70 years for an investment to double in value. Similarly, we find the doubling times for 2%, 3%, and 4% in Table 1.34.

Table 1.34 *Doubling time as a function of interest rate*

i (% annual growth rate)	1	2	3	4
D (doubling time in years)	69.661	35.003	23.45	17.673

(b) We compute $(70/i)$ for $i = 1, 2, 3, 4$. The results are shown in Table 1.35.

Table 1.35 *Approximate doubling time as a function of interest rate: Rule of 70*

i (% annual growth rate)	1	2	3	4
$(70/i)$ (Approximate doubling time in years)	70.000	35.000	23.333	17.500

Comparing Tables 1.34 and 1.35, we see that the quantity $(70/i)$ gives a reasonably accurate approximation to the doubling time, D, for the small interest rates we considered.

Present and Future Value

Many business deals involve payments in the future. For example, when a car is bought on credit, payments are made over a period of time. Being paid $100 in the future is clearly worse than being paid $100 today for many reasons. If we are given the money today, we can do something else with it—for example, put it in the bank, invest it somewhere, or spend it. Thus, even without considering inflation, if we are to accept payment in the future, we would expect to be paid more to compensate for this loss of potential earnings.[92] The question we consider now is, how much more?

To simplify matters, we consider only what we would lose by not earning interest; we do not consider the effect of inflation. Let's look at some specific numbers. Suppose we deposit $100 in an account that earns 7% interest per year compounded annually, so that in a year's time we have $107. Thus, $100 today is worth $107 a year from now. We say that the $107 is the *future value* of the $100, and that the $100 is the *present value* of the $107. In general, we say the following:

- The **future value**, B, of a payment, P, is the amount to which the P would have grown if deposited today in an interest-bearing bank account.
- The **present value**, P, of a future payment, B, is the amount that would have to be deposited in a bank account today to produce exactly B in the account at the relevant time in the future.

[92]This is referred to as the time value of money.

Due to the interest earned, the future value is larger than the present value. The relation between the present and future values depends on the interest rate, as follows.

Suppose B is the *future value* of P and P is the *present value* of B.
If interest is compounded annually at a rate r for t years, then

$$B = P(1 + r)^t, \quad \text{or equivalently,} \quad P = \frac{B}{(1 + r)^t}.$$

If interest is compounded continuously at a rate r for t years, then

$$B = Pe^{rt}, \quad \text{or equivalently,} \quad P = \frac{B}{e^{rt}} = Be^{-rt}.$$

The rate, r, is sometimes called the *discount rate*. The present value is often denoted by PV and the future value by FV.

Example 8   You win the lottery and are offered the choice between $1 million in four yearly installments of $250,000 each, starting now, and a lump-sum payment of $980,000 now. Assuming a 2% interest rate per year, compounded continuously, and ignoring taxes, which should you choose?

Solution   We assume that you pick the option with the largest present value. The first of the four $250,000 payments is made now, so

$$\text{Present value of first payment} = \$250,000.$$

The second payment is made one year from now and so

$$\text{Present value of second payment} = \$250,000e^{-0.02(1)}.$$

Calculating the present value of the third and fourth payments similarly, we find:

$$\begin{aligned}
\text{Total present value} &= \$250,000 + \$250,000e^{-0.02(1)} + \$250,000e^{-0.02(2)} + \$250,000e^{-0.02(3)} \\
&= \$250,000 + \$245,050 + \$240,197 + \$235,441 \\
&= \$970,688.
\end{aligned}$$

Since the present value of the four payments is less than $980,000, you are better off taking the $980,000 now.

Alternatively, we can compare the future values of the two pay schemes. We calculate the future value of both schemes three years from now, on the date of the last $250,000 payment. At that time,

$$\text{Future value of the lump-sum payment} = \$980,000e^{0.02(3)} = \$1,040,600.$$

The future value of the first $250,000 payment is $250,000e^{0.02(3)}$. Calculating the future value of the other payments similarly, we find:

$$\begin{aligned}
\text{Total future value} &= \$250,000e^{0.02(3)} + \$250,000e^{0.02(2)} + \$250,000e^{0.02(1)} + \$250,000 \\
&= \$265,459 + \$260,203 + \$255,050 + \$250,000 \\
&= \$1,030,712.
\end{aligned}$$

As we expect, the future value of the $980,000 payment is greater, so you are better off taking the $980,000 now.[93]

---

[93] If you read the fine print, you will find that many lotteries do not make their payments right away, but often spread them out, sometimes far into the future. This is to reduce the present value of the payments made, so that the value of the prizes is less than it might first appear!

Problems for Section 1.7

■ For Problems **1–2**, find k such that $p = p_0 e^{kt}$ has the given doubling time.

1. 10

2. 0.4

3. Figure 1.79 shows the balances in two bank accounts. Both accounts pay the same interest rate, but one compounds continuously and the other compounds annually. Which curve corresponds to which compounding method? What is the initial deposit in each case?

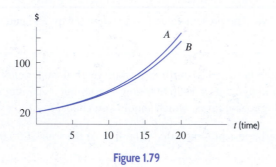

Figure 1.79

4. The exponential function $y(x) = Ce^{\alpha x}$ satisfies the conditions $y(0) = 2$ and $y(1) = 1$. Find the constants C and α. What is $y(2)$?

5. Suppose $1000 is invested in an account paying interest at a rate of 1.5% per year. How much is in the account after 8 years if the interest is compounded

  (a) Annually?   **(b)** Continuously?

6. If you deposit $10,000 in an account earning interest at an 8% annual rate compounded continuously, how much money is in the account after five years?

7. If you need $20,000 in your bank account in 6 years, how much must be deposited now? The interest rate is 10%, compounded continuously.

8. If $15,000 is deposited in an account paying 1.5% interest per year, compounded continuously, how long will it take for the balance to reach $20,000?

9. If a bank pays 1.25% per year interest compounded continuously, how long does it take for the balance in an account to double?

10. Find the doubling time of a quantity that is increasing by 7% per year.

11. Persistent organic pollutants (POPS) are a serious environmental hazard. Figure 1.80 shows their natural decay over time in human fat.[94]

  (a) How long does it take for the concentration to decrease from 100 units to 50 units?

  (b) How long does it take for the concentration to decrease from 50 units to 25 units?

  (c) Explain why your answers to parts (a) and (b) suggest that the decay may be exponential.

  (d) Find an exponential function that models concentration, C, as a function of t, the number of years since 1970.

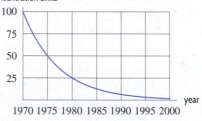

Figure 1.80

12. The half-life of nicotine in the blood is 2 hours. A person absorbs 0.4 mg of nicotine by smoking a cigarette. Fill in the following table with the amount of nicotine remaining in the blood after t hours. Estimate the length of time until the amount of nicotine is reduced to 0.04 mg.

t (hours)	0	2	4	6	8	10
Nicotine (mg)	0.4					

13. World wind energy generating[95] capacity, W, was 371 gigawatts by the end of 2014 and has been increasing at a continuous rate of approximately 16.8% per year. Assume this rate continues. (Generating capacity is the maximum amount of power generated per unit time.)

  (a) Give a formula for W, in gigawatts, as a function of time, t, in years since the end of 2014.

  (b) When is wind capacity predicted to pass 500 gigawatts?

14. Oil consumption in China grew exponentially[96] from 8.938 million barrels per day in 2010 to 10.480 million barrels per day in 2013. Assuming exponential growth continues at the same rate, what will oil consumption be in 2025?

15. From October 2002 to October 2006 the number $N(t)$ of Wikipedia articles[97] was approximated by $N(t) = N_0 e^{t/500}$, where t is the number of days after October 1, 2002. Find the doubling time for the number of Wikipedia articles during this period.

[94]K.C. Jones, P. de Voogt, "Persistent Organic Pollutants (POPs): State of the Science," *Environmental Pollution* 100, 1999, pp. 209–221.

[95]World Wind Energy Association, www.wwindea.org/hyr2015/, accessed September 28, 2016.

[96]Based on www.eia.gov/cfapps/ipdbproject, accessed May 2015.

[97]en.wikipedia.org/wiki/Wikipedia:Modelling_Wikipedia's_growth, accessed January 31, 2017.

16. You want to invest money for your child's education in a certificate of deposit (CD). You want it to be worth $12,000 in 10 years. How much should you invest if the CD pays interest at a 9% annual rate compounded
(a) Annually?   (b) Continuously?

17. According to the EPA, sales of electronic devices in the US doubled between 1997 and 2009, when 438 million electronic devices sold.[98]

(a) Find an exponential function, $S(t)$, to model sales in millions since 1997.
(b) What was the annual percentage growth rate between 1997 and 2009?

18. In 2014, the world's population reached 7.17 billion[99] and was increasing at a rate of 1.1% per year. Assume that this growth rate remains constant. (In fact, the growth rate has decreased since 2008.)

(a) Write a formula for the world population (in billions) as a function of the number of years since 2014.
(b) Estimate the population of the world in the year 2020.
(c) Sketch world population as a function of years since 2014. Use the graph to estimate the doubling time of the population of the world.

19. A cup of coffee contains 100 mg of caffeine, which leaves the body at a continuous rate of 17% per hour.

(a) Write a formula for the amount, A mg, of caffeine in the body t hours after drinking a cup of coffee.
(b) Graph the function from part (a). Use the graph to estimate the half-life of caffeine.
(c) Use logarithms to find the half-life of caffeine.

20. A population, currently 200, is growing at 5% per year.

(a) Write a formula for the population, P, as a function of time, t, years in the future.
(b) Graph P against t.
(c) Estimate the population 10 years from now.
(d) Use the graph to estimate the doubling time of the population.

21. Air pressure, P, decreases exponentially with height, h, above sea level. If P_0 is the air pressure at sea level and h is in meters, then

$$P = P_0 e^{-0.00012h}.$$

(a) At the top of Denali, height 6194 meters (about 20,320 feet), what is the air pressure, as a percent of the pressure at sea level?
(b) The maximum cruising altitude of an ordinary commercial jet is around 12,000 meters (about 39,000 feet). At that height, what is the air pressure, as a percent of the sea level value?

22. The antidepressant fluoxetine (or Prozac) has a half-life of about 3 days. What percentage of a dose remains in the body after one day? After one week?

23. A firm decides to increase output at a constant relative rate from its current level of 20,000 to 30,000 units during the next five years. Calculate the annual percent rate of increase required to achieve this growth.

24. The half-life of a radioactive substance is 12 days. There are 10.32 grams initially.

(a) Write an equation for the amount, A, of the substance as a function of time.
(b) When is the substance reduced to 1 gram?

25. One of the main contaminants of a nuclear accident, such as that at Chernobyl, is strontium-90, which decays exponentially at a rate of approximately 2.4% per year.

(a) Write the percent of strontium-90 remaining, P, as a function of years, t, since the nuclear accident. [Hint: 100% of the contaminant remains at $t = 0$.]
(b) Graph P against t.
(c) Estimate the half-life of strontium-90.
(d) After the Chernobyl disaster, it was predicted that the region would not be safe for human habitation for 100 years. Estimate the percent of original strontium-90 remaining at this time.

26. The number of people living with HIV increased worldwide approximately exponentially from 2.5 million in 1985 to 36.7 million in 2015.[100] (HIV is the virus that causes AIDS.)

(a) Give a formula for the number of people living with HIV, H (in millions), as a function of years, t, since 1985. Use the form $H = H_0 e^{kt}$. Graph this function.
(b) What was the yearly continuous percent change in the number of people living with HIV between 1985 and 2015?

27. (a) Figure 1.81 shows exponential growth. Starting at $t = 0$, estimate the time for the population to double.
(b) Repeat part (a), but this time start at $t = 3$.
(c) Pick any other value of t for the starting point, and notice that the doubling time is the same no matter where you start.

[98] http://www.epa.gov/osw/conserve/materials/ecycling/docs/summarybaselinereport2011.pdf. Accessed March 2015.
[99] www.indexmundi.com, accessed June 14, 2015.
[100] www.unaids.org/en/resources/fact-sheet, accessed September 28, 2016.

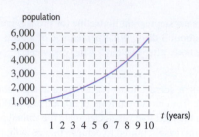

Figure 1.81

28. An exponentially growing animal population numbers 500 at time $t = 0$; two years later, it is 1500. Find a formula for the size of the population in t years and find the size of the population at $t = 5$.

29. If the quantity of a substance decreases by 4% in 10 hours, find its half-life.

30. Pregnant women metabolize some drugs at a slower rate than the rest of the population. The half-life of caffeine is about 4 hours for most people. In pregnant women, it is 10 hours.[101] (This is important because caffeine, like all psychoactive drugs, crosses the placenta to the fetus.) If a pregnant woman and her husband each have a cup of coffee containing 100 mg of caffeine at 8 am, how much caffeine does each have left in the body at 10 pm?

31. The half-life of radioactive strontium-90 is 29 years. In 1960, radioactive strontium-90 was released into the atmosphere during testing of nuclear weapons, and was absorbed into people's bones. How many years does it take until only 10% of the original amount absorbed remains?

32. In 1923, koalas were introduced on Kangaroo Island off the coast of Australia. In 1996, the population was 5000. By 2005, the population had grown to 27,000, prompting a debate on how to control their growth and avoid koalas dying of starvation.[102] Assuming exponential growth, find the (continuous) rate of growth of the koala population between 1996 and 2005. Find a formula for the population as a function of the number of years since 1996, and estimate the population in the year 2020.

33. The population of the US was 281.4 million in 2000 and 316.1 million in 2013.[103] Assuming exponential growth,

(a) In what year is the population expected to go over 350 million?

(b) What population is predicted for the 2020 census?

34. In 2015, the world's population was 7.32 billion, and the population is projected[104] to reach approximately 8 billion by the year 2023. What annual growth rate is projected?

35. A picture supposedly painted by Vermeer (1632–1675) contains 99.5% of its carbon-14 (half-life 5730 years). From this information decide whether the picture is a fake. Explain your reasoning.

36. Food bank usage in Britain has grown dramatically over the past decade. The number of users, in thousands, of the largest food bank in year t is estimated to be $N(t) = 1.3e^{0.81t}$, where t is the number of years since 2006.[105]

(a) What does the 1.3 represent in this context? Give units.

(b) What is the continuous growth rate of users per year?

(c) What is the annual percent growth rate of users per year?

(d) Using only your answer for part (c), decide if the doubling time is more or less than 1 year.

37. In November 2010, a "tiger summit" was held in St. Petersburg, Russia.[106] In 1900, there were 100,000 wild tigers worldwide; in 2010 the number was 3200.

(a) Assuming the tiger population has decreased exponentially, find a formula for $f(t)$, the number of wild tigers t years since 1900.

(b) Between 2000 and 2010, the number of wild tigers decreased by 40%. Is this percentage larger or smaller than the decrease in the tiger population predicted by your answer to part (a)?

38. Tiny marine organisms reproduce at different rates. Phytoplankton doubles in population twice a day, but foraminifera doubles every five days. If the two populations are initially the same size and grow exponentially, how long does it take for

(a) The phytoplankton population to be double the foraminifera population.

(b) The phytoplankton population to be 1000 times the foraminifera population.

39. The world population was 6.9 billion at the end of 2010 and is predicted to reach 9 billion by the end of 2050.[107]

(a) Assuming the population is growing exponentially, what is the continuous growth rate per year?

[101] From Robert M. Julien, *A Primer of Drug Action*, 7th ed., p. 159 (New York: W. H. Freeman, 1995).

[102] news.yahoo.com/s/afp/australiaanimalskoalas, accessed June 1, 2005.

[103] data.worldbank.org, accessed April 1, 2015.

[104] www.geohive.com/earth/his_history3.aspx and www.worldometers.info, accessed September 28, 2016.

[105] Estimates for the Trussell Trust: http://www.bbc.com/news/education-30346060 and http://www.trusselltrust.org/stats. Accessed November 2015.

[106] "Tigers Would be Extinct in Russia if Unprotected," Yahoo! News, Nov. 21, 2010.

[107] "Reviewing the Bidding on the Climate Files", in About Dot Earth, *New York Times*, Nov. 19, 2010.

(b) The United Nations celebrated the "Day of 6 Billion" on October 12, 1999, and "Day of 7 Billion" on October 31, 2011. Using the growth rate in part (a), when is the "Day of 8 Billion" predicted to be?

40. The number of alternative fuel vehicles[108] running on E85, a fuel that is up to 85% plant-derived ethanol, increased exponentially in the US between 2005 and 2010.

    (a) Use this information to complete the missing table values.
    (b) How many E85-powered vehicles were there in the US in 2004?
    (c) By what percent did the number of E85-powered vehicles grow from 2005 to 2009?

Year	2005	2006	2007	2008	2009	2010
No. E85 vehicles	246,363	?	?	?	?	618,505

41. Find the future value in 8 years of a $10,000 payment today, if the interest rate is 3% per year compounded continuously.

42. Find the future value in 15 years of a $20,000 payment today, if the interest rate is 3.8% per year compounded continuously.

43. Find the present value of an $8000 payment to be made in 5 years. The interest rate is 4% per year compounded continuously.

44. Find the present value of a $20,000 payment to be made in 10 years. Assume an interest rate of 3.2% per year compounded continuously.

45. Interest is compounded annually. Consider the following choices of payments to you:

    Choice 1: $1500 now and $3000 one year from now
    Choice 2: $1900 now and $2500 one year from now

    (a) If the interest rate on savings were 5% per year, which would you prefer?
    (b) Is there an interest rate that would lead you to make a different choice? Explain.

46. The island of Manhattan was sold for $24 in 1626. Suppose the money had been invested in an account which compounded interest continuously.

    (a) How much money would be in the account in the year 2012 if the yearly interest rate was
       (i) 5%?   (ii) 7%?
    (b) If the yearly interest rate was 6%, in what year would the account be worth one billion dollars?

47. **(a)** Use the Rule of 70 to predict the doubling time of an investment which is earning 8% interest per year.
    (b) Find the doubling time exactly, and compare your answer to part (a).

48. A business associate who owes you $3000 offers to pay you $2800 now, or else pay you three yearly installments of $1000 each, with the first installment paid now. If you use only financial reasons to make your decision, which option should you choose? Justify your answer, assuming a 3% interest rate per year, compounded continuously.

49. A person is to be paid $2000 for work done over a year. Three payment options are being considered. Option 1 is to pay the $2000 in full now. Option 2 is to pay $1000 now and $1000 in a year. Option 3 is to pay the full $2000 in a year. Assume an annual interest rate of 5% a year, compounded continuously.

    (a) Without doing any calculations, which option is the best option financially for the worker? Explain.
    (b) Find the future value, in one year's time, of all three options.
    (c) Find the present value of all three options.

50. A company is considering whether to buy a new machine, which costs $97,000. The cash flows (adjusted for taxes and depreciation) that would be generated by the new machine are given in the following table:

Year	1	2	3	4
Cash flow	$50,000	$40,000	$25,000	$20,000

    (a) Find the total present value of the cash flows. Treat each year's cash flow as a lump sum at the end of the year and use an interest rate of 7.5% per year, compounded annually.
    (b) Based on a comparison of the cost of the machine and the present value of the cash flows, would you recommend purchasing the machine?

51. Big Tree McGee is negotiating his rookie contract with a professional basketball team. They have agreed to a three-year deal which will pay Big Tree a fixed amount at the end of each of the three years, plus a signing bonus at the beginning of his first year. They are still haggling about the amounts and Big Tree must decide between a big signing bonus and fixed payments per year, or a smaller bonus with payments increasing each year. The two options are summarized in the table. All values are payments in millions of dollars.

	Signing bonus	Year 1	Year 2	Year 3
Option #1	6.0	2.0	2.0	2.0
Option #2	1.0	2.0	4.0	6.0

    (a) Big Tree decides to invest all income in stock funds which he expects to grow at a rate of 10% per year, compounded continuously. He would like to choose the contract option which gives him the greater future value at the end of the three years when the last payment is made. Which option should he choose?
    (b) Calculate the present value of each contract offer.

52. You win \$38,000 in the state lottery to be paid in two installments—\$19,000 now and \$19,000 one year from now. A friend offers you \$36,000 in return for your two lottery payments. Instead of accepting your friend's offer, you take out a one-year loan at an interest rate of 8.25% per year, compounded annually. The loan will be paid back by a single payment of \$19,000 (your second lottery check) at the end of the year. Which is better, your friend's offer or the loan?

53. You are considering whether to buy or lease a machine whose purchase price is \$12,000. Taxes on the machine will be \$580 due in one year, \$464 due in two years, and \$290 due in three years. If you buy the machine, you expect to be able to sell it after three years for \$5,000. If you lease the machine for three years, you make an initial payment of \$2650 and then three payments of \$2650 at the end of each of the next three years. The leasing company will pay the taxes. The interest rate is 7.75% per year, compounded annually. Should you buy or lease the machine? Explain.

54. You are buying a car that comes with a one-year warranty and are considering whether to purchase an extended warranty for \$375. The extended warranty covers the two years immediately after the one-year warranty expires. You estimate that the yearly expenses that would have been covered by the extended warranty are \$150 at the end of the first year of the extension and \$250 at the end of the second year of the extension. The interest rate is 5% per year, compounded annually. Should you buy the extended warranty? Explain.

1.8 NEW FUNCTIONS FROM OLD

We have studied linear and exponential functions and the logarithm function. In this section, we learn how to create new functions by composing, stretching, and shifting functions we already know.

Composite Functions

A drop of water falls onto a paper towel. The area, A of the circular damp spot is a function of r, its radius, which is a function of time, t. We know $A = f(r) = \pi r^2$; suppose $r = g(t) = t + 1$. By substitution, we express A as a function of t:

$$A = f(g(t)) = \pi(t+1)^2.$$

The function $f(g(t))$ is a "function of a function," or a *composite function*, in which there is an *inside function* and an *outside function*. To find $f(g(2))$, we first add one ($g(2) = 2+1 = 3$) and then square and multiply by π. We have

$$f(g(2)) = \pi(2+1)^2 \quad = \quad \pi 3^2 \quad = \quad 9\pi.$$

First calculation   Second calculation

The inside function is $t + 1$ and the outside function is squaring and multiplying by π. In general, the inside function represents the calculation that is done first and the outside function represents the calculation done second.

Example 1   If $f(t) = t^2$ and $g(t) = t + 2$, find

(a) $f(t+1)$ (b) $f(t) + 3$ (c) $f(t+h)$ (d) $f(g(t))$ (e) $g(f(t))$

Solution
(a) Since $t + 1$ is the inside function, $f(t+1) = (t+1)^2$.
(b) Here 3 is added to $f(t)$, so $f(t) + 3 = t^2 + 3$.
(c) Since $t + h$ is the inside function, $f(t+h) = (t+h)^2$.
(d) Since $g(t) = t + 2$, substituting $t + 2$ into f gives $f(g(t)) = f(t+2) = (t+2)^2$.
(e) Since $f(t) = t^2$, substituting t^2 into g gives $g(f(t)) = g(t^2) = t^2 + 2$.

Example 2   If $f(x) = e^x$ and $g(x) = 5x + 1$, find (a) $f(g(x))$ (b) $g(f(x))$

Solution   (a) Substituting $g(x) = 5x + 1$ into f gives $f(g(x)) = f(5x + 1) = e^{5x+1}$.
(b) Substituting $f(x) = e^x$ into g gives $g(f(x)) = g(e^x) = 5e^x + 1$.

Example 3   Using the following table, find $g(f(0))$, $f(g(0))$, $f(g(1))$, and $g(f(1))$.

x	0	1	2	3
$f(x)$	3	1	-1	-3
$g(x)$	0	2	4	6

Solution   To find $g(f(0))$, we first find $f(0) = 3$ from the table. Then we have $g(f(0)) = g(3) = 6$.
For $f(g(0))$, we must find $g(0)$ first. Since $g(0) = 0$, we have $f(g(0)) = f(0) = 3$.
Similar reasoning leads to $f(g(1)) = f(2) = -1$ and $g(f(1)) = g(1) = 2$.

We can write a composite function using a new variable u to represent the value of the inside function. For example,

$$y = (t + 1)^4 \quad \text{is the same as} \quad y = u^4 \quad \text{with} \quad u = t + 1.$$

Other expressions for u, such as $u = (t + 1)^2$, with $y = u^2$, are also possible.

Example 4   Use a new variable u for the inside function to express each of the following as a composite function:
(a) $y = \ln(3t)$ (b) $w = 5(2r + 3)^2$ (c) $P = e^{-0.03t}$

Solution   (a) We take the inside function to be $3t$, so $y = \ln u$ with $u = 3t$.
(b) We take the inside function to be $2r + 3$, so $w = 5u^2$ with $u = 2r + 3$.
(c) We take the inside function to be $-0.03t$, so $P = e^u$ with $u = -0.03t$.

Stretches of Graphs

If the demand function is linear, the graph of a possible revenue function $R = f(p)$ is in Figure 1.82. What does the graph of $R = 3f(p)$ look like? The factor 3 in the function $R = 3f(p)$ stretches each $f(p)$ revenue value by multiplying it by 3. See Figure 1.83. If c is positive, the graph of $R = cf(p)$ is the graph of $R = f(p)$ stretched or shrunk vertically by c units. If c is negative, the function no longer makes sense as a revenue function, but we can still draw the graph. What does the graph of $R = -2f(p)$ look like? The factor -2 in the function $R = -2f(p)$ stretches $f(p)$ by multiplying by 2 and reflecting it across the x-axis. See Figure 1.83.

> Multiplying a function by a constant, c, stretches the graph vertically (if $c > 1$) or shrinks the graph vertically (if $0 < c < 1$). A negative sign (if $c < 0$) reflects the graph across the x-axis, in addition to shrinking or stretching.

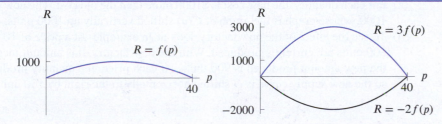

Figure 1.82: Graph of $f(p)$ **Figure 1.83:** Multiples of the function $f(p)$

Shifted Graphs

Consider the function $y = x^2 + 4$. The y-coordinates for this function are exactly 4 units larger than the corresponding y-coordinates of the function $y = x^2$. So the graph of $y = x^2 + 4$ is obtained from the graph of $y = x^2$ by adding 4 to the y-coordinate of each point, that is, by moving the graph of $y = x^2$ up 4 units. (See Figure 1.84.)

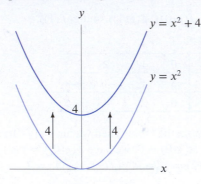

Figure 1.84: Vertical shift: Graphs of $y = x^2$ and $y = x^2 + 4$

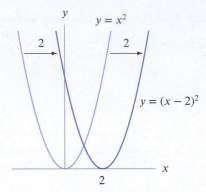

Figure 1.85: Horizontal shift: Graphs of $y = x^2$ and $y = (x - 2)^2$

A graph can also be shifted to the left or to the right. In Figure 1.85, we see that the graph of $y = (x - 2)^2$ is the graph of $y = x^2$ shifted to the right 2 units. In general,

- The graph of $y = f(x) + k$ is the graph of $y = f(x)$ moved up k units (down if k is negative).
- The graph of $y = f(x - k)$ is the graph of $y = f(x)$ moved to the right k units (to the left if k is negative).

Example 5
(a) A cost function, $C(q)$, for a company is shown in Figure 1.86. The fixed cost increases by \$1000. Sketch a graph of the new cost function.
(b) A supply curve, S, for a product is given in Figure 1.87. A new factory opens and produces 100 units of the product no matter what the price. Sketch a graph of the new supply curve.

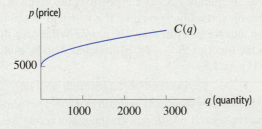

Figure 1.86: A cost function

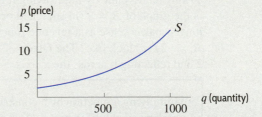

Figure 1.87: A supply function

Solution
(a) For each quantity, the new cost is \$1000 more than the old cost. The new cost function is $C(q) + 1000$, whose graph is the graph of $C(q)$ shifted vertically up 1000 units. (See Figure 1.88.)
(b) To see the effect of the new factory, look at an example. At a price of 10 dollars, approximately 800 units are currently produced. With the new factory, this amount increases by 100 units, so the new amount produced is 900 units. At each price, the quantity produced increases by 100, so the new supply curve is S shifted horizontally to the right by 100 units. (See Figure 1.89.)

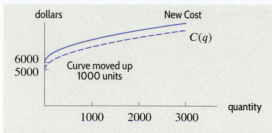

Figure 1.88: New cost function (original curve dashed)

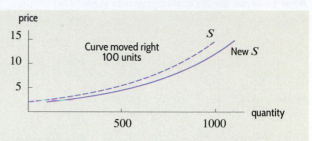

Figure 1.89: New supply curve (original curve dashed)

Problems for Section 1.8

■ In Problems **1–3**, find the following:

(a) $f(g(x))$ (b) $g(f(x))$ (c) $f(f(x))$

1. $f(x) = 5x - 1$ and $g(x) = 3x + 2$
2. $f(x) = x - 2$ and $g(x) = x^2 + 8$
3. $f(x) = 3x$ and $g(x) = e^{2x}$
4. Let $f(x) = x^2$ and $g(x) = 3x - 1$. Find the following:

   (a) $f(2) + g(2)$ (b) $f(2) \cdot g(2)$
   (c) $f(g(2))$ (d) $g(f(2))$

5. For $g(x) = x^2 + 2x + 3$, find and simplify:

   (a) $g(2 + h)$ (b) $g(2)$
   (c) $g(2 + h) - g(2)$

6. If $f(x) = x^2 + 1$, find and simplify:

   (a) $f(t + 1)$ (b) $f(t^2 + 1)$ (c) $f(2)$
   (d) $2f(t)$ (e) $(f(t))^2 + 1$

■ For the functions f and g in Problems **7–10**, find

(a) $f(g(1))$ (b) $g(f(1))$ (c) $f(g(x))$
(d) $g(f(x))$ (e) $f(t)g(t)$

7. $f(x) = x^2, g(x) = x + 1$
8. $f(x) = \sqrt{x + 4}, g(x) = x^2$
9. $f(x) = e^x, g(x) = x^2$
10. $f(x) = 1/x, g(x) = 3x + 4$
11. Use Table 1.36 to find:

    (a) $f(g(1))$ (b) $g(f(1))$ (c) $f(g(4))$
    (d) $g(f(4))$ (e) $f(g(6))$ (f) $g(f(6))$

Table 1.36

x	1	2	3	4	5	6
$f(x)$	5	4	3	3	4	5
$g(x)$	6	5	4	3	2	1

12. Use Table 1.37 to find:

    (a) $f(g(0))$ (b) $f(g(1))$ (c) $f(g(2))$
    (d) $g(f(2))$ (e) $g(f(3))$

Table 1.37

x	0	1	2	3	4	5
$f(x)$	10	6	3	4	7	11
$g(x)$	2	3	5	8	12	15

13. Make a table of values for each of the following functions using Table 1.37:

    (a) $f(x) + 3$ (b) $f(x - 2)$ (c) $5g(x)$
    (d) $-f(x) + 2$ (e) $g(x - 3)$ (f) $f(x) + g(x)$

14. Use the variable u for the inside function to express each of the following as a composite function:

    (a) $y = (5t^2 - 2)^6$ (b) $P = 12e^{-0.6t}$
    (c) $C = 12\ln(q^3 + 1)$

15. Use the variable u for the inside function to express each of the following as a composite function:

    (a) $y = 2^{3x-1}$ (b) $P = \sqrt{5t^2 + 10}$
    (c) $w = 2\ln(3r + 4)$

■ Simplify the quantities in Problems **16–19** using $m(z) = z^2$.

16. $m(z + 1) - m(z)$
17. $m(z + h) - m(z)$
18. $m(z) - m(z - h)$
19. $m(z + h) - m(z - h)$

■ For Problems **20–25**, use the graphs in Figure 1.90.

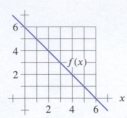
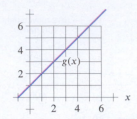

Figure 1.90

20. Estimate $f(g(1))$.   21. Estimate $g(f(1))$.

22. Estimate $f(g(4))$.   23. Estimate $g(f(4))$.

24. Estimate $f(f(2))$.   25. Estimate $g(g(2))$.

■ For Problems **26–29**, use the graphs in Figure 1.91.

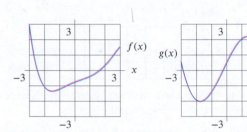

Figure 1.91

26. Estimate $f(g(1))$.   27. Estimate $g(f(2))$.

28. Estimate $f(f(1))$.   29. Estimate $f(g(3))$.

30. Using Table 1.38, create a table of values for $f(g(x))$ and for $g(f(x))$.

Table 1.38

x	-3	-2	-1	0	1	2	3
$f(x)$	0	1	2	3	2	1	0
$g(x)$	3	2	2	0	-2	-2	-3

31. A tree of height y meters has, on average, B branches, where $B = y - 1$. Each branch has, on average, n leaves, where $n = 2B^2 - B$. Find the average number of leaves on a tree as a function of height.

■ In Problems **32–35**, use Figure 1.92 to estimate the function value or explain why it cannot be done.

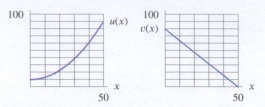

Figure 1.92

32. $u(v(10))$ 33. $u(v(40))$

34. $v(u(10))$ 35. $v(u(40))$

36. The Heaviside step function, H, is graphed in Figure 1.93. Graph the following functions.

(a) $2H(x)$ (b) $H(x) + 1$ (c) $H(x + 1)$
(d) $-H(x)$ (e) $H(-x)$

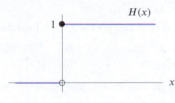

Figure 1.93

■ In Problems **37–42**, use Figure 1.94 to graph the function.

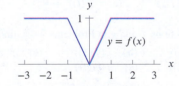

Figure 1.94

37. $y = f(x) + 1$

38. $y = f(x - 2)$

39. $y = 3f(x)$

40. $y = f(x + 1) - 2$

41. $y = -f(x) + 3$

42. $y = -2f(x - 1)$

■ For the functions f in Problems **43–45**, graph:
(a) $f(x + 2)$ (b) $f(x - 1)$ (c) $f(x) - 4$
(d) $f(x + 1) + 3$ (e) $3f(x)$ (f) $-f(x) + 1$

43.

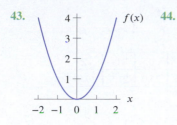

44.

45.

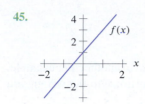

■ In Problems **46–51**, use Figure 1.95 to graph the function.

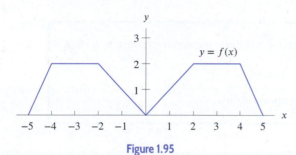

Figure 1.95

46. $y = f(x) + 2$

47. $y = 2f(x)$

48. $y = f(x - 1)$

49. $y = -3f(x)$

50. $y = 2f(x) - 1$

51. $y = 2 - f(x)$

52. Figure 1.96 shows the concentration of a drug in the body, $f(t) = a - be^{kt}$, over time, t:

(a) What does the constant a represent in this context?
(b) What is the sign of the constant a?
(c) What is the relation between the constants a and b?
(d) What is the sign of the constant k?

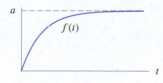

Figure 1.96

53. Morphine, a pain-relieving drug, is administered to a patient intravenously starting at 8 am. The drug saturation curve $Q = f(t)$ in Figure 1.97 gives the quantity, Q, of morphine in the blood t hours after 8 am.

(a) Draw the drug saturation curve $Q = g(t)$ if the IV line is started at noon instead of 8 am.
(b) Is g one of the following transformations of f: vertical shift, vertical stretch, horizontal shift, horizontal stretch? If so, which?
(c) Write $g(t)$ in terms of the function f.

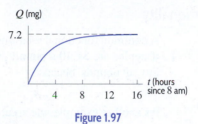

Figure 1.97

54. (a) Write an equation for a graph obtained by vertically stretching the graph of $y = x^2$ by a factor of 2, followed by a vertical upward shift of 1 unit. Sketch it.
(b) What is the equation if the order of the transformations (stretching and shifting) in part (a) is interchanged?
(c) Are the two graphs the same? Explain the effect of reversing the order of transformations.

■ In Problems **55–58** the functions $r = f(t)$ and $V = g(r)$ give the radius and the volume of a commercial hot air balloon being inflated for testing. The variable t is in minutes, r is in feet, and V is in cubic feet. The inflation begins at $t = 0$. In each case, give a mathematical expression that represents the given statement.

55. The volume of the balloon t minutes after inflation began.

56. The volume of the balloon if its radius were twice as big.

57. The time that has elapsed when the radius of the balloon is 30 feet.

58. The time that has elapsed when the volume of the balloon is 10,000 cubic feet.

59. Cyanide is used in solution to isolate gold in a mine.[109] This may result in contaminated groundwater near the mine, requiring the poison be removed, as in the following table, where t is in years since 2012.

(a) Find an exponential model for $c(t)$, the concentration, in parts per million, of cyanide in the groundwater.

---

[109]www.miningfacts.org/environment/what-is-the-role-of-cyanide-in-mining. Accessed June 9, 2015.

(b) Use the model in part (a) to find the number of years it takes for the cyanide concentration to fall to 10 ppm.

(c) The filtering process removing the cyanide is sped up so that the new model is $D(t) = c(2t)$. Find $D(t)$.

(d) If the cyanide removal was started three years earlier, but run at the speed of part (a), find a new model, $E(t)$.

t (years)	0	1	2
$c(t)$ (ppm)	25.0	21.8	19.01

1.9 PROPORTIONALITY AND POWER FUNCTIONS

Proportionality

A common functional relationship occurs when one quantity is *proportional* to another. For example, if apples are \$1.40 a pound, we say the price you pay, p dollars, is proportional to the weight you buy, w pounds, because

$$p = f(w) = 1.40w.$$

As another example, the area, A, of a circle is proportional to the square of the radius, r:

$$A = f(r) = \pi r^2.$$

> We say y is (directly) **proportional** to x if there is a nonzero constant k such that
> $$y = kx.$$
> This k is called the constant of proportionality.

We also say that one quantity is *inversely proportional* to another if one is proportional to the reciprocal of the other. For example, the speed, v, at which you make a 50-mile trip is inversely proportional to the time, t, taken, because v is proportional to $1/t$:

$$v = 50\left(\frac{1}{t}\right) = \frac{50}{t}.$$

Notice that if y is directly proportional to x, then the magnitude of one variable increases (decreases) when the magnitude of the other increases (decreases). If, however, y is inversely proportional to x, then the magnitude of one variable increases when the magnitude of the other decreases.

Example 1   The heart mass of a mammal is proportional to its body mass.[110]

(a) Write a formula for heart mass, H, as a function of body mass, B.

(b) A human with a body mass of 70 kilograms has a heart mass of 0.42 kilograms. Use this information to find the constant of proportionality.

(c) Estimate the heart mass of a horse with a body mass of 650 kg.

Solution   (a) Since H is proportional to B, for some constant k, we have

$$H = kB.$$

(b) We use the fact that $H = 0.42$ when $B = 70$ to solve for k:

$$H = kB$$

---

[110]K. Schmidt-Nielson: *Scaling—Why is Animal Size So Important?* (Cambridge: CUP, 1984).

$$0.42 = k(70)$$
$$k = \frac{0.42}{70} = 0.006.$$

(c) Since $k = 0.006$, we have $H = 0.006B$, so the heart mass of the horse is given by

$$H = 0.006(650) = 3.9 \text{ kilograms.}$$

Example 2   The period of a pendulum, T, is the amount of time required for the pendulum to make one complete swing. For small swings, the period, T, is approximately proportional to the square root of l, the pendulum's length. So

$$T = k\sqrt{l} \quad \text{where } k \text{ is a constant.}$$

Notice that T is not directly proportional to l, but T is proportional to $\sqrt{l}$.

Example 3   An object's weight, w, is inversely proportional to the square of its distance, r, from the earth's center. So, for some constant k,

$$w = \frac{k}{r^2}.$$

Here w is not inversely proportional to r, but to r^2.

Power Functions

In each of the previous examples, one quantity is proportional to the power of another quantity. We make the following definition:

> We say that $Q(x)$ is a **power function** of x if $Q(x)$ is proportional to a constant power of x. If k is the constant of proportionality, and if p is the power, then
>
> $$Q(x) = k \cdot x^p.$$

For example, the function $H = 0.006B$ is a power function with $p = 1$. The function $T = k\sqrt{l} = kl^{1/2}$ is a power function with $p = 1/2$, and the function $w = k/r^2 = kr^{-2}$ is a power function with $p = -2$.

Example 4   Which of the following are power functions? For those which are, write the function in the form $y = kx^p$, and give the coefficient k and the exponent p.

(a) $y = \dfrac{5}{x^3}$ (b) $y = \dfrac{2}{3x}$ (c) $y = \dfrac{5x^2}{2}$

(d) $y = 5 \cdot 2^x$ (e) $y = 3\sqrt{x}$ (f) $y = (3x^2)^3$

Solution   (a) Since $y = 5x^{-3}$, this is a power function with $k = 5$ and $p = -3$.
(b) Since $y = (2/3)x^{-1}$, this is a power function with $k = 2/3$ and $p = -1$.
(c) Since $y = (5/2)x^2$, this is a power function with $k = 5/2 = 2.5$ and $p = 2$.
(d) This is not a power function. It is an exponential function.
(e) Since $y = 3x^{1/2}$, this is a power function with $k = 3$ and $p = 1/2$.
(f) Since $y = 3^3 \cdot (x^2)^3 = 27x^6$, this is a power function with $k = 27$ and $p = 6$.

Graphs of Power Functions

The graph of $y = x^2$ is shown in Figure 1.98. It is decreasing for negative x and increasing for positive x. Notice that it is bending upward, or concave up, for all x. The graph of $y = x^3$ is shown in Figure 1.99. Notice that it is bending downward, or concave down for negative x and bending upward, or concave up for positive x. The graph of $y = \sqrt{x} = x^{1/2}$ is shown in Figure 1.100. Notice that the graph is increasing and concave down.

Since x^2 increases without bound as x increases, we often say that it tends to infinity as x approaches infinity, which we write in symbols as

$$x^2 \to \infty \quad \text{as} \quad x \to \infty.$$

Since x^3 decreases without bound as x decreases, we write

$$x^3 \to -\infty \quad \text{as} \quad x \to -\infty.$$

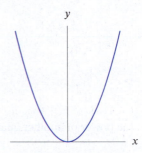

Figure 1.98: Graph of $y = x^2$

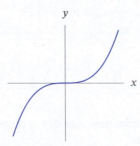

Figure 1.99: Graph of $y = x^3$

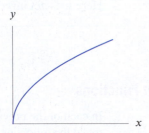

Figure 1.100: Graph of $y = x^{1/2}$

Example 5   If N is the average number of species found on an island and A is the area of the island, observations have shown[111] that N is approximately proportional to the cube root of A. Write a formula for N as a function of A and describe the shape of the graph of this function.

Solution   For some positive constant k, we have

$$N = k\sqrt[3]{A} = kA^{1/3}.$$

It turns out that the value of k depends on the region of the world in which the island is found. The graph of N against A (for $A > 0$) has a shape similar to the graph in Figure 1.100. It is increasing and concave down. Thus, larger islands have more species on them (as we would expect), but the increase slows as the island gets larger.

The function $y = x^0 = 1$ has a graph that is a horizontal line. For negative powers, rewriting

$$y = x^{-1} = \frac{1}{x} \quad \text{and} \quad y = x^{-2} = \frac{1}{x^2}$$

---

[111] *Scientific American*, p. 112 (September, 1989).

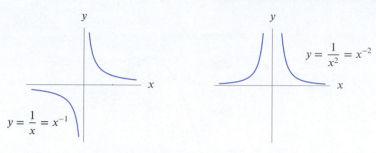

Figure 1.101: Graphs of negative powers of x

makes it clear that as $x > 0$ increases, the denominators increase and the functions decrease. The graphs of $y = x^{-1}$ and $y = x^{-2}$ have both the x- and y-axes as asymptotes. (See Figure 1.101.)

Quadratic Functions and Polynomials

Sums of power functions with nonnegative integer exponents are called *polynomials*. With a_n, a_{n-1}, ..., a_0 constants, polynomials are functions of the form

$$y = p(x) = a_n x^n + a_{n-1} x^{n-1} + \cdots + a_1 x + a_0.$$

Here, n is a nonnegative integer, called the *degree* of the polynomial, and a_n is a nonzero number called the *leading coefficient*. We call $a_n x^n$ the *leading term*.

If $n = 2$, the polynomial is called *quadratic* and has the form $ax^2 + bx + c$ with $a \neq 0$. The graph of a quadratic polynomial is a parabola. It opens up if the leading coefficient a is positive and opens down if a is negative.

Example 6　A company finds that the average number of people attending a concert is 75 if the price is \$50 per person. At a price of \$35 per person, the average number of people in attendance is 120.

(a) Assume that the demand curve is a line. Write the demand, q, as a function of price, p.
(b) Use your answer to part (a) to write the revenue, R, as a function of price, p.
(c) Use a graph of the revenue function to determine what price should be charged to obtain the greatest revenue.

Solution　(a) Two points on the line are $(p, q) = (50, 75)$ and $(p, q) = (35, 120)$. The slope of the line is

$$m = \frac{120 - 75}{35 - 50} = \frac{45}{-15} = -3 \text{ people/dollar}.$$

To find the vertical intercept of the line, we use the slope and one of the points:

$$75 = b + (-3)(50)$$
$$225 = b.$$

The demand function is $q = 225 - 3p$.

(b) Since $R = pq$ and $q = 225 - 3p$, we see that $R = p(225 - 3p) = 225p - 3p^2$.
(c) The revenue function is the quadratic polynomial graphed in Figure 1.102. The maximum revenue occurs at $p = 37.5$. Thus, the company maximizes revenue by charging \$37.50 per person.

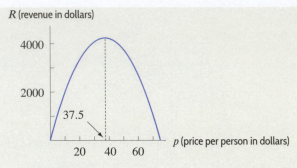

Figure 1.102: Revenue function for concert ticket sales

Problems for Section 1.9

■ In Problems **1–12**, determine whether or not the function is a power function. If it is a power function, write it in the form $y = kx^p$ and give the values of k and p.

1. $y = \dfrac{x}{5}$ **2.** $y = 5\sqrt{x}$ **3.** $y = \dfrac{8}{x}$

4. $y = \dfrac{3}{x^2}$ **5.** $y = 2^x$ **6.** $y = \dfrac{3}{8x}$

7. $y = (3x^5)^2$ **8.** $y = \dfrac{5}{2\sqrt{x}}$ **9.** $y = 3 \cdot 5^x$

10. $y = \dfrac{2x^2}{10}$ **11.** $y = (5x)^3$ **12.** $y = 3x^2 + 4$

■ In Problems **13–16**, write a formula representing the function.

13. The strength, S, of a beam is proportional to the square of its thickness, h.

14. The energy, E, expended by a swimming dolphin is proportional to the cube of the speed, v, of the dolphin.

15. The average velocity, v, for a trip over a fixed distance, d, is inversely proportional to the time of travel, t.

16. The gravitational force, F, between two bodies is inversely proportional to the square of the distance d between them.

17. Use shifts of power functions to find a possible formula for each of the graphs:

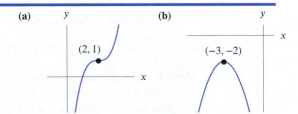

18. The surface area of a mammal, S, satisfies the equation $S = kM^{2/3}$, where M is the body mass and the constant of proportionality k depends on the body shape of the mammal. A human of body mass 70 kilograms has surface area 18,600 cm^2. Find the constant of proportionality for humans. Find the surface area of a human with body mass 60 kilograms.

19. The number of species of lizards, N, found on an island off Baja California is proportional to the fourth root of the area, A, of the island.[112] Write a formula for N as a function of A. Graph this function. Is it increasing or decreasing? Is the graph concave up or concave down? What does this tell you about lizards and island area?

20. The blood mass of a mammal is proportional to its body mass. A rhinoceros with body mass 3000 kilograms has blood mass of 150 kilograms. Find a formula for the blood mass of a mammal as a function of the body mass and estimate the blood mass of a human with body mass 70 kilograms.

21. Kleiber's Law states that the metabolic needs (such as calorie requirements) of a mammal are proportional to its body weight raised to the 0.75 power.[113] Surprisingly, the daily diets of mammals conform to this relation well. Assuming Kleiber's Law holds:

(a) Write a formula for C, daily calorie consumption, as a function of body weight, W.

---

[112]M. L. Rosenzweig, *Species Diversity in Space and Time*, p. 143 (Cambridge: Cambridge University Press, 1995).
[113]S. Strogatz, "Math and the City," *The New York Times,* May 20, 2009. Kleiber originally estimated the exponent as 0.74; it is now believed to be 0.75.

(b) Sketch a graph of this function. (You do not need scales on the axes.)

(c) If a human weighing 150 pounds needs to consume 1800 calories a day, estimate the daily calorie requirement of a horse weighing 700 lbs and of a rabbit weighing 9 lbs.

(d) On a per-pound basis, which animal requires more calories: a mouse or an elephant?

22. Allometry is the study of the relative size of different parts of a body as a consequence of growth. In this problem, you will check the accuracy of an allometric equation: the weight of a fish is proportional to the cube of its length.[114] Table 1.39 relates the weight, y, in gm, of plaice (a type of fish) to its length, x, in cm. Does this data support the hypothesis that (approximately) $y = kx^3$? If so, estimate the constant of proportionality, k.

Table 1.39

x	y	x	y	x	y
33.5	332	37.5	455	41.5	623
34.5	363	38.5	500	42.5	674
35.5	391	39.5	538	43.5	724
36.5	419	40.5	574		

23. Biologists estimate that the number of animal species of a certain body length is inversely proportional to the square of the body length.[115] Write a formula for the number of animal species, N, of a certain body length as a function of the length, L. Are there more species at large lengths or at small lengths? Explain.

24. The specific heat, s, of an element is the number of calories of heat required to raise the temperature of one gram of the element by one degree Celsius. Use the following table to decide if s is proportional or inversely proportional to the atomic weight, w, of the element. If so, find the constant of proportionality.

Element	Li	Mg	Al	Fe	Ag	Pb	Hg
w	6.9	24.3	27.0	55.8	107.9	207.2	200.6
s	0.92	0.25	0.21	0.11	0.056	0.031	.033

25. The circulation time of a mammal (that is, the average time it takes for all the blood in the body to circulate once and return to the heart) is proportional to the fourth root of the body mass of the mammal.

(a) Write a formula for the circulation time, T, in terms of the body mass, B.

(b) If an elephant of body mass 5230 kilograms has a circulation time of 148 seconds, find the constant of proportionality.

(c) What is the circulation time of a human with body mass 70 kilograms?

26. Zipf's Law, developed by George Zipf in 1949, states that in a given country, the population of a city is inversely proportional to the city's rank by size in the country.[116] Assuming Zipf's Law:

(a) Write a formula for the population, P, of a city as a function of its rank, R.

(b) If the constant of proportionality k is 300,000, what is the approximate population of the largest city (rank 1)? The second largest city (rank 2)? The third largest city?

(c) Answer the questions of part (b) if $k = 6$ million.

(d) Interpret the meaning of the constant of proportionality k in this context.

27. The infrastructure needs of a region (for example, the number of miles of electrical cable, the number of miles of roads, the number of gas stations) depend on its population. Cities enjoy economies of scale.[117] For example, the number of gas stations is proportional to the population raised to the power of 0.77.

(a) Write a formula for the number, N, of gas stations in a city as a function of the population, P, of the city.

(b) If city A is 10 times bigger than city B, how do their number of gas stations compare?

(c) Which is expected to have more gas stations per person, a town of 10,000 people or a city of 500,000 people?

28. A sporting goods wholesaler finds that when the price of a product is $25, the company sells 500 units per week. When the price is $30, the number sold per week decreases to 460 units.

(a) Find the demand, q, as a function of price, p, assuming that the demand curve is linear.

(b) Use your answer to part (a) to write revenue as a function of price.

(c) Graph the revenue function in part (b). Find the price that maximizes revenue. What is the revenue at this price?

---

[114]Adapted from R. J. H. Beverton and S. J. Holt, "On the Dynamics of Exploited Fish Populations", *Fishery Investigations*, Series II, 19, 1957.

[115]*US News & World Report*, August 18, 1997, p. 79.

[116]S. Strogatz, "Math and the City," *The New York Times*, May 20, 2009.

[117]S. Strogatz, "Math and the City," *The New York Times*, May 20, 2009.

29. A health club has cost and revenue functions given by $C = 10,000 + 35q$ and $R = pq$, where q is the number of annual club members and p is the price of a one-year membership. The demand function for the club is $q = 3000 - 20p$.

    (a) Use the demand function to write cost and revenue as functions of p.

    (b) Graph cost and revenue as a function of p, on the same axes. (Note that price does not go above $170 and that the annual costs of running the club reach $120,000.)

    (c) Explain why the graph of the revenue function has the shape it does.

    (d) For what prices does the club make a profit?

    (e) Estimate the annual membership fee that maximizes profit. Mark this point on your graph.

1.10 PERIODIC FUNCTIONS

What Are Periodic Functions?

Many functions have graphs that oscillate, resembling a wave. Figure 1.103 shows the number of new housing construction starts (one-family units) in the US, 2012–2015, where t is time in quarter-years.[118] Notice that few new homes begin construction during the first quarter of a year (January, February, and March), whereas many new homes are begun in the second quarter (April, May, and June).

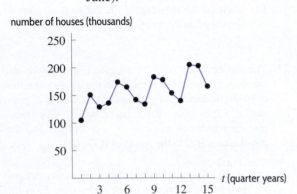

Figure 1.103: New housing construction starts, 2012–2015

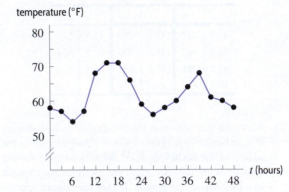

Figure 1.104: Temperature in Phoenix after midnight February 17, 2005

Let's look at another example. Figure 1.104 is a graph of the temperature (in °C) in Phoenix, AZ, in hours after midnight, February 17, 2005. Notice that the maximum is in the afternoon and the minimum is in the early morning.[119] Again, the graph looks like a wave.

Functions whose values repeat at regular intervals are called *periodic*. Many processes, such as the number of housing starts or the temperature, are approximately periodic. The water level in a tidal basin, the blood pressure in a heart, retail sales in the US, and the position of air molecules transmitting a musical note are also all periodic functions of time.

Amplitude and Period

Periodic functions repeat exactly the same cycle forever. If we know one cycle of the graph, we know the entire graph.

> For any periodic function of time:
> - The **amplitude** is half the difference between its maximum and minimum values.
> - The **period** is the time for the function to execute one complete cycle.

---

[118]http://www.census.gov/const/www/quarterly_starts_completions.pdf, accessed September 28, 2016.
[119]http://www.weather.com, accessed February 20, 2005.

Example 1   Estimate the amplitude and period of the new housing starts function shown in Figure 1.103.

Solution   Figure 1.103 is not exactly periodic, since the maximum and minimum are not the same for each cycle. Nonetheless, the minimum is about 300, and the maximum is about 450. The difference between them is 150, so the amplitude is about $\frac{1}{2}(150) = 75$ thousand houses.

    The wave completes a cycle between $t = 1$ and $t = 5$, so the period is $t = 4$ quarter-years, or one year. The business cycle for new housing construction is one year.

Example 2   Figure 1.105 shows the temperature in an unopened freezer. Estimate the temperature in the freezer at 12:30 and at 2:45.

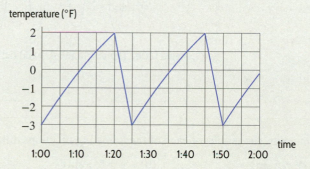

Figure 1.105: Oscillating freezer temperature. Estimate the temperature at 12:30 and 2:45

Solution   The maximum and minimum values each occur every 25 minutes, so the period is 25 minutes. The temperature at 12:30 should be the same as at 12:55 and at 1:20, namely, 2°F. Similarly, the temperature at 2:45 should be the same as at 2:20 and 1:55, or about −1.5°F.

The Sine and Cosine

Many periodic functions are represented using the functions called *sine* and *cosine*. The keys for the sine and cosine on a calculator are usually labeled as ⌈sin⌉ and ⌈cos⌉.
Warning: Your calculator can be in either "degree" mode or "radian" mode. For this book, always use "radian" mode.

Graphs of the Sine and Cosine

The graphs of the sine and the cosine functions are periodic; see Figures 1.106 and 1.107. Notice that the graph of the cosine function is the graph of the sine function, shifted $\pi/2$ to the left.

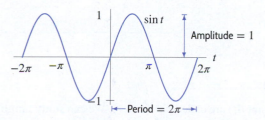

Figure 1.106: Graph of $\sin t$

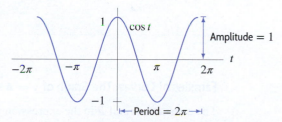

Figure 1.107: Graph of $\cos t$

The maximum and minimum values of $\sin t$ are $+1$ and -1, so the amplitude of the sine function is 1. The graph of $y = \sin t$ completes a cycle between $t = 0$ and $t = 2\pi$; the rest of the graph repeats this portion. The period of the sine function is 2π.

Example 3   Use a graph of $y = 3 \sin 2t$ to estimate the amplitude and period of this function.

Solution   In Figure 1.108, the waves have a maximum of $+3$ and a minimum of -3, so the amplitude is 3. The graph completes one complete cycle between $t = 0$ and $t = \pi$, so the period is π.

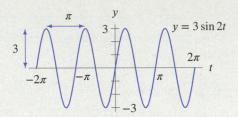

Figure 1.108: The amplitude is 3 and the period is π

Example 4   Explain how the graphs of each of the following functions differ from the graph of $y = \sin t$.

(a) $y = 6 \sin t$ (b) $y = 5 + \sin t$ (c) $y = \sin\left(t + \frac{\pi}{2}\right)$

Solution   (a) The graph of $y = 6 \sin t$ is in Figure 1.109. The maximum and minimum values are $+6$ and -6, so the amplitude is 6. This is the graph of $y = \sin t$ stretched vertically by a factor of 6.

(b) The graph of $y = 5 + \sin t$ is in Figure 1.110. The maximum and minimum values of this function are 6 and 4, so the amplitude is $(6 - 4)/2 = 1$. The amplitude (or size of the wave) is the same as for $y = \sin t$, since this is a graph of $y = \sin t$ shifted up 5 units.

(c) The graph of $y = \sin(t + \pi/2)$ is in Figure 1.111. This has the same amplitude, namely 1, and period, namely 2π, as the graph of $y = \sin t$. It is the graph of $y = \sin t$ shifted $\pi/2$ units to the left. (In fact, this is the graph of $y = \cos t$.)

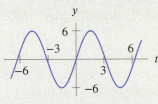

Figure 1.109: Graph of $y = 6 \sin t$

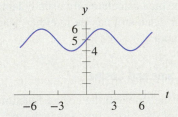

Figure 1.110: Graph of $y = 5 + \sin t$

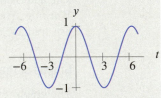

Figure 1.111: Graph of $y = \sin(t + \frac{\pi}{2})$

Families of Curves: The Graph of $y = A \sin(Bt)$

The constants A and B in the expression $y = A \sin(Bt)$ are called *parameters*. We can study families of curves by varying one parameter at a time and studying the result.

Example 5   (a) Graph $y = A \sin t$ for several positive values of A. Describe the effect of A on the graph.

          (b) Graph $y = \sin(Bt)$ for several positive values of B. Describe the effect of B on the graph.

Solution   (a) From the graphs of $y = A \sin t$ for $A = 1, 2, 3$ in Figure 1.112, we see that A is the amplitude.

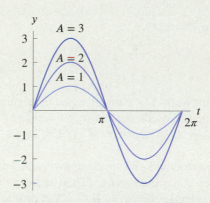

Figure 1.112: Graphs of $y = A \sin t$ with $A = 1, 2, 3$

    (b) The graphs of $y = \sin(Bt)$ for $B = \frac{1}{2}, B = 1$, and $B = 2$ are shown in Figure 1.113. When $B = 1$, the period is 2π; when $B = 2$, the period is π; and when $B = \frac{1}{2}$, the period is 4π. The parameter B affects the period of the function. The graphs suggest that the larger B is, the shorter the period. In fact, the period is $2\pi/B$.

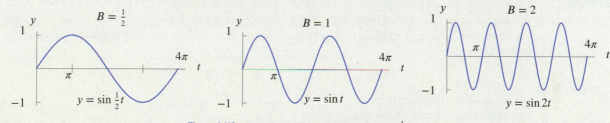

Figure 1.113: Graphs of $y = \sin(Bt)$ with $B = \frac{1}{2}, 1, 2$

In Example 5, the amplitude of $y = A \sin(Bt)$ was determined by the parameter A, and the period was determined by the parameter B. In addition, the oscillations may take place around a midline, given by the vertical shift, C. In general, we have

> The functions $y = A \sin(Bt) + C$ and $y = A \cos(Bt) + C$ are periodic with
>
> $$\text{Amplitude} = |A|, \quad \text{Period} = \frac{2\pi}{|B|}, \quad \text{Vertical shift (midline)} = C$$

Example 6   Find possible formulas for the following periodic functions.

(a) 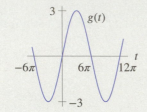   (b) 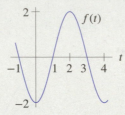   (c)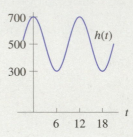

Solution   (a) This function looks like a sine function of amplitude 3, so $g(t) = 3\sin(Bt)$. Since the function executes one full oscillation between $t = 0$ and $t = 12\pi$, when t changes by 12π, the quantity Bt changes by 2π. This means $B \cdot 12\pi = 2\pi$, so $B = 1/6$. Therefore, $g(t) = 3\sin(t/6)$ has the graph shown.

(b) This function looks like an upside-down cosine function with amplitude 2, so $f(t) = -2\cos(Bt)$. The function completes one oscillation between $t = 0$ and $t = 4$. Thus, when t changes by 4, the quantity Bt changes by 2π, so $B \cdot 4 = 2\pi$, or $B = \pi/2$. Therefore, $f(t) = -2\cos(\pi t/2)$ has the graph shown.

(c) This function looks like a cosine function. The maximum is 700 and the minimum is 300, so the amplitude is $\frac{1}{2}(700 - 300) = 200$. The height halfway between the maximum and minimum is 500, so the cosine curve has been shifted up 500 units, so $h(t) = 500 + 200\cos(Bt)$. The period is 12, so $B \cdot 12 = 2\pi$. Thus, $B = \pi/6$. The function $h(t) = 500 + 200\cos(\pi t/6)$ has the graph shown.

Example 7   On October 17, 2017, high tide in Portland, Maine was at midnight.[120] The height of the water (measured in feet above the mean level of the low tide) in the harbor is a periodic function, since it oscillates between high and low tide. If t is in hours since midnight, the height (in feet) is approximated by the formula

$$y = 5 + 6.1\cos\left(\frac{\pi}{6}t\right).$$

(a) Graph this function from $t = 0$ to $t = 24$.

(b) What was the water level at high tide?

(c) When was low tide, and what was the water level at that time?

(d) What is the period of this function, and what does it represent in terms of tides?

(e) What is the amplitude of this function, and what does it represent in terms of tides?

Solution   (a) See Figure 1.114.

(b) The water level at high tide was 11.1 feet (given by the y-intercept on the graph).

(c) Low tide occurs at $t = 6$ (6 am) and at $t = 18$ (6 pm). The water level at this time is -1.1 feet, that is 1.1 feet below the mean level of the low tide.

(d) The period is 12 hours and represents the interval between successive high tides or successive low tides. Of course, there is something wrong with the assumption in the model that the period is 12 hours. If so, the high tide would always be at noon or midnight, instead of progressing slowly through the day, as it in fact does. The interval between successive high tides actually averages about 12 hours 25 minutes, which could be taken into account in a more precise mathematical model.

(e) The maximum is 11.1, and the minimum is -1.1, so the amplitude is $(11.1 - (-1.1))/2 = 12.2/2$, which is 6.1 feet. This represents half the difference between the depths at high and low tide.

---

[120]me.usharbors.com/monthly-tides, accessed January 24, 2017.

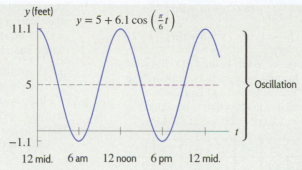

Figure 1.114: Graph of the function approximating the height of the water above mean low tide in Portland, Maine on October 17, 2017

Problems for Section 1.10

■ In Problems 1–6, graph the function. What is the amplitude and period?

1. $y = 3 \sin x$

2. $y = 4 \cos 2x$

3. $y = -3 \sin 2\theta$

4. $y = 3 \sin 2x$

5. $y = 5 - \sin 2t$

6. $y = 4 \cos\left(\frac{1}{2}t\right)$

7. Figure 1.115 shows quarterly US beer production during the period 2013 to 2015. Quarter 1 reflects production during the first three months of each year, etc.[121]

   (a) Explain why a periodic function should be used to model these data.
   (b) Approximately when does the maximum occur? The minimum? Why does this make sense?
   (c) What are the period and amplitude for these data?

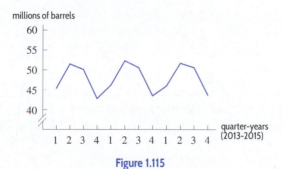

Figure 1.115

8. Sketch a possible graph of sales of sunscreen in the northeastern US over a 3-year period, as a function of months since January 1 of the first year. Explain why your graph should be periodic. What is the period?

9. The following table shows values of a periodic function $f(x)$. The maximum value attained by the function is 5.

   (a) What is the amplitude of this function?

   (b) What is the period of this function?
   (c) Find a formula for this periodic function.

x	0	2	4	6	8	10	12
$f(x)$	5	0	-5	0	5	0	-5

10. Coober Pedy is a town in Australia that mines most of the world's precious opals. Because of the scorching heat, much of the population lives in underground "dugouts." The Coober Pedy monthly high temperatures are shown in Figure 1.116, with $H(n)$ the temperature in month n, where $n = 1$ is January. We fit the data with a function,

$$H(n) = A \cos(Bn) + C.$$

   (a) What does the value of C represent and what is its value?
   (b) What does the value of A represent and what is its value?
   (c) What is the value of B?
   (d) Graph the data and the function.

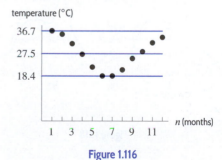

Figure 1.116

11. A person breathes in and out every three seconds. The volume of air in the person's lungs varies between a minimum of 2 liters and a maximum of 4 liters. Which of the following is the best formula for the volume of air in the person's lungs as a function of time?

---

[121] www.ttb.gov/statistics/14beerstats.shtml, accessed September 27, 2016.

(a) $y = 2 + 2\sin\left(\dfrac{\pi}{3}t\right)$ **(b)** $y = 3 + \sin\left(\dfrac{2\pi}{3}t\right)$

(c) $y = 2 + 2\sin\left(\dfrac{2\pi}{3}t\right)$ **(d)** $y = 3 + \sin\left(\dfrac{\pi}{3}t\right)$

12. Values of a function are given in the following table. Explain why this function appears to be periodic. Approximately what are the period and amplitude of the function? Assuming that the function is periodic, estimate its value at $t = 15$, at $t = 75$, and at $t = 135$.

t	20	25	30	35	40	45	50	55	60
$f(t)$	1.8	1.4	1.7	2.3	2.0	1.8	1.4	1.7	2.3

13. Average daily high temperatures in Ottawa, the capital of Canada, range from a low of $-6°$ Celsius on January 1 to a high of $26°$ Celsius on July 1 six months later. See Figure 1.117. Find a formula for H, the average daily high temperature in Ottawa in, $°$C, as a function of t, the number of months since January 1.

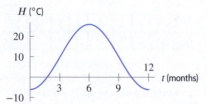

Figure 1.117

14. Figure 1.118 shows the levels of the hormones estrogen and progesterone during the monthly ovarian cycles in females.[122] Is the level of both hormones periodic? What is the period in each case? Approximately when in the monthly cycle is estrogen at a peak? Approximately when in the monthly cycle is progesterone at a peak?

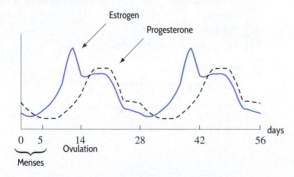

Figure 1.118

15. Delta Cephei is one of the most visible stars in the night sky. Its brightness has periods of 5.4 days, the average brightness is 4.0 and its brightness varies by ± 0.35. Find a formula that models the brightness of Delta Cephei as a function of time, t, with $t = 0$ at peak brightness.

16. Most breeding birds in the northeast US migrate elsewhere during the winter. The number of bird species in an Ohio forest preserve oscillates between a high of 28 in June and a low of 10 in December.[123]

(a) Graph the number of bird species in this preserve as a function of t, the number of months since June. Include at least three years on your graph.

(b) What are the amplitude and period of this function?

(c) Find a formula for the number of bird species, N, as a function of the number of months, t since June.

■ In Problems **17–28**, find a possible formula for the graph.

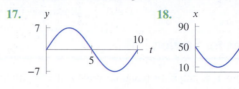

17.

18.

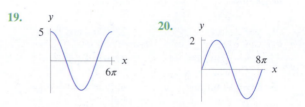

19.

20.

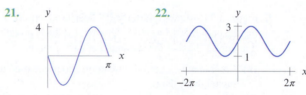

21.

22.

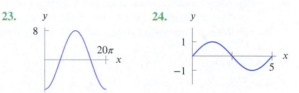

23.

24.

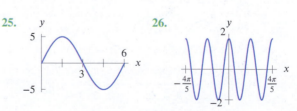

25.

26.

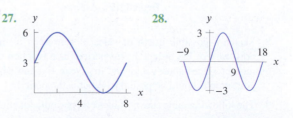

27.

28.

[122] Robert M. Julien, *A Primer of Drug Action*, Seventh Edition, p. 360 (W. H. Freeman and Co., New York: 1995).

[123] M. L. Rosenzweig, *Species Diversity in Space and Time*, p. 71 (Cambridge: Cambridge University Press, 1995).

■ In Problems 29–32, graph the given function on the axes in Figure 1.119.

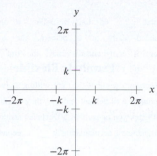

Figure 1.119

29. $y = k \sin x$

30. $y = -k \cos x$

31. $y = k(\cos x) + k$

32. $y = k(\sin x) - k$

33. The Bay of Fundy in Canada has the largest tides in the world. The difference between low and high water levels is 15 meters (nearly 50 feet). At a particular point the depth of the water, y meters, is given as a function of time, t, in hours since midnight by

$$y = D + A \cos(B(t - C)).$$

(a) What is the physical meaning of D?

(b) What is the value of A?

(c) What is the value of B? Assume the time between successive high tides is 12.4 hours.

(d) What is the physical meaning of C?

34. The depth of water in a tank oscillates once every 6 hours. If the smallest depth is 5.5 feet and the largest depth is 8.5 feet, find a possible formula for the depth in terms of time in hours.

35. The desert temperature, H, oscillates daily between 40°F at 5 am and 80°F at 5 pm. Write a possible formula for H in terms of t, measured in hours from 5 am.

36. Table 1.40 gives values for $g(t)$, a periodic function.

(a) Estimate the period and amplitude for this function.

(b) Estimate $g(34)$ and $g(60)$.

Table 1.40

t	0	2	4	6	8	10	12	14
$g(t)$	14	19	17	15	13	11	14	19

t	16	18	20	22	24	26	28	
$g(t)$	17	15	13	11	14	19	17	

37. In Figure 1.120, the blue curve shows monthly mean carbon dioxide (CO_2) concentration, in parts per million (ppm) at Mauna Loa Observatory, Hawaii, as a function of t, in months, since December 2005. The black curve shows the monthly mean concentration adjusted for seasonal CO_2 variation.[124]

(a) Approximately how much did the monthly mean CO_2 increase between December 2005 and December 2010?

(b) Find the average monthly rate of increase of the monthly mean CO_2 between December 2005 and December 2010. Use this information to find a linear function that approximates the black curve.

(c) The seasonal CO_2 variation between December 2005 and December 2010 can be approximated by a sinusoidal function of the form $A \sin Bt$. What is the approximate period of the function? What is its amplitude? Give a formula for the function.

(d) The blue curve may be approximated by a function of the form $h(t) = f(t) + g(t)$, where $f(t)$ is sinusoidal and $g(t)$ is linear. Using your work in parts (b) and (c), find a possible formula for $h(t)$. Graph $h(t)$ using the scale in Figure 1.120.

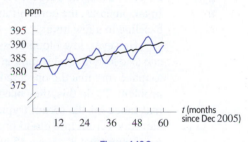

Figure 1.120

PROJECTS FOR CHAPTER ONE

1. Compound Interest

   The newspaper article below is from *The New York Times*, May 27, 1990. Fill in the three blanks. (For the first blank, assume that daily compounding is essentially the same as continuous compounding. For the last blank, assume the interest has been compounded yearly, and give your answer in dollars. Ignore the occurrence of leap years.)

---

[124]www.esrl.noaa.gov/gmd/ccgg/trends/. Accessed March 2011. Monthly means joined by segments to highlight trends.

213 Years After Loan, Uncle Sam Is Dunned

By LISA BELKIN

Special to The New York Times

SAN ANTONIO, May 26 — More than 200 years ago, a wealthy Pennsylvania merchant named Jacob DeHaven lent $450,000 to the Continental Congress to rescue the troops at Valley Forge. That loan was apparently never repaid.

So Mr. DeHaven's descendants are taking the United States Government to court to collect what they believe they are owed.

The total: ____ in today's dollars if the interest is compounded daily at 6 percent, the going rate at the time. If compounded yearly, the bill is only ____.

Family Is Flexible

The descendants say that they are willing to be flexible about the amount of a settlement and that they might even accept a heartfelt thank you or perhaps a DeHaven statue. But they also note that interest is accumulating at ____ a second.

2. Population Center of the US

Since the opening up of the West, the US population has moved westward. To observe this, we look at the "population center" of the US, which is the point at which the country would balance if it were a flat plate with no weight, and every person had equal weight. In 1790 the population center was east of Baltimore, Maryland. It has been moving westward ever since, and in 2000 it was in Edgar Springs, Missouri. During the second half of the 20th century, the population center has moved about 50 miles west every 10 years.

(a) Let us measure position westward from Edgar Springs along the line running through Baltimore. For the years since 2000, express the approximate position of the population center as a function of time in years from 2000.

(b) The distance from Baltimore to Edgar Springs is a bit over 1000 miles. Could the population center have been moving at roughly the same rate for the last two centuries?

(c) Could the function in part (a) continue to apply for the next four centuries? Why or why not? [Hint: You may want to look at a map. Note that distances are in air miles and are not driving distances.]

3. Medical Case Study: Anaphylaxis[125]

During surgery, a patient's blood pressure was observed to be dangerously low. One possible cause is a severe allergic reaction called *anaphylaxis*. A diagnosis of anaphylaxis is based in part on a blood test showing the elevation of the serum *tryptase*, a molecule released by allergic cells. In anaphylaxis, the concentration of tryptase in the blood rises rapidly and then decays back to baseline in a few hours.

However, low blood pressure from an entirely different cause (say from a heart problem) can also lead to an elevation in tryptase. Before diagnosing anaphylaxis, the medical team needs to make sure that the observed tryptase elevation is the result of an allergy problem, not a heart problem. To do this, they need to know the peak level reached by the serum tryptase. The normal range for the serum tryptase is 0–15 ng/ml (nanograms per milliliter). Mild to moderate elevations from low blood pressure are common, but if the peak were three times the normal maximum (that is, above 45 ng/ml), then a diagnosis of anaphylaxis would be made.

The surgeons who resuscitated this patient ran two blood tests to measure T_r, the serum tryptase concentration; the results are in Table 1.41. Use the test results to estimate the peak serum tryptase level at the time of surgery assuming that tryptase decays exponentially. Did this patient experience anaphylaxis?

Table 1.41   *Serum tryptase levels*

t, hours since surgery	4	19.5
T_r, concentration in ng/ml	37	13

[125]From David E. Sloane, M.D., drawing from an actual episode in his clinic.

Chapter 2

RATE OF CHANGE: THE DERIVATIVE

CONTENTS

2.1   INSTANTANEOUS RATE OF CHANGE

Chapter 1 introduced the average rate of change of a function over an interval. In this section, we consider the rate of change of a function at a point. We saw in Chapter 1 that when an object is moving along a straight line, the average rate of change of position with respect to time is the average velocity. If position is expressed as $y = f(t)$, where t is time, then

$$\text{Average rate of change in position between } t = a \text{ and } t = b = \frac{\Delta y}{\Delta t} = \frac{f(b) - f(a)}{b - a}.$$

If you drive 200 miles in 4 hours, your average velocity is $200/4 = 50$ miles per hour. Of course, this does not mean that you travel at exactly 50 mph the entire trip. Your velocity at a given instant during the trip is shown on your speedometer, and this is the quantity that we investigate now.

Instantaneous Velocity

We throw a grapefruit straight upward into the air. Table 2.1 gives its height, y, at time t. What is the velocity of the grapefruit at exactly $t = 1$? We use average velocities to estimate this quantity.

Table 2.1   *Height of the grapefruit above the ground*

t (sec)	0	1	2	3	4	5	6
$y = s(t)$ (feet)	6	90	142	162	150	106	30

The average velocity on the interval $0 \leq t \leq 1$ is 84 ft/sec and the average velocity on the interval $1 \leq t \leq 2$ is 52 ft/sec. Notice that the average velocity before $t = 1$ is larger than the average velocity after $t = 1$ since the grapefruit is slowing down. We expect the velocity *at* $t = 1$ to be between these two average velocities. How can we find the velocity at *exactly* $t = 1$? We look at what happens near $t = 1$ in more detail. Suppose that we find the average velocities on either side of $t = 1$ over smaller and smaller intervals, as in Figure 2.1. Then, for example,

$$\text{Average velocity between } t = 1 \text{ and } t = 1.01 = \frac{\Delta y}{\Delta t} = \frac{s(1.01) - s(1)}{1.01 - 1} = \frac{90.678 - 90}{0.01} = 67.8 \text{ ft/sec.}$$

We expect the instantaneous velocity at $t = 1$ to be between the average velocities on either side of $t = 1$. In Figure 2.1, the values of the average velocity before $t = 1$ and the average velocity after $t = 1$ get closer together as the size of the interval shrinks. For the smallest intervals in Figure 2.1, both velocities are 68.0 ft/sec (to one decimal place), so we say the velocity at $t = 1$ is 68.0 ft/sec (to one decimal place).

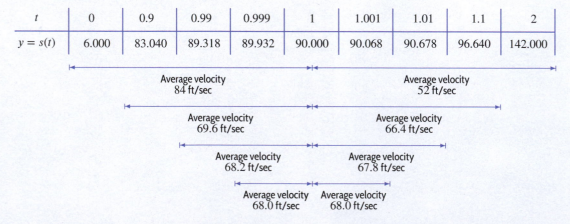

Figure 2.1: Average velocities over intervals on either side of $t = 1$ showing successively smaller intervals

Of course, if we showed more decimal places, the average velocities before and after $t = 1$ would no longer agree. To calculate the velocity at $t = 1$ to more decimal places of accuracy, we take smaller and smaller intervals on either side of $t = 1$ until the average velocities agree to the number of decimal places we want. In this way, we can estimate the velocity at $t = 1$ to any accuracy.

Defining Instantaneous Velocity Using the Idea of a Limit

When we take smaller intervals near $t = 1$, it turns out that the average velocities for the grapefruit are always just above or just below 68 ft/sec. It seems natural, then, to define velocity at the instant $t = 1$ to be 68 ft/sec. This is called the *instantaneous velocity* at this point. Its definition depends on our being convinced that smaller and smaller intervals provide average velocities that come arbitrarily close to 68. This process is referred to as *taking the limit*.

> The **instantaneous velocity** of an object at time t is defined to be the limit of the average velocity of the object over shorter and shorter time intervals containing t.

Notice that the instantaneous velocity seems to be exactly 68, but what if it were 68.000001? How can we be sure that we have taken small enough intervals? Showing that the limit is exactly 68 requires more precise knowledge of how the velocities were calculated and of the limiting process; see the Focus on Theory section.

Instantaneous Rate of Change

We can define the *instantaneous rate of change* of any function $y = f(t)$ at a point $t = a$. We mimic what we did for velocity and look at the average rate of change over smaller and smaller intervals.

> The **instantaneous rate of change** of f at a, also called the **rate of change** of f at a, is defined to be the limit of the average rates of change of f over shorter and shorter intervals around a.

Since the average rate of change is a difference quotient of the form $\Delta y / \Delta t$, the instantaneous rate of change is a limit of difference quotients. In practice, we often approximate a rate of change by one of these difference quotients.

Example 1   The quantity (in mg) of a drug in the blood at time t (in minutes) is given by $Q = 25(0.8)^t$. Estimate the rate of change of the quantity at $t = 3$ and interpret your answer.

Solution   We estimate the rate of change at $t = 3$ by computing the average rate of change over intervals near $t = 3$. We can make our estimate as accurate as we like by choosing our intervals small enough. Let's look at the average rate of change over the interval $3 \leq t \leq 3.01$:

$$\text{Average rate of change} = \frac{\Delta Q}{\Delta t} = \frac{25(0.8)^{3.01} - 25(0.8)^3}{3.01 - 3.00} = \frac{12.7715 - 12.80}{3.01 - 3.00} = -2.85.$$

A reasonable estimate for the rate of change of the quantity at $t = 3$ is -2.85. Since Q is in mg and t in minutes, the units of $\Delta Q / \Delta t$ are mg/minute. Since the rate of change is negative, the quantity of the drug is decreasing. After 3 minutes, the quantity of the drug in the body is decreasing at 2.85 mg/minute.

In Example 1, we estimated the rate of change using an interval to the right of the point ($t = 3$ to $t = 3.01$). In the next section we briefly consider other ways of making estimates.

The Derivative at a Point

The instantaneous rate of change of a function f at a point a is so important that it is given its own name, the *derivative of f at a*, denoted $f'(a)$ (read "f-prime of a"). If we want to emphasize that

$f'(a)$ is the rate of change of $f(x)$ as the variable x increases, we call $f'(a)$ the derivative of f *with respect to x* at $x = a$. Notice that the derivative is just a new name for the rate of change of a function.

> The **derivative of f at a**, written $f'(a)$, is defined to be the instantaneous rate of change of f at the point a.

A definition of the derivative using a formula is given in the Focus on Theory section on page 128.

Example 2   Estimate $f'(2)$ if $f(x) = x^3$.

Solution   Since $f'(2)$ is the derivative, or rate of change, of $f(x) = x^3$ at 2, we look at the average rate of change over intervals near 2. Using the interval $2 \le x \le 2.001$, we see that

$$\begin{array}{l} \text{Average rate of change} \\ \text{on } 2 \le x \le 2.001 \end{array} = \frac{(2.001)^3 - 2^3}{2.001 - 2} = \frac{8.0120 - 8}{0.001} = 12.0.$$

The rate of change of $f(x)$ at $x = 2$ appears to be approximately 12, so we estimate $f'(2) = 12$.

Visualizing the Derivative: Slope of the Graph and Slope of the Tangent Line

Figure 2.2 shows the average rate of change of a function represented by the slope of the secant line joining points A and B. The derivative is found by taking the average rate of change over smaller and smaller intervals. In Figure 2.3, as point B moves toward point A, the secant line approaches the tangent line at point A. Thus, the derivative is represented by the slope of the tangent line to the graph at the point.

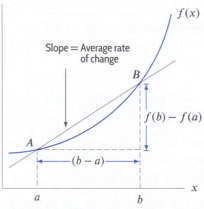

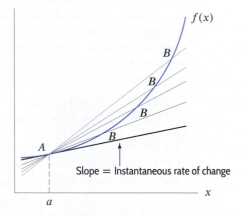

Figure 2.2: Visualizing the average rate of change of f between a and b

Figure 2.3: Visualizing the instantaneous rate of change of f at a

Alternatively, take the graph of a function around a point and "zoom in" to get a close-up view. (See Figure 2.4.) The more we zoom in, the more the graph appears to be straight. We call the slope of this line the *slope of the graph* at the point; it also represents the derivative.

> The derivative of a function at the point A is equal to
>
> - The slope of the graph of the function at A.
> - The slope of the line tangent to the curve at A.

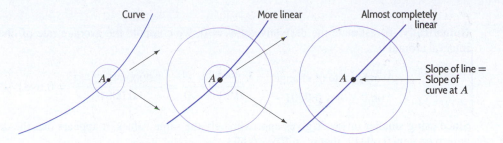

Figure 2.4: Finding the slope of a curve at a point by "zooming in"

The slope interpretation is often useful in gaining rough information about the derivative, as the following examples show.

Example 3 Use a graph of $f(x) = x^2$ to determine whether each of the following quantities is positive, negative, or zero: (a) $f'(1)$ (b) $f'(-1)$ (c) $f'(2)$ (d) $f'(0)$

Solution Figure 2.5 shows tangent line segments to the graph of $f(x) = x^2$ at the points $x = 1$, $x = -1$, $x = 2$, and $x = 0$. Since the derivative is the slope of the tangent line at the point, we have:

(a) $f'(1)$ is positive.
(b) $f'(-1)$ is negative.
(c) $f'(2)$ is positive (and larger than $f'(1)$).
(d) $f'(0) = 0$ since the graph has a horizontal tangent at $x = 0$.

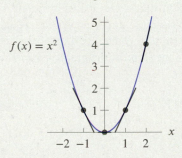

Figure 2.5: Tangent lines showing sign of derivative of $f(x) = x^2$

Example 4 Estimate the derivative of $f(x) = 2^x$ at $x = 0$ graphically and numerically.

Solution Graphically: If we draw a tangent line at $x = 0$ to the exponential curve in Figure 2.6, we see that it has a positive slope between 0.5 and 1.

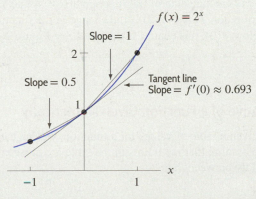

Figure 2.6: Graph of $f(x) = 2^x$ showing the derivative at $x = 0$

Numerically: To estimate the derivative at $x = 0$, we compute the average rate of change on an interval around 0.

$$\text{Average rate of change} \atop \text{on } 0 \le x \le 0.0001 = \frac{2^{0.0001} - 2^0}{0.0001 - 0} = \frac{1.000069317 - 1}{0.0001} = 0.69317.$$

Since using smaller intervals gives approximately the same values, it appears that the derivative is approximately 0.69317; that is, $f'(0) \approx 0.693$.

Example 5   The graph of a function $y = f(x)$ is shown in Figure 2.7. Indicate whether each of the following quantities is positive or negative, and illustrate your answers graphically.

(a) $f'(1)$ (b) $\dfrac{f(3) - f(1)}{3 - 1}$ (c) $f(4) - f(2)$

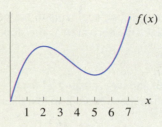

Figure 2.7

Solution   (a) Since $f'(1)$ is the slope of the graph at $x = 1$, we see in Figure 2.8 that $f'(1)$ is positive.

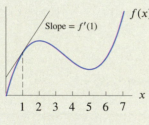

Figure 2.8

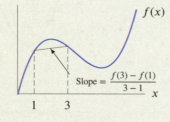

Figure 2.9

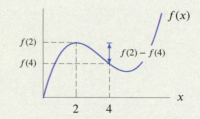

Figure 2.10

(b) The difference quotient $(f(3) - f(1))/(3 - 1)$ is the slope of the secant line between $x = 1$ and $x = 3$. We see from Figure 2.9 that this slope is positive.

(c) Since $f(4)$ is the value of the function at $x = 4$ and $f(2)$ is the value of the function at $x = 2$, the expression $f(4) - f(2)$ is the change in the function between $x = 2$ and $x = 4$. Since $f(4)$ lies below $f(2)$, this change is negative. See Figure 2.10.

Estimating the Derivative of a Function Given Numerically

If we are given a table of values for a function, we can estimate values of its derivative. To do this, we have to assume that the points in the table are close enough together that the function does not change wildly between them.

Example 6   The total acreage of farms in the US[1] since 1992 is in Table 2.2.

(a) What was the average rate of change in farm land between 1997 and 2012?

(b) Estimate $f'(2007)$ and interpret your answer in terms of farm land.

Table 2.2   *Total farm land in million acres*

Year	1992	1997	2002	2007	2012
Farm land (million acres)	946	932	938	922	915

Solution   (a) Between 1997 and 2012,

$$\text{Average rate of change} = \frac{915 - 932}{2012 - 1997} = \frac{-17}{15} = -1.13 \text{ million acres per year.}$$

Between 1997 and 2012, the amount of farm land was decreasing at an average rate of 1.13 million acres per year.

(b) We use the interval from 2007 to 2012 to estimate the instantaneous rate of change at 2007:

$$f'(2007) = \begin{array}{c}\text{Rate of change} \\ \text{in 2007}\end{array} \approx \frac{915 - 922}{2012 - 2007} = \frac{-7}{5} = -1.4 \text{ million acres per year.}$$

In 2007, the amount of farm land was decreasing at a rate of approximately 1.4 million acres per year.

Problems for Section 2.1

■ Problems **1–2** refer to the graph in Figure 2.11.

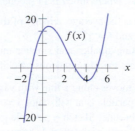

Figure 2.11

1. Does each quantity appear to be positive, negative, or zero?

(a) $f(-2)$ (b) $f(0)$ (c) $f(4)$

(d) $f(5)$ (e) $f'(-2)$ (f) $f'(2)$

(g) $f'(4)$ (h) $f'(5)$

2. Which of the two quantities appears to be larger?

(a) $f(1)$ or $f(2)$ (b) $f(5)$ or $f(6)$

(c) $f(3)$ or $f(4)$ (d) $f'(-2)$ or $f'(2)$

(e) $f'(4)$ or $f'(5)$ (f) $f'(-1)$ or $f'(0)$

■ In Problems **3–6**, which of the functions A, B, C, D in Figure 2.12 satisfy the given condition?

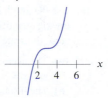
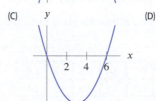
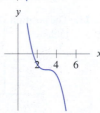

Figure 2.12

3. $f(2) > 0$ **4.** $f(4) < 0$

5. $f'(2) > 0$ **6.** $f'(4) < 0$

[1]www.farmlandinfo.org/statistics and usda.mannlib.cornell.edu/usda/AgCensusImages/1997/01/51/1604/Table-01.pdf, accessed January 10, 2017.

7. Let $s = f(t)$ give an object's height, in feet, above the ground t seconds after it is thrown. After 2 seconds, the object's height is 96 feet, and it is moving up at 16 ft/sec. Fill in the blanks:

   (a) $f(\_\_\_) = \_\_\_\_$
   (b) $f'(\_\_\_) = \_\_\_\_$

8. The quantity, in gallons, of water in a tank after t minutes is $w = f(t)$. After 20 minutes, the tank has 50 gallons of water and the quantity is increasing at 3 gal/min. Fill in the blanks:

   (a) $f(\_\_\_) = \_\_\_\_$
   (b) $f'(\_\_\_) = \_\_\_\_$

9. Figure 2.13 shows $N = f(t)$, the number of farms in the US[2] between 1930 and 2000 as a function of year, t.

   (a) Is $f'(1950)$ positive or negative? What does this tell you about the number of farms?
   (b) Which is more negative: $f'(1960)$ or $f'(1980)$?

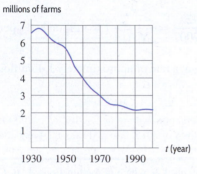

millions of farms

1930   1950   1970   1990 t (year)

Figure 2.13

10. Use the graph in Figure 2.7 to decide if each of the following quantities is positive, negative or approximately zero. Illustrate your answers graphically.

    (a) The average rate of change of $f(x)$ between $x = 3$ and $x = 7$.
    (b) The instantaneous rate of change of $f(x)$ at $x = 3$.

11. The position s of a car at time t is given in the following table.

t (sec)	0	0.2	0.4	0.6	0.8	1.0
s (ft)	0	0.5	1.8	3.8	6.5	9.6

    (a) Find the average velocity over the interval $0 \le t \le 0.2$.
    (b) Find the average velocity over the interval $0.2 \le t \le 0.4$.
    (c) Use the previous answers to estimate the instantaneous velocity of the car at $t = 0.2$.

12. In a time of t seconds, a particle moves a distance of s meters from its starting point, where $s = 4t^2 + 3$.

    (a) Find the average velocity between $t = 1$ and $t = 1 + h$ if:
       (i) $h = 0.1$, (ii) $h = 0.01$, (iii) $h = 0.001$.
    (b) Use your answers to part (a) to estimate the instantaneous velocity of the particle at time $t = 1$.

13. Figure 2.14 shows the cost, $y = f(x)$, of manufacturing x kilograms of a chemical.

    (a) Is the average rate of change of the cost greater between $x = 0$ and $x = 3$, or between $x = 3$ and $x = 5$? Explain your answer graphically.
    (b) Is the instantaneous rate of change of the cost of producing x kilograms greater at $x = 1$ or at $x = 4$? Explain your answer graphically.
    (c) What are the units of these rates of change?

y (thousand $)

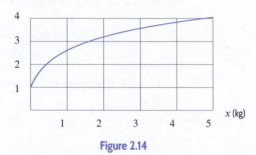

x (kg)

Figure 2.14

14. The distance (in feet) of an object from a point is given by $s(t) = t^2$, where time t is in seconds.

    (a) What is the average velocity of the object between $t = 2$ and $t = 5$?
    (b) By using smaller and smaller intervals around 2, estimate the instantaneous velocity at time $t = 2$.

15. A particle moves along a line with varying velocity. At time t the particle is at a distance $s = f(t)$ from a fixed point on the line. Sketch a possible graph for f if the average velocity of the particle between $t = 0$ and $t = 5$ is the same as its instantaneous velocity at exactly two times between $t = 0$ and $t = 5$.

16. (a) Using Table 2.3, find the average rate of change in the world's population, P, between 1980 and 2015.[3] Give units.
    (b) If $P = f(t)$ with t in years, estimate $f'(2010)$ and give units.

Table 2.3   *World population, in billions of people*

Year	1980	1985	1990	1995	2000	2005	2010	2015
Population	4.45	4.86	5.32	5.74	6.13	6.51	6.92	7.32

---

[2] www.nass.usda.gov:81/ipedb/farmnum.htm, accessed April 11, 2005.
[3] www.geohive.com/earth/his_history3.aspx, accessed September 5, 2016.

17. The size, S, of a tumor (in cubic millimeters) is given by $S = 2^t$, where t is the number of months since the tumor was discovered. Give units with your answers.

    (a) What is the total change in the size of the tumor during the first six months?
    (b) What is the average rate of change in the size of the tumor during the first six months?
    (c) Estimate the rate at which the tumor is growing at $t = 6$. (Use smaller and smaller intervals.)

18. Let $g(x) = 4^x$. Use small intervals to estimate $g'(1)$.

19. (a) Let $g(t) = (0.8)^t$. Use a graph to determine whether $g'(2)$ is positive, negative, or zero.
    (b) Use a small interval to estimate $g'(2)$.

20. For the function shown in Figure 2.15, at what labeled points is the slope of the graph positive? Negative? At which labeled point does the graph have the greatest (i.e., most positive) slope? The least slope (i.e., negative and with the largest magnitude)?

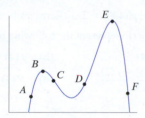

Figure 2.15

21. Match the points labeled on the curve in Figure 2.16 with the given slopes.

Slope	Point
−3	
−1	
0	
1/2	
1	
2	

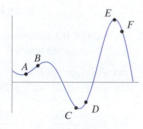

Figure 2.16

22. For the function $f(x) = 3^x$, estimate $f'(1)$. From the graph of $f(x)$, would you expect your estimate to be greater than or less than the true value of $f'(1)$?

23. The table gives $P = f(t)$, the number of households, in millions, in the US with cable television t years since 1998.[4]

    (a) Does $f'(2)$ appear to be positive or negative? What does this tell you about the number of households with cable television?
    (b) Estimate $f'(2)$. Estimate $f'(10)$. Explain what each is telling you, in terms of cable television.

t	0	2	4	6	8	10	12
P	64.650	66.250	66.472	65.727	65.319	64.274	60.958

24. The following table gives the percent of the US population living in urban areas as a function of year.[5]

    (a) Find the average rate of change of the percent of the population living in urban areas between 1890 and 1990.
    (b) Estimate the rate at which this percent is increasing for the year 1990.
    (c) Estimate the rate of change of this function for the year 1830 and explain what it is telling you.

Year	1800	1830	1860	1890	1920
Percent	6.0	9.0	19.8	35.1	51.2
Year	1950	1980	1990	2000	2005
Percent	64.0	73.7	75.2	79.0	79.0

25. World population was 6.13 billion in 2000 and was 7.32 billion in 2015, which means the population grew, on average, by 0.079 billion people per year during that time. The percent rate of growth of the population is declining, but the absolute rate is increasing. At the start of 2000, the population was growing by 0.076 billion people per year, while at the start of 2015, it was growing by 0.082 billion people per year.[6] Is each of the numbers (i) an average rate of change, (ii) an instantaneous rate of change, or (iii) not a rate of change of world population?

    (a) 6.13   (b) 7.32   (c) 0.079
    (d) 0.076   (e) 0.082

26. The height, in feet, of a tomato, t seconds after it is dropped from a 200-ft balcony, is $s = f(t) = -16t^2 + 200$.

    (a) Find the average velocity between $t = 2$ and $t = 3$.
    (b) Find the average velocity between $t = 2$ and $t = 2 + h$ if:
        (i) $h = 0.1$,   (ii) $h = 0.01$,   (iii) $h = 0.001$.
    (c) Use your answers to part (b) to estimate the instantaneous velocity at $t = 2$.
    (d) What is $f(2)$ and what is an estimated value for $f'(2)$?

---

[4] www.census.gov/2010census/, accessed December 12, 2016.
[5] *Statistical Abstracts of the US*, 1985, and http://www.census.gov, accessed September 2012.
[6] http://hwww.geohive.com/earth/his_history3.aspx. Accessed December 2016.

27. (a) Graph $f(x) = x^2$ and $g(x) = x^2 + 3$ on the same axes. What can you say about the slopes of the tangent lines to the two graphs at the point $x = 0$? $x = 1$? $x = 2$? $x = a$, where a is any value?
    (b) Explain why adding a constant to any function will not change the value of the derivative at any point.

28. Figure 2.17 shows $f(t)$ and $g(t)$, the positions of two cars with respect to time, t, in minutes.

    (a) Describe how the velocity of each car is changing during the time shown.
    (b) Find an interval over which the cars have the same average velocity.
    (c) Which of the following statements are true?
        (i) Sometime in the first half minute, the two cars are traveling at the same instantaneous velocity.
        (ii) During the second half minute (from $t = 1/2$ to $t = 1$), there is a time that the cars are traveling at the same instantaneous velocity.
        (iii) The cars are traveling at the same velocity at $t = 1$ minute.
        (iv) There is no time during the period shown that the cars are traveling at the same velocity.

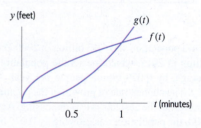

Figure 2.17

29. The function in Figure 2.18 has $f(4) = 25$ and $f'(4) = 1.5$. Find the coordinates of the points A, B, C.

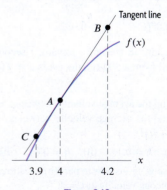

Figure 2.18

30. Use Figure 2.19 to fill in the blanks in the following statements about the function f at point A.
    (a) $f(\underline{}) = \underline{}$ (b) $f'(\underline{}) = \underline{}$

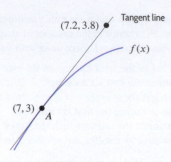

Figure 2.19

■ For Problems 31–34, estimate the change in y for the given change in x.

31. $y = f(x)$, $f'(100) = 0.4$, x increases from 100 to 101

32. $y = f(x)$, $f'(12) = 30$, x increases from 12 to 12.2

33. $y = g(x)$, $g'(250) = -0.5$, x increases from 250 to 251.5

34. $y = p(x)$, $p'(400) = 2$, x decreases from 400 to 398

35. Show how to represent the following on Figure 2.20.
    (a) $f(4)$ (b) $f(4) - f(2)$
    (c) $\dfrac{f(5) - f(2)}{5 - 2}$ (d) $f'(3)$

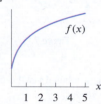

Figure 2.20

36. For each of the following pairs of numbers, use Figure 2.20 to decide which is larger. Explain your answer.
    (a) $f(3)$ or $f(4)$?
    (b) $f(3) - f(2)$ or $f(2) - f(1)$?
    (c) $\dfrac{f(2) - f(1)}{2 - 1}$ or $\dfrac{f(3) - f(1)}{3 - 1}$?
    (d) $f'(1)$ or $f'(4)$?

37. Use Figure 2.21 to decide which is larger in each of the following pairs.
    (a) Average rate of change between $x = 0$ and $x = 2$ or between $x = 2$ and $x = 4$?
    (b) $g(1)$ or $g(4)$?
    (c) $g'(2)$ or $g'(4)$?

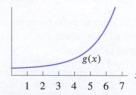

Figure 2.21

38. Estimate the instantaneous rate of change of the function $f(x) = x \ln x$ at $x = 1$ and at $x = 2$. What do these values suggest about the concavity of the graph between 1 and 2?

39. The population, $P(t)$, of China, in billions, can be approximated by[7]

$$P(t) = 1.394(1.006)^t,$$

where t is the number of years since the start of 2014. According to this model, how fast was the population growing at the start of 2014 and at the start of 2015? Give your answers in millions of people per year.

40. The US population[8] officially reached 300 million on October 17, 2006 and was gaining 1 person each 11 seconds. If $f(t)$ is the US population in millions t years after October 17, 2006, find $f(0)$ and $f'(0)$.

41. The following table shows the number of hours worked in a week, $f(t)$, hourly earnings, $g(t)$, in dollars, and weekly earnings, $h(t)$, in dollars, of production workers as functions of t, the year.[9]

(a) Indicate whether each of the following derivatives is positive, negative, or zero: $f'(t)$, $g'(t)$, $h'(t)$. Interpret each answer in terms of hours or earnings.

(b) Estimate each of the following derivatives, and interpret your answers:

   (i) $f'(1970)$ and $f'(1995)$

   (ii) $g'(1970)$ and $g'(1995)$

   (iii) $h'(1970)$ and $h'(1995)$

t	1970	1975	1980	1985	1990	1995	2000
$f(t)$	37.0	36.0	35.2	34.9	34.3	34.3	34.3
$g(t)$	3.40	4.73	6.84	8.73	10.09	11.64	14.00
$h(t)$	125.80	170.28	240.77	304.68	349.29	399.53	480.41

2.2 THE DERIVATIVE FUNCTION

In Section 2.1 we looked at the derivative of a function at a point. In general, the derivative takes on different values at different points and is itself a function. Recall that the derivative is the slope of the tangent line to the graph at the point.

Finding the Derivative of a Function Given Graphically

Example 1   Estimate the derivative of the function $f(x)$ graphed in Figure 2.22 at $x = -2, -1, 0, 1, 2, 3, 4, 5$.

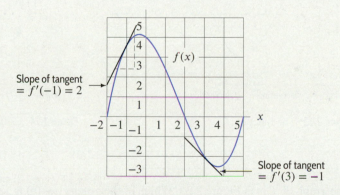

Figure 2.22: Estimating the derivative graphically as the slope of a tangent line

Solution   From the graph, we estimate the derivative at any point by placing a straightedge so that it forms the tangent line at that point, and then using the grid to estimate the slope of the tangent line. For example, the tangent at $x = -1$ is drawn in Figure 2.22, and has a slope of about 2, so $f'(-1) \approx 2$. Notice that the slope at $x = -2$ is positive and fairly large; the slope at $x = -1$ is positive but smaller. At $x = 0$, the slope is negative, by $x = 1$ it has become more negative, and so on. Some estimates of

[7] www.worldometers.info, accessed April 1, 2015.

[8] www.today.com/id/15298443/ns/today/t/us-population-hits-million-mark/#.VsuG1hgVmV0, accessed February 2015.

[9] *The World Almanac and Book of Facts 2005*, p. 151 (New York). Production workers include nonsupervisory workers in mining, manufacturing, construction, transportation, public utilities, wholesale and retail trade, finance, insurance, real estate, and services.

the derivative, to the nearest integer, are listed in Table 2.4. You should check these values yourself. Is the derivative positive where you expect? Negative?

Table 2.4   *Estimated values of derivative of function in Figure 2.22*

x	−2	−1	0	1	2	3	4	5
Derivative at x	6	2	−1	−2	−2	−1	1	4

The important point to notice is that for every x-value, there is a corresponding value of the derivative. The derivative, therefore, is a function of x.

> For a function f, we define the **derivative function**, f', by
>
> $$f'(x) = \text{Instantaneous rate of change of } f \text{ at } x.$$

Example 2   Plot the values of the derivative function calculated in Example 1. Compare the graphs of f' and f.

Solution   Graphs of f and f' are in Figures 2.23 and 2.24, respectively. Notice that f' is positive (its graph is above the x-axis) where f is increasing, and f' is negative (its graph is below the x-axis) where f is decreasing. The value of $f'(x)$ is 0 where f has a maximum or minimum value (at approximately $x = -0.4$ and $x = 3.7$).

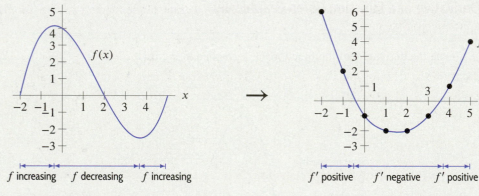

Figure 2.23: The function f

Figure 2.24: Estimates of the derivative, f'

Example 3   The graph of f is in Figure 2.25. Which of the graphs (a)–(c) is a graph of the derivative, f'?

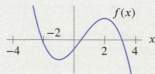

Figure 2.25

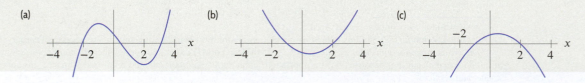

Solution   Since the tangent of $f(x)$ is horizontal at $x = -1$ and $x = 2$, the derivative is zero there. Therefore, the graph of $f'(x)$ has x-intercepts at $x = -1$ and $x = 2$.

The function f is decreasing for $x < -1$, increasing for $-1 < x < 2$, and decreasing for $x > 2$. The derivative is positive (its graph is above the x-axis) where f is increasing, and the derivative is negative (its graph is below the x-axis) where f is decreasing. The correct graph is (c).

What Does the Derivative Tell Us Graphically?

Where the derivative, f', of a function is positive, the tangent to the graph of f is sloping up; where f' is negative, the tangent is sloping down. If $f' = 0$ everywhere, then the tangent is horizontal everywhere and so f is constant. The sign of the derivative f' tells us whether the function f is increasing or decreasing.

> If $f' > 0$ on an interval, then f is *increasing* on that interval.
> If $f' < 0$ on an interval, then f is *decreasing* on that interval.
> If $f' = 0$ on an interval, then f is *constant* on that interval.

The magnitude of the derivative gives us the magnitude of the rate of change of f. If f' is large in magnitude, then the graph of f is steep (up if f' is positive or down if f' is negative); if f' is small in magnitude, the graph of f is gently sloping.

Estimating the Derivative of a Function Given Numerically

If we are given a table of function values instead of a graph of the function, we can estimate values of the derivative.

Example 4   Table 2.5 gives values of $c(t)$, the concentration (mg/cc) of a drug in the bloodstream at time t (min). Construct a table of estimated values for $c'(t)$, the rate of change of $c(t)$ with respect to t.

Table 2.5   *Concentration of a drug as a function of time*

t (min)	0	0.1	0.2	0.3	0.4	0.5	0.6	0.7	0.8	0.9	1.0
$c(t)$ (mg/cc)	0.84	0.89	0.94	0.98	1.00	1.00	0.97	0.90	0.79	0.63	0.41

Solution   To estimate the derivative of c using the values in the table, we assume that the data points are close enough together that the concentration does not change wildly between them. From the table, we see that the concentration is increasing between $t = 0$ and $t = 0.4$, so we expect a positive derivative there. From $t = 0.5$ to $t = 1.0$, the concentration starts to decrease, and the rate of decrease gets larger and larger, so we would expect the derivative to be negative and of greater and greater magnitude.

We estimate the derivative for each value of t using a difference quotient. For example,

$$c'(0) \approx \frac{c(0.1) - c(0)}{0.1 - 0} = \frac{0.89 - 0.84}{0.1} = 0.5 \text{ (mg/cc) per minute.}$$

Similarly, we get the estimates

$$c'(0.1) \approx \frac{c(0.2) - c(0.1)}{0.2 - 0.1} = \frac{0.94 - 0.89}{0.1} = 0.5,$$

$$c'(0.2) \approx \frac{c(0.3) - c(0.2)}{0.3 - 0.2} = \frac{0.98 - 0.94}{0.1} = 0.4,$$

and so on. These values are tabulated in Table 2.6. Notice that the derivative has small positive values up until $t = 0.4$, and then it gets more and more negative, as we expected.

Table 2.6 *Derivative of concentration*

t	0	0.1	0.2	0.3	0.4	0.5	0.6	0.7	0.8	0.9
$c'(t)$	0.5	0.5	0.4	0.2	0.0	−0.3	−0.7	−1.1	−1.6	−2.2

Improving Numerical Estimates for the Derivative

In the previous example, our estimate for the derivative of $c(t)$ at $t = 0.2$ used the point to the right. We found the average rate of change between $t = 0.2$ and $t = 0.3$. However, we could equally well have gone to the left and used the rate of change between $t = 0.1$ and $t = 0.2$ to approximate the derivative at 0.2. For a more accurate result, we could average these slopes, getting the approximation

$$c'(0.2) \approx \frac{1}{2} \left(\begin{array}{c} \text{Slope to left} \\ \text{of 0.2} \end{array} + \begin{array}{c} \text{Slope to right} \\ \text{of 0.2} \end{array} \right) = \frac{0.5 + 0.4}{2} = 0.45.$$

Each of these methods of approximating the derivative gives a reasonable answer. We will usually estimate the derivative by going to the right.

Finding the Derivative of a Function Given by a Formula

If we are given a formula for a function f, can we come up with a formula for f'? Using the definition of the derivative, we often can. Indeed, much of the power of calculus depends on our ability to find formulas for the derivatives of all the familiar functions. This is explained in detail in Chapter 3. In the next example, we see how to guess a formula for the derivative.

Example 5   Guess a formula for the derivative of $f(x) = x^2$.

Solution   We use difference quotients to estimate the values of $f'(1)$, $f'(2)$, and $f'(3)$. Then we look for a pattern in these values which we use to guess a formula for $f'(x)$.
Near $x = 1$, we have

$$f'(1) \approx \frac{1.001^2 - 1^2}{0.001} = \frac{1.002 - 1}{0.001} = \frac{0.002}{0.001} = 2.$$

Similarly,

$$f'(2) \approx \frac{2.001^2 - 2^2}{0.001} = \frac{4.004 - 4}{0.001} = \frac{0.004}{0.001} = 4$$

$$f'(3) \approx \frac{3.001^2 - 3^2}{0.001} = \frac{9.006 - 9}{0.001} = \frac{0.006}{0.001} = 6.$$

Knowing the value of f' at specific points cannot tell us the formula for f', but it can be suggestive: knowing $f'(1) \approx 2$, $f'(2) \approx 4$, $f'(3) \approx 6$ suggests that $f'(x) = 2x$. In Chapter 3, we show that this is indeed the case.

Problems for Section 2.2

1. The graph of $f(x)$ is given in Figure 2.26. Draw tangent lines to the graph at $x = -2$, $x = -1$, $x = 0$, and $x = 2$. Estimate $f'(-2)$, $f'(-1)$, $f'(0)$, and $f'(2)$.

Figure 2.26

2. The graph of $f(x)$ is given in Figure 2.27. Estimate $f'(1)$, $f'(2)$, $f'(3)$, $f'(4)$, and $f'(5)$.

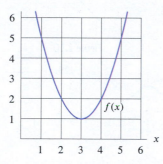

Figure 2.27

■ For Problems 3–8, graph the derivative of the given function.

3.
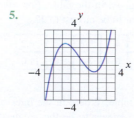

4.
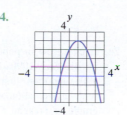

5.
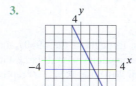

6.
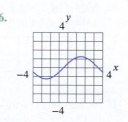

7.
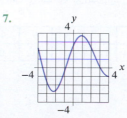

8.
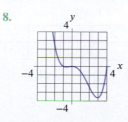

9. In the graph of f in Figure 2.28, at which of the labeled x-values is

   (a) $f(x)$ greatest?   **(b)** $f(x)$ least?

   (c) $f'(x)$ greatest?   **(d)** $f'(x)$ least?

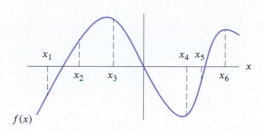

Figure 2.28

10. Find approximate values for $f'(x)$ at each of the x-values given in the following table.

x	0	5	10	15	20
$f(x)$	100	70	55	46	40

11. Using slopes to the left and right of 0, estimate $R'(0)$ if $R(x) = 100(1.1)^x$.

■ For Problems 12–17, sketch the graph of $f'(x)$.

12.

13.
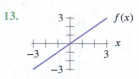

14.
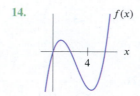

15.
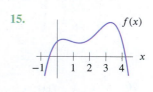

16.
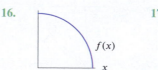

17.

■ Match the functions in Problems **18–21** with one of the derivatives in Figure 2.29.

■ In Problems **22–25**, match f' with the corresponding f in Figure 2.30.

(I)

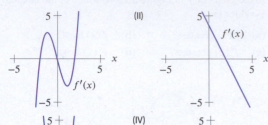

(II)

$f'(x)$

(III)

$f'(x)$

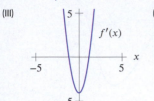

(IV)

(V)

$f'(x)$

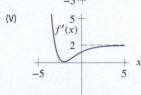

(VI)

$f'(x)$

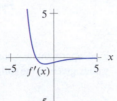

(VII)

$f'(x)$

(VIII)

$f'(x)$

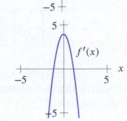

Figure 2.29

(I)

(II)

(III)

(IV)

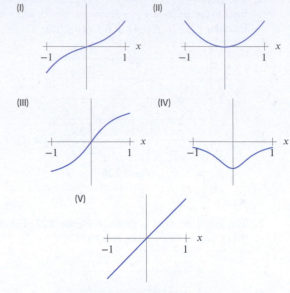

(V)

Figure 2.30

22.

$f'(x)$

23.

$f'(x)$

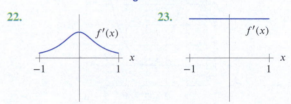

24.

$f'(x)$

25.

$f'(x)$

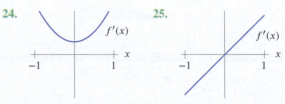

18.

$f(x)$

19.

$f(x)$

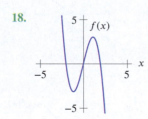
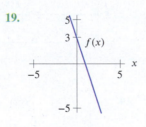

20.

$f(x)$

21.

$f(x)$

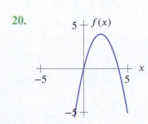
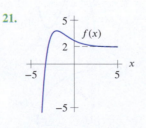

26. A city grew in population throughout the 1980s and into the early 1990s. The population was at its largest in 1995, and then shrank until 2010. Let $P = f(t)$ represent the population of the city t years since 1980. Sketch graphs of $f(t)$ and $f'(t)$, labeling the units on the axes.

27. Values of x and $g(x)$ are given in the table. For what value of x does $g'(x)$ appear to be closest to 3?

x	2.7	3.2	3.7	4.2	4.7	5.2	5.7	6.2
$g(x)$	3.4	4.4	5.0	5.4	6.0	7.4	9.0	11.0

28. Values of $f(x)$ are in the table. Where in the interval $-12 \leq x \leq 9$ does $f'(x)$ appear to be the greatest? Least?

x	−12	−9	−6	−3	0	3	6	9
$f(x)$	1.02	1.05	1.12	1.14	1.15	1.14	1.12	1.06

29. Draw a possible graph of $y = f(x)$ given the following information about its derivative.

   - $f'(x) > 0$ for $x < -1$
   - $f'(x) < 0$ for $x > -1$
   - $f'(x) = 0$ at $x = -1$

30. Draw a possible graph of a continuous function $y = f(x)$ that satisfies the following three conditions:

   - $f'(x) > 0$ for $1 < x < 3$
   - $f'(x) < 0$ for $x < 1$ and $x > 3$
   - $f'(x) = 0$ at $x = 1$ and $x = 3$

31. A vehicle moving along a straight road has distance $f(t)$ from its starting point at time t. Which of the graphs in Figure 2.31 could be $f'(t)$ for the following scenarios? (Assume the scales on the vertical axes are all the same.)

   (a) A bus on a popular route, with no traffic
   (b) A car with no traffic and all green lights
   (c) A car in heavy traffic conditions

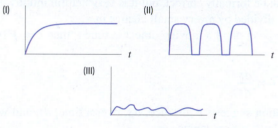

Figure 2.31

32. (a) Let $f(x) = \ln x$. Use small intervals to estimate $f'(1)$, $f'(2)$, $f'(3)$, $f'(4)$, and $f'(5)$.
   (b) Use your answers to part (a) to guess a formula for the derivative of $f(x) = \ln x$.

33. Suppose $f(x) = \frac{1}{3}x^3$. Estimate $f'(2)$, $f'(3)$, and $f'(4)$. What do you notice? Can you guess a formula for $f'(x)$?

34. Match each property (a)–(d) with one or more of graphs (I)–(IV) of functions.

   (a) $f'(x) = 1$ for all $0 \leq x \leq 4$

   (b) $f'(x) > 0$ for all $0 \leq x \leq 4$
   (c) $f'(2) = 1$
   (d) $f'(1) = 2$

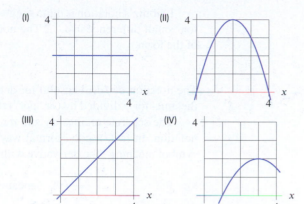

35. A child inflates a balloon, admires it for a while and then lets the air out at a constant rate. If $V(t)$ gives the volume of the balloon at time t, then Figure 2.32 shows $V'(t)$ as a function of t. At what time does the child:

   (a) Begin to inflate the balloon?
   (b) Finish inflating the balloon?
   (c) Begin to let the air out?
   (d) What would the graph of $V'(t)$ look like if the child had alternated between pinching and releasing the open end of the balloon, instead of letting the air out at a constant rate?

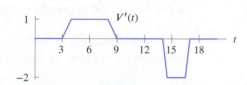

Figure 2.32

2.3   INTERPRETATIONS OF THE DERIVATIVE

We have seen the derivative interpreted as a slope and as a rate of change. In this section, we see other interpretations. The purpose of these examples is not to make a catalog of interpretations but to illustrate the process of obtaining them. There is another notation for the derivative that is often helpful.

An Alternative Notation for the Derivative

So far we have used the notation f' to stand for the derivative of the function f. An alternative notation for derivatives was introduced by the German mathematician Gottfried Wilhelm Leibniz (1646–1716) when calculus was first being developed. We know that $f'(x)$ is approximated by the average rate of change over a small interval. If $y = f(x)$, then the average rate of change is given by $\Delta y / \Delta x$. For small Δx, we have

$$f'(x) \approx \frac{\Delta y}{\Delta x}.$$

Leibniz's notation for the derivative, dy/dx, is meant to remind us of this. If $y = f(x)$, then we write

$$f'(x) = \frac{dy}{dx}.$$

Leibniz's notation is quite suggestive, especially if we think of the letter d in dy/dx as standing for "small difference in … ." The notation dy/dx reminds us that the derivative is a limit of ratios of the form

$$\frac{\text{Difference in } y\text{-values}}{\text{Difference in } x\text{-values}}.$$

The notation dy/dx is useful for determining the units for the derivative: the units for dy/dx are the units for y divided by (or "per") the units for x.

The separate entities dy and dx officially have no independent meaning: they are part of one notation. In fact, a good formal way to view the notation dy/dx is to think of d/dx as a single symbol meaning "the derivative with respect to x of …". Thus, dy/dx could be viewed as

$$\frac{d}{dx}(y), \quad \text{meaning "the derivative with respect to } x \text{ of } y\text{."}$$

On the other hand, many scientists and mathematicians really do think of dy and dx as separate entities representing "infinitesimally" small differences in y and x, even though it is difficult to say exactly how small "infinitesimal" is. It may not be formally correct, but it is very helpful intuitively to think of dy/dx as a very small change in y divided by a very small change in x.

For example, recall that if $s = f(t)$ is the position of a moving object at time t, then $v = f'(t)$ is the velocity of the object at time t. Writing

$$v = \frac{ds}{dt}$$

reminds us that v is a velocity since the notation suggests a distance, ds, over a time, dt, and we know that distance over time is velocity. Similarly, we recognize

$$\frac{dy}{dx} = f'(x)$$

as the slope of the graph of $y = f(x)$ by remembering that slope is vertical rise, dy, over horizontal run, dx.

The disadvantage of the Leibniz notation is that it is awkward to specify the value at which a derivative is evaluated. To specify $f'(2)$, for example, we have to write

$$\left.\frac{dy}{dx}\right|_{x=2}.$$

Using Units to Interpret the Derivative

Suppose a body moves along a straight line. If $s = f(t)$ gives the position in meters of the body from a fixed point on the line as a function of time, t, in seconds, then knowing that

$$\frac{ds}{dt} = f'(2) = 10 \text{ meters/sec}$$

tells us that when $t = 2$ sec, the body is moving at a velocity of 10 meters/sec. If the body continues to move at this velocity for a whole second (from $t = 2$ to $t = 3$), it would move an additional 10 meters.

In other words, if Δs represents the change in position during a time interval Δt, and if the body continues to move at this velocity, we have $\Delta s = 10\Delta t$, so

$$\Delta s = f'(2)\,\Delta t = 10\,\Delta t.$$

If the velocity is varying, this relationship is no longer exact.

For small values of Δt, we have:

$$\Delta s \approx f'(t)\Delta t.$$

Notice that the derivative acts as a multiplier: For a given positive Δt, a large derivative gives large change in s; a small derivative gives a small change in s. In general:

- The units of the derivative of a function are the units of the dependent variable divided by the units of the independent variable. In other words, the units of dA/dB are the units of A divided by the units of B.
- If the derivative of a function is not changing rapidly near a point, then the derivative is approximately equal to the change in the function when the independent variable increases by 1 unit.

The following examples illustrate how useful units can be in suggesting interpretations of the derivative.

Example 1 The cost C (in dollars) of building a house A square feet in area is given by the function $C = f(A)$. What are the units and the practical interpretation of the function $f'(A)$?

Solution In the Leibniz notation,

$$f'(A) = \frac{dC}{dA}.$$

This is a cost divided by an area, so it is measured in dollars per square foot. You can think of dC as the extra cost of building an extra dA square feet of house. So if you are planning to build a house with area A square feet, $f'(A)$ is approximately the cost per square foot of the *extra* area involved in building a house 1 square foot larger, and is called the *marginal cost*.

Example 2 The cost of extracting T tons of ore from a copper mine is $C = f(T)$ dollars. What does it mean to say that $f'(2000) = 100$?

Solution In the Leibniz notation,

$$f'(2000) = \left.\frac{dC}{dT}\right|_{T=2000}.$$

Since C is measured in dollars and T is measured in tons, dC/dT is measured in dollars per ton. You can think of dC as the extra cost of extracting an extra dT tons of ore. So the statement

$$\left.\frac{dC}{dT}\right|_{T=2000} = 100$$

says that when 2000 tons of ore have already been extracted from the mine, the cost of extracting the next ton is approximately \$100. Another way of saying this is that it costs about \$100 to extract the 2001st ton.

Example 3 If $q = f(p)$ gives the number of thousands of tons of zinc produced when the price is p dollars per ton, then what are the units and the meaning of

$$\left.\frac{dq}{dp}\right|_{p=900} = 0.2?$$

Solution   The units of dq/dp are the units of q over the units of p, or thousands of tons per dollar. You can think of dq as the extra zinc produced when the price increases by dp. The statement

$$\left.\frac{dq}{dp}\right|_{p=900} = f'(900) = 0.2 \text{ thousand tons per dollar}$$

tells us that the instantaneous rate of change of q with respect to p is 0.2 when $p = 900$. This means that when the price is \$900, the quantity produced increases by about 0.2 thousand tons, or 200 tons more for a one-dollar increase in price.

Example 4   The time, L (in hours), that a drug stays in a person's system is a function of the quantity administered, q, in mg, so $L = f(q)$.

(a) Interpret the statement $f(10) = 6$. Give units for the numbers 10 and 6.
(b) Write the derivative of the function $L = f(q)$ in Leibniz notation. If $f'(10) = 0.5$, what are the units of the 0.5?
(c) Interpret the statement $f'(10) = 0.5$ in terms of dose and duration.

Solution   (a) We know that $f(q) = L$. In the statement $f(10) = 6$, we have $q = 10$ and $L = 6$, so the units are 10 mg and 6 hours. The statement $f(10) = 6$ tells us that a dose of 10 mg lasts 6 hours.
(b) Since $L = f(q)$, we see that L depends on q. The derivative of this function is dL/dq. Since L is in hours and q is in mg, the units of the derivative are hours per mg. In the statement $f'(10) = 0.5$, the 0.5 is the derivative and the units are hours per mg.
(c) The statement $f'(10) = 0.5$ tells us that, at a dose of 10 mg, the instantaneous rate of change of duration is 0.5 hour per mg. In other words, if we increase the dose by 1 mg, the drug stays in the body approximately 30 minutes longer.

In the previous example, notice that $f'(10) = 0.5$ tells us that a 1 mg increase in dose leads to about a 0.5-hour increase in duration. If, on the other hand, we had had $f'(10) = 20$, we would have known that a 1-mg increase in dose leads to about a 20-hour increase in duration. Thus the derivative is the multiplier relating changes in dose to changes in duration. The magnitude of the derivative tells us how sensitive the time is to changes in dose.

We define the derivative of velocity, dv/dt, as *acceleration*.

Example 5   If the velocity of a body at time t seconds is measured in meters/sec, what are the units of the acceleration?

Solution   Since acceleration, dv/dt, is the derivative of velocity, the units of acceleration are units of velocity divided by units of time, or (meters/sec)/sec, written meters/sec^2.

Using the Derivative to Estimate Values of a Function

Since the derivative tells us how fast the value of a function is changing, we can use the derivative at a point to estimate values of the function at nearby points.

Example 6   Fertilizers can improve agricultural production. A Cornell University study[10] on maize (corn) production in Kenya found that the average value, $y = f(x)$, in Kenyan shillings of the yearly maize production from an average plot of land is a function of the quantity, x, of fertilizer used in kilograms.

(The shilling is the Kenyan unit of currency.)

(a) Interpret the statements $f(5) = 11,500$ and $f'(5) = 350$.

(b) Use the statements in part (a) to estimate $f(6)$ and $f(10)$.

(c) The value of the derivative, f', is increasing for x near 5. Which estimate in part (b) is more reliable?

Solution

(a) The statement $f(5) = 11,500$ tells us that $y = 11,500$ when $x = 5$. This means that if 5 kg of fertilizer are applied, maize worth 11,500 Kenyan shillings is produced. Since the derivative is dy/dx, the statement $f'(5) = 350$ tells us that

$$\frac{dy}{dx} = 350 \quad \text{when } x = 5.$$

This means that if the amount of fertilizer used is 5 kg and increases by 1 kg, then maize production value increases by about 350 Kenyan shillings.

(b) We want to estimate $f(6)$, that is the production value when 6 kg of fertilizer are used. If instead of 5 kg of fertilizer, one more kilogram is used, giving 6 kg altogether, we expect production value, in Kenyan shillings, to increase from 11,500 by about 350. Thus,

$$f(6) \approx 11,500 + 350 = 11,850.$$

Similarly, if 5 kg more fertilizer is used, so 10 kg are used altogether, we expect production value to increase by about $5 \cdot 350 = 1750$ Kenyan shillings, so the value is approximately

$$f(10) \approx 11,500 + 1750 = 13,250.$$

(c) To estimate $f(6)$, we assume that production value increases at rate of 350 Kenyan shillings per kilogram between $x = 5$ and $x = 6$ kg. To estimate $f(10)$, we assume that production value continues to increase at the same rate all the way from $x = 5$ to $x = 10$ kg. Since the derivative is increasing for x near 5, the estimate of $f(6)$ is more reliable.

In Example 6, representing the change in y by Δy and the change in x by Δx, we used the result introduced earlier in this section:

Tangent Line Approximation: Local Linearity

If $y = f(x)$ and Δx is near 0, then $\Delta y \approx f'(x)\Delta x$. For x near a, we have $\Delta y = f(x) - f(a)$ and $\Delta x = x - a$, so

$$f(x) \approx f(a) + f'(a)\Delta x.$$

Relative Rate of Change

In Section 1.5, we saw that an exponential function has a constant percent rate of change. Now we link this idea to derivatives. Analogous to the relative change, we look at the rate of change as a fraction of the original quantity.

The **relative rate of change** of $y = f(t)$ at $t = a$ is defined to be

$$\text{Relative rate of change of } y \text{ at } a = \frac{dy/dt}{y} = \frac{f'(a)}{f(a)}.$$

---

[10]P. Marenya and C. Barrett, "State-conditional Fertilizer Yield Response on Western Kenyan Farms", *Social Science Research Network*, abstract = 1141937, 2009.

We see in Section 3.3 that an exponential function has a constant relative rate of change. If the independent variable is time, the relative rate is often given as a percent change per unit time.

Example 7   Annual world soybean production, $W = f(t)$, in million tons, is a function of t years since the start of 2000.

(a) Interpret the statements $f(8) = 253$ and $f'(8) = 17$ in terms of soybean production.
(b) Calculate the relative rate of change of W at $t = 8$; interpret it in terms of soybean production.

Solution   (a) The statement $f(8) = 253$ tells us that 253 million tons of soybeans were produced in the year 2008. The statement $f'(8) = 17$ tells us that in 2008 annual soybean production was increasing at a rate of 17 million tons per year.

(b) We have

$$\text{Relative rate of change of soybean production} = \frac{f'(8)}{f(8)} = \frac{17}{253} = 0.067.$$

In 2008, annual soybean production was increasing at a continuous rate of 6.7% per year.

Example 8   Solar photovoltaic (PV) cells are the world's fastest growing energy source. At time t in years since 2010, peak PV energy-generating capacity worldwide was approximately $E = 50.15e^{0.32t}$ gigawatts.[11]

Estimate the relative rate of change of PV energy-generating capacity in 2018 using this model with

(a) $\Delta t = 1$ (b) $\Delta t = 0.1$ (c) $\Delta t = 0.01$

Solution   Let $E = f(t)$. In 2018 we have $t = 8$. The relative rate of change of f in 2018 is $f'(8)/f(8)$. We estimate $f'(8)$ using a difference quotient.

(a) Estimating the relative rate of change using $\Delta t = 1$ at $t = 8$, we have

$$\frac{dE/dt}{E} = \frac{f'(8)}{f(8)} \approx \frac{1}{f(8)} \frac{f(9) - f(8)}{1} = 0.377 = 37.7\% \text{ per year.}$$

(b) With $\Delta t = 0.1$ and $t = 8$, we have

$$\frac{dE/dt}{E} = \frac{f'(8)}{f(8)} \approx \frac{1}{f(8)} \frac{f(8.1) - f(8)}{0.1} = 0.325 = 32.5\% \text{ per year.}$$

(c) With $\Delta t = 0.01$ and $t = 8$, we have

$$\frac{dE/dt}{E} = \frac{f'(8)}{f(8)} \approx \frac{1}{f(8)} \frac{f(8.01) - f(8)}{0.01} = 0.321 = 32.1\% \text{ per year.}$$

The relative rate of change is approximately 32% per year. From Section 1.6, we know that the exponential function $E = 50.15e^{0.32t}$ has a continuous rate of change of 32% per year for all t, so the exact relative rate of change is 32%.

Example 9   In April 2009, the US Bureau of Economic Analysis announced that the US gross domestic product (GDP) was decreasing at an annual rate of 6.1%. The GDP of the US at that time was 13.84 trillion dollars. Calculate the annual rate of change of the US GDP in April 2009.

---

[11]Based on en.wikipedia.org/wiki/Growth_of_photovoltaics, accessed January 14, 2017. A gigawatt measures power and can be used to measure energy-generating capacity.

Solution   The Bureau of Economic Analysis is reporting the relative rate of change. In April 2009, the relative rate of change of GDP was -0.061 per year. To find the rate of change, we use:

$$\text{Relative rate of change in April 2009} = \frac{\text{Rate of change in April 2009}}{\text{GDP in April 2009}}$$

$$-0.061 = \frac{\text{Rate of change}}{13.84}$$

$$\text{Rate of change in April 2009} = -0.061 \cdot 13.84 = -0.84424 \text{ trillion dollars per year.}$$

The GDP of the US was decreasing at a continuous rate of 844.24 billion dollars per year in April 2009.

Problems for Section 2.3

■ In Problems 1–4, write the Leibniz notation for the derivative of the given function and include units.

1. The distance to the ground, D, in feet, of a skydiver is a function of the time t in minutes since the skydiver jumped out of the airplane.

2. The cost, C, of a steak, in dollars, is a function of the weight, W, of the steak, in pounds.

3. The number, N, of gallons of gas left in a gas tank is a function of the distance, D, in miles, the car has been driven.

4. An employee's pay, P, in dollars, for a week is a function of the number of hours worked, H.

5. The time for a chemical reaction, T (in minutes), is a function of the amount of catalyst present, a (in milliliters), so $T = f(a)$.

   (a) If $f(5) = 18$, what are the units of 5? What are the units of 18? What does this statement tell us about the reaction?
   (b) If $f'(5) = -3$, what are the units of 5? What are the units of -3? What does this statement tell us?

6. An economist is interested in how the price of a certain item affects its sales. At a price of $\$p$, a quantity, q, of the item is sold. If $q = f(p)$, explain the meaning of each of the following statements:

   (a) $f(150) = 2000$ (b) $f'(150) = -25$

7. The Arctic Sea ice extent, the area of sea covered by ice, grows over the winter months, typically from November to March. Let $F(t)$ be the Arctic Sea ice extent, in million of square kilometers, as a function of time, t, in days since November 1, 2014. Then $F'(t) = 0.073$ on January 1, 2015.[12]

   (a) Give the units of the 0.073, and interpret the number in practical terms.

(b) Estimate ΔF between January 1 and January 6, 2015. Explain what this tells us about Arctic Sea ice.

8. The cost, $C = f(w)$, in dollars of buying a chemical is a function of the weight bought, w, in pounds.

   (a) In the statement $f(12) = 5$, what are the units of the 12? What are the units of the 5? Explain what this is saying about the cost of buying the chemical.
   (b) Do you expect the derivative f' to be positive or negative? Why?
   (c) In the statement $f'(12) = 0.4$, what are the units of the 12? What are the units of the 0.4? Explain what this is saying about the cost of buying the chemical.

9. Figure 2.33 shows world solar energy output, in megawatts, as a function of years since 1990.[13] Estimate $f'(6)$. Give units and interpret your answer.

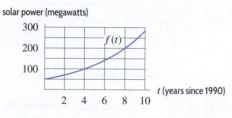

solar power (megawatts)

Figure 2.33

10. Figure 2.34 shows the power output, $p(w)$, of a wind turbine as a function of wind speed, w.

    (a) At what wind speed does the turbine generate the most power?
    (b) The derivative $p'(7)$ is positive but the derivative $p'(15)$ is negative. What does this tell you?

---

[12] Sea Ice Index data from nsidc.org/data/seaice_index/archives.html. Accessed March 2015.
[13] The Worldwatch Institute, *Vital Signs* 2001, p. 47 (New York: W.W. Norton, 2001).

(c) Give units and interpret $p(7) = 17,500$ and $p'(7) = 8000$.

(d) Give units and interpret $p(15) = 50,000$ and $p'(15) = -800$.

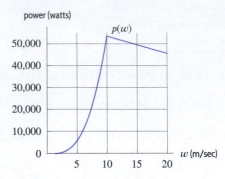

Figure 2.34

11. When you breathe, a muscle (called the diaphragm) reduces the pressure around your lungs and they expand to fill with air. The table shows the volume of a lung as a function of the reduction in pressure from the diaphragm. Pulmonologists (lung doctors) define the *compliance* of the lung as the derivative of this function.[14]

(a) What are the units of compliance?
(b) Estimate the maximum compliance of the lung.
(c) Explain why the compliance gets small when the lung is nearly full (around 1 liter).

Pressure reduction (cm of water)	Volume (liters)
0	0.20
5	0.29
10	0.49
15	0.70
20	0.86
25	0.95
30	1.00

12. Average leaf width, w (in mm), in tropical Australia[15] is a function of the average annual rainfall, r (in mm), so $w = f(r)$. We have $f'(1500) = 0.0218$.

(a) What are the units of the 1500?
(b) What are the units of the 0.0218?
(c) About how much difference in average leaf width would you find in two forests whose average annual rainfalls are near 1500 mm but differ by 200 mm?

13. The depth, h (in mm), of the water runoff down a slope during a steady rain is a function of the distance, x (in meters), from the top of the slope,[16] so $h = f(x)$. We have $f'(15) = 0.02$.

(a) What are the units of the 15?
(b) What are the units of the 0.02?
(c) About how much difference in runoff depth is there between two points around 15 meters down the slope if one of them is 4 meters farther from the top of the slope than the other?

14. When an ice dam for a glacial lake breaks, the maximal outflow rate, Q in meters^3/sec is a function of V, the volume of the lake (in millions of meters^3).

(a) What are the units of dQ/dV?
(b) Observation shows that $dQ/dV|_{V=12} = 22$. About how much is the difference in maximal outflow when dams break in two lakes with volumes near $12 \cdot 10^6$ meters^3 if one of them has volume 500,000 meters^3 greater than the other?

15. On May 9, 2007, CBS Evening News had a 4.3 point rating. (Ratings measure the number of viewers.) News executives estimated that a 0.1 drop in the ratings for the CBS Evening News corresponds to a $5.5 million drop in revenue.[17] Express this information as a derivative. Specify the function, the variables, the units, and the point at which the derivative is evaluated.

16. Let $S(t)$ be the amount of water, measured in acre-feet,[18] that is stored in a reservoir in week t.

(a) What are the units of $S'(t)$?
(b) What is the practical meaning of $S'(t) > 0$? What circumstances might cause this situation?

17. A yam has just been taken out of the oven and is cooling off before being eaten. The temperature, T, of the yam (measured in degrees Fahrenheit) is a function of how long it has been out of the oven, t (measured in minutes). Thus, we have $T = f(t)$.

(a) Is $f'(t)$ positive or negative? Why?
(b) What are the units for $f'(t)$?

18. Let $f(x)$ be the elevation in feet of the Mississippi River x miles from its source. What are the units of $f'(x)$? What can you say about the sign of $f'(x)$?

19. Meteorologists define the temperature lapse rate to be $-dT/dz$ where T is the air temperature in Celsius at altitude z kilometers above the ground.

(a) What are the units of the lapse rate?
(b) What is the practical meaning of a lapse rate of 6.5?

[14]en.wikipedia.org, accessed April 3, 2015.
[15]H. Shugart, *Terrestrial Ecosystems in Changing Environments* (Cambridge: CUP, 1998), p. 145.
[16]R. S. Anderson and S. P. Anderson, *Geomorphology* (Cambridge: Cambridge University Press, 2010), p. 369.
[17]*OC Register*, May 9, 2007; *The New York Times*, May 14, 2007.
[18]An acre-foot is the amount of water it takes to cover one acre of area with 1 foot of water.

20. Investing $1000 at an annual interest rate of $r\%$, compounded continuously, for 10 years gives you a balance of B, where $B = g(r)$. Give a financial interpretation of the statements:

(a) $g(2) \approx 1221$.

(b) $g'(2) \approx 122$. What are the units of $g'(2)$?

21. In April 2015 in the US, there was one birth every 8 seconds, one death every 12 seconds, and one new international migrant every 32 seconds.[19]

(a) Let $f(t)$ be the population of the US, where t is time in seconds measured from the start of April 2015. Find $f'(0)$. Give units.

(b) To the nearest second, how long did it take for the US population to add one person in April 2015?

22. Table 2.7 shows world gold production,[20] $G = f(t)$, as a function of year, t.

(a) Does $f'(t)$ appear to be positive or negative? What does this mean in terms of gold production?

(b) In which time interval does $f'(t)$ appear to be greatest?

(c) Estimate $f'(2015)$. Give units and interpret your answer in terms of gold production.

(d) Use the estimated value of $f'(2015)$ to estimate $f(2016)$ and $f(2020)$, and interpret your answers.

Table 2.7 *World gold production*

t (year)	2011	2012	2013	2014	2015
G (metric tons)	2660	2690	2800	2990	3000

23. The average weight, W, in pounds, of an adult is a function, $W = f(c)$, of the average number of Calories per day, c, consumed.

(a) Interpret the statements $f(1800) = 155$ and $f'(2000) = 0$ in terms of diet and weight.

(b) What are the units of $f'(c) = dW/dc$?

24. The cost, C (in dollars), to produce g gallons of a chemical can be expressed as $C = f(g)$. Using units, explain the meaning of the following statements in terms of the chemical:

(a) $f(200) = 1300$ (b) $f'(200) = 6$

25. Let G be annual US government purchases, T be annual US tax revenues, and Y be annual US output of all goods and services. All three quantities are given in dollars. Interpret the statements about the two derivatives, called fiscal policy multipliers.

(a) $dY/dG = 0.60$ (b) $dY/dT = -0.26$

26. The weight, W, in lbs, of a child is a function of its age, a, in years, so $W = f(a)$.

(a) Do you expect $f'(a)$ to be positive or negative? Why?

(b) What does $f(8) = 45$ tell you? Give units for the numbers 8 and 45.

(c) What are the units of $f'(a)$? Explain what $f'(a)$ tells you in terms of age and weight.

(d) What does $f'(8) = 4$ tell you about age and weight?

(e) As a increases, do you expect $f'(a)$ to increase or decrease? Explain.

27. A recent study reports that men who retired late developed Alzheimer's at a later stage than those who stopped work earlier. Each additional year of employment was associated with about a six-week later age of onset. Express these results as a statement about the derivative of a function. State clearly what function you use, including the units of the dependent and independent variables.

28. The thickness, P, in mm, of pelican eggshells depends on the concentration, c, of PCBs in the eggshell, measured in ppm (parts per million); that is, $P = f(c)$.

(a) The derivative $f'(c)$ is negative. What does this tell you?

(b) Give units and interpret $f(200) = 0.28$ and $f'(200) = -0.0005$ in terms of PCBs and eggshells.

29. Suppose that $f(t)$ is a function with $f(25) = 3.6$ and $f'(25) = -0.2$. Estimate $f(26)$ and $f(30)$.

30. For a function $f(x)$, we know that $f(20) = 68$ and $f'(20) = -3$. Estimate $f(21)$, $f(19)$ and $f(25)$.

■ In Problems **31–36**, use the tangent line approximation.

31. Given $f(4) = 5$, $f'(4) = 7$, approximate $f(4.02)$.

32. Given $f(4) = 5$, $f'(4) = 7$, approximate $f(3.92)$.

33. Given $f(5) = 3$, $f'(5) = -2$, approximate $f(5.03)$.

34. Given $f(2) = -4$, $f'(2) = -3$, approximate $f(1.95)$.

35. Given $f(-3) = -4$, $f'(-3) = 2$, approximate $f(-2.99)$.

36. Given $f(3) = -4$, $f'(3) = -2$ approximate $f(2.99)$.

37. Let $\left.\dfrac{dV}{dr}\right|_{r=2} = 16$.

(a) For small Δr, write an approximate equation relating ΔV and Δr near $r = 2$.

(b) Estimate ΔV if $\Delta r = 0.1$.

(c) Let $V = 32$ when $r = 2$. Estimate V when $r = 2.1$.

---

[19]www.census.gov, accessed April 8, 2015.
[20]minerals.usgs.gov/minerals/pubs/commodity/gold/, accessed September 2016.

38. Let $R = f(S)$ and $f'(10) = 3$.

    (a) For small ΔS, write an approximate equation relating ΔR and ΔS near $S = 10$.

    (b) Estimate the change in R if S changes from $S = 10$ to $S = 10.2$.

    (c) Let $f(10) = 13$. Estimate $f(10.2)$.

■ Problems 39–42 concern $g(t)$ in Figure 2.35, which gives the weight of a human fetus as a function of its age.

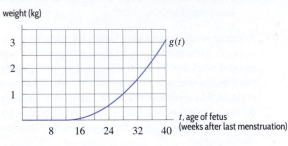

Figure 2.35

39. (a) What are the units of $g'(24)$?
    (b) What is the biological meaning of $g'(24) = 0.096$?

40. (a) Which is greater, $g'(20)$ or $g'(36)$?
    (b) What does your answer say about fetal growth?

41. Is the instantaneous weight growth rate greater or less than the average rate of change of weight over the 40-week period

    (a) At week 16?   **(b)** At week 36?

42. Estimate **(a)** $g'(20)$ **(b)** $g'(36)$

    (c) The average rate of change of weight for the entire 40-week gestation.

43. Annual net sales, in billion of dollars, for the Hershey Company, the largest US producer of chocolate, is a function $S = f(t)$ of time, t, in years since 2010.

    (a) Interpret the statements $f(5) = 7.39$ and $f'(5) = -0.03$ in terms of Hershey sales.[21]

    (b) Estimate $f(8)$ and interpret it in terms of Hershey sales.

44. For some painkillers, the size of the dose, D, given depends on the weight of the patient, W. Thus, $D = f(W)$, where D is in milligrams and W is in pounds.

    (a) Interpret the statements $f(140) = 120$ and $f'(140) = 3$ in terms of this painkiller.

    (b) Use the information in the statements in part (a) to estimate $f(145)$.

45. US beef[22] production, $M = f(t)$, in billion pounds, is a function of t, years since 2010.

    (a) Interpret $f(5) = 23.7$ and $f'(5) = -0.1$ in terms of beef production.

    (b) Estimate $f(8)$ and interpret it in terms of beef production.

46. The quantity, Q mg, of nicotine in the body t minutes after a cigarette is smoked is given by $Q = f(t)$.

    (a) Interpret the statements $f(20) = 0.36$ and $f'(20) = -0.002$ in terms of nicotine. What are the units of the numbers 20, 0.36, and -0.002?

    (b) Use the information given in part (a) to estimate $f(21)$ and $f(30)$. Justify your answers.

47. A mutual fund is currently valued at \$80 per share and its value per share is increasing at a rate of \$0.50 a day. Let $V = f(t)$ be the value of the share t days from now.

    (a) Express the information given about the mutual fund in term of f and f'.

    (b) Assuming that the rate of growth stays constant, estimate and interpret $f(10)$.

48. Figure 2.36 shows how the contraction velocity, $v(x)$, of a muscle changes as the load on it changes.

    (a) Find the slope of the line tangent to the graph of contraction velocity at a load of 2 kg. Give units.

    (b) Using your answer to part (a), estimate the change in the contraction velocity if the load is increased from 2 kg by adding 50 grams.

    (c) Express your answer to part (a) as a derivative of $v(x)$.

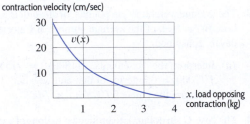

Figure 2.36

49. Figure 2.37 shows how the pumping rate of a person's heart changes after bleeding.

    (a) Find the slope of the line tangent to the graph at time 2 hours. Give units.

    (b) Using your answer to part (a), estimate how much the pumping rate increases during the minute beginning at time 2 hours.

    (c) Express your answer to part (a) as a derivative of $g(t)$.

---

[21]www.statista.com/statistics/235932/total-global-chocolate-sales-of-the-hershey-company/, accessed September 4, 2016.
[22]www.ers.usda.gov/topics/animal-products/cattle-beef/statistics-information.aspx, accessed September 2016.

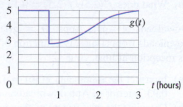

Figure 2.37

50. Suppose $C(r)$ is the total cost of paying off a car loan borrowed at an annual interest rate of $r\%$. What are the units of $C'(r)$? What is the practical meaning of $C'(r)$? What is its sign?

■ Problems 51–55 refer to Figure 2.38, which shows the depletion of food stores in the human body during starvation.

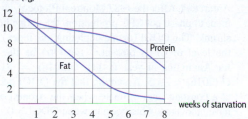

Figure 2.38

51. Which is being consumed at a greater rate, fat or protein, during the

(a) Third week?   (b) Seventh week?

52. The fat storage graph is linear for the first four weeks. What does this tell you about the use of stored fat?

53. Estimate the rate of fat consumption after

(a) 3 weeks   (b) 6 weeks   (c) 8 weeks

54. What seems to happen during the sixth week? Why do you think this happens?

55. Figure 2.39 shows the derivatives of the protein and fat storage functions. Which graph is which?

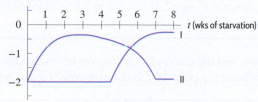

Figure 2.39

^{23}www.arctic.noaa.gov. Accessed April 3, 2015.

56. A person with a certain liver disease first exhibits larger and larger concentrations of certain enzymes (called SGOT and SGPT) in the blood. As the disease progresses, the concentration of these enzymes drops, first to the predisease level and eventually to zero (when almost all of the liver cells have died). Monitoring the levels of these enzymes allows doctors to track the progress of a patient with this disease. If $C = f(t)$ is the concentration of the enzymes in the blood as a function of time,

(a) Sketch a possible graph of $C = f(t)$.
(b) Mark on the graph the intervals where $f' > 0$ and where $f' < 0$.
(c) What does $f'(t)$ represent, in practical terms?

57. A company's revenue from car sales, C (in thousands of dollars), is a function of advertising expenditure, a, in thousands of dollars, so $C = f(a)$.

(a) What does the company hope is true about the sign of f'?
(b) What does the statement $f'(100) = 2$ mean in practical terms? How about $f'(100) = 0.5$?
(c) Suppose the company plans to spend about $100,000 on advertising. If $f'(100) = 2$, should the company spend more or less than $100,000 on advertising? What if $f'(100) = 0.5$?

58. A company making solar panels spends x dollars on materials, and the revenue from the sale of the solar panels is $f(x)$ dollars.

(a) What does the statement $f'(80,000) = 2$ mean in practical terms? How about $f'(80,000) = 0.5$?
(b) Suppose the company plans to spend about $80,000 on materials. If $f'(80,000) = 2$, should the company spend more than $80,000 on materials? What if $f'(100) = 0.5$?

59. For a new type of biofuel, scientists estimate that it takes $A = f(g)$ gallons of gasoline to produce the raw materials to generate g gallons of biofuel. Assume the biofuel is equal in efficiency to the gasoline.

(a) At a certain level of production, we have $dA/dg = 1.3$. Interpret this in practical terms. Is this level of production sustainable? Explain.
(b) Repeat part (a) if $dA/dg = 0.2$.

60. Analysis of satellite data indicates that the Greenland ice sheet lost approximately 2900 gigatons (gt) of mass between March 2002 and September 2014. The mean mass loss rate for 2013–14 was 6 gt/year; the rate for 2012–13 was 474 gt/year.^{23}

(a) What derivative does this tell us about? Define the function and give units for each variable.
(b) What numerical statement can you make about the derivative? Give units.

61. The area of the Amazon's rain forest, $R = f(t)$, in thousand square kilometers,[24] is a function of the number of years, t, since 2010.

   (a) Interpret $f(5) = 5500$ and $f'(5) = -10.9$ in terms of the Amazon's rain forests.

   (b) Find and interpret the relative rate of change of $f(t)$ when $t = 5$.

62. The number of active Facebook users hit 1.55 billion at the end of September 2015 and 1.59 billion[25] at the end of December. With t in months since the start of 2015, let $f(t)$ be the number of active users in billions. Estimate $f(12)$ and $f'(12)$ and the relative rate of change of f at $t = 12$. Interpret your answers in terms of Facebook users.

63. Estimate the relative rate of change of $f(t) = t^2$ at $t = 4$. Use $\Delta t = 0.01$.

64. Estimate the relative rate of change of $f(t) = t^2$ at $t = 10$. Use $\Delta t = 0.01$.

65. The world population in billions is approximately $P = 7.4e^{0.0107t}$ where t is in years since 2016.[26] Estimate the relative rate of change of population in 2020 using this model and

   (a) $\Delta t = 1$ **(b)** $\Delta t = 0.1$ **(c)** $\Delta t = 0.01$

66. The weight, w, in kilograms, of a baby is a function $f(t)$ of her age, t, in months.

   (a) What does $f(2.5) = 5.67$ tell you?
   (b) What does $f'(2.5)/f(2.5) = 0.13$ tell you?

67. Downloads of Apple Apps, $D = g(t)$, in billions of downloads from iTunes, is a function of t months since its inception in June, 2008.[27]

   (a) Interpret the statements $g(36) = 15$ and $g'(36) = 0.93$ in terms of App downloads.

   (b) Calculate the relative rate of change of D at $t = 36$; interpret it in terms of App downloads.

68. The number of barrels of oil produced from North Dakota oil wells since January 2014 is estimated to be $B = 29.05e^{0.0214t}$ million barrels, where t is in months since January 2014.[28] Estimate the relative rate of change of oil production in March 2015 using

   (a) $\Delta t = 1$ **(b)** $\Delta t = 0.1$ **(c)** $\Delta t = 0.01$

69. During the 1970s and 1980s, the buildup of chlorofluorocarbons (CFCs) created a hole in the ozone layer over Antarctica. After the 1987 Montreal Protocol, an agreement to phase out CFC production, the ozone hole has shrunk. The ODGI (ozone depleting gas index) shows the level of CFCs present.[29] Let $O(t)$ be the ODGI for Antarctica in year t; then $O(2016) = 81.4$ and $O'(2016) = -1.5$. Assuming that the ODGI decreases at a constant rate, estimate when the ozone hole will have recovered, which occurs when ODGI $= 0$.

2.4   THE SECOND DERIVATIVE

Since the derivative is itself a function, we can calculate its derivative. For a function f, the derivative of its derivative is called the *second derivative*, and written f''. If $y = f(x)$, the second derivative can also be written as $\dfrac{d^2y}{dx^2}$, which means $\dfrac{d}{dx}\left(\dfrac{dy}{dx}\right)$, the derivative of $\dfrac{dy}{dx}$.

What Does the Second Derivative Tell Us?

Recall that the derivative of a function tells us whether the function is increasing or decreasing:

   If $f' > 0$ on an interval, then f is increasing on that interval.
   If $f' < 0$ on an interval, then f is decreasing on that interval.

Since f'' is the derivative of f', we have

   If $f'' > 0$ on an interval, then f' is increasing on that interval.
   If $f'' < 0$ on an interval, then f' is decreasing on that interval.

So the question becomes: What does it mean for f' to be increasing or decreasing? The case in which f' is increasing is shown in Figure 2.40, where the graph of f is bending upward, or is *concave up*. In the case when f' is decreasing, shown in Figure 2.41, the graph is bending downward, or is *concave down*. We have the following result:

> If $f'' > 0$ on an interval, then f' is increasing and the graph of f is concave up on that interval.
> If $f'' < 0$ on an interval, then f' is decreasing and the graph of f is concave down on that interval.

---

[24] en.wikipedia.org/wiki/Amazon_rainforest, accessed September 5, 2016.
[25] techcrunch.com, accessed September 5, 2016.
[26] http://www.worldometers.info/world-population/, accessed September 6, 2016.
[27] www.asymco.com/2011/07/13/itunes-app-total-downloads-finally-overtook-song-downloads, accessed September 6, 2016.
[28] www.dmr.nd.gov/oilgas/stats/statisticsvw.asp, accessed September 6, 2016.
[29] www.esrl.noaa.gov/gmd/odgi, accessed April, 2017.

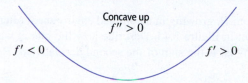

Figure 2.40: Meaning of f'': The slope increases from negative to positive as you move from left to right, so f'' is positive and f is concave up

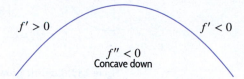

Figure 2.41: Meaning of f'': The slope decreases from positive to negative as you move from left to right, so f'' is negative and f is concave down

Example 1   For the functions whose graphs are given in Figure 2.42, decide where their second derivatives are positive and where they are negative.

Solution   From the graphs it appears that

(a) $f'' > 0$ everywhere, because the graph of f is concave up everywhere.
(b) $g'' < 0$ everywhere, because the graph is concave down everywhere.
(c) $h'' > 0$ for $x > 0$, because the graph of h is concave up there; $h'' < 0$ for $x < 0$, because the graph of h is concave down there.

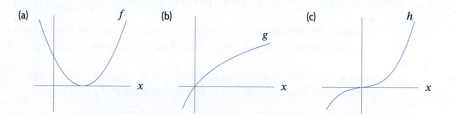

Figure 2.42: What signs do the second derivatives have?

Interpretation of the Second Derivative as a Rate of Change

If we think of the derivative as a rate of change, then the second derivative is a rate of change of a rate of change. If the second derivative is positive, the rate of change is increasing; if the second derivative is negative, the rate of change is decreasing.

The second derivative is often a matter of practical concern. In 1985 a newspaper headline reported the Secretary of Defense as saying that Congress and the Senate had cut the defense budget. As his opponents pointed out, however, Congress had merely cut the rate at which the defense budget was increasing.[30] In other words, the derivative of the defense budget was still positive (the budget was increasing), but the second derivative was negative (the budget's rate of increase had slowed).

---

[30] In the *Boston Globe*, March 13, 1985, Representative William Gray (D–Pa.) was reported as saying: "It's confusing to the American people to imply that Congress threatens national security with reductions when you're really talking about a reduction in the increase."

Example 2 A population, P, growing in a confined environment often follows a *logistic* growth curve, like the graph shown in Figure 2.43. Describe how the rate at which the population is increasing changes over time. What is the sign of the second derivative d^2P/dt^2? What is the practical interpretation of t^* and L?

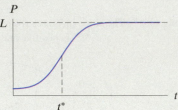

Figure 2.43: Logistic growth curve

Solution Initially, the population is increasing, and at an increasing rate. So, initially dP/dt is increasing and $d^2P/dt^2 > 0$. At t^*, the rate at which the population is increasing is a maximum; the population is growing fastest then. Beyond t^*, the rate at which the population is growing is decreasing, so $d^2P/dt^2 < 0$. At t^*, the graph changes from concave up to concave down and $d^2P/dt^2 = 0$.

The quantity L represents the limiting value of the population that is approached as t tends to infinity; L is called the *carrying capacity* of the environment and represents the maximum population that the environment can support.

Example 3 Table 2.8 shows the number of abortions per year, A, in thousands, reported in the US in the year t.[31]

Table 2.8 *Abortions reported in the US (1972–2013)*

Year, t	1972	1975	1980	1985	1990	1995	2000	2005	2010	2013
Abortions, A (1000s)	587	1034	1554	1589	1609	1359	1313	1206	1103	1000

(a) Calculate the average rate of change for the time intervals shown between 1972 and 2013.
(b) What can you say about the sign of d^2A/dt^2 during the period 1972–1995?

Solution (a) For each time interval we can calculate the average rate of change of the number of abortions per year over this interval. For example, between 1972 and 1975

$$\text{Average rate of change} = \frac{\Delta A}{\Delta t} = \frac{1034 - 587}{1975 - 1972} = \frac{447}{3} = 149.$$

Thus, between 1972 and 1975, there were approximately 149,000 more abortions reported each year. Values of $\Delta A/\Delta t$ are listed in Table 2.9.

Table 2.9 *Rate of change of number of abortions reported*

Time	1972–1975	1975–1980	1980–1985	1985–1990	1990–1995	1995–2000	2000–2005	2005–2010	2010–2013
Average rate of change, $\Delta A/\Delta t$ (1000s/year)	149	104	7	4	−50	−9.2	−21.4	−42.0	−34.3

(b) We assume the data lies on a smooth curve. Since the values of $\Delta A/\Delta t$ are decreasing dramatically for 1975–1995, we can be pretty certain that dA/dt also decreases, so d^2A/dt^2 is negative for this period. For 1972–1975, the sign of d^2A/dt^2 is less clear; abortion data from 1968 would help. Figure 2.44 confirms this; the graph appears to be concave down for 1975–1995. The fact

[31] www.johnstonsarchive.net/policy/abortion/graphusabrate.html, accessed January 9th, 2017.

that dA/dt is positive during the period 1972–1980 corresponds to the fact that the number of abortions reported increased from 1972 to 1980. The fact that dA/dt is negative during the period 1990–2013 corresponds to the fact that the number of abortions reported decreased from 1990 to 2013. The fact that d^2A/dt^2 is negative for 1975–1995 reflects the fact that the rate of increase slowed over this period.

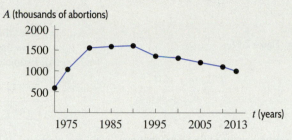

Figure 2.44: How the number of reported abortions in the US is changing with time

Problems for Section 2.4

1. For the function $g(x)$ graphed in Figure 2.45, are the following nonzero quantities positive or negative?

    (a) $g'(0)$ **(b)** $g''(0)$

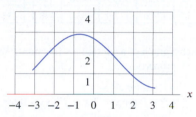

Figure 2.45

2. At one of the labeled points on the graph in Figure 2.46 both dy/dx and d^2y/dx^2 are negative. Which is it?

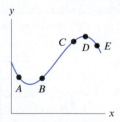

Figure 2.46

3. Graph the functions described in parts (a)–(d).

    (a) First and second derivatives everywhere positive.
    (b) Second derivative everywhere negative; first derivative everywhere positive.
    (c) Second derivative everywhere positive; first derivative everywhere negative.
    (d) First and second derivatives everywhere negative.

■ For Problems **4–9**, give the signs of the first and second derivatives for the function. Each derivative is either positive everywhere, zero everywhere, or negative everywhere.

4.

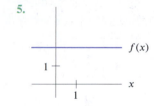

5.

6.

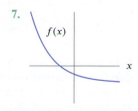

7.

8.

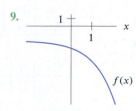

9.

■ In Problems **10–12**, use Figure 2.47 to determine which of the two values is greater.

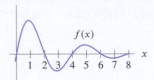

Figure 2.47

10. $f'(0)$ or $f'(4)$?   **11.** $f'(2)$ or $f'(6)$?

12. $f''(1)$ or $f''(3)$?

■ In Problems **13–14**, use the graph given for each function.
(a) Estimate the intervals on which the derivative is positive and the intervals on which the derivative is negative.
(b) Estimate the intervals on which the second derivative is positive and the intervals on which the second derivative is negative.

13.

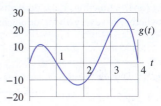

14.

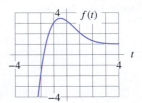

■ In Problems **15–16**, use the values given for each function.
(a) Does the derivative of the function appear to be positive or negative over the given interval? Explain.
(b) Does the second derivative of the function appear to be positive or negative over the given interval? Explain.

15.

t	100	110	120	130	140
$w(t)$	10.7	6.3	4.2	3.5	3.3

16.

t	0	1	2	3	4	5
$s(t)$	12	14	17	20	31	55

17. Sketch the graph of a function whose first derivative is everywhere negative and whose second derivative is positive for some x-values and negative for other x-values.

18. IBM-Peru uses second derivatives to assess the relative success of various advertising campaigns. They assume that all campaigns produce some increase in sales. If a graph of sales against time shows a positive second derivative during a new advertising campaign, what does this suggest to IBM management? Why? What does a negative second derivative suggest?

19. Values of $f(t)$ are given in the following table.

(a) Does this function appear to have a positive or negative first derivative? Second derivative? Explain.
(b) Estimate $f'(2)$ and $f'(8)$.

t	0	2	4	6	8	10
$f(t)$	150	145	137	122	98	56

20. The table gives the number of passenger cars, $C = f(t)$, in millions,[32] in the US in the year t.

(a) Do $f'(t)$ and $f''(t)$ appear to be positive or negative during the period 1975–1990?
(b) Do $f'(t)$ and $f''(t)$ appear to be positive or negative during the period 1990–2000?
(c) Estimate $f'(2005)$. Using units, interpret your answer in terms of passenger cars.

t	1975	1980	1985	1990	1995	2000	2005
C	106.7	121.6	127.9	133.7	128.4	133.6	136.6

21. Sketch a graph of a continuous function f with the following properties:

- $f'(x) > 0$ for all x
- $f''(x) < 0$ for $x < 2$ and $f''(x) > 0$ for $x > 2$.

22. Sketch the graph of a function f such that $f(2) = 5$, $f'(2) = 1/2$, and $f''(2) > 0$.

23. At exactly two of the labeled points in Figure 2.48, the derivative f' is 0; the second derivative f'' is not zero at any of the labeled points. On a copy of the table, give the signs of f, f', f'' at each marked point.

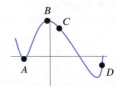

Figure 2.48

Point	f	f'	f''
A			
B			
C			
D			

24. For three minutes the temperature of a feverish person has had positive first derivative and negative second derivative. Which of the following is correct?

(a) The temperature rose in the last minute more than it rose in the minute before.

[32]www.bts.gov/publications/national_transportation_statistics/html/table_01_11.html. Accessed April 27, 2011.

(b) The temperature rose in the last minute, but less than it rose in the minute before.

(c) The temperature fell in the last minute but less than it fell in the minute before.

(d) The temperature rose for two minutes but fell in the last minute.

25. Yesterday's temperature at t hours past midnight was $f(t)$ °C. At noon the temperature was 20°C. The first derivative, $f'(t)$, decreased all morning, reaching a low of 2°C/hour at noon, then increased for the rest of the day. Which one of the following must be correct?

   (a) The temperature fell in the morning and rose in the afternoon.

   (b) At 1 pm the temperature was 18°C.

   (c) At 1 pm the temperature was 22°C.

   (d) The temperature was lower at noon than at any other time.

   (e) The temperature rose all day.

26. A function f has $f(5) = 20$, $f'(5) = 2$, and $f''(x) < 0$, for $x \geq 5$. Which of the following are possible values for $f(7)$ and which are impossible?

   (a) 26   **(b)** 24   **(c)** 22

27. An industry is being charged by the Environmental Protection Agency (EPA) with dumping unacceptable levels of toxic pollutants in a lake. Over a period of several months, an engineering firm makes daily measurements of the rate at which pollutants are being discharged into the lake. The engineers produce a graph similar to either Figure 2.49(a) or Figure 2.49(b). For each case, give an idea of what argument the EPA might make in court against the industry and in the industry's defense.

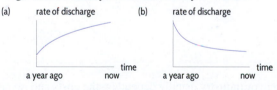

Figure 2.49

28. "Winning the war on poverty" has been described cynically as slowing the rate at which people are slipping below the poverty line. Assuming that this is happening:

   (a) Graph the total number of people in poverty against time.

   (b) If N is the number of people below the poverty line at time t, what are the signs of dN/dt and d^2N/dt^2? Explain.

29. A headline in the New York Times on December 14, 2014, read:[33]

   "A Steep Slide in Law School Enrollment Accelerates"

   (a) What function is the author talking about?

   (b) Draw a possible graph for the function.

   (c) In terms of derivatives, what is the headline saying?

30. In economics, *total utility* refers to the total satisfaction from consuming some commodity. According to the economist Samuelson:[34]

   > As you consume more of the same good, the total (psychological) utility increases. However, ... with successive new units of the good, your total utility will grow at a slower and slower rate because of a fundamental tendency for your psychological ability to appreciate more of the good to become less keen.

   (a) Sketch the total utility as a function of the number of units consumed.

   (b) In terms of derivatives, what is Samuelson saying?

31. Let $P(t)$ represent the price of a share of stock of a corporation at time t. What does each of the following statements tell us about the signs of the first and second derivatives of $P(t)$?

   (a) "The price of the stock is rising faster and faster."

   (b) "The price of the stock is close to bottoming out."

32. Each of the graphs in Figure 2.50 shows the position of a particle moving along the x-axis as a function of time, $0 \leq t \leq 5$. The vertical scales of the graphs are the same. During this time interval, which particle has

   (a) Constant velocity?

   (b) The greatest initial velocity?

   (c) The greatest average velocity?

   (d) Zero average velocity?

   (e) Zero acceleration?

   (f) Positive acceleration throughout?

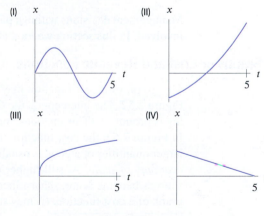

Figure 2.50

---

[33]http://www.nytimes.com, accessed January 8, 2015.
[34]From Paul A. Samuelson, *Economics*, 11th edition (New York: McGraw-Hill, 1981).

33. The Arctic Sea ice extent, the area of the sea covered by ice, grows seasonally over the winter months each year, typically from November to March, and is modeled by $G(t)$, in millions of square kilometers, t months after November 1, 2014.[35]

(a) What is the sign of $G'(t)$ for $0 < t < 4$?

(b) Suppose $G''(t) < 0$ for $0 < t < 4$. What does this tell us about how the Arctic Sea ice extent grows?

(c) Sketch a graph of $G(t)$ for $0 \le t \le 4$, given that $G(0) = 10.3$, $G(4) = 14.4$, and G'' is as in part (b). Label your axes, including units.

34. In 2009, a study was done on the impact of sea-level rise in the mid-Atlantic states.[36] Let $a(t)$ be the depth of the sea in millimeters (mm) at a typical point on the Atlantic Coast, and let $m(t)$ be the depth of the sea in mm at a typical point on the Gulf of Mexico, with time t in years since data collection started.

(a) The study reports "Sea level is rising and there is evidence that the rate is accelerating." What does this statement tell us about $a(t)$ and $m(t)$?

(b) The study also reports "The Atlantic Coast and the Gulf of Mexico experience higher rates of sea-level rise (2 to 4 mm per year and 2 to 10 mm per year, respectively) than the current global average (1.7 mm per year)." What does this tell us about $a(t)$ and $m(t)$?

(c) Assume the rates at which the sea level rises on the Atlantic Coast and the Gulf of Mexico are constant for a century and within the ranges given in the report.

(i) What is the largest amount the sea could rise on the Atlantic Coast during a century? Your answer should be a range of values.

(ii) What is the shortest amount of time in which the sea level in the Gulf of Mexico could rise 1 meter?

35. (a) Using Table 2.10, calculate the average rate of change of the number of Facebook users, N (in millions), per month for each of the 3-month intervals.[37]

(b) What can you say about the sign of d^2N/dt^2 during the period September 2014–September 2015?

Table 2.10

Month, t	Sep 2014	Dec 2014	Mar 2015	June 2015	Sep 2015
N	1350	1393	1441	1490	1545

36. In 1913, Carlson[38] conducted the classic experiment in which he grew yeast, *Saccharomyces cerevisiae*, in laboratory cultures and collected data every hour for 18 hours. Table 2.11 shows the yeast population, P, at representative times t in hours.

(a) Calculate the average rate of change of P per hour for the time intervals shown between 0 and 18 hours.

(b) What can you say about the sign of d^2P/dt^2 during the period 0–18 hours?

Table 2.11

t	0	2	4	6	8	10	12	14	16	18
P	9.6	29.0	71.1	174.6	350.7	513.3	594.8	640.8	655.9	661.8

2.5   MARGINAL COST AND REVENUE

Management decisions within a particular firm or industry usually depend on the costs and revenues involved. In this section we look at the cost and revenue functions.

Graphs of Cost and Revenue Functions

The graph of a cost function may be linear, as in Figure 2.51, or it may have the shape shown in Figure 2.52. The intercept on the C-axis represents the fixed costs, which are incurred even if nothing is produced. (This includes, for instance, the cost of the machinery needed to begin production.) In Figure 2.52, the cost function increases quickly at first and then more slowly because producing larger quantities of a good is usually more efficient than producing smaller quantities—this is called *economy of scale*. At still higher production levels, the cost function increases faster again as resources become scarce; sharp increases may occur when new factories have to be built. Thus, the graph of a cost function, C, may start out concave down and become concave up later on.

---

[35]Data on the Arctic Sea ice extent was recorded daily in 2014 and is archived at nsidc.org/arcticseaicenews/2015/03/. Accessed September 2015.

[36]www.epa.gov/climatechange/effects/coastal/sap4-1.html, *Coastal Sensitivity to Sea-Level Rise: A Focus on the Mid-Atlantic Region,* US Climate Change Science Program, January 2009.

[37]www.statista.com/statistics/264810/number-of-monthly-active-facebook-users-worldwide/, accessed December 12, 2016.

[38]T. Carlson, "Über Geschwindigkeit und Grösse der Hefevermehrung in Würze", *Biochem. Z.*, 1913.

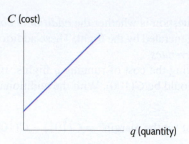

Figure 2.51: A linear cost function

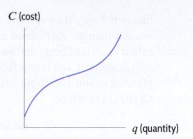

Figure 2.52: A nonlinear cost function

The revenue function is $R = pq$, where p is price and q is quantity. If the price, p, is a constant, the graph of R against q is a straight line through the origin with slope equal to the price. (See Figure 2.53.) In practice, for large values of q, the market may become glutted, causing the price to drop and giving R the shape in Figure 2.54.

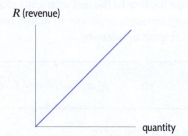

Figure 2.53: Revenue: Constant price

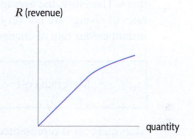

Figure 2.54: Revenue: Decreasing price

Example 1 If cost, C, and revenue, R, are given by the graph in Figure 2.55, for what production quantities does the firm make a profit?

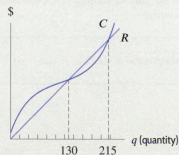

Figure 2.55: Costs and revenues for Example 1

Solution The firm makes a profit whenever revenues are greater than costs, that is, when $R > C$. The graph of R is above the graph of C approximately when $130 < q < 215$. Production between 130 units and 215 units will generate a profit.

Marginal Analysis

Many economic decisions are based on an analysis of the costs and revenues "at the margin." Let's look at this idea through an example.

Suppose you are running an airline and you are trying to decide whether to offer an additional flight. How should you decide? We'll assume that the decision is to be made purely on financial grounds: if the flight will make money for the company, it should be added. Obviously you need to consider the costs and revenues involved. Since the choice is between adding this flight and leaving things the way they are, the crucial question is whether the *additional costs* incurred are greater or smaller than the *additional revenues* generated by the flight. These additional costs and revenues are called *marginal costs* and *marginal revenues*.

Suppose $C(q)$ is the function giving the cost of running q flights. If the airline had originally planned to run 100 flights, its costs would be $C(100)$. With the additional flight, its costs would be $C(101)$. Therefore,

$$\text{Additional cost "at the margin"} = C(101) - C(100).$$

Now

$$C(101) - C(100) = \frac{C(101) - C(100)}{101 - 100},$$

and this quantity is the average rate of change of cost between 100 and 101 flights. In Figure 2.56 the average rate of change is the slope of the secant line. If the graph of the cost function is not curving too fast near the point, the slope of the secant line is close to the slope of the tangent line there. Therefore, the average rate of change is close to the instantaneous rate of change. Since these rates of change are not very different, many economists choose to define marginal cost, MC, as the instantaneous rate of change of cost with respect to quantity:

$$\boxed{\text{Marginal cost} = MC = C'(q) \qquad \text{so} \qquad \text{Marginal cost} \approx C(q+1) - C(q).}$$

Marginal cost is represented by the slope of the cost curve.

Similarly if the revenue generated by q flights is $R(q)$ and the number of flights increases from 100 to 101, then

$$\text{Additional revenue "at the margin"} = R(101) - R(100).$$

Now $R(101) - R(100)$ is the average rate of change of revenue between 100 and 101 flights. As before, the average rate of change is approximately equal to the instantaneous rate of change, so economists often define

$$\boxed{\text{Marginal revenue} = MR = R'(q) \qquad \text{so} \qquad \text{Marginal revenue} \approx R(q+1) - R(q).}$$

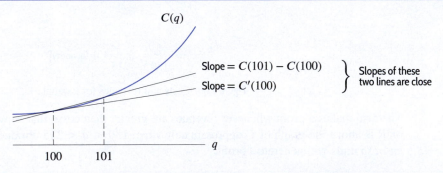

Figure 2.56: Marginal cost: Slope of one of these lines

Example 2   If $C(q)$ and $R(q)$ for the airline are given in Figure 2.57, should the company add the 101st flight?

Solution   The marginal revenue is the slope of the revenue curve at $q = 100$. The marginal cost is the slope of the graph of C at $q = 100$. Figure 2.57 suggests that the slope at point A is smaller than the slope at B, so $MC < MR$ for $q = 100$. This means that the airline will make more in extra revenue than it will spend in extra costs if it runs another flight, so it should go ahead and run the 101st flight.

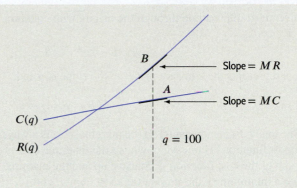

Figure 2.57: Cost and revenue for Example 2

Example 3   The graph of a cost function is given in Figure 2.58. Does it cost more to produce the 500th item or the 2000th? Does it cost more to produce the 3000th item or the 4000th? At approximately what production level is marginal cost smallest? What is the total cost at this production level?

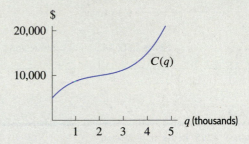

Figure 2.58: Estimating marginal cost: Where is marginal cost smallest?

Solution   The cost to produce an additional item is the marginal cost, which is represented by the slope of the cost curve. Since the slope of the cost function in Figure 2.58 is greater at $q = 0.5$ (when the quantity produced is 0.5 thousand, or 500) than at $q = 2$, it costs more to produce the 500th item than the 2000th item. Since the slope is greater at $q = 4$ than $q = 3$, it costs more to produce the 4000th item than the 3000th item.

The slope of the cost function is close to zero at $q = 2$, and is positive everywhere else, so the slope is smallest at $q = 2$. The marginal cost is smallest at a production level of 2000 units. Since $C(2) \approx 10,000$, the total cost to produce 2000 units is about \$10,000.

Example 4   If the revenue and cost functions, R and C, are given by the graphs in Figure 2.59, sketch graphs of the marginal revenue and marginal cost functions, MR and MC.

Figure 2.59: Total revenue and total cost for Example 4

Solution   The revenue graph is a line through the origin, with equation

$$R = pq$$

where p represents the constant price, so the slope is p and

$$MR = R'(q) = p.$$

The total cost is increasing, so the marginal cost is always positive. For small q values, the graph of the cost function is concave down, so the marginal cost is decreasing. For larger q, say $q > 100$, the graph of the cost function is concave up and the marginal cost is increasing. Thus, the marginal cost has a minimum at about $q = 100$. (See Figure 2.60.)

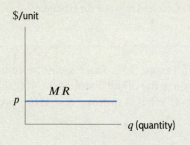

Figure 2.60: Marginal revenue and costs for Example 4

Problems for Section 2.5

1. It costs \$4800 to produce 1295 items and it costs \$4830 to produce 1305 items. What is the approximate marginal cost at a production level of 1300 items?

2. The function $C(q)$ gives the cost in dollars to produce q barrels of olive oil.

    (a) What are the units of marginal cost?
    (b) What is the practical meaning of the statement $MC = 3$ for $q = 100$?

3. In Figure 2.61, estimate the marginal revenue when the level of production is 600 units and interpret it.

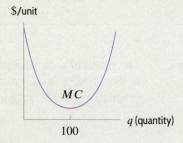

Figure 2.62

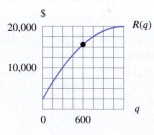

Figure 2.61

4. In Figure 2.62, is marginal cost greater at $q = 5$ or at $q = 30$? At $q = 20$ or at $q = 40$? Explain.

5. In Figure 2.63, estimate the marginal cost when the level of production is 10,000 units and interpret it.

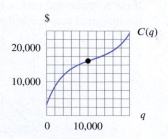

Figure 2.63

6. For q units of a product, a manufacturer's cost is $C(q)$ dollars and revenue is $R(q)$ dollars, with $C(500) = 7200$, $R(500) = 9400$, $MC(500) = 15$, and $MR(500) = 20$.

   (a) What is the profit or loss at $q = 500$?
   (b) If production is increased from 500 to 501 units, by approximately how much does profit change?

7. The cost of recycling q tons of paper is given in the following table. Estimate the marginal cost at $q = 2000$. Give units and interpret your answer in terms of cost. At approximately what production level does marginal cost appear smallest?

q (tons)	1000	1500	2000	2500	3000	3500
$C(q)$ (dollars)	2500	3200	3640	3825	3900	4400

8. Total cost is $C = 8500 + 4.65q$ and total revenue is $R = 5.15q$, both in dollars, where q represents the quantity produced.

   (a) What is the fixed cost?
   (b) What is the marginal cost per item?
   (c) What is the price at which this item is sold?
   (d) For what production levels does this company make a profit?
   (e) How much does the company make for each additional unit sold?

9. When production is 4500, marginal revenue is \$8 per unit and marginal cost is \$9.25 per unit. Do you expect maximum profit to occur at a production level above or below 4500? Explain.

10. Figure 2.64 shows part of the graph of cost and revenue for a car manufacturer. Which is greater, marginal cost or marginal revenue, at

   (a) q_1?   (b) q_2?

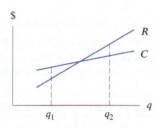

Figure 2.64

11. Let $C(q)$ represent the total cost of producing q items. Suppose $C(15) = 2300$ and $C'(15) = 108$. Estimate the total cost of producing: (a) 16 items (b) 14 items.

12. To produce 1000 items, the total cost is \$5000 and the marginal cost is \$25 per item. Estimate the costs of producing 1001 items, 999 items, and 1100 items.

13. Let $C(q)$ represent the cost and $R(q)$ represent the revenue, in dollars, of producing q items.

   (a) If $C(50) = 4300$ and $C'(50) = 24$, estimate $C(52)$.
   (b) If $C'(50) = 24$ and $R'(50) = 35$, approximately how much profit is earned by the 51st item?
   (c) If $C'(100) = 38$ and $R'(100) = 35$, should the company produce the 101st item? Why or why not?

14. Figure 2.65 shows cost and revenue for producing q units. For the production levels in (a)–(d), is the company making or losing money? Should the company be increasing or decreasing production to increase profits?

   (a) $q = 75$ (b) $q = 150$
   (c) $q = 225$ (d) $q = 300$

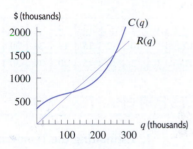

Figure 2.65

15. Cost and revenue functions for a charter bus company are shown in Figure 2.66. Should the company add a 50th bus? How about a 90th? Explain your answers using marginal revenue and marginal cost.

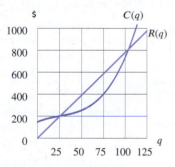

Figure 2.66

16. An industrial production process costs $C(q)$ million dollars to produce q million units; these units then sell for $R(q)$ million dollars. If $C(2.1) = 5.1$, $R(2.1) = 6.9$, $MC(2.1) = 0.6$, and $MR(2.1) = 0.7$, calculate

   (a) The profit earned by producing 2.1 million units
   (b) The approximate change in revenue if production increases from 2.1 to 2.14 million units.
   (c) The approximate change in revenue if production decreases from 2.1 to 2.05 million units.
   (d) The approximate change in profit in parts (b) and (c).

17. A company's cost of producing q liters of a chemical is $C(q)$ dollars; this quantity can be sold for $R(q)$ dollars. Suppose $C(2000) = 5930$ and $R(2000) = 7780$.

   (a) What is the profit at a production level of 2000?
   (b) If $MC(2000) = 2.1$ and $MR(2000) = 2.5$, what is the approximate change in profit if q is increased from 2000 to 2001? Should the company increase or decrease production from $q = 2000$?
   (c) If $MC(2000) = 4.77$ and $MR(2000) = 4.32$, should the company increase or decrease production from $q = 2000$?

18. Table 2.12 shows the cost, $C(q)$, and revenue, $R(q)$, in terms of quantity q. Estimate the marginal cost, $C'(q)$, and marginal revenue, $R'(q)$, for q between 0 and 7.

   Table 2.12

q	0	1	2	3	4	5	6	7
$C(q)$	9	10	12	15	19	24	30	37
$R(q)$	0	5	10	15	20	25	30	35

19. Table 2.13 shows the cost, $C(q)$, and revenue, $R(q)$, in terms of quantity q. Estimate the marginal cost, $MC(q)$, and marginal revenue, $MR(q)$, for q between 0 and 6.

   Table 2.13

q	1	2	3	4	5	6
$C(q)$	20	60	120	200	300	420
$R(q)$	100	220	330	410	450	480

PROJECTS FOR CHAPTER TWO

1. **Estimating the Temperature of a Yam**

   You put a yam in a hot oven, maintained at a constant temperature of 200°C. As the yam picks up heat from the oven, its temperature rises.[39]

   (a) Draw a possible graph of the temperature T of the yam against time t (minutes) since it is put into the oven. Explain any interesting features of the graph, and in particular explain its concavity.
   (b) At $t = 30$, the temperature T of the yam is 120° and increasing at the (instantaneous) rate of 2°/min. Using this information and the shape of the graph, estimate the temperature at time $t = 40$.
   (c) In addition, you are told that at $t = 60$, the temperature of the yam is 165°. Can you improve your estimate of the temperature at $t = 40$?
   (d) Assuming all the data given so far, estimate the time at which the temperature of the yam is 150°.

2. **Temperature and Illumination**

   Alone in your dim, unheated room, you light a single candle rather than curse the darkness. Depressed with the situation, you walk directly away from the candle, sighing. The temperature (in degrees Fahrenheit) and illumination (in % of one candle power) decrease as your distance (in feet) from the candle increases. In fact, you have tables showing this information.

Distance (feet)	Temperature (°F)	Distance (feet)	Illumination (%)
0	55	0	100
1	54.5	1	85
2	53.5	2	75
3	52	3	67
4	50	4	60
5	47	5	56
6	43.5	6	53

[39] From Peter D. Taylor, *Calculus: The Analysis of Functions* (Toronto: Wall & Emerson, Inc., 1992).

You are cold when the temperature is below 40°. You are in the dark when the illumination is at most 50% of one candle power.

(a) Two graphs are shown in Figures 2.67 and 2.68. One is temperature as a function of distance and one is illumination as a function of distance. Which is which? Explain.

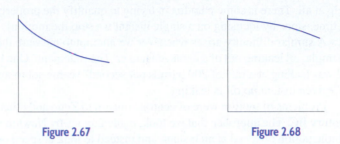

Figure 2.67   Figure 2.68

(b) What is the average rate at which the temperature is changing when the illumination drops from 75% to 56%?

(c) You can still read your watch when the illumination is about 65%. Can you still read your watch at 3.5 feet? Explain.

(d) Suppose you know that at 6 feet the instantaneous rate of change of the temperature is $-4.5°$F/ft and the instantaneous rate of change of illumination is -3% candle power/ft. Estimate the temperature and the illumination at 7 feet.

(e) Are you in the dark before you are cold, or vice versa?

3. **Chlorofluorocarbons (CFCs) in the Atmosphere**

Chlorofluorocarbons (CFCs) were used as propellants in spray cans until their buildup in the atmosphere started destroying the ozone, which protects us from ultraviolet rays. Since the 1987 Montreal Protocol (an agreement to curb CFCs), the CFCs in the atmosphere above the US have been reduced from a high of 1915 parts per trillion (ppt) in 2000 to 1640 ppt in 2014.[40] The reduction has been approximately linear. Let $C(t)$ be the concentration of CFCs in ppt in year t.

(a) Find $C(2000)$ and $C(2014)$.

(b) Estimate $C'(2000)$ and $C'(2014)$.

(c) Assuming $C(t)$ is linear, find a formula for $C(t)$.

(d) When is $C(t)$ expected to reach 1500 ppt, the level before CFCs were introduced?

(e) If you were told that in the future, $C(t)$ would not be exactly linear, and that $C''(t) > 0$, would your answer to part (d) be too early or too late?

---

[40]www.esrl.noaa.gov, accessed April 27, 2015.

FOCUS ON THEORY

LIMITS, CONTINUITY, AND THE DEFINITION OF THE DERIVATIVE

The velocity at a single instant in time is surprisingly difficult to define precisely. Consider the statement "At the instant it crossed the finish line, the horse was traveling at 42 mph." How can such a claim be substantiated? A photograph taken at that instant will show the horse motionless—it is no help at all. There is some paradox in trying to quantify the property of motion at a particular instant in time, since by focusing on a single instant we stop the motion!

A similar difficulty arises whenever we attempt to measure the rate of change of anything—for example, oil leaking out of a damaged tanker. The statement "One hour after the ship's hull ruptured, oil was leaking at a rate of 200 barrels per second" seems not to make sense. We could argue that at any given instant *no* oil is leaking.

Problems of motion were of central concern to Zeno and other philosophers as early as the fifth century BC. The approach that we took, made famous by Newton's calculus, is to stop looking for a simple notion of speed at an instant, and instead to look at speed over small intervals containing the instant. This method sidesteps the philosophical problems mentioned earlier but brings new ones of its own.

Definition of the Derivative Using Average Rates

In Section 2.1, we defined the derivative as the instantaneous rate of change of a function. We can estimate a derivative by computing average rates of change over smaller and smaller intervals. We use this idea to give a symbolic definition of the derivative. Letting h represent the size of the interval, we have

$$\text{Average rate of change between } x \text{ and } x + h = \frac{f(x + h) - f(x)}{(x + h) - x} = \frac{f(x + h) - f(x)}{h}.$$

To find the derivative, or instantaneous rate of change at the point x, we use smaller and smaller intervals. To find the derivative exactly, we take the limit as h, the size of the interval, shrinks to zero, so we say

$$\text{Derivative} = \text{Limit, as } h \text{ approaches zero, of } \frac{f(x + h) - f(x)}{h}.$$

Finally, instead of writing the phrase "limit, as h approaches 0," we use the notation $\lim_{h \to 0}$. This leads to the following symbolic definition:

For any function f, we define the **derivative function**, f', by

$$f'(x) = \lim_{h \to 0} \frac{f(x + h) - f(x)}{h},$$

provided the limit exists. The function f is said to be **differentiable** at any point x at which the derivative function is defined.

Notice that we have replaced the original difficulty of computing velocity at a point by an argument that the average rates of change approach a number as the time intervals shrink in size. In a sense, we have traded one hard question for another, since we don't yet have any idea how to be certain what number the average velocities are approaching.

The Idea of a Limit

We used a limit to define the derivative. Now we look a bit more at the idea of the limit of a function at the point c. Provided the limit exists:

> We write $\lim\limits_{x \to c} f(x)$ to represent the number approached by $f(x)$ as x approaches c.

Example 1 Investigate $\lim\limits_{x \to 2} x^2$.

Solution Notice that we can make x^2 as close to 4 as we like by taking x sufficiently close to 2. (Look at the values of 1.9^2, 1.99^2, 1.999^2, and 2.1^2, 2.01^2, 2.001^2 in Table 2.14; they seem to be approaching 4.) We write

$$\lim_{x \to 2} x^2 = 4,$$

which is read "the limit, as x approaches 2, of x^2 is 4." Notice that the limit does not ask what happens *at* $x = 2$, so it is not sufficient to substitute 2 to find the answer. The limit describes behavior of a function *near* a point, not *at* the point.

Table 2.14   *Values of x^2 near $x = 2$*

x	1.9	1.99	1.999	2.001	2.01	2.1
x^2	3.61	3.96	3.996	4.004	4.04	4.41

Example 2 Use a graph to estimate $\lim\limits_{x \to 0} \dfrac{2^x - 1}{x}$.

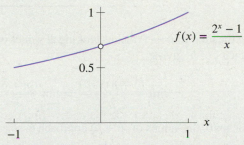

Figure 2.69: Find the limit as $x \to 0$ of $\dfrac{2^x - 1}{x}$

Solution Notice that the expression $\dfrac{2^x - 1}{x}$ is undefined at $x = 0$. To find out what happens to this expression as x approaches 0, look at a graph of $f(x) = \dfrac{2^x - 1}{x}$. Figure 2.69 shows that as x approaches 0 from either side, the value of $\dfrac{2^x - 1}{x}$ appears to approach 0.7. If we zoom in on the graph near $x = 0$, we can estimate the limit with greater accuracy, giving

$$\lim_{x \to 0} \frac{2^x - 1}{x} \approx 0.693.$$

Example 3   Estimate $\lim_{h \to 0} \dfrac{(3 + h)^2 - 9}{h}$ numerically.

Solution   The limit is the value approached by this expression as h approaches 0. The values in Table 2.15 seem to be approaching 6 as $h \to 0$. So it is a reasonable guess that

$$\lim_{h \to 0} \frac{(3 + h)^2 - 9}{h} = 6.$$

However, we cannot be sure that the limit is *exactly* 6 by looking at the table. To calculate the limit exactly requires algebra.

Table 2.15   *Values of* $\left((3 + h)^2 - 9\right)/h$

h	-0.1	-0.01	-0.001	0.001	0.01	0.1
$\left((3 + h)^2 - 9\right)/h$	5.9	5.99	5.999	6.001	6.01	6.1

Example 4   Use algebra to find $\lim_{h \to 0} \dfrac{(3 + h)^2 - 9}{h}$.

Solution   Expanding the numerator gives

$$\frac{(3 + h)^2 - 9}{h} = \frac{9 + 6h + h^2 - 9}{h} = \frac{6h + h^2}{h}.$$

Since taking the limit as $h \to 0$ means looking at values of h near, but not equal, to 0, we can cancel a common factor of h, giving

$$\lim_{h \to 0} \frac{(3 + h)^2 - 9}{h} = \lim_{h \to 0} \frac{6h + h^2}{h} = \lim_{h \to 0}(6 + h).$$

As h approaches 0, the values of $(6 + h)$ approach 6, so

$$\lim_{h \to 0} \frac{(3 + h)^2 - 9}{h} = \lim_{h \to 0}(6 + h) = 6.$$

Continuity

Roughly speaking, a function is said to be *continuous* on an interval if its graph has no breaks, jumps, or holes in that interval. A continuous function has a graph that can be drawn without lifting the pencil from the paper.

Example: The function $f(x) = 3x^2 - x^2 + 2x + 1$ is continuous on any interval. (See Figure 2.70.)

Example: The function $f(x) = 1/x$ is not defined at $x = 0$. It is continuous on any interval not containing the origin. (See Figure 2.71.)

Example: A company rents cars for $7 per hour or fraction thereof, so it costs $7 for a trip of one hour or less, $14 for a trip between one and two hours, and so on. If $p(x)$ is the price of trip lasting x hours, then its graph (in Figure 2.72) is a series of steps. This function is not continuous on intervals such as $(0, 2)$ because the graph jumps at $x = 1$.

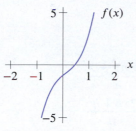

Figure 2.70: The graph of $f(x) = 3x^3 - x^2 + 2x - 1$

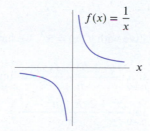

Figure 2.71: Graph of $f(x) = 1/x$: Not defined at 0

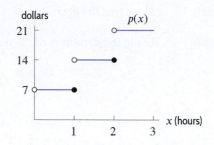

Figure 2.72: Cost of renting a car

What Does Continuity Mean Numerically?

Continuity is important in practical work because it means that small errors in the independent variable lead to small errors in the value of the function.

Example: Suppose that $f(x) = x^2$ and that we want to compute $f(\pi)$. Knowing f is continuous tells us that taking $x = 3.14$ should give a good approximation to $f(\pi)$, and that we can get a better approximation to $f(\pi)$ by using more decimals of π.

Example: If $p(x)$ is the price of renting a car graphed in Figure 2.72, then $p(0.99) = p(1) = \$7$, whereas $p(1.01) = \$14$, because as soon as we pass one hour, the price jumps to $14. So a small difference in time can lead to a significant difference in the cost. Hence p is not continuous at $x = 1$.

Definition of Continuity

We now define continuity using limits. The idea of continuity rules out breaks, jumps, or holes by demanding that the behavior of a function *near* a point be consistent with its behavior *at* the point:

The function f is **continuous** at $x = c$ if f is defined at $x = c$ and

$$\lim_{x \to c} f(x) = f(c).$$

The function is **continuous on an interval** (a, b) if it is continuous at every point in the interval.

Which Functions Are Continuous?

Requiring a function to be continuous on an interval is not asking very much, as any function whose graph is an unbroken curve over the interval is continuous. For example, exponential functions, polynomials, and sine and cosine are continuous on every interval. Functions created by adding, multiplying, or composing continuous functions are also continuous.

Using the Definition to Calculate Derivatives

By estimating the derivative of the function $f(x) = x^2$ at several points, we guessed in Example 5 of Section 2.2 that the derivative of x^2 is $f'(x) = 2x$. In order to show that this formula is correct, we have to use the symbolic definition of the derivative given earlier in this section.

In evaluating the expression

$$\lim_{h \to 0} \frac{f(x+h) - f(x)}{h},$$

we simplify the difference quotient first, and then take the limit as h approaches zero.

Example 5   Show that the derivative of $f(x) = x^2$ is $f'(x) = 2x$.

Solution   Using the definition of the derivative with $f(x) = x^2$, we have

$$f'(x) = \lim_{h \to 0} \frac{f(x+h) - f(x)}{h} = \lim_{h \to 0} \frac{(x+h)^2 - x^2}{h}$$

$$= \lim_{h \to 0} \frac{x^2 + 2xh + h^2 - x^2}{h} = \lim_{h \to 0} \frac{2xh + h^2}{h}$$

$$= \lim_{h \to 0} \frac{h(2x + h)}{h}.$$

To take the limit, look at what happens when h is close to 0, but do not let $h = 0$. Since $h \neq 0$, we cancel the common factor of h, giving

$$f'(x) = \lim_{h \to 0} \frac{h(2x + h)}{h} = \lim_{h \to 0} (2x + h) = 2x,$$

because as h gets close to zero, $2x + h$ gets close to $2x$. So

$$f'(x) = \frac{d}{dx}(x^2) = 2x.$$

Example 6   Show that if $f(x) = 3x - 2$, then $f'(x) = 3$.

Solution   Since the slope of the linear function $f(x) = 3x - 2$ is 3 and the derivative is the slope, we see that $f'(x) = 3$. We can also use the definition to get this result:

$$f'(x) = \lim_{h \to 0} \frac{f(x+h) - f(x)}{h} = \lim_{h \to 0} \frac{(3(x+h) - 2) - (3x - 2)}{h}$$

$$= \lim_{h \to 0} \frac{3x + 3h - 2 - 3x + 2}{h} = \lim_{h \to 0} \frac{3h}{h}.$$

To find the limit, look at what happens when h is close to, but not equal to, 0. Simplifying, we get

$$f'(x) = \lim_{h \to 0} \frac{3h}{h} = \lim_{h \to 0} 3 = 3.$$

Problems on Limits and the Definition of the Derivative

1. On Figure 2.73, mark lengths that represent the quantities in parts (a)–(e). (Pick any h, with $h > 0$.)

   (a) $a + h$ **(b)** h **(c)** $f(a)$
   (d) $f(a + h)$ **(e)** $f(a+h)-f(a)$

   (f) Using your answers to parts (a)–(e), show how the quantity $\dfrac{f(a + h) - f(a)}{h}$ can be represented as the slope of a line on the graph.

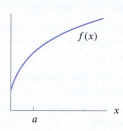

   Figure 2.73

HINT INAPPROPRIATE

2. On Figure 2.74, mark lengths that represent the quantities in parts (a)–(e). (Pick any h, with $h > 0$.)

   (a) $a + h$ **(b)** h **(c)** $f(a)$
   (d) $f(a + h)$ **(e)** $f(a+h)-f(a)$

   (f) Using your answers to parts (a)–(e), represent the quantity $\dfrac{f(a + h) - f(a)}{h}$ as the slope of a line on the graph.

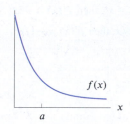

   Figure 2.74

■ For the expressions in Problems 3–6,
   (a) What is the result when we substitute $h = 0$?
   (b) Find the values when we substitute $h = 0.1$, $h = 0.01$, $h = 0.001$, and $h = 0.0001$.
   (c) Use the results of part (b) to estimate the limit as h goes to zero.

3. $\dfrac{e^{5h} - 1}{h}$

4. $\dfrac{2e^{3h} - 2}{h}$

5. $\dfrac{\ln(h + 1)}{h}$

6. $\dfrac{\sin(3h)}{h}$ (in radians)

■ Use a graph to estimate the limits in Problems 7–8.

7. $\lim\limits_{x \to 0} \dfrac{5^x - 1}{x}$

8. $\lim\limits_{x \to 0} \dfrac{\sin x}{x}$ (with x in radians)

■ In Problems 9–12, estimate the limit by substituting smaller and smaller values of h. For trigonometric functions, use radians. Give answers to one decimal place.

9. $\lim\limits_{h \to 0} \dfrac{7^h - 1}{h}$

10. $\lim\limits_{h \to 0} \dfrac{(3 + h)^3 - 27}{h}$

11. $\lim\limits_{h \to 0} \dfrac{\cos h - 1}{h}$

12. $\lim\limits_{h \to 0} \dfrac{e^{1+h} - e}{h}$

■ In Problems 13–16, does the function $f(x)$ appear to be continuous on the interval $0 \le x \le 2$? If not, what about on the interval $0 \le x \le 0.5$?

13.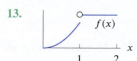

14.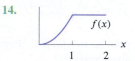

15.

16.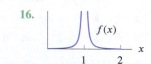

17. The graph of $y = f(x)$ is in Figure 2.75. Does the function $f(x)$ appear to be continuous on the given interval?

   (a) $1 \le x \le 3$
   (b) $0.5 \le x \le 1.5$
   (c) $3 \le x \le 5$
   (d) $2 \le x \le 6$

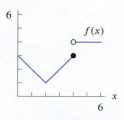

   Figure 2.75

18. The graph of $y = f(x)$ is in Figure 2.76. Does the function $f(x)$ appear to be continuous on the given interval?

(a) $1 \leq x \leq 3$
(b) $0.5 \leq x \leq 1.5$
(c) $3 \leq x \leq 5$
(d) $4 \leq x \leq 5$

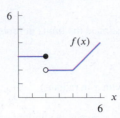

Figure 2.76

■ In Problems 19–22, find the limit by simplifying the expression.

19. $\lim\limits_{h \to 0} \dfrac{(5 + 2h) - 5}{h}$

20. $\lim\limits_{h \to 0} \dfrac{(3h - 7) + 7}{h}$

21. $\lim\limits_{h \to 0} \dfrac{(h + 1)^2 - 1}{h}$

22. $\lim\limits_{h \to 0} \dfrac{(h - 2)^2 - 4}{h}$

■ Are the functions in Problems 23–28 continuous on the given intervals?

23. $f(x) = x + 2$ on $-3 \leq x \leq 3$

24. $f(x) = 2^x$ on $0 \leq x \leq 10$

25. $f(x) = x^2 + 2$ on $0 \leq x \leq 5$

26. $f(x) = \dfrac{1}{x - 1}$ on $2 \leq x \leq 3$

27. $f(x) = \dfrac{1}{x - 1}$ on $0 \leq x \leq 2$

28. $f(x) = \dfrac{1}{x^2 + 1}$ on $0 \leq x \leq 2$

■ Which of the functions described in Problems 29–33 are continuous?

29. The number of people in a village as a function of time.

30. The distance traveled by a car in stop-and-go traffic as a function of time.

31. The weight of a baby as a function of time during the second month of the baby's life.

32. The number of pairs of pants as a function of the number of yards of cloth from which they are made. Each pair requires 3 yards.

33. You start in North Carolina and go westward on Interstate 40 toward California. Consider the function giving the local time of day as a function of your distance from your starting point.

■ Use the definition of the derivative to show how the formulas in Problems 34–43 are obtained.

34. If $f(x) = 5x$, then $f'(x) = 5$.

35. If $f(x) = 3x - 2$, then $f'(x) = 3$.

36. If $f(x) = x^2 + 4$, then $f'(x) = 2x$.

37. If $f(x) = 3x^2$, then $f'(x) = 6x$.

38. If $f(x) = x - x^2$, then $f'(x) = 1 - 2x$.

39. If $f(x) = 5x^2 + 1$, then $f'(x) = 10x$.

40. If $f(x) = 2x^2 + x$, then $f'(x) = 4x + 1$.

41. If $f(x) = -2x^3$, then $f'(x) = -6x^2$.

42. If $f(x) = 1 - x^3$, then $f'(x) = -3x^2$.

43. If $f(x) = 1/x$, then $f'(x) = -1/x^2$.

Chapter 3

SHORTCUTS TO DIFFERENTIATION

CONTENTS

3.1   DERIVATIVE FORMULAS FOR POWERS AND POLYNOMIALS

The derivative of a function at a point represents a slope and a rate of change. In Chapter 2, we learned how to estimate values of the derivative of a function given by a graph or by a table. Now, we learn how to find a formula for the derivative of a function given by a formula.

Derivative of a Constant Function

The graph of a constant function $f(x) = k$ is a horizontal line, with a slope of 0 everywhere. Therefore, its derivative is 0 everywhere. (See Figure 3.1.)

$$\text{If } f(x) = k, \text{ then } f'(x) = 0.$$

For example, $\dfrac{d}{dx}(5) = 0$.

Figure 3.1: A constant function

Derivative of a Linear Function

We already know that the slope of a line is constant. This tells us that the derivative of a linear function is constant.

$$\text{If } f(x) = b + mx, \text{ then } f'(x) = \text{Slope} = m.$$

For example, $\dfrac{d}{dx}\left(5 - \dfrac{3}{2}x\right) = -\dfrac{3}{2}$.

Derivative of a Constant Times a Function

Figure 3.2 shows the graph of $y = f(x)$ and of three multiples: $y = 3f(x)$, $y = \frac{1}{2}f(x)$, and $y = -2f(x)$. How are the derivatives of these functions related? In other words, for a particular x-value, how are the slopes of these graphs related?

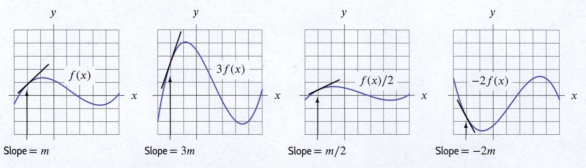

Figure 3.2: A function and its multiples: Derivative of multiple is multiple of derivative

Multiplying by a constant stretches or shrinks the graph (and reflects it across the x-axis if the constant is negative). This changes the slope of the curve at each point. If the graph has been stretched, the "rises" have all been increased by the same factor, whereas the "runs" remain the same. Thus, the slopes are all steeper by the same factor. If the graph has been shrunk, the slopes are all smaller by the same factor. If the graph has been reflected across the x-axis, the slopes will all have their signs reversed. Thus, if a function is multiplied by a constant, c, so is its derivative:

Derivative of a Constant Multiple

If c is a constant,

$$\frac{d}{dx}[cf(x)] = cf'(x).$$

Derivatives of Sums and Differences

Values of two functions, $f(x)$ and $g(x)$, and their sum $f(x) + g(x)$ are listed in Table 3.1.

Table 3.1   *Sum of functions*

x	$f(x)$	$g(x)$	$f(x) + g(x)$
0	100	0	100
1	110	0.2	110.2
2	130	0.4	130.4
3	160	0.6	160.6
4	200	0.8	200.8

We see that adding the increments of $f(x)$ and the increments of $g(x)$ gives the increments of $f(x) + g(x)$. For example, as x increases from 0 to 1, $f(x)$ increases by 10 and $g(x)$ increases by 0.2, while $f(x) + g(x)$ increases by $110.2 - 100 = 10.2$. Similarly, as x increases from 3 to 4, $f(x)$ increases by 40 and $g(x)$ by 0.2, while $f(x) + g(x)$ increases by $200.8 - 160.6 = 40.2$.

This example suggests that the rate at which $f(x) + g(x)$ is increasing is the sum of the rates at which $f(x)$ and $g(x)$ are increasing. Similar reasoning applies to the difference, $f(x) - g(x)$. In terms of derivatives:

Derivative of Sum and Difference

$$\frac{d}{dx}[f(x) + g(x)] = f'(x) + g'(x) \qquad \text{and} \qquad \frac{d}{dx}[f(x) - g(x)] = f'(x) - g'(x).$$

Powers of x

We start by looking at $f(x) = x^2$ and $g(x) = x^3$. We show in the Focus on Theory section at the end of this chapter that

$$f'(x) = \frac{d}{dx}\left(x^2\right) = 2x \quad \text{and} \quad g'(x) = \frac{d}{dx}\left(x^3\right) = 3x^2.$$

The graphs of $f(x) = x^2$ and $g(x) = x^3$ and their derivatives are shown in Figures 3.3 and 3.4. Notice $f'(x) = 2x$ has the behavior we expect. It is negative for $x < 0$ (when f is decreasing), zero for $x = 0$, and positive for $x > 0$ (when f is increasing). Similarly, $g'(x) = 3x^2$ is zero when $x = 0$, but positive everywhere else, as g is increasing everywhere else. These examples are special cases of the power rule.

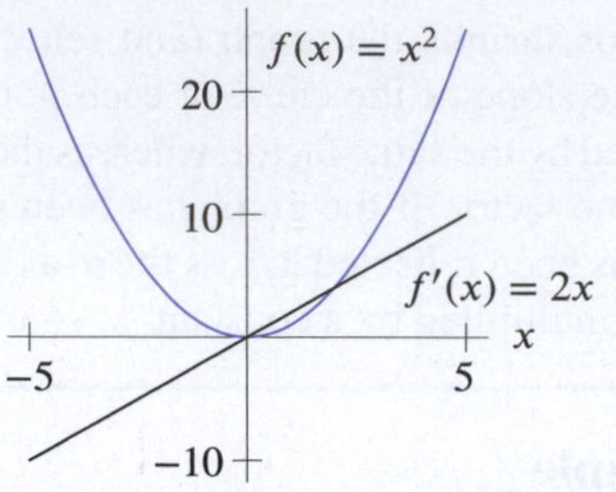

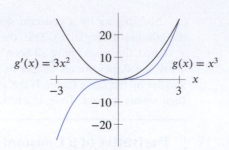

Figure 3.3: Graphs of $f(x) = x^2$ and its derivative $f'(x) = 2x$

Figure 3.4: Graphs of $g(x) = x^3$ and its derivative $g'(x) = 3x^2$

The Power Rule

For any constant real number n,

$$\frac{d}{dx}(x^n) = nx^{n-1}.$$

Example 1   Find the derivative of (a) $h(x) = x^8$ (b) $P(t) = t^7$.

Solution   (a) $h'(x) = 8x^7$.   (b) $P'(t) = 7t^6$.

We can use the power rule to differentiate negative and fractional powers.

Example 2   Use the power rule to differentiate (a) $\dfrac{1}{x^3}$ (b) $\sqrt{x}$ (c) $2t^{4.5}$.

Solution   (a) For $n = -3$: $\dfrac{d}{dx}\left(\dfrac{1}{x^3}\right) = \dfrac{d}{dx}(x^{-3}) = -3x^{-3-1} = -3x^{-4} = -\dfrac{3}{x^4}$.

(b) For $n = 1/2$: $\dfrac{d}{dx}\left(\sqrt{x}\right) = \dfrac{d}{dx}\left(x^{1/2}\right) = \dfrac{1}{2}x^{(1/2)-1} = \dfrac{1}{2}x^{-1/2} = \dfrac{1}{2\sqrt{x}}$.

(c) For $n = 4.5$: $\dfrac{d}{dt}\left(2t^{4.5}\right) = 2\left(4.5t^{4.5-1}\right) = 9t^{3.5}$.

Derivatives of Polynomials

Using the derivatives of powers, constant multiples, and sums, we can differentiate any polynomial.

Example 3   Differentiate:

(a) $A(t) = 3t^5$

(b) $r(p) = p^5 + p^3$

(c) $f(x) = 5x^2 - 7x^3$

(d) $g(t) = \dfrac{t^2}{4} + 3$

Solution   (a) Using the constant multiple rule: $A'(t) = \dfrac{d}{dt}(3t^5) = 3\dfrac{d}{dt}(t^5) = 3 \cdot 5t^4 = 15t^4$.

(b) Using the sum rule: $r'(p) = \dfrac{d}{dp}(p^5 + p^3) = \dfrac{d}{dp}(p^5) + \dfrac{d}{dp}(p^3) = 5p^4 + 3p^2$.

(c) Using all three rules together:

$$f'(x) = \frac{d}{dx}(5x^2 - 7x^3) = \frac{d}{dx}(5x^2) - \frac{d}{dx}(7x^3) \quad \text{Derivative of difference}$$

$$= 5\frac{d}{dx}(x^2) - 7\frac{d}{dx}(x^3) \quad \text{Derivative of multiple}$$

$$= 5(2x) - 7(3x^2) = 10x - 21x^2. \quad \text{Power rule}$$

(d) Using both rules:

$$g'(t) = \frac{d}{dt}\left(\frac{t^2}{4} + 3\right) = \frac{1}{4}\frac{d}{dt}(t^2) + \frac{d}{dt}(3)$$

$$= \frac{1}{4}(2t) + 0 = \frac{t}{2}. \quad \text{Since the derivative of a constant, } \frac{d}{dt}(3), \text{ is zero}$$

Using the Derivative Formulas

Since the slope of the tangent line to a curve is given by the derivative, we use differentiation to find the equation of the tangent line.

Example 4   Find an equation for the tangent line at $x = 1$ to the graph of

$$y = x^3 + 2x^2 - 5x + 7.$$

Sketch the graph of the curve and its tangent line on the same axes.

Solution   Differentiating gives

$$\frac{dy}{dx} = 3x^2 + 2(2x) - 5(1) + 0 = 3x^2 + 4x - 5,$$

so the slope of the tangent line at $x = 1$ is

$$m = \frac{dy}{dx}\bigg|_{x=1} = 3(1)^2 + 4(1) - 5 = 2.$$

When $x = 1$, we have $y = 1^3 + 2(1^2) - 5(1) + 7 = 5$, so the point $(1, 5)$ lies on the tangent line. Using the formula $y - y_0 = m(x - x_0)$ gives

$$y - 5 = 2(x - 1)$$
$$y = 3 + 2x.$$

The equation of the tangent line is $y = 3 + 2x$. See Figure 3.5.

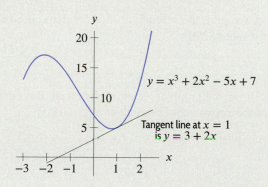

Figure 3.5: Find the equation for this tangent line

Example 5   Find and interpret the second derivatives of (a) $f(x) = x^2$ (b) $g(x) = x^3$.

Solution   (a) Differentiating $f(x) = x^2$ gives $f'(x) = 2x$, so $f''(x) = \dfrac{d}{dx}(2x) = 2$. Since f'' is always positive, the graph of f is concave up, as expected for a parabola opening upward. (See Figure 3.6.)

(b) Differentiating $g(x) = x^3$ gives $g'(x) = 3x^2$, so $g''(x) = \dfrac{d}{dx}(3x^2) = 3\dfrac{d}{dx}(x^2) = 3 \cdot 2x = 6x$.
Since $6x$ is positive for $x > 0$ and negative for $x < 0$, the graph of $g(x) = x^3$ is concave up for $x > 0$ and concave down for $x < 0$. (See Figure 3.7.)

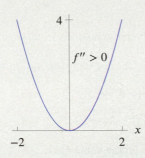

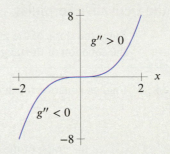

Figure 3.6: Graph of $f(x) = x^2$ with $f''(x) = 2$ Figure 3.7: Graph of $g(x) = x^3$ with $g''(x) = 6x$

Example 6   The revenue (in dollars) from producing q units of a product is given by

$$R(q) = 1000q - 3q^2.$$

Find $R(125)$ and $R'(125)$. Give units and interpret your answers.

Solution   We have
$$R(125) = 1000 \cdot 125 - 3 \cdot 125^2 = 78{,}125 \text{ dollars.}$$
Since $R'(q) = 1000 - 6q$, we have
$$R'(125) = 1000 - 6 \cdot 125 = 250 \text{ dollars per unit.}$$

If 125 units are sold, the revenue is 78,125 dollars. If an additional unit is then sold, the revenue increases by about $250.

Example 7   Figure 3.8 shows the graph of a cubic polynomial. Both graphically and algebraically, describe the behavior of the derivative of this cubic.

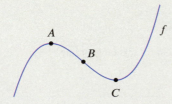

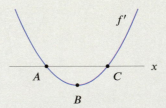

Figure 3.8: The cubic of Example 7   Figure 3.9: Derivative of the cubic of Example 7

Solution	Graphical approach: Suppose we move along the curve from left to right. To the left of A, the slope is positive; it starts very positive and decreases until the curve reaches A, where the slope is 0. Between A and C the slope is negative. Between A and B the slope is decreasing (getting more negative); it is most negative at B. Between B and C the slope is negative but increasing; at C the slope is zero. From C to the right, the slope is positive and increasing. The graph of the derivative function is shown in Figure 3.9.

Algebraic approach: f is a cubic that goes to $+\infty$ as $x \to +\infty$, so

$$f(x) = ax^3 + bx^2 + cx + d$$

with $a > 0$. Hence,

$$f'(x) = 3ax^2 + 2bx + c,$$

whose graph is a parabola opening upward because the coefficient of x^2 is positive, as in Figure 3.9.

Problems for Section 3.1

■ For Problems 1–38, find the derivative. Assume a, b, c, k are constants.

1. $y = 3x$

2. $y = 5$

3. $y = x^{-12}$

4. $y = x^{12}$

5. $y = 8t^3$

6. $y = x^{4/3}$

7. $y = 5x + 13$

8. $y = 3t^4 - 2t^2$

9. $f(q) = q^3 + 10$

10. $f(x) = \dfrac{1}{x^4}$

11. $y = 6x^3 + 4x^2 - 2x$

12. $y = x^2 + 5x + 9$

13. $y = 8t^3 - 4t^2 + 12t - 3$

14. $y = 3x^2 + 7x - 9$

15. $y = -3x^4 - 4x^3 - 6x$

16. $y = 4.2q^2 - 0.5q + 11.27$

17. $f(z) = -\dfrac{1}{z^{6.1}}$

18. $g(t) = \dfrac{1}{t^5}$

19. $y = \sqrt{x}$

20. $y = \dfrac{1}{r^{7/2}}$

21. $f(x) = \sqrt{\dfrac{1}{x^3}}$

22. $h(\theta) = \dfrac{1}{\sqrt[3]{\theta}}$

23. $z = (t-1)(t+1)$

24. $R = (s^2 + 1)^2$

25. $y = z^2 + \dfrac{1}{2z}$

26. $y = 3t^5 - 5\sqrt{t} + \dfrac{7}{t}$

27. $h(t) = \dfrac{3}{t} + \dfrac{4}{t^2}$

28. $y = 3t^2 + \dfrac{12}{\sqrt{t}} - \dfrac{1}{t^2}$

29. $h(\theta) = \theta(\theta^{-1/2} - \theta^{-2})$

30. $y = \sqrt{x}(x+1)$

31. $y = ax^2 + bx + c$

32. $f(x) = kx^2$

33. $v = at^2 + \dfrac{b}{t^2}$

34. $Q = aP^2 + bP^3$

35. $V = \frac{4}{3}\pi r^2 b$

36. $P = a + b\sqrt{t}$

37. $h(x) = \dfrac{ax + b}{c}$

38. $w = 3ab^2 q$

39. (a) Use a graph of $P(q) = 6q - q^2$ to determine whether each of the following derivatives is positive, negative, or zero: $P'(1)$, $P'(3)$, $P'(4)$. Explain.
    (b) Find $P'(q)$ and the three derivatives in part (a).

40. Let $f(x) = x^3 - 4x^2 + 7x - 11$. Find $f'(0)$, $f'(2)$, $f'(-1)$.

41. Let $f(t) = t^2 - 4t + 5$.
    (a) Find $f'(t)$.
    (b) Find $f'(1)$ and $f'(2)$.
    (c) Use a graph of $f(t)$ to check that your answers to part (b) are reasonable. Explain.

42. Find the rate of change of a population of size $P(t) = t^3 + 4t + 1$ at time $t = 2$.

43. With t in years since 2016, the height of a sand dune (in centimeters) is $f(t) = 700 - 3t^2$. Find $f(5)$ and $f'(5)$. Using units, explain what each means in terms of the sand dune.

■ In Problems 44–45, find the relative rate of change $f'(t)/f(t)$ at the given value of t. Assume t is in years and give your answer as a percent.

44. $f(t) = 3t + 2$; $t = 5$

45. $f(t) = 2t^3 + 10$; $t = 4$

■ In Problems 46–47, use the tangent line approximation.

46. Given $f(x) = x^4 - x^2 + 3$ approximate $f(1.04)$.

47. Given $f(x) = x^3 + x^2 - 6$, approximate $f(0.97)$.

48. The number, N, of acres of harvested land in a region is given by

$$N = f(t) = 120\sqrt{t},$$

where t is the number of years since farming began in the region. Find $f(9)$, $f'(9)$, and the relative rate of change f'/f at $t = 9$. Interpret your answers in terms of harvested land.

49. Zebra mussels are freshwater shellfish that first appeared in the St. Lawrence River in the early 1980s and have spread throughout the Great Lakes. Suppose that t months after they appeared in a small bay, the number of zebra mussels is given by $Z(t) = 300t^2$. How many zebra mussels are in the bay after four months? At what rate is the population growing at that time? Give units.

50. The quantity, Q, in tons, of material at a municipal waste site is a function of the number of years since 2000, with

$$Q = f(t) = 3t^2 + 100.$$

Find $f(10)$, $f'(10)$, and the relative rate of change f'/f at $t = 10$. Interpret your answers in terms of waste.

51. If $f(t) = 2t^3 - 4t^2 + 3t - 1$, find $f'(t)$ and $f''(t)$.

52. If $f(t) = t^4 - 3t^2 + 5t$, find $f'(t)$ and $f''(t)$.

■ For the functions in Problems 53–56:

  (a) Find the derivative at $x = -1$.

  (b) Find the second derivative at $x = -1$.

  (c) Use your answers to parts (a) and (b) to match the function to one of the graphs in Figure 3.10, each of which is shown centered on the point $(-1, -1)$.

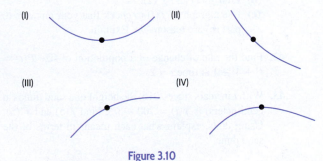

Figure 3.10

53. $k(x) = x^3 - x - 1$

54. $f(x) = 2x^3 + 3x^2 - 2$

55. $g(x) = x^4 - x^2 - 2x - 3$

56. $h(x) = 2x^4 + 8x^3 + 15x^2 + 14x + 4$

57. (a) Use Figure 3.11 to rank the quantities $f'(-1)$, $f'(0)$, $f'(1)$, $f'(4)$ from smallest to largest.

(b) Confirm your answer by calculating the quantities using the formula, $f(x) = x^3 - 3x^2 + 2x + 10$.

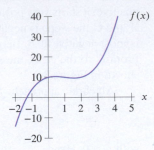

Figure 3.11

58. Find the equation of the line tangent to the graph of $f(x) = 2x^3 - 5x^2 + 3x - 5$ at $x = 1$.

59. For $f(x) = x^2 + 5x + 2$, find the equation of the tangent line at

  (a) $x = 1$ (b) $x = -1$

60. Find the equation of the line tangent to the graph of $f(t) = 6t - t^2$ at $t = 4$. Sketch the graph of $f(t)$ and the tangent line on the same axes.

61. (a) Find the equation of the tangent line to $f(x) = x^3$ at the point where $x = 2$.

  (b) Graph the tangent line and the function on the same axes. If the tangent line is used to estimate values of the function near $x = 2$, will the estimates be overestimates or underestimates?

62. For a certain type of fish, the weight y, in grams, is related to the length x, in cm, by $y = f(x) = 0.009x^3$.

  (a) Find $f'(x)$.

  (b) Assume $f(A) = B$ and $f'(C) = D$. For each of A, B, C, and D, which of the following are the correct units?

  grams,   cm,   cm/gram,   grams/cm

63. A jökulhlaup is the rapid draining of a glacial lake when an ice dam bursts. The maximum outflow rate, Q (in m^3/sec), during a jökulhlaup is given[1] in terms of its volume, v (in km^3), before the dam-break by $Q = 7700v^{0.67}$.

  (a) Find $\dfrac{dQ}{dv}$.

  (b) Evaluate $\dfrac{dQ}{dv}\Big|_{v=0.1}$. Include units. What does this derivative mean for glacial lakes?

64. The time, T, in seconds for one complete oscillation of a pendulum is given by $T = f(L) = 1.111\sqrt{L}$, where L is the length of the pendulum in feet. Find the following quantities, with units, and interpret in terms of the pendulum.

  (a) $f(100)$ (b) $f'(100)$.

---

[1]J. J. Clague and W. H. Mathews, "The Magnitude of Jökulhlaups." *Journal of Glaciology*, 12, no. 66, pp. 501–504.

65. Kleiber's Law states that the daily calorie requirement, $C(w)$, of a mammal is proportional to the mammal's body weight w raised to the 0.75 power.[2] If body weight is measured in pounds, the constant of proportionality is approximately 42.

    (a) Give formulas for $C(w)$ and $C'(w)$.
    (b) Find and interpret
        (i) $C(10)$ and $C'(10)$
        (ii) $C(100)$ and $C'(100)$
        (iii) $C(1000)$ and $C'(1000)$

66. If you are outdoors, the wind may make it feel a lot colder than the thermometer reads. You feel the wind-chill temperature, which, if the air temperature is 20°F, is given in °F by $W(v) = 48.17 - 27.2v^{0.16}$, where v is the wind velocity in mph for $5 \leq v \leq 60$.[3]

    (a) If the air temperature is 20°F, and the wind is blow-ing at 40 mph, what is the windchill temperature, to the nearest degree?
    (b) Find $W'(40)$, and explain what this means in terms of windchill.

67. Circulation time is the average amount of time it takes for all the blood in the body to circulate once and return to the heart. For a mammal with mass m kilograms, the circulation time C, in seconds, is

$$C = f(m) = 17.4m^{0.25}.$$

    (a) Find $f'(m)$.
    (b) For a human with mass 70 kg, find and interpret $f(70)$ and $f'(70)$.

68. An island of A square kilometers[4] in the Florida Keys has N species of beetles, given by

$$N = f(A) = 8.3A^{0.25}.$$

    (a) Find $f'(A)$.
    (b) At 40 km^2, Key Largo is the largest island in the Keys. Find $f(40)$ and $f'(40)$.
    (c) Interpret $f(40)$ in terms of beetles and Key Largo.
    (d) Global warming is causing the oceans to rise, re-ducing land mass. Interpret $f'(40)$ in terms of Key Largo, shrinking land mass, and beetles.

69. (a) Use the formula for the area of a circle of radius r, $A = \pi r^2$, to find dA/dr.
    (b) The result from part (a) should look familiar. What does dA/dr represent geometrically?
    (c) Use the difference quotient to explain the observa-tion you made in part (b).

70. Suppose W is proportional to r^3. The derivative dW/dr is proportional to what power of r?

71. Show that for any power function $f(x) = x^n$, we have $f'(1) = n$.

72. The cost to produce q items is $C(q) = 1000 + 2q^2$ dol-lars. Find the marginal cost of producing the 25th item. Interpret your answer in terms of costs.

73. The demand curve for a product is given by $q = 300 - 3p$, where p is the price of the product and q is the quan-tity that consumers buy at this price.

    (a) Write the revenue as a function, $R(p)$, of price.
    (b) Find $R'(10)$ and interpret your answer in terms of revenue.
    (c) For what prices is $R'(p)$ positive? For what prices is it negative?

74. A ball is dropped from the top of the Empire State Building. The height, y, of the ball above the ground (in feet) is given as a function of time, t (in seconds), by

$$y = 1250 - 16t^2.$$

    (a) Find the velocity of the ball at time t. What is the sign of the velocity? Why is this to be expected?
    (b) When does the ball hit the ground, and how fast is it going at that time? Give your answer in feet per second and in miles per hour (1 ft/sec = 15/22 mph).

75. The yield, Y, of an apple orchard (measured in bushels of apples per acre) is a function of the amount x of fer-tilizer in pounds used per acre. Suppose

$$Y = f(x) = 320 + 140x - 10x^2.$$

    (a) What is the yield if 5 pounds of fertilizer is used per acre?
    (b) Find $f'(5)$. Give units with your answer and inter-pret it in terms of apples and fertilizer.
    (c) Given your answer to part (b), should more or less fertilizer be used? Explain.

76. The demand for a product is given, for $p, q \geq 0$, by

$$p = f(q) = 50 - 0.03q^2.$$

    (a) Find the p- and q-intercepts for this function and interpret them in terms of demand for this product.
    (b) Find $f(20)$ and give units with your answer. Ex-plain what it tells you in terms of demand.
    (c) Find $f'(20)$ and give units with your answer. Ex-plain what it tells you in terms of demand.

77. The cost (in dollars) of producing q items is given by $C(q) = 0.08q^3 + 75q + 1000$.

    (a) Find the marginal cost function.
    (b) Find $C(50)$ and $C'(50)$. Give units with your an-swers and explain what each is telling you about costs of production.

---

[2] S. Strogatz, "Math and the City," *The New York Times*, May 20, 2009.
[3] en.wikipedia.org/wiki/Wind_chill, accessed December 13, 2016.
[4] J. Brown and S. B. Peck, "The Long-horned Beetle of South Florida," *Canadian Journal of Zoology*, 74: 2154–2169, 1996.

78. Let $f(x) = x^3 - 6x^2 - 15x + 20$. Find $f'(x)$ and all values of x for which $f'(x) = 0$. Explain the relationship between these values of x and the graph of $f(x)$.

79. If the demand curve is a line, we can write $p = b + mq$,

where p is the price of the product, q is the quantity sold at that price, and b and m are constants.

(a) Write the revenue as a function of quantity sold.

(b) Find the marginal revenue function.

3.2 EXPONENTIAL AND LOGARITHMIC FUNCTIONS

The Exponential Function

What do we expect the graph of the derivative of the exponential function $f(x) = a^x$ to look like? The graph of an exponential function with $a > 1$ is shown in Figure 3.12. The function increases slowly for $x < 0$ and more rapidly for $x > 0$, so the values of f' are small for $x < 0$ and larger for $x > 0$. Since the function is increasing for all values of x, the graph of the derivative must lie above the x-axis. In fact, the graph of f' resembles the graph of f itself. We will see how this observation holds for $f(x) = 2^x$ and $g(x) = 3^x$.

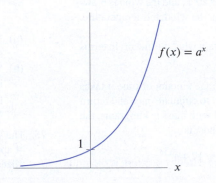

Figure 3.12: $f(x) = a^x$, with $a > 1$

The Derivatives of 2^x and 3^x

In Section 2.1, we estimated the derivative of $f(x) = 2^x$ at $x = 0$:

$$f'(0) \approx 0.693.$$

By estimating the derivative at other values of x, we obtain the graph in Figure 3.13. Since the graph of f' looks like the graph of f shrunk vertically, we assume that f' is a multiple of f. Since $f'(0) \approx 0.693 = 0.693 \cdot 1 = 0.693 f(0)$, the multiplier is approximately 0.693, which suggests that

$$\frac{d}{dx}(2^x) = f'(x) \approx (0.693)2^x.$$

Similarly, in Figure 3.14, the derivative of $g(x) = 3^x$ is a multiple of g, with multiplier $g'(0) \approx 1.099$. So

$$\frac{d}{dx}(3^x) = g'(x) \approx (1.099)3^x.$$

The Derivative of a^x and the Number e

The calculation of the derivative of $f(x) = a^x$, for $a > 0$, is similar to that of 2^x and 3^x. The derivative is again proportional to the original function. When $a = 2$, the constant of proportionality (0.693) is less than 1, and the derivative is smaller than the original function. When $a = 3$, the constant of proportionality (1.099) is greater than 1, and the derivative is greater than the original function. Is there an in-between case, when derivative and function are exactly equal? In other words:

Is there a value of a that makes $\dfrac{d}{dx}(a^x) = a^x$?

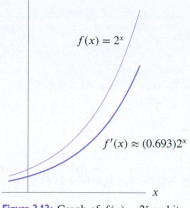

Figure 3.13: Graph of $f(x) = 2^x$ and its derivative

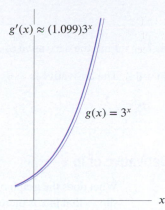

Figure 3.14: Graph of $g(x) = 3^x$ and its derivative

The answer is yes: the value is $a \approx 2.718 \ldots$, the number e introduced in Chapter 1. This means that the function e^x is its own derivative:

$$\frac{d}{dx}(e^x) = e^x.$$

It turns out that the constants involved in the derivatives of 2^x and 3^x are natural logarithms. In fact, since $0.693 \approx \ln 2$ and $1.099 \approx \ln 3$, we (correctly) guess that

$$\frac{d}{dx}(2^x) = (\ln 2)2^x \quad \text{and} \quad \frac{d}{dx}(3^x) = (\ln 3)3^x.$$

In the Focus on Theory section at the end of this chapter, we show that, in general:

The Exponential Rule

For any positive constant a,

$$\frac{d}{dx}(a^x) = (\ln a)a^x.$$

Since $\ln a$ is a constant, the derivative of a^x is proportional to a^x. Many quantities have rates of change that are proportional to themselves; for example, the simplest model of population growth has this property. The fact that the constant of proportionality is 1 when $a = e$ makes e a particularly useful base for exponential functions.

Example 1 Differentiate $2 \cdot 3^x + 5e^x$.

Solution We have $\dfrac{d}{dx}(2 \cdot 3^x + 5e^x) = 2\dfrac{d}{dx}(3^x) + 5\dfrac{d}{dx}(e^x) = 2\ln 3 \cdot 3^x + 5e^x.$

The Derivative of e^{kt}

Since functions of the form e^{kt} where k is a constant are often useful, we calculate the derivative of e^{kt}. In Section 3.3 on the chain rule, we see that if k is a constant,

$$\frac{d}{dt}(e^{kt}) = ke^{kt}.$$

Example 2   Find the derivative of $P = 5 + 3x^2 - 7e^{-0.2x}$.

Solution   The derivative is

$$\frac{dP}{dx} = 0 + 3(2x) - 7(-0.2e^{-0.2x}) = 6x + 1.4e^{-0.2x}.$$

The Derivative of ln x

What does the graph of the derivative of the logarithmic function $f(x) = \ln x$ look like? Figure 3.15 shows that $\ln x$ is increasing, so its derivative is positive. The graph of $f(x) = \ln x$ is concave down, so the derivative is decreasing. Furthermore, the slope of $f(x) = \ln x$ is very large near $x = 0$ and very small for large x, so the derivative tends to $+\infty$ for x near 0 and tends to 0 for very large x. See Figure 3.16. It turns out that

$$\boxed{\frac{d}{dx}(\ln x) = \frac{1}{x}.}$$

We give an algebraic justification for this rule in the Focus on Theory section.

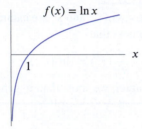

Figure 3.15: Graph of $f(x) = \ln x$

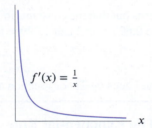

Figure 3.16: Graph of the derivative of $f(x) = \ln x$

Example 3   Differentiate $y = 5 \ln t + 7e^t - 4t^2 + 12$.

Solution   We have

$$\frac{d}{dt}(5 \ln t + 7e^t - 4t^2 + 12) = 5\frac{d}{dt}(\ln t) + 7\frac{d}{dt}(e^t) - 4\frac{d}{dt}(t^2) + \frac{d}{dt}(12)$$

$$= 5\left(\frac{1}{t}\right) + 7(e^t) - 4(2t) + 0$$

$$= \frac{5}{t} + 7e^t - 8t.$$

Using the Derivative Formulas

Example 4   With t is in years since the start of 2000, the population of Nevada, P, in millions, can be approximated by

$$P = 2.020(1.036)^t.$$

At what rate was the population growing at the beginning of 2009? Give units with your answer.

Solution   The instantaneous rate of growth is the derivative, so we want dP/dt when $t = 9$. We have:

$$\frac{dP}{dt} = \frac{d}{dt}(2.020(1.036)^t) = 2.020(\ln 1.036)(1.036)^t = 0.0714(1.036)^t.$$

Substituting $t = 9$ gives

$$\frac{dP}{dt} = 0.0714(1.036)^9 = 0.0982.$$

The population of Nevada was growing at a rate of about 0.0982 million, or 98,200, people per year at the start of 2009.

Example 5   Find the equation of the tangent line to the graph of $f(x) = \ln x$ at the point where $x = 2$. Draw a graph with $f(x)$ and the tangent line on the same axes.

Solution   Since $f'(x) = 1/x$, the slope of the tangent line at $x = 2$ is $f'(2) = 1/2 = 0.5$. When $x = 2$, $y = \ln 2 = 0.693$, so a point on the tangent line is $(2, 0.693)$. Substituting into the equation for a line, we have:

$$y - 0.693 = 0.5(x - 2)$$
$$y = -0.307 + 0.5x.$$

The equation of the tangent line is $y = -0.307 + 0.5x$. See Figure 3.17.

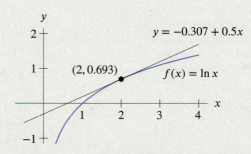

Figure 3.17: Graph of $f(x) = \ln x$ and a tangent line

Example 6   Suppose $1000 is deposited into a bank account that pays 8% annual interest, compounded continuously.

(a) Find a formula $f(t)$ for the balance t years after the initial deposit.
(b) Find $f(10)$ and $f'(10)$ and explain what your answers mean in terms of money.

Solution   (a) The balance is $f(t) = 1000e^{0.08t}$.
(b) Substituting $t = 10$ gives

$$f(10) = 1000e^{(0.08)(10)} = 2225.54.$$

This means that the balance is $2225.54 after 10 years.
    To find $f'(10)$, we compute $f'(t) = 1000(0.08e^{0.08t}) = 80e^{0.08t}$. Therefore,

$$f'(10) = 80e^{(0.08)(10)} = 178.04.$$

This means that after 10 years, the balance is growing at the rate of about $178 per year.

Problems for Section 3.2

■ Differentiate the functions in Problems 1–28. Assume that A, B, and C are constants.

1. $P = 3t^3 + 2e^t$

2. $f(x) = 2e^x + x^2$

3. $f(x) = x^3 + 3^x$

4. $y = 5t^2 + 4e^t$

5. $y = 5 \cdot 5^t + 6 \cdot 6^t$

6. $y = 2^x + \dfrac{2}{x^3}$

7. $y = 4 \cdot 10^x - x^3$

8. $f(x) = 2^x + 2 \cdot 3^x$

9. $y = 5 \cdot 2^x - 5x + 4$

10. $y = 3x - 2 \cdot 4^x$

11. $y = e^{0.7t}$

12. $f(t) = e^{3t}$

13. $P = e^{-0.2t}$

14. $y = e^{-4t}$

15. $P = 200e^{0.12t}$

16. $P = 50e^{-0.6t}$

17. $P(t) = 12.41(0.94)^t$

18. $P(t) = 3000(1.02)^t$

19. $y = B + Ae^t$

20. $P(t) = Ce^t$.

21. $y = 10^x + \dfrac{10}{x}$

22. $f(x) = Ae^x - Bx^2 + C$

23. $D = 10 - \ln p$

24. $R = 3 \ln q$

25. $R(q) = q^2 - 2 \ln q$

26. $y = t^2 + 5 \ln t$

27. $f(t) = Ae^t + B \ln t$

28. $y = x^2 + 4x - 3 \ln x$

■ In Problems 29–34, find the relative rate of change, $f'(t)/f(t)$, of the function $f(t)$.

29. $f(t) = 15t + 12$

30. $f(t) = 10t + 5$

31. $f(t) = 30e^{-7t}$

32. $f(t) = 8e^{5t}$

33. $f(t) = 35t^{-4}$

34. $f(t) = 6t^2$

35. For $f(t) = 4 - 2e^t$, find $f'(-1)$, $f'(0)$, and $f'(1)$. Graph $f(t)$, and draw tangent lines at $t = -1$, $t = 0$, and $t = 1$. Do the slopes of the lines match the derivatives you found?

36. (a) Use Figure 3.18 to rank the quantities $f'(1), f'(2), f'(3)$ from smallest to largest.
(b) Confirm your answer by calculating the quantities using the formula, $f(x) = 2e^x - 3x^2 \sqrt{x}$.

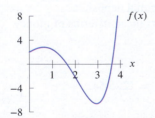

Figure 3.18

37. Find the equation of the tangent line to the graph of $y = 3^x$ at $x = 1$. Check your work by sketching a graph of the function and the tangent line on the same axes.

38. Find the equation of the tangent line to $y = e^{-2t}$ at $t = 0$. Check by sketching the graphs of $y = e^{-2t}$ and the tangent line on the same axes.

39. Find the equation of the tangent line to $f(x) = 10e^{-0.2x}$ at $x = 4$.

40. A fish population is approximated by $P(t) = 10e^{0.6t}$, where t is in months. Calculate and use units to explain what each of the following tells us about the population:

(a) $P(12)$ **(b)** $P'(12)$

41. If \$1000 is deposited in a bank account that pays 3% interest compounded continuously, the balance B after t years is
$$B = f(t) = 1000e^{0.03t}.$$
(a) Find $f'(t)$.
(b) Find $f(10)$ and $f'(10)$. Give units.

42. The amount, A mg, of caffeine in the body t hours after drinking a cup of coffee can be approximated by
$$A = f(t) = 120e^{-0.17t}.$$
(a) Find $f'(t)$.
(b) Find $f(2)$ and $f'(2)$. Give units.

43. The world's population[5] is about $f(t) = 7.17e^{0.011t}$ billion, where t is time in years since July 2014. Find $f(0)$, $f'(0)$, $f(10)$, and $f'(10)$. Using units, interpret your answers in terms of population.

44. The demand curve for a product is given by
$$q = f(p) = 10{,}000e^{-0.25p},$$
where q is the quantity sold and p is the price of the product, in dollars. Find $f(2)$ and $f'(2)$. Explain in economic terms what information each of these answers gives you.

45. The value of an automobile purchased in 2014 can be approximated by the function $V(t) = 30(0.85)^t$, where t is the time, in years, from the date of purchase, and $V(t)$ is the value, in thousands of dollars.

(a) Evaluate and interpret $V(4)$, including units.
(b) Find an expression for $V'(t)$, including units.
(c) Evaluate and interpret $V'(4)$, including units.
(d) Use $V(t)$, $V'(t)$, and any other considerations you think are relevant to write a paragraph in support of or in opposition to the following statement: "From a monetary point of view, it is best to keep this vehicle as long as possible."

[5]www.indexmundi.com, accessed April 9, 2015.

46. A new DVD is available for sale in a store one week after its release. The cumulative revenue, R, from sales of the DVD in this store in week t after its release is

$$R = f(t) = 350 \ln t \quad \text{with} \quad t > 1.$$

Find $f(5)$, $f'(5)$, and the relative rate of change f'/f at $t = 5$. Interpret your answers in terms of revenue.

47. The population of Hungary[6] has been modeled by

$$P = 9.919(0.998)^t,$$

where P is in millions and t is years since July 2014.

   (a) What does this model predict for the population of Hungary in July 2020?
   (b) How fast (in people/year) does this model predict Hungary's population will be decreasing in 2020?

48. With t in years since January 1, 2010, the population P of Slim Chance is predicted by

$$P = 35,000(0.98)^t.$$

At what rate will the population be changing on January 1, 2023?

49. With a yearly inflation rate of 2%, prices are given by

$$P = P_0(1.02)^t,$$

where P_0 is the price in dollars when $t = 0$ and t is time in years. Suppose $P_0 = 1$. How fast (in cents/year) are prices rising when $t = 10$?

50. Find the value of c in Figure 3.19, where the line l tangent to the graph of $y = 2^x$ at $(0, 1)$ intersects the x-axis.

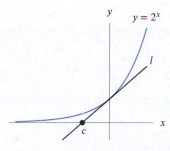

Figure 3.19

51. At a time t hours after it was administered, the concentration of a drug in the body is $f(t) = 27e^{-0.14t}$ ng/ml. What is the concentration 4 hours after it was administered? At what rate is the concentration changing at that time?

52. The cost of producing a quantity, q, of a product is given by

$$C(q) = 1000 + 30e^{0.05q} \quad \text{dollars.}$$

Find the cost and the marginal cost when $q = 50$. Interpret these answers in economic terms.

53. Carbon-14 is a radioactive isotope used to date objects. If A_0 represents the initial amount of carbon-14 in the object, then the quantity remaining at time t, in years, is

$$A(t) = A_0 e^{-0.000121t}.$$

   (a) A tree, originally containing 185 micrograms of carbon-14, is now 500 years old. At what rate is the carbon-14 decaying now?
   (b) In 1988, scientists found that the Shroud of Turin, which was reputed to be the burial cloth of Jesus, contained 91% of the amount of carbon-14 in freshly made cloth of the same material.[7] According to this data, how old was the Shroud of Turin in 1988?

54. For the cost function $C = 1000 + 300 \ln q$ (in dollars), find the cost and the marginal cost at a production level of 500. Interpret your answers in economic terms.

55. In 2012, the population of Mexico was 115 million and growing 1.09% annually, while the population of the US was 314 million and growing 0.9% annually.[8]

   (a) Find the Mexican growth rate in people/year in 2012.
   (b) Find the US growth rate, measured the same way, and use it determine which population was growing faster in 2012.

56. In 2009, the population, P, of India was 1.166 billion and growing at 1.5% annually.[9]

   (a) Give a formula for P in terms of time, t, measured in years since 2009.
   (b) Find $\dfrac{dP}{dt}$, $\dfrac{dP}{dt}\Big|_{t=0}$, and $\dfrac{dP}{dt}\Big|_{t=25}$. What do each of these represent in practical terms?

57. **(a)** Find the equation of the tangent line to $y = \ln x$ at $x = 1$.
   (b) Use it to calculate approximate values for $\ln(1.1)$ and $\ln(2)$.
   (c) Using a graph, explain whether the approximate values are smaller or larger than the true values. Would the same result have held if you had used the tangent line to estimate $\ln(0.9)$ and $\ln(0.5)$? Why?

[6]www.cia.gov, accessed April 10, 2015.
[7]*The New York Times,* October 18, 1988.
[8]www.indexmundi.com, accessed April 9, 2015.
[9]https://www.cia.gov/library/publications/the-world-factbook/print/in.html, accessed April 14, 2009.

58. With t in years since the start of 2014, worldwide annual extraction of copper is $17.9(1.025)^t$ million tonnes.[10]

    (a) How fast is the annual extraction changing at time t? Give units.

    (b) How fast is the annual extraction changing at the start of 2025?

    (c) Suppose annual extraction changes at the rate found in part (b) for the five years 2025–2030. By how much does the annual extraction change over this period?

    (d) Is your answer to part (c) larger or smaller than the change in annual extraction predicted by the model $17.9(1.025)^t$?

59. Food bank usage in Britain has grown dramatically over the past decade. The number of users, in thousands, of the largest bank is estimated[11] to be $N(t) = 1.3(2.25)^t$, where t is the number of years since 2006.

    (a) At what rate is the number of food bank users changing at time t? Give units.

    (b) Does this rate of change increase or decrease with time?

60. Find the quadratic polynomial $g(x) = ax^2 + bx + c$ which best fits the function $f(x) = e^x$ at $x = 0$, in the sense that

$$g(0) = f(0), \text{ and } g'(0) = f'(0), \text{ and } g''(0) = f''(0).$$

Using a computer or calculator, sketch graphs of f and g on the same axes. What do you notice?

3.3 THE CHAIN RULE

We now see how to differentiate composite functions such as $f(t) = \ln(3t)$ and $g(x) = e^{-x^2}$.

The Derivative of a Composition of Functions

Suppose $y = f(z)$ with $z = g(t)$ for some inside function g and outside function f, where f and g are differentiable. A small change in t, called Δt, generates a small change in z, called Δz. In turn, Δz generates a small change in y, called Δy. Provided Δt and Δz are not zero, we can say

$$\frac{\Delta y}{\Delta t} = \frac{\Delta y}{\Delta z} \cdot \frac{\Delta z}{\Delta t}.$$

Since the derivative $\dfrac{dy}{dt}$ is the limit of the quotient $\dfrac{\Delta y}{\Delta t}$ as Δt gets smaller and smaller, this suggests

> ### The Chain Rule
>
> If $y = f(z)$ and $z = g(t)$ are differentiable, then the derivative of $y = f(g(t))$ is given by
>
> $$\frac{dy}{dt} = \frac{dy}{dz} \cdot \frac{dz}{dt}.$$
>
> In words, the derivative of a composite function is the derivative of the outside function times the derivative of the inside function:
>
> $$\frac{d}{dt}(f(g(t))) = f'(g(t)) \cdot g'(t).$$

The following example shows us how to interpret the chain rule in practical terms.

Example 1

The amount of gas, G, in gallons, consumed by a car depends on the distance traveled, s, in miles, and s depends on the time, t, in hours. If 0.05 gallons of gas are consumed for each mile traveled, and the car is traveling at 30 miles/hr, how fast is gas being consumed? Give units.

---

[10]Data from http://minerals.usgs.gov/minerals/pubs/commodity/ Accessed February 8, 2015.

[11]Estimates for the Trussell Trust http://www.bbc.com/news/education-30346060 and http://www.trusselltrust.org/stats. Accessed November 2015.

Solution   We expect the rate of gas consumption to be in gallons/hr. We are told that

$$\text{Rate gas is consumed with respect to distance} = \frac{dG}{ds} = 0.05 \text{ gallons/mile}$$

$$\text{Rate distance is increasing with respect to time} = \frac{ds}{dt} = 30 \text{ miles/hr.}$$

We want to calculate the rate at which gas is being consumed with respect to time, or dG/dt. We think of G as a function of s, and s as a function of t. By the chain rule we know that

$$\frac{dG}{dt} = \frac{dG}{ds} \cdot \frac{ds}{dt} = \left(0.05 \frac{\text{gallons}}{\text{mile}}\right) \cdot \left(30 \frac{\text{miles}}{\text{hour}}\right) = 1.5 \text{ gallons/hour.}$$

Thus, gas is being consumed at a rate of 1.5 gallons/hour.

The Chain Rule for Functions Given by Formulas

In order to use the chain rule to differentiate a composite function, we first rewrite the function using a new variable z to represent the inside function:

$$y = (t+1)^4 \quad \text{is the same as} \quad y = z^4 \quad \text{where} \quad z = t+1.$$

Example 2   Use a new variable z for the inside function to express each of the following as a composite function:
(a) $y = \ln(3t)$ (b) $P = e^{-0.03t}$ (c) $w = 5(2r+3)^2$.

Solution   (a) The inside function is $3t$, so we have $y = \ln z$ with $z = 3t$.
(b) The inside function is $-0.03t$, so we have $P = e^z$ with $z = -0.03t$.
(c) The inside function is $2r+3$, so we have $w = 5z^2$ with $z = 2r+3$.

Example 3   Find derivatives of the following functions: (a) $y = (4t^2+1)^7$ (b) $P = e^{kt}$ for constant k.

Solution   (a) Here $z = 4t^2+1$ is the inside function; $y = z^7$ is the outside function. Since $dy/dz = 7z^6$ and $dz/dt = 8t$, we have

$$\frac{dy}{dt} = \frac{dy}{dz} \cdot \frac{dz}{dt} = (7z^6)(8t) = 7(4t^2+1)^6(8t) = 56t(4t^2+1)^6.$$

(b) Let $z = kt$ and $P = e^z$. Then $dP/dz = e^z$ and, since k is constant, $dz/dt = k$, so

$$\frac{dP}{dt} = \frac{dP}{dz} \cdot \frac{dz}{dt} = e^z \cdot k = e^{kt} \cdot k = ke^{kt}.$$

Notice that the derivative formula for e^{kt} introduced in Section 3.2 and justified in Example 3(b) is just a special case of the chain rule.

The derivative rules give us

$$\frac{d}{dt}(t^n) = nt^{n-1} \qquad \frac{d}{dt}(e^t) = e^t \qquad \frac{d}{dt}(\ln t) = \frac{1}{t}.$$

Using the chain rule in addition, we have the following results.

If z is a differentiable function of t, then

$$\frac{d}{dt}(z^n) = nz^{n-1}\frac{dz}{dt}, \qquad \frac{d}{dt}(e^z) = e^z\frac{dz}{dt}, \qquad \frac{d}{dt}(\ln z) = \frac{1}{z}\frac{dz}{dt}.$$

Example 4   Differentiate (a) $(3t^3 - t)^5$ (b) $\ln(q^2 + 1)$ (c) e^{-x^2}.

Solution   (a) Let $z = 3t^3 - t$, giving

$$\frac{d}{dt}(3t^3 - t)^5 = \frac{d}{dt}(z^5) = 5z^4\frac{dz}{dt} = 5(3t^3 - t)^4(9t^2 - 1).$$

(b) We have $z = q^2 + 1$, so

$$\frac{d}{dq}(\ln(q^2 + 1)) = \frac{d}{dq}(\ln z) = \frac{1}{z}\frac{dz}{dq} = \frac{1}{q^2 + 1}(2q).$$

(c) Taking $z = -x^2$, the derivative is

$$\frac{d}{dx}(e^{-x^2}) = \frac{d}{dx}(e^z) = e^z\frac{dz}{dx} = e^{-x^2}(-2x) = -2xe^{-x^2}.$$

As we see in the following example, it is often faster to use the chain rule without introducing the new variable, z.

Example 5   Differentiate
(a) $(x^2 + 4)^3$ (b) $5\ln(2t^2 + 3)$ (c) $\sqrt{1 + 2e^{5t}}$

Solution   (a) We have

$$\frac{d}{dx}\left((x^2 + 4)^3\right) = 3(x^2 + 4)^2 \cdot \frac{d}{dx}\left(x^2 + 4\right)$$
$$= 3(x^2 + 4)^2 \cdot 2x$$
$$= 6x(x^2 + 4)^2.$$

(b) We have

$$\frac{d}{dt}\left(5\ln(2t^2 + 3)\right) = 5 \cdot \frac{1}{2t^2 + 3} \cdot \frac{d}{dt}\left(2t^2 + 3\right)$$
$$= 5 \cdot \frac{1}{2t^2 + 3} \cdot 4t$$
$$= \frac{20t}{2t^2 + 3}.$$

(c) Here we use the chain rule twice, giving

$$\frac{d}{dt}\left((1 + 2e^{5t})^{1/2}\right) = \frac{1}{2}(1 + 2e^{5t})^{-1/2} \cdot \frac{d}{dt}\left(1 + 2e^{5t}\right)$$
$$= \frac{1}{2}(1 + 2e^{5t})^{-1/2} \cdot 2e^{5t} \cdot \frac{d}{dt}\left(5t\right)$$
$$= \frac{1}{2}(1 + 2e^{5t})^{-1/2} \cdot 2e^{5t} \cdot 5$$
$$= \frac{5e^{5t}}{\sqrt{1 + 2e^{5t}}}.$$

Example 6   Let $h(x) = f(g(x))$ and $k(x) = g(f(x))$. Evaluate   (a) $h'(1)$ (b) $k'(1)$ given that

$$f(1) = 1, \quad f'(0) = 0, \quad f'(1) = 3, \quad g(1) = 0, \quad g'(1) = 1.$$

Solution   (a) The chain rule tells us that $h'(x) = f'(g(x)) \cdot g'(x)$, so

$$
\begin{aligned}
h'(1) &= f'(g(1)) \cdot g'(1) \\
&= f'(0) \cdot g'(1) \\
&= 0 \cdot 1 \\
&= 0.
\end{aligned}
$$

(b) The chain rule tells us that $k'(x) = g'(f(x)) \cdot f'(x)$, so

$$
\begin{aligned}
k'(1) &= g'(f(1)) \cdot f'(1) \\
&= g'(1) \cdot f'(1) \\
&= 1 \cdot 3 \\
&= 3.
\end{aligned}
$$

Example 7   Let $h(x) = f(g(x))$ and $k(x) = g(f(x))$. Use Figure 3.20 to estimate: (a) $h'(1)$ (b) $k'(2)$

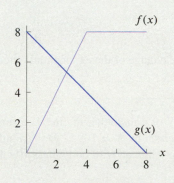

Figure 3.20: Graphs of f and g for Example 7

Solution   (a) The chain rule tells us that $h'(x) = f'(g(x)) \cdot g'(x)$, so

$$
\begin{aligned}
h'(1) &= f'(g(1)) \cdot g'(1) \\
&= f'(7) \cdot g'(1) \\
&= 0 \cdot (-1) \\
&= 0.
\end{aligned}
$$

We use the slopes of the lines in Figure 3.20 to find the derivatives $f'(7) = 0$ and $g'(1) = -1$.
(b) The chain rule tells us that $k'(x) = g'(f(x)) \cdot f'(x)$, so

$$
\begin{aligned}
k'(2) &= g'(f(2)) \cdot f'(2) \\
&= g'(4) \cdot f'(2) \\
&= (-1) \cdot 2 \\
&= -2.
\end{aligned}
$$

We use slopes to compute the derivatives $g'(4) = -1$ and $f'(2) = 2$.

Relative Rates and Logarithms

In Section 2.3 we defined the relative rate of change of a function $z = f(t)$ to be

$$\text{Relative rate of change} = \frac{f'(t)}{f(t)} = \frac{1}{z}\frac{dz}{dt}.$$

Since

$$\frac{d}{dt}(\ln z) = \frac{1}{z}\frac{dz}{dt},$$

we have the following result:

For any positive function $f(t)$,

$$\begin{array}{c}\text{Relative rate of change}\\ \text{of } f(t)\end{array} = \frac{d}{dt}(\ln f(t)).$$

Just as linear functions have constant rates of change, in the following example we see that exponential functions have constant relative rates of change.

Example 8   Find the relative rate of change of the exponential function $z = P_0 e^{kt}$.

Solution   Since

$$\ln z = \ln(P_0 e^{kt}) = \ln P_0 + \ln(e^{kt}) = \ln P_0 + kt,$$

we have

$$\frac{d}{dt}(\ln z) = k.$$

The relative rate of change of the exponential function $P_0 e^{kt}$ is the constant k.

Example 9   The surface area S of a mammal, in cm^2, is a function of the body mass, M, of the mammal, in kilograms, and is given by $S = 1095 \cdot M^{2/3}$. Find the relative rate of change of S with respect to M and evaluate for a human with body mass 70 kilograms. Interpret your answer.

Solution   We have

$$\ln S = \ln(1095 \cdot M^{2/3}) = \ln(1095) + \ln(M^{2/3}) = \ln(1095) + \frac{2}{3}\ln M.$$

Thus,

$$\begin{aligned}\text{Relative rate of change} &= \frac{d}{dM}(\ln S)\\ &= \frac{d}{dM}\left(\ln(1095) + \frac{2}{3}\ln M\right)\\ &= \frac{2}{3}\cdot\frac{1}{M}.\end{aligned}$$

For a human with body mass $M = 70$ kilograms, we have

$$\text{Relative rate} = \frac{2}{3}\frac{1}{70} = 0.0095 = 0.95\% \text{ per kg.}$$

The surface area of a human with body mass 70 kilograms increases by about 0.95% if body mass increases by 1 kilogram.

Problems for Section 3.3

■ Find the derivative of the functions in Problems 1–27.

1. $f(x) = (x + 1)^{99}$
2. $g(x) = (4x^2 + 1)^7$
3. $w = (t^2 + 1)^{100}$
4. $R = (q^2 + 1)^4$
5. $w = (5r - 6)^3$
6. $f(x) = (x^3 + x^2)^{-90}$
7. $y = 12 - 3x^2 + 2e^{3x}$
8. $y = \sqrt{s^3 + 1}$
9. $f(x) = 6e^{5x} + e^{-x^2}$
10. $C = 12(3q^2 - 5)^3$
11. $< w = e^{-3t^2}$
12. $y = 5e^{5t+1}$
13. $y = \ln(5t + 1)$
14. $w = e^{\sqrt{s}}$
15. $f(t) = \ln(t^2 + 1)$
16. $f(x) = \ln(1 - x)$
17. $f(x) = \ln(e^x + 1)$
18. $f(x) = \ln(1 - e^{-x})$
19. $f(x) = \ln(\ln x)$
20. $f(x) = (\ln x)^3$
21. $y = 5 + \ln(3t + 2)$
22. $y = (5 + e^x)^2$
23. $y = 5x + \ln(x + 2)$
24. $y = \sqrt{e^x + 1}$
25. $P = (1 + \ln x)^{0.5}$
26. $f(\theta) = (e^\theta + e^{-\theta})^{-1}$
27. $f(x) = \sqrt{2 + \sqrt{x}}$

■ In Problems 28–29, find the relative rate of change $f'(t)/f(t)$ at the given value of t. Assume t is in years and give your answer as a percent.

28. $f(t) = 2e^{0.3t}$; $t = 7$
29. $f(t) = \ln(t^2 + 1)$; $t = 2$

■ In Problems 30–33, find the relative rate of change of $f(t)$ using the formula $\frac{d}{dt} \ln f(t)$.

30. $f(t) = 6.8e^{-0.5t}$
31. $f(t) = 5e^{1.5t}$
32. $f(t) = 4.5t^{-4}$
33. $f(t) = 3t^2$

34. For $f(x) = (2x - 3)^3$, find the equation of the tangent line at
    (a) $x = 0$
    (b) $x = 2$

35. Find the equation of the tangent line to $f(x) = (x - 1)^3$ at the point where $x = 2$.

36. If you invest P dollars in a bank account at an annual interest rate of $r\%$, then after t years you will have B dollars, where
$$B = P\left(1 + \frac{r}{100}\right)^t.$$
    (a) Find dB/dt, assuming P and r are constant. In terms of money, what does dB/dt represent?
    (b) Find dB/dr, assuming P and t are constant. In terms of money, what does dB/dr represent?

37. A company estimates that the total revenue, R, in dollars, received from the sale of q items is $R = \ln(1 + 1000q^2)$. Calculate and interpret the marginal revenue if $q = 10$.

38. For t in years since 2010, daily oil consumption in China, in thousands of barrels, was approximated by[12]
$$B = 8938e^{0.05t}.$$
    (a) Is daily oil consumption increasing or decreasing with time?
    (b) How fast is oil consumption changing at time t?

39. For $t \geq 0$ in minutes, the temperature, H, of a pot of soup in degrees Celsius is[13]
$$H = 5 + 95e^{-0.054t}.$$
    (a) Is the temperature increasing or decreasing with time?
    (b) How fast is the temperature changing at time t? Give units.

40. The distance, s, of a moving body from a fixed point is given as a function of time by $s = 20e^{t/2}$. Find the velocity, v, of the body as a function of t.

■ For Problems 41–44, let $h(x) = f(g(x))$ and $k(x) = g(f(x))$. Use Figure 3.21 to estimate the derivatives.

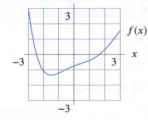
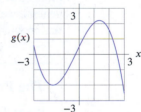

Figure 3.21

41. $h'(1)$
42. $k'(1)$
43. $h'(2)$
44. $k'(2)$

■ In Problems 45–50, use Figure 3.22 to evaluate the derivative.

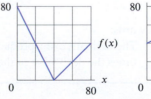

Figure 3.22

45. $\frac{d}{dx} f(g(x))|_{x=30}$
46. $\frac{d}{dx} f(g(x))|_{x=70}$
47. $\frac{d}{dx} g(f(x))|_{x=30}$
48. $\frac{d}{dx} g(f(x))|_{x=70}$
49. $\frac{d}{dx} f(g(x))|_{x=20}$
50. $\frac{d}{dx} g(f(x))|_{x=60}$

[12]Based on /www.eia.gov/cfapps/ipdbproject/ Accessed May 2015.
[13]Based on http://www.ugrad.math.ubc.ca/coursedoc/math100/notes/diffeqs/cool.html. Accessed May 2015.

51. Given $y = f(x)$ with $f(1) = 4$ and $f'(1) = 3$, find

    (a) $g'(1)$ if $g(x) = \sqrt{f(x)}$.

    (b) $h'(1)$ if $h(x) = f\left(\sqrt{x}\right)$.

52. Some economists suggest that an extra year of education increases a person's wages, on average, by about 14%. Assume you could make \$10 per hour with your current level of education and that inflation increases wages at a continuous rate of 3.5% per year.

    (a) If you had had four more years of education, how

much would you make per hour?

    (b) What is the difference between your wages in 20 years' time with and without the additional four years of education?

    (c) Is the difference you found in part (b) increasing with time? If so, at what rate? (Assume the number of additional years of education stays fixed at four.)

53. Show that if the graphs of $f(t)$ and $h(t) = Ae^{kt}$ are tangent at $t = a$, then k is the relative rate of change of f at $t = a$.

3.4   THE PRODUCT AND QUOTIENT RULES

This section shows how to find the derivatives of products and quotients of functions.

The Product Rule

Suppose we know the derivatives of $f(x)$ and $g(x)$ and want to calculate the derivative of the product, $f(x)g(x)$. We start by looking at an example. Let $f(x) = x$ and $g(x) = x^2$. Then

$$f(x)g(x) = x \cdot x^2 = x^3,$$

so the derivative of the product is $3x^2$. Notice that the derivative of the product is *not* equal to the product of the derivatives, since $f'(x) = 1$ and $g'(x) = 2x$, so $f'(x)g'(x) = (1)(2x) = 2x$. In general, we have the following rule, which is justified in the Focus on Theory section at the end of this chapter.

The Product Rule

If $u = f(x)$ and $v = g(x)$ are differentiable functions, then

$$(fg)' = f'g + fg'.$$

The product rule can also be written

$$\frac{d(uv)}{dx} = \frac{du}{dx} \cdot v + u \cdot \frac{dv}{dx}.$$

In words:
    The derivative of a product is the derivative of the first times the second, plus the first times the derivative of the second.

We check that this rule gives the correct answers for $f(x) = x$ and $g(x) = x^2$. The derivative of $f(x)g(x)$ is

$$f'(x)g(x) + f(x)g'(x) = 1(x^2) + x(2x) = x^2 + 2x^2 = 3x^2.$$

This is the answer we expect for the derivative of $f(x)g(x) = x \cdot x^2 = x^3$.

Example 1   Differentiate (a) $x^2 e^{2x}$ (b) $t^3 \ln(t + 1)$ (c) $(3x^2 + 5x)e^x$.

Solution   (a) Using the product rule, we have

$$\frac{d}{dx}(x^2 e^{2x}) = \frac{d}{dx}(x^2) \cdot e^{2x} + x^2 \frac{d}{dx}(e^{2x})$$
$$= (2x)e^{2x} + x^2(2e^{2x})$$
$$= 2xe^{2x} + 2x^2 e^{2x}.$$

(b) Differentiating using the product rule gives

$$\frac{d}{dt}(t^3 \ln(t+1)) = \frac{d}{dt}(t^3) \cdot \ln(t+1) + t^3 \frac{d}{dt}(\ln(t+1))$$
$$= (3t^2)\ln(t+1) + t^3\left(\frac{1}{t+1}\right)$$
$$= 3t^2 \ln(t+1) + \frac{t^3}{t+1}.$$

(c) The product rule gives

$$\frac{d}{dx}((3x^2 + 5x)e^x) = \left(\frac{d}{dx}(3x^2 + 5x)\right)e^x + (3x^2 + 5x)\frac{d}{dx}(e^x)$$
$$= (6x + 5)e^x + (3x^2 + 5x)e^x$$
$$= (3x^2 + 11x + 5)e^x.$$

Example 2   Find the derivative of $C = \dfrac{e^{2t}}{t}$.

Solution   We write $C = e^{2t}t^{-1}$ and use the product rule:

$$\frac{d}{dt}(e^{2t}t^{-1}) = \frac{d}{dt}(e^{2t}) \cdot t^{-1} + e^{2t}\frac{d}{dt}(t^{-1})$$
$$= (2e^{2t}) \cdot t^{-1} + e^{2t}(-1)t^{-2}$$
$$= \frac{2e^{2t}}{t} - \frac{e^{2t}}{t^2}.$$

Example 3   A demand curve for a product has the equation $p = 80e^{-0.003q}$, where p is price and q is quantity sold.

(a) Find the revenue as a function of quantity sold.
(b) Find the marginal revenue function.

Solution   (a) Since Revenue = Price × Quantity, we have $R = pq = (80e^{-0.003q})q = 80qe^{-0.003q}$.
(b) The marginal revenue function is the derivative of revenue with respect to quantity. The product rule gives

$$\text{Marginal Revenue} = \frac{d}{dq}(80qe^{-0.003q})$$
$$= \left(\frac{d}{dq}(80q)\right)e^{-0.003q} + 80q\left(\frac{d}{dq}(e^{-0.003q})\right)$$
$$= (80)e^{-0.003q} + 80q(-0.003e^{-0.003q})$$
$$= (80 - 0.24q)e^{-0.003q}.$$

The Quotient Rule

Suppose we want to differentiate a function of the form $Q(x) = f(x)/g(x)$. (Of course, we have to avoid points where $g(x) = 0$.) We want a formula for Q' in terms of f' and g'. We have the following rule, which is justified in the Focus on Theory section at the end of this chapter.

The Quotient Rule

If $u = f(x)$ and $v = g(x)$ are differentiable functions, then

$$\left(\frac{f}{g}\right)' = \frac{f'g - fg'}{g^2},$$

or equivalently,

$$\frac{d}{dx}\left(\frac{u}{v}\right) = \frac{\dfrac{du}{dx} \cdot v - u \cdot \dfrac{dv}{dx}}{v^2}.$$

In words:

>   The derivative of a quotient is the derivative of the numerator times the denominator minus the numerator times the derivative of the denominator, all over the denominator squared.

Example 4   Differentiate (a) $\dfrac{5x^2}{x^3 + 1}$ (b) $\dfrac{1}{1 + e^x}$ (c) $\dfrac{e^x}{x^2}$.

Solution   (a) Using the quotient rule, we have

$$\frac{d}{dx}\left(\frac{5x^2}{x^3 + 1}\right) = \frac{\left(\dfrac{d}{dx}(5x^2)\right)(x^3 + 1) - 5x^2\dfrac{d}{dx}(x^3 + 1)}{(x^3 + 1)^2} = \frac{10x(x^3 + 1) - 5x^2(3x^2)}{(x^3 + 1)^2}$$

$$= \frac{-5x^4 + 10x}{(x^3 + 1)^2}.$$

(b) Differentiating using the quotient rule yields

$$\frac{d}{dx}\left(\frac{1}{1 + e^x}\right) = \frac{\left(\dfrac{d}{dx}(1)\right)(1 + e^x) - 1\dfrac{d}{dx}(1 + e^x)}{(1 + e^x)^2} = \frac{0(1 + e^x) - 1(0 + e^x)}{(1 + e^x)^2}$$

$$= \frac{-e^x}{(1 + e^x)^2}.$$

(c) The quotient rule gives

$$\frac{d}{dx}\left(\frac{e^x}{x^2}\right) = \frac{\left(\dfrac{d}{dx}(e^x)\right)x^2 - e^x\left(\dfrac{d}{dx}(x^2)\right)}{(x^2)^2} = \frac{e^x x^2 - e^x(2x)}{x^4}$$

$$= e^x\left(\frac{x^2 - 2x}{x^4}\right) = e^x\left(\frac{x - 2}{x^3}\right).$$

Example 5   If $f(2) = 1$, $f'(2) = 5$, $g(2) = 3$ and $g'(2) = 6$, find

(a) $h'(2)$ if $h(x) = f(x)g(x)$

(b) $k'(2)$ if $k(x) = f(x)/g(x)$.

Solution   (a) Differentiating using the product rule yields

$$h'(x) = f'(x)g(x) + f(x)g'(x),$$

so

$$h'(2) = f'(2)g(2) + f(2)g'(2) = 5 \cdot 3 + 1 \cdot 6 = 21.$$

(b) Differentiating using the quotient rule yields

$$k'(x) = \frac{f'(x)g(x) - f(x)g'(x)}{(g(x))^2},$$

so

$$k'(2) = \frac{f'(2)g(2) - f(2)g'(2)}{(g(2))^2} = \frac{5 \cdot 3 - 1 \cdot 6}{3^2} = 1.$$

Example 6   Let $h(x) = f(x)g(x)$ and $k(x) = f(x)/g(x)$. Use Figure 3.23 to estimate: (a) $h'(2)$ (b) $k'(6)$

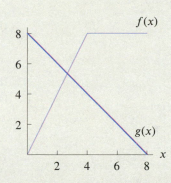

Figure 3.23: Graphs of f and g for Example 6

Solution   (a) The product rule tells us that $h'(x) = f'(x)g(x) + f(x)g'(x)$, so

$$h'(2) = f'(2)g(2) + f(2)g'(2)$$
$$= 2 \cdot 6 + 4 \cdot (-1)$$
$$= 8.$$

We use the slopes of the lines in Figure 3.23 to find the derivatives $f'(2) = 2$ and $g'(2) = -1$.

(b) The quotient rule tells us that $k'(x) = (f'(x)g(x) - f(x)g'(x))/(g(x))^2$, so

$$k'(6) = \frac{f'(6)g(6) - f(6)g'(6)}{(g(6))^2}$$
$$= \frac{0 \cdot 2 - 8 \cdot (-1)}{2^2}$$
$$= 2.$$

We use slopes to compute the derivatives $f'(6) = 0$ and $g'(6) = -1$.

Problems for Section 3.4

1. If $f(x) = x^2(x^3 + 5)$, find $f'(x)$ two ways: by using the product rule and by multiplying out before taking the derivative. Do you get the same result? Should you?

2. If $f(x) = (2x + 1)(3x - 2)$, find $f'(x)$ two ways: by using the product rule and by multiplying out. Do you get the same result?

■ For Problems **3–32**, find the derivative. Assume that a, b, c, and k are constants.

3. $f(t) = te^{-2t}$

4. $f(x) = xe^x$

5. $y = t^2(3t + 1)^3$

6. $y = 5xe^{x^2}$

7. $y = x \ln x$

8. $y = (t^2 + 3)e^t$

9. $y = (t^3 - 7t^2 + 1)e^t$

10. $z = (3t + 1)(5t + 2)$

11. $R = 3qe^{-q}$

12. $P = t^2 \ln t$

13. $f(x) = \dfrac{x^2 + 3}{x}$

14. $f(z) = \sqrt{z}e^{-z}$

15. $y = te^{-t^2}$

16. $f(t) = te^{5-2t}$

17. $g(p) = p \ln(2p + 1)$

18. $y = x \cdot 2^x$

19. $f(w) = (5w^2 + 3)e^{w^2}$

20. $w = (t^3 + 5t)(t^2 - 7t + 2)$

21. $z = (te^{3t} + e^{5t})^9$

22. $f(x) = \dfrac{x}{e^x}$

23. $z = \dfrac{1 - t}{1 + t}$

24. $w = \dfrac{3z}{1 + 2z}$

25. $w = \dfrac{3y + y^2}{5 + y}$

26. $y = \dfrac{e^x}{1 + e^x}$

27. $f(x) = \dfrac{ax + b}{cx + k}$

28. $y = \dfrac{1 + z}{\ln z}$

29. $f(x) = axe^{-bx}$

30. $f(x) = (ax^2 + b)^3$

31. $g(\alpha) = e^{\alpha e^{-2\alpha}}$

32. $f(t) = ae^{bt}$

33. If $f(x) = (3x + 8)(2x - 5)$, find $f'(x)$ and $f''(x)$.

34. Find the equation of the tangent line to the graph of $f(x) = \dfrac{2x - 5}{x + 1}$ at the point at which $x = 0$.

35. Find the equation of the tangent line to the graph of $f(x) = 5xe^x$ at the point at which $x = 0$.

36. Find the equation of the tangent line to the graph of $f(x) = x^2e^{-x}$ at $x = 0$. Check by graphing this function and the tangent line on the same axes.

37. Find the equation of the tangent line to the graph of $f(x) = x^3e^x$ at the point at which $x = 2$.

38. If $\dfrac{d}{dt}(tf(t)) = 1 + f(t)$, what is $f'(t)$?

39. The concentration, C in ng/ml, of nicotine in the body t minutes after starting to smoke a cigarette can be approximated by

$$C = f(t) = 4te^{-0.08t}.$$

  (a) Find $f'(t)$.
  (b) Find $f(15)$ and $f'(15)$. Give units.

40. The quantity of a drug, Q mg, present in the body t hours after an injection of the drug is given is

$$Q = f(t) = 100te^{-0.5t}.$$

Find $f(1)$, $f'(1)$, $f(5)$, and $f'(5)$. Give units and interpret the answers.

41. A drug concentration curve is given by $C = f(t) = 20te^{-0.04t}$, with C in mg/ml and t in minutes.

  (a) Graph C against t. Is $f'(15)$ positive or negative? Is $f'(45)$ positive or negative? Explain.
  (b) Find $f(30)$ and $f'(30)$ analytically. Interpret them in terms of the concentration of the drug in the body.

42. If p is price in dollars and q is quantity, demand for a product is given by

$$q = 5000e^{-0.08p}.$$

  (a) What quantity is sold at a price of \$10?
  (b) Find the derivative of demand with respect to price when the price is \$10 and interpret your answer in terms of demand.

43. The demand for a product is given in Problem 42. Find the revenue and the derivative of revenue with respect to price at a price of \$10. Interpret your answers in economic terms.

44. The quantity demanded of a certain product, q, is given in terms of p, the price, by

$$q = 1000e^{-0.02p}$$

  (a) Write revenue, R, as a function of price.
  (b) Find the rate of change of revenue with respect to price.
  (c) Find the revenue and rate of change of revenue with respect to price when the price is \$10. Interpret your answers in economic terms.

45. The quantity, q, of a skateboard sold depends on the selling price, p, in dollars, so we write $q = f(p)$. You are given that $f(140) = 15{,}000$ and $f'(140) = -100$.

   (a) What do $f(140) = 15{,}000$ and $f'(140) = -100$ tell you about the sales of skateboards?
   (b) The total revenue, R, earned by the sale of skateboards is given by $R = pq$. Find $\left.\dfrac{dR}{dp}\right|_{p=140}$.
   (c) What is the sign of $\left.\dfrac{dR}{dp}\right|_{p=140}$? If the skateboards are currently selling for \$140, what happens to revenue if the price is increased to \$141?

46. A patient's total cholesterol level, $T(t)$, and good cholesterol level, $G(t)$, at t weeks after January 1, 2016, are measured in milligrams per deciliter of blood (mg/dl). The cholesterol ratio, $R(t) = G(t)/T(t)$, is used to gauge the safety of a patient's cholesterol, with risk of cholesterol-related illnesses being minimized when $R(t) > 1/5$ (that is, good cholesterol is at least $1/5$ of total cholesterol).

   (a) Explain how it is possible for total cholesterol of the patient to increase but the cholesterol ratio to remain constant.
   (b) On January 1, the patient's total cholesterol level is 120 mg/dl and good cholesterol level is 30 mg/dl. Though $R > 1/5$, the doctor prefers that the patient's good cholesterol increase to 40 mg/dl, so prescribes a diet starting January 1 which increases good cholesterol by 1 mg/dl per week without changing the cholesterol ratio. What is the rate of change of total cholesterol the first week of the diet?

47. If C (in units of 10^{-4} molar) is the concentration of glucose in a solution, then $E.$ $coli$ bacteria in the solution grow at a rate, R (in cell divisions per hour), given by[14]

   $$R = \frac{1.35C}{0.22 + C} \text{ cell divisions per hour.}$$

   (a) Find the growth rate of bacteria growing in $2 \cdot 10^{-4}$ molar glucose solution. Include units.
   (b) Find dR/dC.
   (c) Find $dR/dC|_{C=2}$. Include units.
   (d) Find the tangent line approximation of the growth rate for bacteria growing in glucose concentrations near $2 \cdot 10^{-4}$ molar.

   (e) Use the tangent line approximation to estimate the growth rate in a $2.2 \cdot 10^{-4}$ molar glucose solution. Compare it with the growth rate from the original model.

48. Show that the relative rate of change of a product fg is the sum of the relative rates of change of f and g.

49. Show that the relative rate of change of a quotient f/g is the difference between the relative rates of change of f and g.

50. If $h = f^n$, show that
   $$\frac{(f^n)'}{f^n} = n\frac{f'}{f}.$$

51. For positive constants c and k, the *Monod growth curve* describes the growth of a population, P, as a function of the available quantity of a resource, r:
   $$P = \frac{cr}{k + r}.$$
   Find dP/dr and interpret it in terms of the growth of the population.

52. If a someone is lost in the wilderness, the search and rescue team identifies the boundaries of the search area and then uses probabilities to help optimize the chances of finding the person, assuming the subject is immobile. The probability, O, of the person being outside the search area after the search has begun and the person has not been found is given by
   $$O(E) = \frac{I}{1 - (1 - I)E},$$
   where I is the probability of the person being outside the search area at the start of the search and E is the search effort, a measure of how well the search area has been covered by the resources in the field.

   (a) If there was a 20% chance that the subject was not in the search area at the start of the search, and the search effort was 80%, what is the current probability of the person being outside the search area? (Probabilities are between 0 and 1, so 20% = 0.2 and 80% = 0.8.)
   (b) In practical terms, what does $I = 1$ mean? Is this realistic?
   (c) Evaluate $O'(E)$. Is it positive or negative? What does that tell you about O as E increases?

3.5 DERIVATIVES OF PERIODIC FUNCTIONS

Since the sine and cosine functions are periodic, their derivatives must be periodic also. (Why?) Let's look at the graph of $f(x) = \sin x$ in Figure 3.24 and estimate the derivative function graphically.

[14]Jacques Monod, "The Growth of Bacterial Cultures," *Annual Review of Microbiology,* 1949 vol. 3, pp. 371–394.

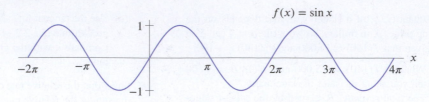

Figure 3.24: The sine function

First we might ask ourselves where the derivative is zero. (At $x = \pm\pi/2, \pm3\pi/2, \pm5\pi/2$, etc.) Then ask where the derivative is positive and where it is negative. (Positive for $-\pi/2 < x < \pi/2$; negative for $\pi/2 < x < 3\pi/2$, etc.) Since the largest positive slopes are at $x = 0, 2\pi$, and so on, and the largest negative slopes are at $x = \pi, 3\pi$, and so on, we get something like the graph in Figure 3.25.

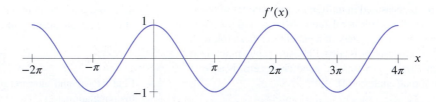

Figure 3.25: Derivative of $f(x) = \sin x$

The graph of the derivative in Figure 3.25 looks suspiciously like the graph of the cosine function. This might lead us to conjecture, quite correctly, that the derivative of the sine is the cosine. However, we cannot be sure just from the graphs.

One thing we can do is to check that the derivative function in Figure 3.25 has amplitude 1 (as it must if it is the cosine). That means we have to convince ourselves that the derivative of $f(x) = \sin x$ is 1 when $x = 0$. The next example suggests that this is true when x is in radians.

Example 1

Using a calculator, estimate the derivative of $f(x) = \sin x$ at $x = 0$. Make sure your calculator is set in radians.

Solution

We use the average rate of change of $\sin x$ on the small interval $0 \le x \le 0.01$ to compute

$$f'(0) \approx \frac{\sin(0.01) - \sin(0)}{0.01 - 0} = \frac{0.0099998 - 0}{0.01} = 0.99998 \approx 1.0.$$

The derivative of $f(x) = \sin x$ at $x = 0$ is approximately 1.0.

Warning: It is important to notice that in the previous example x was in *radians*; any conclusions we have drawn about the derivative of $\sin x$ are valid *only* when x is in radians.

Example 2

Starting with the graph of the cosine function, sketch a graph of its derivative.

Solution

The graph of $g(x) = \cos x$ is in Figure 3.26(a). Its derivative is 0 at $x = 0, \pm\pi, \pm2\pi$, and so on; it is positive for $-\pi < x < 0, \pi < x < 2\pi$, and so on, and it is negative for $0 < x < \pi, 2\pi < x < 3\pi$, and so on. The derivative is in Figure 3.26(b).

(a)

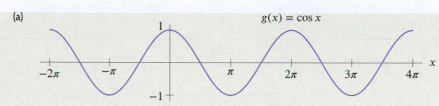

(b)

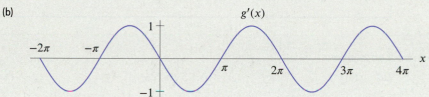

Figure 3.26: $g(x) = \cos x$ and its derivative, $g'(x)$

As we did with the sine, we'll use the graphs to make a conjecture. The derivative of the cosine in Figure 3.26(b) looks exactly like the graph of sine, except reflected across the x-axis. It turns out that the derivative of $\cos x$ is $-\sin x$.

> For x in radians,
>
> $$\frac{d}{dx}(\sin x) = \cos x \quad \text{and} \quad \frac{d}{dx}(\cos x) = -\sin x.$$

Example 3   Differentiate   (a) $5 \sin t - 8 \cos t$ (b) $5 - 3 \sin x + x^3$

Solution   (a) Differentiating gives

$$\frac{d}{dt}(5\sin t - 8\cos t) = 5\frac{d}{dt}(\sin t) - 8\frac{d}{dt}(\cos t) = 5(\cos t) - 8(-\sin t) = 5\cos t + 8\sin t.$$

(b) We have

$$\frac{d}{dx}(5 - 3\sin x + x^3) = \frac{d}{dx}(5) - 3\frac{d}{dx}(\sin x) + \frac{d}{dx}(x^3) = 0 - 3(\cos x) + 3x^2 = -3\cos x + 3x^2.$$

The chain rule tells us how to differentiate composite functions involving the sine and cosine. Suppose $y = \sin(3t)$, so $y = \sin z$ and $z = 3t$, so

$$\frac{dy}{dt} = \frac{dy}{dz} \cdot \frac{dz}{dt} = \cos z \frac{dz}{dt} = \cos(3t) \cdot 3 = 3\cos(3t).$$

In general,

> If z is a differentiable function of t, then
>
> $$\frac{d}{dt}(\sin z) = \cos z \frac{dz}{dt} \quad \text{and} \quad \frac{d}{dt}(\cos z) = -\sin z \frac{dz}{dt}.$$

In many applications, $z = kt$ for some constant k. Then we have:

If k is a constant, then

$$\frac{d}{dt}(\sin kt) = k \cos kt \quad \text{and} \quad \frac{d}{dt}(\cos kt) = -k \sin kt.$$

Example 4   Differentiate:

(a) $\sin(t^2)$　　　　　(b) $5\cos(2t)$　　　　　(c) $t \sin t$.

Solution   (a) We have $y = \sin z$ with $z = t^2$, so

$$\frac{d}{dt}(\sin(t^2)) = \frac{d}{dt}(\sin z) = \cos z \frac{dz}{dt} = \cos(t^2) \cdot 2t = 2t \cos(t^2).$$

(b) We have $y = 5\cos z$ with $z = 2t$, so

$$\frac{d}{dt}(5\cos(2t)) = 5\frac{d}{dt}(\cos(2t)) = 5(-2\sin(2t)) = -10\sin(2t).$$

(c) We use the product rule:

$$\frac{d}{dt}(t \sin t) = \frac{d}{dt}(t) \cdot \sin t + t\frac{d}{dt}(\sin t) = 1 \cdot \sin t + t(\cos t) = \sin t + t \cos t.$$

Problems for Section 3.5

■ Differentiate the functions in Problems **1–20**. Assume that A and B are constants.

1. $P = 3 + \cos t$
2. $y = 5 \sin x$
3. $y = B + A \sin t$
4. $y = t^2 + 5 \cos t$
5. $y = 5 \sin x - 5x + 4$
6. $R(q) = q^2 - 2 \cos q$
7. $R = \sin(5t)$
8. $f(x) = \sin(3x)$
9. $y = 2 \cos(5t)$
10. $W = 4 \cos(t^2)$
11. $y = A \sin(Bt)$
12. $y = \sin(x^2)$
13. $y = 6 \sin(2t) + \cos(4t)$
14. $z = \cos(4\theta)$
15. $f(x) = 2x \sin(3x)$
16. $f(x) = x^2 \cos x$
17. $z = \dfrac{e^{t^2} + t}{\sin(2t)}$
18. $f(\theta) = \theta^3 \cos \theta$
19. $f(\theta) = \dfrac{\sin \theta}{\theta}$
20. $f(t) = \dfrac{t^2}{\cos t}$

21. Find the equation of the tangent line to the graph of $y = \sin x$ at $x = \pi$. Graph the function and the tangent line on the same axes.

22. Find the line tangent to $f(x) = 3x + \cos(5x)$ at the point where $x = 0$.

23. Find the equations of the tangent lines to the graph of $f(x) = \sin x$ at $x = 0$ and at $x = \pi/3$. Use each tangent line to approximate $\sin(\pi/6)$. Would you expect these results to be equally accurate, since they are taken equally far away from $x = \pi/6$ but on opposite sides? If the accuracy is different, can you account for the difference?

24. Is the graph of $y = \sin(x^4)$ increasing or decreasing when $x = 10$? Is it concave up or concave down?

25. If t is the number of months since June, the number of bird species, N, found in an Ohio forest oscillates approximately according to the formula

$$N = f(t) = 19 + 9 \cos\left(\frac{\pi}{6}t\right).$$

(a) Graph $f(t)$ for $0 \le t \le 24$ and describe what it shows. Use the graph to decide whether $f'(1)$ and $f'(10)$ are positive or negative.
(b) Find $f'(t)$.
(c) Find and interpret $f(1)$, $f'(1)$, $f(10)$, and $f'(10)$.

26. The average adult takes about 12 breaths per minute. As a patient inhales, the volume of air in the lung increases. As the patient exhales, the volume of air in the lung decreases. For t in seconds since start of the breathing cycle, the volume of air inhaled or exhaled since $t = 0$ is given,[15] in hundreds of cubic centimeters, by

$$A(t) = -2\cos\left(\frac{2\pi}{5}t\right) + 2.$$

(a) How long is one breathing cycle?
(b) Find $A'(1)$ and explain what it means.

■ In Problems **27–30**, find and interpret the value of the expression in practical terms. Let $C(t)$ be the concentration of carbon dioxide in parts per million (ppm) in the air as a function of time, t, in months since December 1, 2005:[16]

$$C(t) = 3.5\sin\left(\frac{\pi t}{6}\right) + 381 + \frac{t}{6}.$$

27. $C'(36)$ **28.** $C'(60)$

29. $C'(30)$ **30.** $\dfrac{C(60) - C(0)}{60}$

31. The concentration of carbon dioxide[17] in the air, $C(t)$, in parts per million at time, t, in months, since December 1, 2005 is given by

$$C(t) = 3.5\sin\left(\frac{\pi t}{6}\right) + 381 + \frac{t}{6}.$$

(a) Is $C(t)$ periodic? How about $C'(t)$?
(b) Which of expressions (i)–(iii) best supports the claim that the concentration of carbon dioxide in the air went up between Dec. 1, 2006 and Dec. 1, 2010?

(i) $C'(60) - C'(12)$ (ii) $\dfrac{C'(60) - C'(12)}{48}$ (iii) $\dfrac{C(60) - C(12)}{48}$

32. Normal human body temperature fluctuates with a rhythm tied to our sleep cycle.[18] If $H(t)$ is body temperature in degrees Celsius at time t in hours since 9 am, then $H(t)$ may be modeled by

$$H(t) = 36.8 + 0.6\sin\left(\frac{\pi}{12}t\right).$$

(a) Calculate $H'(t)$ and give units.
(b) Calculate $H'(4)$ and $H'(12)$, then interpret the meaning of your answers in everyday terms.

33. A company's monthly sales, $S(t)$, are seasonal and given as a function of time, t, in months, by

$$S(t) = 2000 + 600\sin\left(\frac{\pi}{6}t\right).$$

(a) Graph $S(t)$ for $t = 0$ to $t = 12$. What is the maximum monthly sales? What is the minimum monthly sales? If $t = 0$ is January 1, when during the year are sales highest?
(b) Find $S(2)$ and $S'(2)$. Interpret in terms of sales.

34. A boat at anchor is bobbing up and down in the sea. The vertical distance, y, in feet, between the sea floor and the boat is given as a function of time, t, in minutes, by

$$y = 15 + \sin(2\pi t).$$

(a) Find the vertical velocity, v, of the boat at time t.
(b) Make rough sketches of y and v against t.

35. The depth of the water, y, in meters, in the Bay of Fundy, Canada, is given as a function of time, t, in hours after midnight, by the function

$$y = 10 + 7.5\cos(0.507t).$$

How quickly is the tide rising or falling (in meters/hour) at each of the following times?

(a) 6:00 am   **(b)** 9:00 am
(c) Noon   **(d)** 6:00 pm

36. Paris, France, has a latitude of approximately 49° N. If t is the number of days since the start of 2009, the number of hours of daylight in Paris can be approximated by

$$D(t) = 4\cos\left(\frac{2\pi}{365}(t - 172)\right) + 12.$$

(a) Find $D(40)$ and $D'(40)$. Explain what this tells about daylight in Paris.
(b) Find $D(172)$ and $D'(172)$. Explain what this tells about daylight in Paris.

37. On September 6th, 2017, there was a full moon. If t is the number of days since September 6th, the percent of moon illuminated can be represented by

$$H(t) = 50 + 50\cos\left(\frac{\pi}{15}t\right).$$

(a) Find $H'(t)$. Explain what this tells about the moon.
(b) For $0 \le t \le 30$, when is $H'(t) = 0$? What does this tell us about the moon?
(c) For $0 \le t \le 30$, when is $H'(t)$ negative? When is it positive? Explain what positive and negative values of $H'(t)$ tell us about the moon.

---

[15]Based upon information obtained from Dr. Gadi Avshalomov on August 14, 2008.
[16]Based on data from Mauna Loa, Hawaii, at esrl.noaa.gov/gmd/ccgg/trends/. Accessed March 2011.
[17]Based on data from Mauna Loa, Hawaii, at esrl.noaa.gov/gmd/ccgg/trends/. Accessed March 2011.
[18]Model based on data from circadian.org. Accessed January 2014.

PROJECTS FOR CHAPTER THREE

1. **Coroner's Rule of Thumb**

   Coroners estimate time of death using the rule of thumb that a body cools about 2°F during the first hour after death and about 1°F for each additional hour. Assuming an air temperature of 68°F and a living body temperature of 98.6°F, the temperature $T(t)$ in °F of a body at a time t hours since death is given by

   $$T(t) = 68 + 30.6e^{-kt}.$$

   (a) For what value of k will the body cool by 2°F in the first hour?

   (b) Using the value of k found in part (a), after how many hours will the temperature of the body be decreasing at a rate of 1°F per hour?

   (c) Using the value of k found in part (a), show that, 24 hours after death, the coroner's rule of thumb gives approximately the same temperature as the formula.

2. **Air Pressure and Altitude**

   Air pressure at sea level is 30 inches of mercury. At an altitude of h feet above sea level, the air pressure, P, in inches of mercury, is given by

   $$P = 30e^{-3.23 \times 10^{-5} h}$$

   (a) Sketch a graph of P against h.

   (b) Find the equation of the tangent line at $h = 0$.

   (c) A rule of thumb used by travelers is that air pressure drops about 1 inch for every 1000-foot increase in height above sea level. Write a formula for the air pressure given by this rule of thumb.

   (d) What is the relation between your answers to parts (b) and (c)? Explain why the rule of thumb works.

   (e) Are the predictions made by the rule of thumb too large or too small? Why?

3. **Relative Growth Rates: Population, GDP, and GDP per Capita**

   (a) Let Y be the world's annual production (GDP). The world GDP per capita is given by Y/P where P is world population. Figure 3.27 shows relative growth rate of GDP and GDP per capita for 1952–2000.[19]

   (i) Explain why the vertical distance between the two curves gives the relative rate of growth of the world population.

   (ii) Estimate the relative rate of population growth in 1970 and 2000.

   (b) In 2006 the relative rate of change of the GDP in developing countries was 4.5% per year, and the relative rate of change of the population was 1.2%. What was the relative rate of change of the per capita GDP in developing countries?

   (c) In 2006 the relative rate of change of the world's total production (GDP) was 3.8% per year, and the relative rate of change of the world's per capita production was 2.6% per year. What was the relative rate of change of the world population?

---

[19] Angus Maddison, *The World Economy: Historical Statistics*, OECD, 2003.

relative growth
rate per year

Figure 3.27

4. Keeling Curve: Atmospheric Carbon Dioxide

Since the 1950s, the carbon dioxide concentration in the air has been recorded at the Mauna Loa Observatory in Hawaii.[20] A graph of this data is called the Keeling Curve, after Charles Keeling, who started recording the data. With t in years since 1950, fitting functions to the data gives three models for the carbon dioxide concentration in parts per million (ppm):

$$f(t) = 303 + 1.3t$$
$$g(t) = 304e^{0.0038t}$$
$$h(t) = 0.0135t^2 + 0.5133t + 310.5.$$

(a) What family of function is used in each model?

(b) Find the rate of change of carbon dioxide in 2010 in each of the three models. Give units.

(c) Arrange the three models in increasing order of the rates of change they give for 2010. (Which model predicts the largest rate of change in 2010? Which predicts the smallest?)

(d) Consider the same three models for all positive time t. Will the ordering in part (c) remain the same for all t? If not, how will it change?

---

[20]www.esrl.noaa.gov/gmd/ccgg/, accessed March 10, 2013.

FOCUS ON THEORY

ESTABLISHING THE DERIVATIVE FORMULAS

The graph of $f(x) = x^2$ suggests that the derivative of x^2 is $f'(x) = 2x$. However, as we saw in the Focus on Theory section in Chapter 2, to be sure that this formula is correct, we have to use the definition:

$$f'(x) = \lim_{h \to 0} \frac{f(x+h) - f(x)}{h}.$$

As in Chapter 2, we simplify the difference quotient and then take the limit as h approaches zero.

Example 1   Confirm that the derivative of $g(x) = x^3$ is $g'(x) = 3x^2$.

Solution   Using the definition, we calculate $g'(x)$:

$$g'(x) = \lim_{h \to 0} \frac{g(x+h) - g(x)}{h} = \lim_{h \to 0} \frac{(x+h)^3 - x^3}{h}$$

$$\text{Multiplying out} \longrightarrow = \lim_{h \to 0} \frac{x^3 + 3x^2 h + 3xh^2 + h^3 - x^3}{h}$$

$$= \lim_{h \to 0} \frac{3x^2 h + 3xh^2 + h^3}{h}$$

$$\text{Simplifying} \longrightarrow = \lim_{h \to 0} (3x^2 + 3xh + h^2) = 3x^2.$$

$$\text{Looking at what happens as } h \to 0$$

So $g'(x) = \dfrac{d}{dx}(x^3) = 3x^2$.

Example 2   Give an informal justification that the derivative of $f(x) = e^x$ is $f'(x) = e^x$.

Solution   Using $f(x) = e^x$, we have

$$f'(x) = \lim_{h \to 0} \frac{f(x+h) - f(x)}{h} = \lim_{h \to 0} \frac{e^{x+h} - e^x}{h}$$

$$= \lim_{h \to 0} \frac{e^x e^h - e^x}{h} = \lim_{h \to 0} e^x \left(\frac{e^h - 1}{h} \right).$$

What is the limit of $\dfrac{e^h - 1}{h}$ as $h \to 0$? The graph of $\dfrac{e^h - 1}{h}$ in Figure 3.28 suggests that $\dfrac{e^h - 1}{h}$ approaches 1 as $h \to 0$. In fact, it can be proved that the limit equals 1, so

$$f'(x) = \lim_{h \to 0} e^x \left(\frac{e^h - 1}{h} \right) = e^x \cdot 1 = e^x.$$

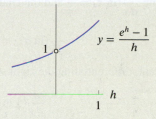

Figure 3.28: What is $\lim\limits_{h \to 0} \frac{e^h - 1}{h}$?

Example 3   Show that if $f(x) = 2x^2 + 1$, then $f'(x) = 4x$.

Solution   We use the definition of the derivative with $f(x) = 2x^2 + 1$:

$$f'(x) = \lim_{h \to 0} \frac{f(x+h) - f(x)}{h} = \lim_{h \to 0} \frac{(2(x+h)^2 + 1) - (2x^2 + 1)}{h}$$

$$= \lim_{h \to 0} \frac{2(x^2 + 2xh + h^2) + 1 - 2x^2 - 1}{h} = \lim_{h \to 0} \frac{2x^2 + 4xh + 2h^2 + 1 - 2x^2 - 1}{h}$$

$$= \lim_{h \to 0} \frac{4xh + 2h^2}{h} = \lim_{h \to 0} \frac{h(4x + 2h)}{h}.$$

To find the limit, look at what happens when h is close to 0, but $h \neq 0$. Simplifying, we have

$$f'(x) = \lim_{h \to 0} \frac{h(4x + 2h)}{h} = \lim_{h \to 0}(4x + 2h) = 4x$$

because as h gets close to 0, we know that $4x + 2h$ gets close to $4x$.

Using the Chain Rule to Establish Derivative Formulas

We use the chain rule to justify the formulas for derivatives of $\ln x$ and of a^x.

Derivative of $\ln x$

We'll differentiate an identity that involves $\ln x$. In Section 1.6, we have $e^{\ln x} = x$. Differentiating gives

$$\frac{d}{dx}(e^{\ln x}) = \frac{d}{dx}(x) = 1.$$

On the left side, since e^x is the outside function and $\ln x$ is the inside function, the chain rule gives

$$\frac{d}{dx}(e^{\ln x}) = e^{\ln x} \cdot \frac{d}{dx}(\ln x).$$

Thus, solving for $\frac{d}{dx}(\ln x)$, we have the result in Section 3.2,

$$\frac{d}{dx}(\ln x) = \frac{1}{e^{\ln x}} = \frac{1}{x}.$$

Derivative of a^x

Graphical arguments suggest that the derivative of a^x is proportional to a^x. Now we show that the constant of proportionality is $\ln a$. For $a > 0$, we use the identity from Section 1.6:

$$\ln(a^x) = x \ln a.$$

On the left side, using $\dfrac{d}{dx}(\ln x) = \dfrac{1}{x}$ and the chain rule gives

$$\frac{d}{dx}(\ln a^x) = \frac{1}{a^x} \cdot \frac{d}{dx}(a^x).$$

Since $\ln a$ is a constant, differentiating the right side gives

$$\frac{d}{dx}(x \ln a) = \ln a.$$

Since the two sides are equal, we have

$$\frac{1}{a^x}\frac{d}{dx}(a^x) = \ln a.$$

Solving for $\dfrac{d}{dx}(a^x)$ gives the result of Section 3.2. For $a > 0$,

$$\frac{d}{dx}(a^x) = (\ln a)a^x.$$

The Product Rule

Suppose we want to calculate the derivative of the product of differentiable functions, $f(x)g(x)$, using the definition of the derivative. Notice that in the second step below, we are adding and subtracting the same quantity: $f(x)g(x + h)$.

$$\frac{d[f(x)g(x)]}{dx} = \lim_{h \to 0} \frac{f(x + h)g(x + h) - f(x)g(x)}{h}$$

$$= \lim_{h \to 0} \frac{f(x + h)g(x + h) - f(x)g(x + h) + f(x)g(x + h) - f(x)g(x)}{h}$$

$$= \lim_{h \to 0} \left[\frac{f(x + h) - f(x)}{h} \cdot g(x + h) + f(x) \cdot \frac{g(x + h) - g(x)}{h} \right].$$

Taking the limit as $h \to 0$ gives the product rule:

$$(f(x)g(x))' = f'(x) \cdot g(x) + f(x) \cdot g'(x).$$

The Quotient Rule

Let $Q(x) = f(x)/g(x)$ be the quotient of differentiable functions. Assuming that $Q(x)$ is differentiable, we can use the product rule on $f(x) = Q(x)g(x)$:

$$f'(x) = Q'(x)g(x) + Q(x)g'(x).$$

Substituting for $Q(x)$ gives

$$f'(x) = Q'(x)g(x) + \frac{f(x)}{g(x)}g'(x).$$

Solving for $Q'(x)$ gives

$$Q'(x) = \frac{f'(x) - \dfrac{f(x)}{g(x)}g'(x)}{g(x)}.$$

Multiplying the top and bottom by $g(x)$ to simplify gives the quotient rule:

$$\left(\frac{f(x)}{g(x)} \right)' = \frac{f'(x)g(x) - f(x)g'(x)}{(g(x))^2}.$$

Problems on Establishing the Derivative Formulas

■ For Problems 1–7, use the definition of the derivative to obtain the following results.

1. If $f(x) = 2x + 1$, then $f'(x) = 2$.

2. If $f(x) = 5x^2$, then $f'(x) = 10x$.

3. If $f(x) = 2x^2 + 3$, then $f'(x) = 4x$.

4. If $f(x) = x^2 + x$, then $f'(x) = 2x + 1$.

5. If $f(x) = 4x^2 + 1$, then $f'(x) = 8x$.

6. If $f(x) = x^4$, then $f'(x) = 4x^3$. [Hint: $(x + h)^4 = x^4 + 4x^3h + 6x^2h^2 + 4xh^3 + h^4$.]

7. If $f(x) = x^5$, then $f'(x) = 5x^4$. [Hint: $(x + h)^5 = x^5 + 5x^4h + 10x^3h^2 + 10x^2h^3 + 5xh^4 + h^5$.]

8. (a) Use a graph of $g(h) = \dfrac{2^h - 1}{h}$ to explain why we believe that $\lim\limits_{h \to 0} \dfrac{2^h - 1}{h} \approx 0.6931$.

(b) Use the definition of the derivative and the result from part (a) to explain why, if $f(x) = 2^x$, we believe that $f'(x) \approx (0.6931)2^x$.

9. Use the definition of the derivative to show that if $f(x) = C$, where C is a constant, then $f'(x) = 0$.

10. Use the definition of the derivative to show that if $f(x) = b + mx$, for constants m and b, then $f'(x) = m$.

11. Use the definition of the derivative to show that if $f(x) = k \cdot u(x)$, where k is a constant and $u(x)$ is a function, then $f'(x) = k \cdot u'(x)$.

12. Use the definition of the derivative to show that if $f(x) = u(x) + v(x)$, for functions $u(x)$ and $v(x)$, then $f'(x) = u'(x) + v'(x)$.

FOCUS ON PRACTICE

■ Find derivatives for the functions in Problems 1–63. Assume $a, b, c,$ and k are constants.

1. $f(t) = t^2 + t^4$

2. $g(x) = 5x^4$

3. $y = 5x^3 + 7x^2 - 3x + 1$

4. $s(t) = 6t^{-2} + 3t^3 - 4t^{1/2}$

5. $f(x) = \dfrac{1}{x^2} + 5\sqrt{x} - 7$

6. $P(t) = 100e^{0.05t}$

7. $f(x) = 5e^{2x} - 2 \cdot 3^x$

8. $P(t) = 1{,}000(1.07)^t$

9. $D(p) = e^{p^2} + 5p^2$

10. $y = t^2 e^{5t}$

11. $y = x^2 \sqrt{x^2 + 1}$

12. $f(x) = \ln(x^2 + 1)$

13. $s(t) = 8\ln(2t + 1)$

14. $g(w) = w^2 \ln(w)$

15. $f(x) = 2^x + x^2 + 1$

16. $P(t) = \sqrt{t^2 + 4}$

17. $C(q) = (2q + 1)^3$

18. $g(x) = 5x(x + 3)^2$

19. $P(t) = be^{kt}$

20. $f(x) = ax^2 + bx + c$

21. $y = x^2 \ln(2x + 1)$

22. $f(t) = (e^t + 4)^3$

23. $f(x) = 5\sin(2x)$

24. $W(r) = r^2 \cos r$

25. $g(t) = 3\sin(5t) + 4$

26. $y = e^{3t} \sin(2t)$

27. $y = 2e^x + 3\sin x + 5$

28. $f(t) = 3t^2 - 4t + 1$

29. $y = 17x + 24x^{1/2}$

30. $g(x) = -\tfrac{1}{2}(x^5 + 2x - 9)$

31. $f(x) = 5x^4 + \dfrac{1}{x^2}$

32. $y = \dfrac{e^{2x}}{x^2 + 1}$

33. $f(x) = \dfrac{x^2 + 3x + 2}{x + 1}$

34. $y = \left(\dfrac{x^2 + 2}{3}\right)^2$

35. $g(x) = \sin(2 - 3x)$

36. $f(z) = \dfrac{z^2 + 1}{3z}$

37. $q(r) = \dfrac{3r}{5r + 2}$

38. $y = x\ln x - x + 2$

39. $j(x) = \ln(e^{ax} + b)$

40. $g(t) = \dfrac{t - 4}{t + 4}$

41. $h(w) = (w^4 - 2w)^5$

42. $h(w) = w^3 \ln(10w)$

43. $f(x) = \ln(\sin x + \cos x)$

44. $w(r) = \sqrt{r^4 + 1}$

45. $h(w) = -2w^{-3} + 3\sqrt{w}$

46. $h(x) = \sqrt{\dfrac{x^2 + 9}{x + 3}}$

47. $v(t) = t^2 e^{-ct}$

48. $f(x) = \dfrac{x}{1 + \ln x}$

49. $g(\theta) = e^{\sin\theta}$

50. $p(t) = e^{4t+2}$

51. $j(x) = \dfrac{x^3}{a} + \dfrac{a}{b}x^2 - cx$

52. $f(z) = \dfrac{z^2 + 1}{\sqrt{z}}$

53. $h(r) = \dfrac{r^2}{2r + 1}$

54. $g(x) = 2x - \dfrac{1}{\sqrt[3]{x}} + 3^x - e$

55. $f(t) = 2te^t - \dfrac{1}{\sqrt{t}}$

56. $w = \dfrac{5 - 3z}{5 + 3z}$

57. $f(x) = \dfrac{x^3}{9}(3\ln x - 1)$

58. $g(x) = \dfrac{x^2 + \sqrt{x} + 1}{x^{3/2}}$

59. $y = (x^2 + 5)^3 (3x^3 - 2)^2$

60. $f(x) = \dfrac{a^2 - x^2}{a^2 + x^2}$

61. $w(r) = \dfrac{ar^2}{b + r^3}$

62. $H(t) = (at^2 + b)e^{-ct}$

63. $g(w) = \dfrac{5}{(a^2 - w^2)^2}$

Chapter 4

USING THE DERIVATIVE

CONTENTS

4.1   LOCAL MAXIMA AND MINIMA

What Derivatives Tell Us About a Function and Its Graph

As we saw in Chapter 2, values of a function and its derivatives are related as follows:

- If $f' > 0$ on an interval, then f is increasing on that interval.
- If $f' < 0$ on an interval, then f is decreasing on that interval.
- If $f'' > 0$ on an interval, then the graph of f is concave up on that interval.
- If $f'' < 0$ on an interval, then the graph of f is concave down on that interval.

We now use these principles in conjunction with the derivative formulas from Chapter 3. For example, when we graph a function on a computer or calculator, we often see only part of the picture. The first and second derivatives can help identify regions with interesting behavior.

Example 1   Use a computer or calculator to sketch a useful graph of the function

$$f(x) = x^3 - 9x^2 - 48x + 52.$$

Solution   Since f is a cubic polynomial, we expect a graph that is roughly S-shaped. Graphing this function with $-10 \leq x \leq 10$, $-10 \leq y \leq 10$ gives the two nearly vertical lines in Figure 4.1. We know that there is more going on than this, but how do we know where to look?

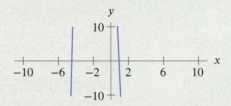

Figure 4.1: Unhelpful graph of $f(x) = x^3 - 9x^2 - 48x + 52$

We use the derivative to determine where the function is increasing and where it is decreasing. The derivative of f is

$$f'(x) = 3x^2 - 18x - 48.$$

To find where $f' > 0$ or $f' < 0$, we first find where $f' = 0$; that is, where $3x^2 - 18x - 48 = 0$. Factoring gives

$$3(x - 8)(x + 2) = 0,$$

so $x = -2$ or $x = 8$. Since $f' = 0$ *only* at $x = -2$ and $x = 8$, and since f' is continuous, f' cannot change sign on any of the three intervals $x < -2$, or $-2 < x < 8$, or $8 < x$.

How can we tell the sign of f' on each of these intervals? The easiest way is to pick a point and substitute into f'. For example, since $f'(-3) = 33 > 0$, we know f' is positive for $x < -2$, so f is increasing for $x < -2$. Similarly, since $f'(0) = -48$ and $f'(10) = 72$, we know that f decreases between $x = -2$ and $x = 8$ and increases for $x > 8$. Summarizing:

	$x = -2$		$x = 8$	
f increasing ↗		f decreasing ↘		f increasing ↗
$f' > 0$	$f' = 0$	$f' < 0$	$f' = 0$	$f' > 0$

We find that $f(-2) = 104$ and $f(8) = -396$. Hence, on the interval $-2 < x < 8$ the function decreases from a high of 104 to a low of -396. (Now we see why not much showed up in our first calculator graph.)

One more point on the graph is easy to get: the y-intercept, $f(0) = 52$. With just these three points we can get a much more helpful graph. By setting the plotting window to $-10 \leq x \leq 20$ and

$-400 \leq y \leq 400$, we get Figure 4.2, which gives much more insight into the behavior of $f(x)$ than the graph in Figure 4.1.

In Figure 4.2, we see that part of the graph is concave up and part is concave down. We can use the second derivative to analyze concavity. We have

$$f''(x) = 6x - 18.$$

Thus, $f''(x) < 0$ when $x < 3$ and $f''(x) > 0$ when $x > 3$, so the graph of f is concave down for $x < 3$ and concave up for $x > 3$. At $x = 3$, we have $f''(x) = 0$. See Figure 4.2. Summarizing:

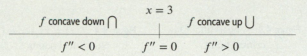

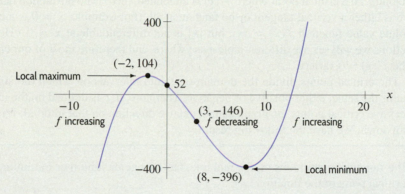

Figure 4.2: Useful graph of $f(x) = x^3 - 9x^2 - 48x + 52$. Notice that the scales on the x- and y-axes are different.

Local Maxima and Minima

We are often interested in points such as those marked local maximum and local minimum in Figure 4.2. We have the following definition:

> Suppose p is a point in the domain of f:
> - f has a **local minimum** at p if $f(p)$ is less than or equal to the values of f for points near p.
> - f has a **local maximum** at p if $f(p)$ is greater than or equal to the values of f for points near p.

We use the adjective "local" because we are describing only what happens near p. Local maxima and minima are sometimes called local extrema.

How Do We Detect a Local Maximum or Minimum?

In the preceding example, the points $x = -2$ and $x = 8$, where $f'(x) = 0$, played a key role in leading us to local maxima and minima. We give a name to such points:

> For any function f, a point p in the domain of f where $f'(p) = 0$ or $f'(p)$ is undefined is called a **critical point** of the function. In addition, the point $(p, f(p))$ on the graph of f is also called a critical point. A **critical value** of f is the value, $f(p)$, at a critical point, p.

Notice that "critical point of f" can refer either to points in the domain of f or to points on the graph of f. You will know which meaning is intended from the context. A function may have any number of critical points or none at all. (See Figures 4.3–4.5.)

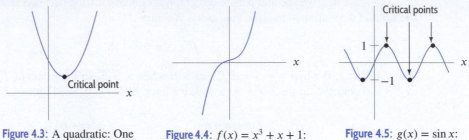

Figure 4.3: A quadratic: One critical point

Figure 4.4: $f(x) = x^3 + x + 1$: No critical points

Figure 4.5: $g(x) = \sin x$: Many critical points

Geometrically, at a critical point where $f'(p) = 0$, the line tangent to the graph of f at p is horizontal. At a critical point where $f'(p)$ is undefined, there is no horizontal tangent to the graph—there is either a vertical tangent or no tangent at all. (For example, $x = 0$ is a critical point for the absolute value function $f(x) = |x|$, but $|x|$ is not differentiable at $x = 0$.) However, most of the functions we will see are differentiable everywhere, and therefore most of our critical points will be of the $f'(p) = 0$ variety.

The critical points divide the domain of f into intervals on which the sign of the derivative remains the same, either positive or negative. Therefore, if f is defined on the interval between two successive critical points, its graph cannot change direction on that interval; it is either going up or going down. We have the following result:

> If a function, continuous on the real line, has a local maximum or minimum at p, then p is a critical point of the function.

If a continuous function f has domain the interval $a \leq x \leq b$, then f may have local maxima or minima at the endpoints $x = a$ and $x = b$, even if these points are not critical points of f.

Testing for Local Maxima and Minima

If f' has different signs on either side of a critical point p with $f'(p) = 0$, then the graph changes direction at p and looks like one of those in Figure 4.6. We have the following result:

> ### The First-Derivative Test for Local Maxima and Minima
>
> Suppose p is a critical point of a continuous function f. Moving from left to right:
> - If f' changes from negative to positive at p, then f has a local minimum at p.
> - If f' changes from positive to negative at p, then f has a local maximum at p.

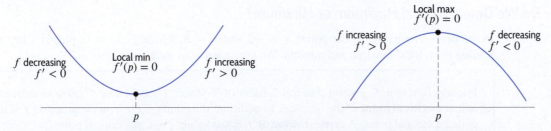

Figure 4.6: Changes in direction at a critical point, p: Local maxima and minima

The second derivative of a function can also be useful in testing if a critical point is a local maximum or a local minimum. Suppose p is a critical point of f, with $f'(p) = 0$, so that the graph of f has a horizontal tangent line at p. If the graph is concave up at p, then f has a local minimum at p. Likewise, if the graph is concave down, f has a local maximum. See Figure 4.7. This suggests:

The Second-Derivative Test for Local Maxima and Minima

Suppose p is a critical point of a continuous function f, and $f'(p) = 0$.

- If $f''(p) > 0$, then f has a local minimum at p.
- If $f''(p) < 0$, then f has a local maximum at p.
- If $f''(p) = 0$, then the test tells us nothing.

Figure 4.7: Local maxima and minima and concavity

Example 2   Use the second-derivative test to confirm that $f(x) = x^3 - 9x^2 - 48x + 52$ has a local maximum at $x = -2$ and a local minimum at $x = 8$.

Solution   In Example 1, we calculated $f'(x) = 3x^2 - 18x - 48 = 3(x - 8)(x + 2)$, so $f'(8) = f'(-2) = 0$. Differentiating again gives $f''(x) = 6x - 18$. Since

$$f''(8) = 6 \cdot 8 - 18 = 30 \qquad \text{and} \qquad f''(-2) = 6(-2) - 18 = -30,$$

the second-derivative test confirms that $x = 8$ is a local minimum and $x = -2$ is a local maximum.

Example 3   (a) Graph a function f with the following properties:
- $f(x)$ has critical points at $x = 2$ and $x = 5$;
- $f'(x)$ is positive to the left of 2 and positive to the right of 5;
- $f'(x)$ is negative between 2 and 5.

(b) Identify the critical points as local maxima, local minima, or neither.

Solution   (a) We know that $f(x)$ is increasing when $f'(x)$ is positive, and $f(x)$ is decreasing when $f'(x)$ is negative. The function is increasing to the left of 2 and increasing to the right of 5, and it is decreasing between 2 and 5. A possible sketch is given in Figure 4.8.

(b) We see that the function has a local maximum at $x = 2$ and a local minimum at $x = 5$.

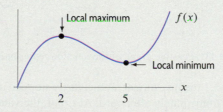

Figure 4.8: A function with critical points at $x = 2$ and $x = 5$

Warning!

Not every critical point of a function is a local maximum or minimum. For instance, consider $f(x) = x^3$, graphed in Figure 4.9. The derivative is $f'(x) = 3x^2$ so $x = 0$ is a critical point. But $f'(x) = 3x^2$ is positive on both sides of $x = 0$, so f increases on both sides of $x = 0$. There is neither a local maximum nor a local minimum for $f(x)$ at $x = 0$.

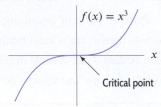

Figure 4.9: A critical point that is neither a local maximum nor minimum.

Example 4   The value of an investment at time t is given by $S(t)$. The rate of change, $S'(t)$, of the value of the investment is shown in Figure 4.10.

(a) What are the critical points of the function $S(t)$?

(b) Identify each critical point as a local maximum, a local minimum, or neither.

(c) Explain the financial significance of each of the critical points.

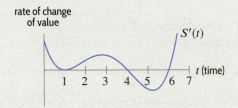

Figure 4.10: Graph of $S'(t)$, the rate of change of the value of the investment

Solution   (a) The critical points of S occur at times t when $S'(t) = 0$. We see in Figure 4.10 that $S'(t) = 0$ at $t = 1$, 4, and 6, so the critical points occur at $t = 1$, 4, and 6.

(b) In Figure 4.10, we see that $S'(t)$ is positive to the left of 1 and between 1 and 4, that $S'(t)$ is negative between 4 and 6, and that $S'(t)$ is positive to the right of 6. Therefore $S(t)$ is increasing to the left of 1 and between 1 and 4 (with a slope of zero at 1), decreasing between 4 and 6, and increasing again to the right of 6. A possible sketch of $S(t)$ is given in Figure 4.11. We see that S has neither a local maximum nor a local minimum at the critical point $t = 1$, but that it has a local maximum at $t = 4$ and a local minimum at $t = 6$.

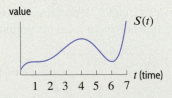

Figure 4.11: Possible graph of the function representing the value of the investment at time t

(c) At time $t = 1$ the investment momentarily stopped increasing in value, though it started increasing again immediately afterward. At $t = 4$, the value peaked and began to decline. At $t = 6$, it started increasing again.

Example 5   Find the critical point of the function $f(x) = x^2 + bx + c$. What is its graphical significance?

Solution   Since $f'(x) = 2x + b$, the critical point x satisfies the equation $2x + b = 0$. Thus, the critical point is at $x = -b/2$. The graph of f is a parabola and the critical point is its vertex. See Figure 4.12.

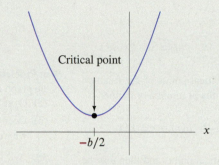

Figure 4.12: Critical point of the parabola $f(x) = x^2 + bx + c$. (Sketched with $b, c > 0$)

Problems for Section 4.1

■ In Problems 1–4, indicate all critical points of the function
f. How many critical points are there? Identify each critical
point as a local maximum, a local minimum, or neither.

1.

2.

3.

4.

5. (a) Graph a function with two local minima and one
local maximum.
(b) Graph a function with two critical points. One of
these critical points should be a local minimum,
and the other should be neither a local maximum
nor a local minimum.

6. Graph two continuous functions f and g, each of which
has exactly five critical points, the points $A-E$ in Figure 4.13, and that satisfy the following conditions:
(a) $f(x) \to \infty$ as $x \to -\infty$ and
$f(x) \to \infty$ as $x \to \infty$
(b) $g(x) \to -\infty$ as $x \to -\infty$ and
$g(x) \to 0$ as $x \to \infty$

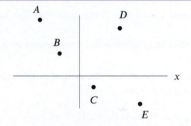

Figure 4.13

7. During an illness a person ran a fever. His temperature rose steadily for eighteen hours, then went steadily
down for twenty hours. When was there a critical point
for his temperature as a function of time?

■ In Problems 8–13,
(a) Use the derivative to find all critical points.
(b) Use a graph to classify each critical point as a local minimum, a local maximum, or neither.

8. $f(x) = 5x - x^2 + 8$ **9.** $f(x) = x^3 - 75x$

10. $f(x) = 9x^2 - x^3$ **11.** $f(x) = x^4 - 8x^2$

12. $f(x) = x^4 - 4x^3$ **13.** $f(x) = x^5 - 15x^3$

■ In Problems 14–19, graph the function and describe in words the interesting features of the graph, including the location of the critical points and where the function is monotonic (that is, increasing or decreasing). Then use the derivative and algebra to explain the shape of the graph.

14. $f(x) = x^3 + 6x + 1$

15. $f(x) = x^3 - 6x + 1$

16. $f(x) = 3x^5 - 5x^3$

17. $f(x) = e^x - 10x$

18. $f(x) = x \ln x, \quad x > 0$

19. $f(x) = x + 2 \sin x$

■ In Problems 20–21, find the critical points of the function and classify them as local maxima or minima or neither.

20. $g(x) = xe^{-3x}$

21. $h(x) = x + 1/x$

■ In Problems 22–25, find all critical points and then use the first-derivative test to determine local maxima and minima. Check your answer by graphing.

22. $f(x) = 3x^4 - 4x^3 + 6$

23. $f(x) = (x^2 - 4)^7$

24. $f(x) = (x^3 - 8)^4$

25. $f(x) = \dfrac{x}{x^2 + 1}$

■ In Problems 26–29, the continuous function has a critical point.
(a) Is the critical point a local maximum or a local minimum?
(b) Sketch the graph near the critical point. Label the coordinates of the critical point.

26. $f(1) = 5, f'(1) = 0, f''(1) = -2$

27. $g(-5) = 4, g'(-5) = 0, g''(-5) = 2$

28. $h(2) = -5, h'(2) = 0, h''(2) = -4$

29. $j(3) = 5, j'(3)$ is undefined, $j'(x) = -1$ for $x < 3$ and $j'(x) = 1$ for $x > 3$.

30. The function $f(x) = x^4 - 4x^3 + 8x$ has a critical point at $x = 1$. Use the second-derivative test to identify it as a local maximum or local minimum.

31. Find and classify the critical points of $f(x) = x^3(1-x)^4$ as local maxima and minima.

■ In Problems 32–35, use the derivative to find all critical points. Assume that A, B, and C are positive constants.

32. $f(x) = Ax^2 + Bx + C$

33. $f(x) = Ax^3 - Bx$

34. $f(x) = C + Ax^2 - Bx^3$

35. $f(x) = Ax^4 - Bx^2 + C$

36. If U and V are positive constants, find all critical points of

$$F(t) = Ue^t + Ve^{-t}.$$

37. Figure 4.14 is the graph of a derivative f'. On the graph, mark the x-values that are critical points of f. At which critical points does f have local maxima, local minima, or neither?

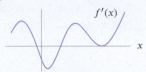

Figure 4.14

■ In Problems 38–41, the function f is defined for all x. Use the graph of f' to decide:
(a) Over what intervals is f increasing? Decreasing?
(b) Does f have local maxima or minima? If so, which, and where?

38.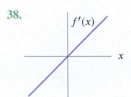

39.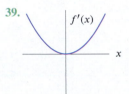

40.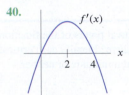

41.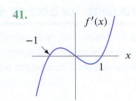

42. Figure 4.15 is a graph of f'. For what values of x does f have a local maximum? A local minimum?

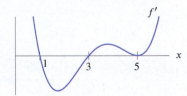

Figure 4.15: Graph of f' (not f)

43. Consumer demand for a product is changing over time, and the rate of change of demand, $f'(t)$, in units/week, is given, in week t, for $0 \le t \le 10$, in the following table.

(a) When is the demand for this product increasing? When is it decreasing?
(b) Approximately when is demand at a local maximum? A local minimum?

t	0	1	2	3	4	5	6	7	8	9	10
$f'(t)$	12	10	4	-2	-3	-1	3	7	11	15	10

44. Suppose f has a continuous derivative whose values are given in the following table.

   (a) Estimate the x-coordinates of critical points of f for $0 \leq x \leq 10$.

   (b) For each critical point, indicate if it is a local maximum of f, local minimum, or neither.

x	0	1	2	3	4	5	6	7	8	9	10
$f'(x)$	5	2	1	−2	−5	−3	−1	2	3	1	−1

45. The derivative of $f(t)$ is given by $f'(t) = t^3 - 6t^2 + 8t$ for $0 \leq t \leq 5$. Graph $f'(t)$, and describe how the function $f(t)$ changes over the interval $t = 0$ to $t = 5$. When is $f(t)$ increasing and when is it decreasing? Where does $f(t)$ have a local maximum and where does it have a local minimum?

46. **(a)** If a is a positive constant, find all critical points of $f(x) = x^3 - ax$.

   (b) Find the value of a so that f has local extrema at $x = \pm 2$.

47. **(a)** If a is a constant, find all critical points of $f(x) = 5ax - 2x^2$.

   (b) Find the value of a so that f has a local maximum at $x = 6$.

■ In Problems 48–49, find constants a and b so that the minimum for the parabola $f(x) = x^2 + ax + b$ is at the given point. [Hint: Begin by finding the critical point in terms of a.]

48. $(3, 5)$ **49.** $(-2, -3)$

50. Find the value of a so that $f(x) = xe^{ax}$ has a critical point at $x = 3$.

51. For what values of a and b does $f(x) = a(x - b \ln x)$ have a local minimum at the point $(2, 5)$? Figure 4.16 shows a graph of $f(x)$ with $a = 1$ and $b = 1$.

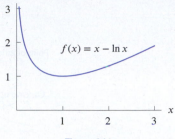

Figure 4.16

52. Sketch several members of the family $y = x^3 - ax^2$ on the same axes. Discuss the effect of the parameter a on the graph. Find all critical points for this function.

53. **(a)** For a a positive constant, find all critical points of $f(x) = x - a\sqrt{x}$.

   (b) What value of a gives a critical point at $x = 5$? Does $f(x)$ have a local maximum or a local minimum at this critical point?

54. Find values of a and b so that $f(x) = axe^{bx}$ has $f(1/3) = 1$ and f has a local maximum at $x = 1/3$.

■ In Problems 55–57, investigate the one-parameter family of functions. Assume that a is positive.

   (a) Graph $f(x)$ using three different values for a.

   (b) Using your graph in part (a), describe the critical points of f and how they appear to move as a increases.

   (c) Find a formula for the x-coordinates of the critical point(s) of f in terms of a.

55. $f(x) = (x - a)^2$ **56.** $f(x) = x^3 - ax$

57. $f(x) = x^2 e^{-ax}$

58. If $m, n \geq 2$ are integers, find and classify the critical points of $f(x) = x^m (1 - x)^n$.

4.2   INFLECTION POINTS

Concavity and Inflection Points

A study of the points on the graph of a function where the slope changes sign led us to critical points. Now we will study the points on the graph where the concavity changes, either from concave up to concave down, or from concave down to concave up.

> A point at which the graph of a function f changes concavity is called an **inflection point** of f.

The words "inflection point of f" can refer either to a point in the domain of f or to a point on the graph of f. The context of the problem will tell you which is meant.

How Do You Locate an Inflection Point?

Since the concavity of the graph of f changes at an inflection point, the sign of f'' changes there: it is positive on one side of the inflection point and negative on the other. Thus, at the inflection point, f'' is zero or undefined. (See Figure 4.17.)

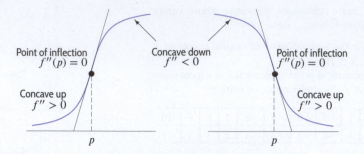

Figure 4.17: Change in concavity (from positive to negative or vice versa) at point p

---

Suppose f'' is defined on both sides of a point p:

- If f'' is zero or undefined at p, then p is a possible inflection point.

- To test whether p is an inflection point, check whether f'' changes sign at p.

Example 1   Find the inflection points of $f(x) = x^3 - 9x^2 - 48x + 52$.

Solution   In Figure 4.18, part of the graph of f is concave up and part is concave down, so the function must have an inflection point. However, it is difficult to locate the inflection point accurately by examining the graph. To find the inflection point exactly, we calculate where the second derivative is zero. Since $f'(x) = 3x^2 - 18x - 48$,

$$f''(x) = 6x - 18 \qquad \text{so} \qquad f''(x) = 0 \quad \text{when} \quad x = 3.$$

We can see that the graph of $f(x)$ changes concavity at $x = 3$, so $x = 3$ is an inflection point.

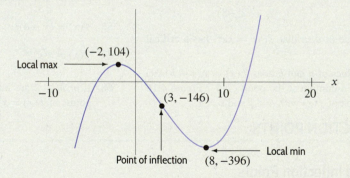

Figure 4.18: Graph of $f(x) = x^3 - 9x^2 - 48x + 52$ showing the inflection point at $x = 3$

Example 2   Graph a function f with the following properties: f has a critical point at $x = 4$ and an inflection point at $x = 8$; the value of f' is negative to the left of 4 and positive to the right of 4; the value of f'' is positive to the left of 8 and negative to the right of 8.

Solution   Since f' is negative to the left of 4 and positive to the right of 4, the value of $f(x)$ is decreasing to the left of 4 and increasing to the right of 4. The values of f'' tell us that the graph of $f(x)$ is concave up to the left of 8 and concave down to the right of 8. A possible sketch is given in Figure 4.19.

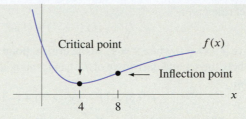

Figure 4.19: A function with a critical point at $x = 4$ and an inflection point at $x = 8$

Example 3   Figure 4.20 shows a population growing toward a limiting population, L. There is an inflection point on the graph at the point where the population reaches $L/2$. What is the significance of the inflection point to the population?

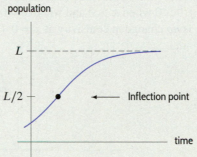

Figure 4.20: Inflection point on graph of a population growing toward a limiting population, L

Solution   At times before the inflection point, the population is increasing faster every year. At times after the inflection point, the population is increasing slower every year. At the inflection point, the population is growing fastest.

Example 4   (a) How many critical points and how many inflection points does the function $f(x) = xe^{-x}$ have?
(b) Use derivatives to find the critical points and inflection points exactly.

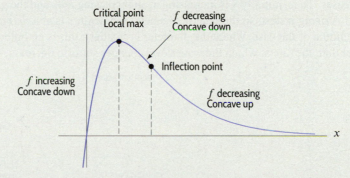

Figure 4.21: Graph of $f(x) = xe^{-x}$

Solution   (a) Figure 4.21 shows the graph of $f(x) = xe^{-x}$. It appears to have one critical point, which is a local maximum. Are there any inflection points? Since the graph of the function is concave down at the critical point and concave up for large x, the graph of the function changes concavity, so there must be an inflection point to the right of the critical point.
(b) To find the critical point, find the point where the first derivative of f is zero or undefined. The

product rule gives

$$f'(x) = x(-e^{-x}) + (1)(e^{-x}) = (1-x)e^{-x}.$$

We have $f'(x) = 0$ when $x = 1$, so the critical point is at $x = 1$. To find the inflection point, we find where the second derivative of f changes sign. Using the product rule on the first derivative, we have

$$f''(x) = (1-x)(-e^{-x}) + (-1)(e^{-x}) = (x-2)e^{-x}.$$

We have $f''(x) = 0$ when $x = 2$. Since $f''(x) > 0$ for $x > 2$ and $f''(x) < 0$ for $x < 2$, the concavity changes sign at $x = 2$. So the inflection point is at $x = 2$.

Warning!

Not every point x where $f''(x) = 0$ (or f'' is undefined) is an inflection point (just as not every point where $f' = 0$ is a local maximum or minimum). For instance, $f(x) = x^4$ has $f''(x) = 12x^2$ so $f''(0) = 0$, but $f'' > 0$ when $x > 0$ and when $x < 0$, so the graph of f is concave up on both sides of $x = 0$. There is *no* change in concavity at $x = 0$. (See Figure 4.22.)

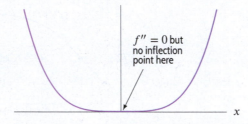

$f'' = 0$ but no inflection point here

Figure 4.22: Graph of $f(x) = x^4$

Example 5   Water is being poured into the vase in Figure 4.23 at a constant rate measured in liters per minute. Graph $y = f(t)$, the depth of the water against time, t. Explain the concavity, and indicate the inflection points.

Solution   Notice that the volume of water in the vase increases at a constant rate.

At first the water level, y, rises slowly because the base of the vase is wide, so it takes a lot of water to make the depth increase. However, as the vase narrows, the rate at which the water level rises increases. Thus, initially y is increasing at an increasing rate, and the graph is concave up. The water level is rising fastest, so the rate of change of the depth y is at a maximum, when the water reaches the middle of the vase, where the diameter is smallest; this is an inflection point. (See Figure 4.24.) After that, the rate at which the water level changes starts to decrease, so the graph is concave down.

Figure 4.23: A vase

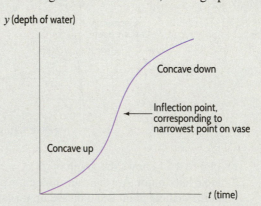

y (depth of water)

Concave down

Inflection point, corresponding to narrowest point on vase

Concave up

t (time)

Figure 4.24: Graph of depth of water in the vase, y, against time, t

Example 6   What is the concavity of the graph of $f(x) = ax^2 + bx + c$?

Solution   We have $f'(x) = 2ax + b$ and $f''(x) = 2a$. The second derivative of f has the same sign as a. If $a > 0$, the graph is concave up everywhere, an upward-opening parabola. If $a < 0$, the graph is concave down everywhere, a downward-opening parabola. (See Figure 4.25.) If $a = 0$, the function is linear and the graph is a straight line.

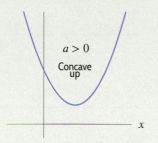
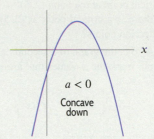

Figure 4.25: Concavity of $f(x) = ax^2 + bx + c$

Problems for Section 4.2

■ In Problems 1–4, indicate the approximate locations of all inflection points. How many inflection points are there?

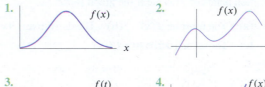

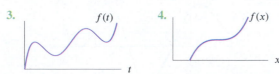

5. Graph a function with only one critical point (at $x = 5$) and one inflection point (at $x = 10$). Label the critical point and the inflection point on your graph.

6. (a) Graph a polynomial with two local maxima and two local minima.
   (b) What is the least number of inflection points this function must have? Label the inflection points.

7. Graph a function which has a critical point and an inflection point at the same place.

8. During a flood, the water level in a river first rose faster and faster, then rose more and more slowly until it reached its highest point, then went back down to its pre-flood level. Consider water depth as a function of time.
   (a) Is the time of highest water level a critical point or an inflection point of this function?
   (b) Is the time when the water first began to rise more slowly a critical point or an inflection point?

9. As I left home in the morning, I put on a light jacket because, although the temperature was dropping, it seemed that the temperature would not go much lower. But I was wrong. Around noon a northerly wind blew up and the temperature began to drop faster and faster. The worst was around 6 pm when, fortunately, the temperature started going back up.
   (a) When was there a critical point in the graph of temperature as a function of time?
   (b) When was there an inflection point in the graph of temperature as a function of time?

10. For $f(x) = x^3 - 18x^2 - 10x + 6$, find the inflection point algebraically. Graph the function with a calculator or computer and confirm your answer.

11. Find the inflection points of $f(x) = x^4 + x^3 - 3x^2 + 2$.

■ In Problems 12–21, use the first derivative to find all critical points and use the second derivative to find all inflection points. Use a graph to identify each critical point as a local maximum, a local minimum, or neither.

12. $f(x) = x^2 - 5x + 3$

13. $f(x) = x^3 - 3x + 10$

14. $f(x) = 2x^3 + 3x^2 - 36x + 5$

15. $f(x) = \dfrac{x^3}{6} + \dfrac{x^2}{4} - x + 2$

16. $f(x) = x^4 - 2x^2$

17. $f(x) = 3x^4 - 4x^3 + 6$

18. $f(x) = x^4 - 8x^2 + 5$

19. $f(x) = x^4 - 4x^3 + 10$

20. $f(x) = x^5 - 5x^4 + 35$

21. $f(x) = 3x^5 - 5x^3$

22. **(a)** Use a graph to estimate the x-values of any critical points and inflection points of $f(x) = e^{-x^2}$.
    (b) Use derivatives to find the x-values of any critical points and inflection points exactly.

23. **(a)** Find all critical points and all inflection points of the function $f(x) = x^4 - 2ax^2 + b$. Assume a and b are positive constants.
    (b) Find values of the parameters a and b if f has a critical point at the point $(2, 5)$.
    (c) If there is a critical point at $(2, 5)$, where are the inflection points?

24. Figure 4.26 is the graph of a second derivative f''. On the graph, mark the x-values that are inflection points of f.

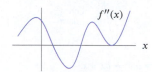

Figure 4.26

25. **(a)** Figure 4.27 shows the graph of f. Which of the x-values A, B, C, D, E, F, and G appear to be critical points of f?
    (b) Which appear to be inflection points of f?
    (c) How many local maxima does f appear to have? How many local minima?

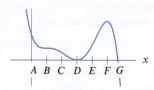

Figure 4.27

26. Figure 4.27 shows the graph of the *derivative*, f'.
    (a) Which of the x-values A, B, C, D, E, F, and G appear to be critical points of f?
    (b) Which appear to be inflection points of f?
    (c) How many local maxima does f appear to have? How many local minima?

■ For Problems **27–30**, sketch a possible graph of $y = f(x)$, using the given information about the derivatives $y' = f'(x)$ and $y'' = f''(x)$. Assume that the function is defined and continuous for all real x.

27.

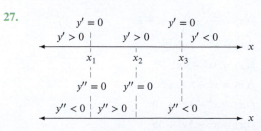

28.

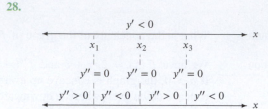

29.

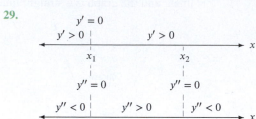

30.

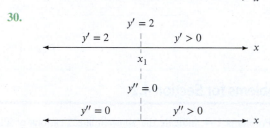

31. Indicate on Figure 4.28 approximately where the inflection points of $f(x)$ are if the graph shows

    (a) The function $f(x)$ **(b)** The derivative $f'(x)$
    (c) The second derivative $f''(x)$

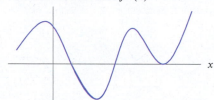

Figure 4.28

■ Problems **32–35** concern $f(t)$ in Figure 4.29, which gives the length of a human fetus as a function of its age.

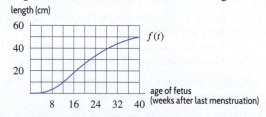

Figure 4.29

32. **(a)** What are the units of $f'(24)$?
    (b) What is the biological meaning of $f'(24) = 1.6$?

33. **(a)** Which is greater, $f'(20)$ or $f'(36)$?
    (b) What does your answer say about fetal growth?

34. **(a)** At what time does the inflection point occur?
    (b) What is the biological significance of this point?

35. Estimate

    (a) $f'(20)$ **(b)** $f'(36)$

    (c) The average rate of change of length over the 40 weeks shown.

36. **(a)** Water is flowing at a constant rate (i.e., constant volume per unit time) into a cylindrical container standing vertically. Sketch a graph showing the depth of water against time.

    (b) Water is flowing at a constant rate into a cone-shaped container standing on its point. Sketch a graph showing the depth of the water against time.

37. If water is flowing at a constant rate (that is, constant volume per unit time) into the Grecian urn in Figure 4.30, sketch a graph of the depth of the water against time. Mark on the graph the time at which the water reaches the widest point of the urn.

Figure 4.30

38. The vase in Figure 4.31 is filled with water at a constant rate (that is, constant volume per unit time).

    (a) Graph $y = f(t)$, the depth of the water, against time, t. Show on your graph the points at which the concavity changes.

(b) At what depth is $y = f(t)$ growing most quickly? Most slowly? Estimate the ratio between the growth rates at these two depths.

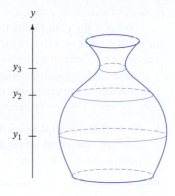

Figure 4.31

■ Find formulas for the functions described in Problems **39–41**.

39. A cubic polynomial, $ax^3 + bx^2 + cx + d$, with a critical point at $x = 2$, an inflection point at $(1, 4)$, and a leading coefficient of 1.

40. A function of the form $y = \dfrac{a}{1 + be^{-t}}$ with y-intercept 2 and an inflection point at $t = 1$.

41. A curve of the form $y = e^{-(x-a)^2/b}$ for $b > 0$ with a local maximum at $x = 2$ and points of inflection at $x = 1$ and $x = 3$.

4.3 GLOBAL MAXIMA AND MINIMA

Global Maxima and Minima

The techniques for finding maximum and minimum values make up the field called *optimization*. Local maxima and minima occur where a function takes larger or smaller values than at nearby points. However, we are often interested in where a function is larger or smaller than at all other points. For example, a firm may want to minimize costs in order to maximize profit. We define:

For any function f:

- f has a **global minimum** at p if $f(p)$ is less than or equal to all values of f.
- f has a **global maximum** at p if $f(p)$ is greater than or equal to all values of f.

How Do We Find Global Maxima and Minima?

Figure 4.32 shows a continuous function, f, defined on an interval $a \leq x \leq b$ (including its endpoints). Notice that the global maximum and minimum of f occur at a local maximum or a local minimum, which could be at an endpoint, $x = a$ or $x = b$.

To find the global maximum and minimum of a continuous function on an interval including endpoints: Compare values of the function at all the critical points in the interval and at the endpoints.

What if the continuous function is defined on an interval $a < x < b$ (excluding its endpoints), or on the entire real line, which has no endpoints? The function shown in Figure 4.33 has no global maximum because the function has no largest value. The global minimum of this function coincides with one of the local minima and is marked. A function defined on the entire real line or on an interval excluding endpoints may or may not have a global maximum or a global minimum.

To find the global maximum and minimum of a continuous function on an interval excluding endpoints or on the entire real line: Find the values of the function at all the critical points and sketch a graph.

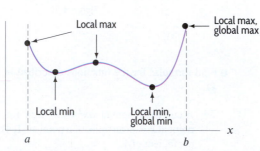

Figure 4.32: Global maximum and minimum on an interval domain, $a \le x \le b$

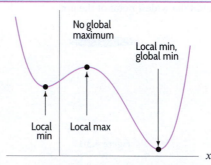

Figure 4.33: Global maximum and minimum on the entire real line

Example 1   Find the global maximum and minimum of $f(x) = x^3 - 9x^2 - 48x + 52$ on the interval $-5 \le x \le 14$.

Solution   We have calculated the critical points of this function previously using

$$f'(x) = 3x^2 - 18x - 48 = 3(x+2)(x-8),$$

so $x = -2$ and $x = 8$ are critical points. Since the global maxima and minima occur at a critical point or at an endpoint of the interval, we evaluate f at these four points:

$$f(-5) = -58, \qquad f(-2) = 104, \qquad f(8) = -396, \qquad f(14) = 360.$$

Comparing these four values, we see that the global maximum is 360 and occurs at $x = 14$, and that the global minimum is -396 and occurs at $x = 8$. See Figure 4.34.

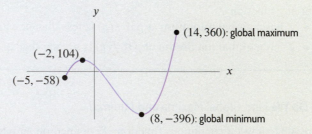

Figure 4.34: Global maximum and minimum on the interval $-5 \le x \le 14$

Example 2   For time, $t \geq 0$, in days, the rate at which photosynthesis takes place in the leaf of a plant, represented by the rate at which oxygen is produced, is approximated by[1]

$$p(t) = 100(e^{-0.02t} - e^{-0.1t}).$$

When is photosynthesis occurring fastest? What is the rate at that time?

Solution   Photosynthesis is fastest at the global maximum of $p(t)$. To find this maximum value, we first find critical points. We differentiate, set equal to zero, and solve for t:

$$p'(t) = 100(-0.02e^{-0.02t} + 0.1e^{-0.1t}) = 0$$
$$-0.02e^{-0.02t} = -0.1e^{-0.1t}$$
$$\frac{e^{-0.02t}}{e^{-0.1t}} = \frac{0.1}{0.02}$$
$$e^{-0.02t+0.1t} = 5$$
$$e^{0.08t} = 5$$
$$0.08t = \ln 5$$
$$t = \frac{\ln 5}{0.08} = 20.12 \text{ days.}$$

Differentiating again gives

$$p''(t) = 100(0.0004e^{-0.02t} - 0.01e^{-0.1t})$$

and substituting $t = 20.12$ gives $p''(20.12) = -0.107$, so $t = 20.12$ is a local maximum.

Since $t = 20.12$ is the only critical point and is a local maximum, $t = 20.12$ must give the global maximum. See Figure 4.35. The maximum rate is

$$p(20.12) = 100\left(e^{-0.02(20.12)} - e^{-0.1(20.12)}\right) = 53.50.$$

Photosynthesis occurs fastest after about 20 days. The maximum rate is about 53.5 units of oxygen per day.

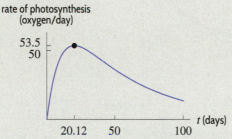

Figure 4.35: Maximum rate of photosynthesis

A Graphical Example: Minimizing Gas Consumption

Next we look at an example in which a function is given graphically and the maximum and minimum values are read from a graph. We already know how to estimate the maximum and minimum values of $f(x)$ from a graph of $f(x)$—read off the highest and lowest values. In this example, we see how to estimate the minimum value of the quantity $f(x)/x$ from a graph of $f(x)$ against x.

---

[1]Examples adapted from Rodney Gentry, *Introduction to Calculus for the Biological and Health Sciences* (Reading: Addison-Wesley, 1978).

We investigate what driving speed maximizes fuel efficiency.[2] Gas consumption, g (in gallons/hour), is a function of velocity, v (in mph) as shown in Figure 4.36. We want to minimize the gas consumption per *mile*, not the gas consumption per hour. Notice that

$$\text{Units of } \frac{g}{v} = \frac{\text{gallons/hour}}{\text{miles/hour}} = \frac{\text{gallons}}{\text{miles}},$$

suggesting we minimize $G = g/v$.

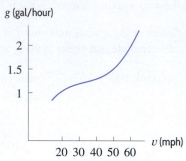

Figure 4.36: Gas consumption versus velocity

Example 3   Using Figure 4.36, estimate the velocity which minimizes $G = g/v$.

Solution   We want to find the minimum value of $G = g/v$ when g and v are related by the graph in Figure 4.36. We could use Figure 4.36 to sketch a graph of G against v and estimate a critical point. But there is an easier way. Figure 4.37 shows that g/v is the slope of the line from the origin to the point P. Where on the curve should P be to make the slope a minimum? From Figure 4.37, we see that the slope of the line is both a local and global minimum when the line is tangent to the curve. From Figure 4.38, we see that the velocity at this point is about 50 mph. Thus, to minimize gas consumption per mile, we should drive about 50 mph.

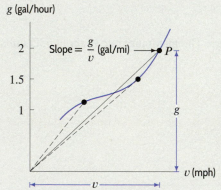

Figure 4.37: Graphical representation of gas consumption per mile, $G = g/v$

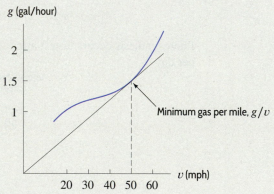

Figure 4.38: Velocity for maximum gas efficiency

Problems for Section 4.3

■ For Problems **1–2**, indicate all critical points on the given graphs. Determine which correspond to local minima, local maxima, global minima, global maxima, or none of these. (Note that the graphs are on closed intervals.)

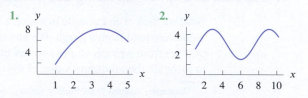

---

[2] Adapted from Peter D. Taylor, *Calculus: The Analysis of Functions* (Toronto: Wall & Emerson, 1992).

■ In Problems **3–8**, for $f(x)$ in Figure 4.39, find the x-values of the global maximum and global minimum on the given domain.

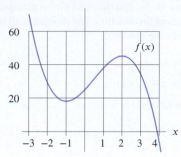

Figure 4.39

3. Domain: $0 \le x \le 4$

4. Domain: $-3 \le x \le 0$

5. Domain: $-2 \le x \le 3$

6. Domain: $-2 \le x \le 1$

7. Domain: $0 \le x \le 3$

8. Domain: $-3 \le x \le 4$

9. For each interval, use Figure 4.40 to choose the statement that gives the location of the global maximum and global minimum of f on the interval.

   (a) $4 \le x \le 12$ **(b)** $11 \le x \le 16$

   (c) $4 \le x \le 9$ **(d)** $8 \le x \le 18$

   (I)   Maximum at right endpoint, minimum at left endpoint.

   (II)   Maximum at right endpoint, minimum at critical point.

   (III)   Maximum at left endpoint, minimum at right endpoint.

   (IV)   Maximum at left endpoint, minimum at critical point.

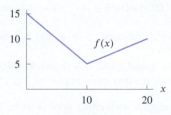

Figure 4.40

■ In Problems **10–13**, graph a function with the given properties.

10. Has local minimum and global minimum at $x = 3$ but no local or global maximum.

11. Has local minimum at $x = 3$, local maximum at $x = 8$, but no global maximum or minimum.

12. Has no local or global maxima or minima.

13. Has local and global minimum at $x = 3$, local and global maximum at $x = 8$.

14. True or false? Give an explanation for your answer. The global maximum of $f(x) = x^2$ on every closed interval is at one of the endpoints of the interval.

15. Plot the graph of $f(x) = x^3 - e^x$ using a graphing calculator or computer to find all local and global maxima and minima for: **(a)** $-1 \le x \le 4$ **(b)** $-3 \le x \le 2$

■ In Problems **16–19**, sketch the graph of a function on the interval $0 \le x \le 10$ with the given properties.

16. Has local minimum at $x = 3$, local maximum at $x = 8$, but global maximum and global minimum at the endpoints of the interval.

17. Has local and global maximum at $x = 3$, local and global minimum at $x = 10$.

18. Has global maximum at $x = 0$, global minimum at $x = 10$, and no other local maxima or minima.

19. Has local and global minimum at $x = 3$, local and global maximum at $x = 8$.

20. The function $y = t(x)$ is positive and continuous with a global maximum at the point $(3, 3)$. Graph $t(x)$ if $t'(x)$ and $t''(x)$ have the same sign for $x < 3$, but opposite signs for $x > 3$.

21. Figure 4.41 shows the rate at which photosynthesis is taking place in a leaf.

   (a) At what time, approximately, is photosynthesis proceeding fastest for $t \ge 0$?

   (b) If the leaf grows at a rate proportional to the rate of photosynthesis, for what part of the interval $0 \le t \le 200$ is the leaf growing? When is it growing fastest?

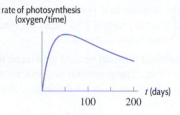

Figure 4.41

■ For the functions in Problems **22–25**, do the following:

(a) Find f' and f''.

(b) Find the critical points of f.

(c) Find any inflection points of f.

(d) Evaluate f at its critical points and at the endpoints of the given interval. Identify local and global maxima and minima of f in the interval.

(e) Graph f.

22. $f(x) = x^3 - 3x^2$ $(-1 \le x \le 3)$

23. $f(x) = 2x^3 - 9x^2 + 12x + 1$ $(-0.5 \le x \le 3)$

24. $f(x) = x^3 - 3x^2 - 9x + 15$ $(-5 \le x \le 4)$

25. $f(x) = x + \sin x$ $(0 \le x \le 2\pi)$

26. Find the value of x that maximizes $y = 12 + 18x - 5x^2$ and the corresponding value of y, by

    (a) Estimating the values from a graph of y.
    (b) Finding the values using calculus.

27. Find the value(s) of x that give critical points of $y = ax^2 + bx + c$, where a, b, c are constants. Under what conditions on a, b, c is the critical value a maximum? A minimum?

■ In Problems 28–31, the continuous function has exactly one critical point. Find the x-values at which the global maximum and the global minimum occur in the interval given.

28. $f'(1) = 0$, $f''(1) = -2$ on $1 \leq x \leq 3$

29. $g'(-5) = 0$, $g''(-5) = 2$ on $-6 \leq x \leq -5$

30. $h'(2)$ is undefined, $h'(x) = -1$ for $x < 2$ and $h'(x) = 1$ for $x > 2$, on $1 \leq x \leq 4$

31. $j'(3)$ is undefined, $j'(x) = 2$ for $x < 3$ and $j'(x) = -2$ for $x > 3$, on $1 \leq x \leq 5$

32. Figure 4.42 shows a function f. Does f have a global maximum? A global minimum? If so, where? Assume that $f(x)$ is defined for all x and that the graph does not change concavity outside the window shown.

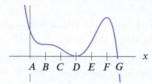

A B C D E F G

Figure 4.42

33. Figure 4.42 shows a *derivative*, f'. Does the function f have a global maximum? A global minimum? If so, where? Assume that $f(x)$ and $f'(x)$ are defined for all x and that the graph of $f'(x)$ does not change concavity outside the window shown.

34. A grapefruit is tossed straight up with an initial velocity of 50 ft/sec. The grapefruit is 5 feet above the ground when it is released. Its height, in feet, at time t seconds is given by
$$y = -16t^2 + 50t + 5.$$
How high does it go before returning to the ground?

35. The sum of two nonnegative numbers is 100. What is the maximum value of the product of these two numbers?

36. The product of two positive numbers is 784. What is the minimum value of their sum?

37. The sum of three nonnegative numbers is 36, and one of the numbers is twice one of the other numbers. What is the maximum value of the product of these three numbers?

38. The perimeter of a rectangle is 64 cm. Find the lengths of the sides of the rectangle giving the maximum area.

■ In Problems 39–46, find the exact global maximum and minimum values of the function. The domain is all real numbers unless otherwise specified.

39. $g(x) = 4x - x^2 - 5$

40. $f(x) = x + 1/x$ for $x > 0$

41. $g(t) = te^{-t}$ for $t > 0$

42. $f(x) = 2e^x + 3e^{-x}$

43. $f(x) = e^{3x} - e^{2x}$

44. $f(x) = x - \ln x$ for $x > 0$

45. $f(t) = \dfrac{t}{1 + t^2}$

46. $f(t) = (\sin^2 t + 2) \cos t$

47. What value of w minimizes S if $S - 5pw = 3qw^2 - 6pq$ and p and q are positive constants?

48. The energy expended by a bird per day, E, depends on the time spent foraging for food per day, F hours. Foraging for a shorter time requires better territory, which then requires more energy for its defense.[3] Find the foraging time that minimizes energy expenditure if
$$E = 0.25F + \frac{1.7}{F^2}.$$

49. The impact of a drug is a measure of its effect, for example, the reduction in blood pressure, loss of weight, or the duration of a headache. The impact, I, generally depends on the dose, D, given.[4] For positive constants a and b, two possible impact functions, valid for $0 \leq D \leq 300$, are
$$I = f(D) = aD\sqrt{b + D} \qquad I = g(D) = aD\sqrt{b - D}.$$

    (a) Which function, f or g, has a critical point for $D > 0$?
    (b) For f and g and $a = 1$ and $b = 300$, what dose gives the maximum impact on $0 \leq D \leq 300$?

50. A rectangular swimming pool is to be built with an area of 1800 square feet. The owner wants 5-foot-wide decks along either side and 10-foot-wide decks at the two ends. Find the dimensions of the smallest piece of property on which the pool can be built satisfying these conditions.

51. If you have 100 feet of fencing and want to enclose a rectangular area up against a long, straight wall, what is the largest area you can enclose?

---

[3] Adapted from Graham Pyke, reported by J. R. Krebs and N. B. Davies, *An Introduction to Behavioural Ecology* (Oxford: Blackwell, 1987).
[4] www.brynmawr.edu/math/people/vandiver/documents/Optimization.pdf and prezi.com/fqwwg6tcuqqp/calculus-in-medicine/. Accessed December 2016.

52. An apple tree produces, on average, 400 kg of fruit each season. However, if more than 200 trees are planted per km^2, crowding reduces the yield by 1 kg for each tree over 200.

(a) Express the total yield, y, from one square kilometer as a function of the number of trees on it. Graph this function.

(b) How many trees should a farmer plant on each square kilometer to maximize yield?

53. On the west coast of Canada, crows eat whelks (a shellfish). To open the whelks, the crows drop them from the air onto a rock. If the shell does not smash the first time, the whelk is dropped again.[5] The average number of drops, n, needed when the whelk is dropped from a height of x meters is approximated by

$$n(x) = 1 + \frac{27}{x^2}.$$

(a) Give the total vertical distance the crow travels upward to open a whelk as a function of drop height, x.

(b) Crows are observed to drop whelks from the height that minimizes the total vertical upward distance traveled per whelk. What is this height?

54. During a flu outbreak in a school of 763 children, the number of infected children, I, was expressed in terms of the number of susceptible (but still healthy) children, S, by the function[6]

$$I = 192 \ln\left(\frac{S}{762}\right) - S + 763.$$

What is the maximum possible number of infected children?

55. The number of offspring in a population may not be a linear function of the number of adults. The Ricker curve, used to model fish populations, claims that $y = axe^{-bx}$, where x is the number of adults, y is the number of offspring, and a and b are positive constants.

(a) Find and classify all critical points of the Ricker curve.

(b) Is there a global maximum? What does this imply about populations?

56. The oxygen supply, S, in the blood depends on the hematocrit, H, the percentage of red blood cells in the blood:

$$S = aHe^{-bH} \quad \text{for positive constants } a, b.$$

(a) What value of H maximizes the oxygen supply? What is the maximum oxygen supply?

(b) How does increasing the value of the constants a and b change the maximum value of S?

57. The quantity of a drug in the bloodstream t hours after a tablet is swallowed is given, in mg, by

$$q(t) = 20(e^{-t} - e^{-2t}).$$

(a) How much of the drug is in the bloodstream at time $t = 0$?

(b) When is the maximum quantity of drug in the bloodstream? What is that maximum?

(c) In the long run, what happens to the quantity?

58. When birds lay eggs, they do so in clutches of several at a time. When the eggs hatch, each clutch gives rise to a brood of baby birds. We want to determine the clutch size which maximizes the number of birds surviving to adulthood per brood. If the clutch is small, there are few baby birds in the brood; if the clutch is large, there are so many baby birds to feed that most die of starvation. The number of surviving birds per brood as a function of clutch size is shown by the benefit curve in Figure 4.43.[7]

(a) Estimate the clutch size which maximizes the number of survivors per brood.

(b) Suppose also that there is a biological cost to having a larger clutch: the female survival rate is reduced by large clutches. This cost is represented by the dotted line in Figure 4.43. If we take cost into account by assuming that the optimal clutch size in fact maximizes the vertical distance between the curves, what is the new optimal clutch size?

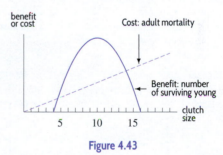

Figure 4.43

59. Let $f(v)$ be the amount of energy consumed by a flying bird, measured in joules per second (a joule is a unit of energy), as a function of its speed v (in meters/sec). Let $a(v)$ be the amount of energy consumed by the same bird, measured in joules per meter.

(a) Suggest a reason in terms of the way birds fly for the shape of the graph of $f(v)$ in Figure 4.44.

[5] Adapted from Reto Zach, reported by J. R. Krebs and N. B. Davies, *An Introduction to Behavioural Ecology* (Oxford: Blackwell, 1987).

[6] Data from Communicable Disease Surveillance Centre (UK), reported in "Influenza in a Boarding School", *British Medical Journal*, March 4, 1978.

[7] Data from C. M. Perrins and D. Lack, reported by J. R. Krebs and N. B. Davies, *An Introduction to Behavioural Ecology* (Oxford: Blackwell, 1987).

(b) What is the relationship between $f(v)$ and $a(v)$?

(c) Where on the graph is $a(v)$ a minimum?

(d) Should the bird try to minimize $f(v)$ or $a(v)$ when it is flying? Why?

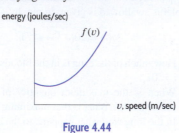

Figure 4.44

60. A person's blood pressure, p, in millimeters of mercury (mm Hg) is given, for t in seconds, by

$$p = 100 + 20 \sin(2.5\pi t).$$

(a) What are the maximum and minimum values of blood pressure?

(b) What is the time between successive maxima?

(c) Show your answers on a graph of blood pressure against time.

4.4 PROFIT, COST, AND REVENUE

Maximizing Profit

A fundamental issue for a producer of goods is how to maximize profit. For a quantity, q, the profit $\pi(q)$ is the difference between the revenue, $R(q)$, and the cost, $C(q)$, of supplying that quantity. Thus, $\pi(q) = R(q) - C(q)$. The marginal cost, $MC = C'$, is the derivative of C; marginal revenue is $MR = R'$.

Now we look at how to maximize total profit, given functions for revenue and cost. The next example suggests a criterion for identifying the optimal production level.

Example 1   Estimate the maximum profit if the revenue and cost are given by the curves R and C, respectively, in Figure 4.45.

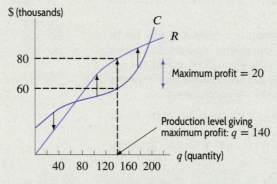

Figure 4.45: Maximum profit at $q = 140$

Solution   Since profit is revenue minus cost, the profit is represented by the vertical distance between the cost and revenue curves, marked by the vertical arrows in Figure 4.45. When revenue is below cost, the company is taking a loss; when revenue is above cost, the company is making a profit. The maximum profit must occur between about $q = 70$ and $q = 200$, which is the interval in which the company is making a profit. Profit is maximized when the vertical distance between the curves is largest (and revenue is above cost). This occurs at approximately $q = 140$.

The profit accrued at $q = 140$ is the vertical distance between the curves, so the maximum profit $= \$80,000 - \$60,000 = \$20,000$.

Maximum Profit Can Occur Where $MR = MC$

We now analyze the marginal costs and marginal revenues near the optimal point. Zooming in on Figure 4.45 around $q = 140$ gives Figure 4.46.

At a production level q_1 to the left of 140 in Figure 4.46, marginal cost is less than marginal revenue. The company would make more money by producing more units, so production should be increased (toward a production level of 140). At any production level q_2 to the right of 140, marginal cost is greater than marginal revenue. The company would lose money by producing more units and would make more money by producing fewer units. Production should be adjusted down toward 140.

What about the marginal revenue and marginal cost at $q = 140$? Since $MC < MR$ to the left of 140, and $MC > MR$ to the right of 140, we expect $MC = MR$ at 140. In this example, profit is maximized at the point where the slopes of the cost and revenue graphs are equal.

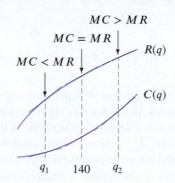

Figure 4.46: Example 1: Maximum profit occurs where $MC = MR$

We can get the same result analytically. Global maxima and minima of a function can only occur at critical points of the function or at the endpoints of the interval. To find critical points of π, look for zeros of the derivative:

$$\pi'(q) = R'(q) - C'(q) = 0.$$

So

$$R'(q) = C'(q),$$

that is, the slopes of the graphs of $R(q)$ and $C(q)$ are equal at q. In economic language,

---

The maximum (or minimum) profit can occur where

$$\text{Marginal profit} = 0,$$

that is, where

$$\text{Marginal revenue} = \text{Marginal cost.}$$

Of course, maximum or minimum profit does not *have* to occur where $MR = MC$; either one could occur at an endpoint. Example 2 shows how to visualize maxima and minima of the profit on a graph of marginal revenue and marginal cost.

Example 2   The total revenue and total cost curves for a product are given in Figure 4.47.

(a) Sketch the marginal revenue and marginal cost, MR and MC, on the same axes. Mark the two quantities where marginal revenue equals marginal cost. What is the significance of these two quantities? At which quantity is profit maximized?

(b) Graph the profit function $\pi(q)$.

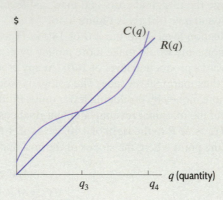

Figure 4.47: Total revenue and total cost

Solution   (a) Since $R(q)$ is a straight line with positive slope, the graph of its derivative, MR, is a horizontal line. (See Figure 4.48.) Since $C(q)$ is always increasing, its derivative, MC, is always positive. As q increases, the cost curve changes from concave down to concave up, so the derivative of the cost function, MC, changes from decreasing to increasing. (See Figure 4.48.) The local minimum on the marginal cost curve corresponds to the inflection point of $C(q)$.

Where is profit maximized? We know that the maximum profit can occur when Marginal revenue = Marginal cost, that is where the curves in Figure 4.48 cross at q_1 and q_2. Do these points give the maximum profit?

We first consider q_1. To the left of q_1, we have $MR < MC$, so $\pi' = MR - MC$ is negative and the profit function is decreasing there. To the right of q_1, we have $MR > MC$, so π' is positive and the profit function is increasing. This behavior, decreasing and then increasing, means that the profit function has a local minimum at q_1. This is certainly not the production level we want.

What happens at q_2? To the left of q_2, we have $MR > MC$, so π' is positive and the profit function is increasing. To the right of q_2, we have $MR < MC$, so π' is negative and the profit function is decreasing. This behavior, increasing and then decreasing, means that the profit function has a local maximum at q_2. The global maximum profit occurs either at the production level q_2 or at an endpoint (the largest and smallest possible production levels). Since the profit is negative at the endpoints (see Figure 4.47), the global maximum occurs at q_2.

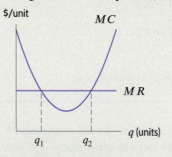

Figure 4.48: Marginal revenue and marginal cost

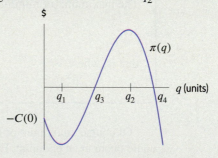

Figure 4.49: Profit function

(b) The graph of the profit function is in Figure 4.49. At the maximum and minimum, the slope of the profit curve is zero:

$$\pi'(q_1) = \pi'(q_2) = 0.$$

Note that since $R(0) = 0$ and $C(0)$ represents the fixed costs of production, we have

$$\pi(0) = R(0) - C(0) = -C(0).$$

Therefore the vertical intercept of the profit function is a negative number, equal in magnitude to the size of the fixed cost.

Example 3   Find the quantity which maximizes profit if the total revenue and total cost (in dollars) are given by

$$R(q) = 5q - 0.003q^2$$
$$C(q) = 300 + 1.1q,$$

where q is quantity and $0 \leq q \leq 1000$ units. What production level gives the minimum profit?

Solution   We begin by looking for production levels that give Marginal revenue = Marginal cost. Since

$$MR = R'(q) = 5 - 0.006q$$
$$MC = C'(q) = 1.1,$$

$MR = MC$ leads to

$$5 - 0.006q = 1.1$$
$$q = \frac{3.9}{0.006} = 650 \text{ units.}$$

Does this represent a local maximum or minimum of the profit π? To decide, look to the left and right of 650 units.

When $q = 649$, we have $MR = \$1.106$ per unit, which is greater than $MC = \$1.10$ per unit.

Thus, producing one more unit (the 650th) brings in more revenue than it costs, so profit increases.

When $q = 651$, we have $MR = \$1.094$ per unit, which is less than $MC = \$1.10$ per unit.

It is not profitable to produce the 651st unit. We conclude that $q = 650$ gives a local maximum for the profit function π.

To check whether $q = 650$ gives a global maximum, we compare the profit at the endpoints, $q = 0$ and $q = 1000$, with the profit at $q = 650$.

At $q = 0$, the only cost is $300 (the fixed costs) and there is no revenue, so $\pi(0) = -\$300$.
At $q = 1000$, we have $R(1000) = \$2000$ and $C(1000) = \$1400$, so $\pi(1000) = \$600$.
At $q = 650$, we have $R(650) = \$1982.50$ and $C(650) = \$1015$, so $\pi(650) = \$967.50$.

Therefore, the maximum profit is obtained at a production level of $q = 650$ units. The minimum profit (a loss) occurs when $q = 0$ and there is no production at all.

Maximizing Revenue

For some companies, costs do not depend on the number of items sold. For example, a city bus company with a fixed schedule has the same costs no matter how many people ride the buses. In such a situation, profit is maximized by maximizing revenue.

Example 4   At a price of $80 for a half-day trip, a white-water rafting company attracts 300 customers. Every $5 decrease in price attracts an additional 30 customers.

(a) Find the demand equation.
(b) Express revenue as a function of price.
(c) What price should the company charge per trip to maximize revenue?

Solution   (a) We first find the equation relating price to demand. If price, p, is 80, the number of trips sold, q, is 300. If p is 75, then q is 330, and so on. See Table 4.1. Because demand changes by a constant (30 people) for every $5 drop in price, q is a linear function of p. Then

$$\text{Slope} = \frac{300 - 330}{80 - 75} = -\frac{30}{5} = -6 \text{ people/dollar,}$$

so the demand equation is $q = -6p + b$. Since $p = 80$ when $q = 300$, we have

$$300 = -6 \cdot 80 + b$$
$$b = 300 + 6 \cdot 80 = 780.$$

The demand equation is $q = -6p + 780$.

(b) Since revenue $R = p \cdot q$, revenue as a function of price is

$$R(p) = p(-6p + 780) = -6p^2 + 780p.$$

(c) Figure 4.50 shows this revenue function has a maximum. To find it, we differentiate:

$$R'(q) = -12p + 780 = 0$$
$$p = \frac{780}{12} = 65.$$

The maximum revenue is achieved when the price is $65.

Table 4.1 *Demand for rafting trips*

Price, p	Number of trips sold, q
80	300
75	330
70	360
65	390
...	...

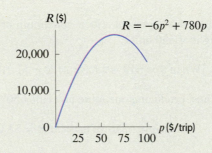

Figure 4.50: Revenue for a rafting company as a function of price

Problems for Section 4.4

1. Figure 4.51 shows cost and revenue. For what production levels is the profit function positive? Negative? Estimate the production at which profit is maximized.

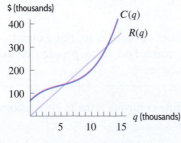

Figure 4.51

2. The revenue from selling q items is $R(q) = 500q - q^2$, and the total cost is $C(q) = 150 + 10q$. Write a function that gives the total profit earned, and find the quantity which maximizes the profit.

3. Revenue is given by $R(q) = 450q$ and cost is given by $C(q) = 10{,}000 + 3q^2$. At what quantity is profit maximized? What is the total profit at this production level?

4. Using the cost and revenue graphs in Figure 4.52, sketch the following functions. Label the points q_1 and q_2.

(a) Total profit   (b) Marginal cost

(c) Marginal revenue

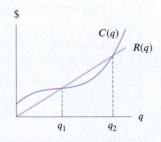

Figure 4.52

5. Figure 4.48 in Section 4.4 shows the points, q_1 and q_2, where marginal revenue equals marginal cost.

   (a) On the graph of the corresponding total cost and total revenue functions in Figure 4.53, label the points q_1 and q_2. Using slopes, explain the significance of these points.
   (b) Explain in terms of profit why one is a local minimum and one is a local maximum.

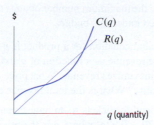

Figure 4.53

6. Let $C(q)$ be the total cost of producing a quantity q of a certain product. See Figure 4.54.

   (a) What is the meaning of $C(0)$?
   (b) Describe in words how the marginal cost changes as the quantity produced increases.
   (c) Explain the concavity of the graph (in terms of economics).
   (d) Explain the economic significance (in terms of marginal cost) of the point at which the concavity changes.
   (e) Do you expect the graph of $C(q)$ to look like this for all types of products?

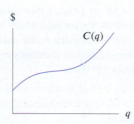

Figure 4.54

7. Let $C(q)$ represent the cost, $R(q)$ the revenue, and $\pi(q)$ the total profit, in dollars, of producing q items.

   (a) If $C'(50) = 75$ and $R'(50) = 84$, approximately how much profit is earned by the 51st item?
   (b) If $C'(90) = 71$ and $R'(90) = 68$, approximately how much profit is earned by the 91st item?
   (c) If $\pi(q)$ is a maximum when $q = 78$, how do you think $C'(78)$ and $R'(78)$ compare? Explain.

8. Table 4.2 shows cost, $C(q)$, and revenue, $R(q)$.

   (a) At approximately what production level, q, is profit maximized? Explain your reasoning.
   (b) What is the price of the product?
   (c) What are the fixed costs?

Table 4.2

q	0	500	1000	1500	2000	2500	3000
$R(q)$	0	1500	3000	4500	6000	7500	9000
$C(q)$	3000	3800	4200	4500	4800	5500	7400

9. Table 4.3 shows marginal cost, MC, and marginal revenue, MR.

   (a) Use the marginal cost and marginal revenue at a production of $q = 5000$ to determine whether production should be increased or decreased from 5000.
   (b) Estimate the production level that maximizes profit.

Table 4.3

q	5000	6000	7000	8000	9000	10,000
MR	60	58	56	55	54	53
MC	48	52	54	55	58	63

10. Figure 4.55 shows graphs of marginal cost and marginal revenue. Estimate the production levels that could maximize profit. Explain your reasoning.

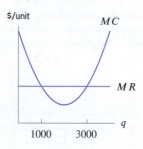

Figure 4.55

11. The marginal cost and marginal revenue of a company are $MC(q) = 0.03q^2 - 1.4q + 34$ and $MR(q) = 30$, where q is the number of items manufactured. To increase profits, should the company increase or decrease production from each of the following levels?

    (a) 25 items   (b) 50 items   (c) 80 items

12. A manufacturing process has marginal costs given in the table; the item sells for $30 per unit. At how many quantities, q, does the profit appear to be a maximum? In what intervals do these quantities appear to lie?

q	0	10	20	30	40	50	60
MC ($/unit)	34	23	18	19	26	39	58

13. Cost and revenue functions are given in Figure 4.56. Approximately what quantity maximizes profits?

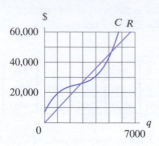

Figure 4.56

14. Cost and revenue functions are given in Figure 4.56.

(a) At a production level of $q = 3000$, is marginal cost or marginal revenue greater? Explain what this tells you about whether production should be increased or decreased.
(b) Answer the same questions for $q = 5000$.

15. When production is 2000, marginal revenue is $4 per unit and marginal cost is $3.25 per unit. Do you expect maximum profit to occur at a production level above or below 2000? Explain.

16. Revenue and cost functions for a company are given in Figure 4.57.

(a) Estimate the marginal cost at $q = 400$.
(b) Should the company produce the 500$^\text{th}$ item? Why?
(c) Estimate the quantity which maximizes profit.

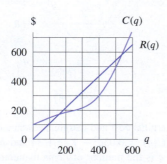

Figure 4.57

17. A company has 100 units to spend for equipment and labor combined. The company spends x on equipment and $100 - x$ on labor, enabling it to produce Q items where

$$Q = 5x^{0.3}(100 - x)^{0.8}.$$

How much should the company spend on equipment to maximize production Q? On labor? What is the maximum production Q?

18. A manufacturing process with a $12 million budget uses x kilograms of one raw material and y kilograms of a second raw material to make $Q = 3\ln(x + 1) + 2\ln(y + 1)$ units of product. The first raw material costs $6 million per kilogram and the second costs $3 million per kilogram.

(a) If the budget is fully spent, find an expression for y in terms of x. For what values of x does this expression make sense?
(b) Find the maximum number of units that can be produced under this budget.

19. The demand equation for a product is $p = 45 - 0.01q$. Write the revenue as a function of q and find the quantity that maximizes revenue. What price corresponds to this quantity? What is the total revenue at this price?

20. The demand for tickets to an amusement park is given by $p = 70 - 0.02q$, where p is the price of a ticket in dollars and q is the number of people attending at that price.

(a) What price generates an attendance of 3000 people? What is the total revenue at that price? What is the total revenue if the price is $20?
(b) Write the revenue function as a function of attendance, q, at the amusement park.
(c) What attendance maximizes revenue?
(d) What price should be charged to maximize revenue?
(e) What is the maximum revenue? Can we determine the corresponding profit?

21. An ice cream company finds that at a price of $4.00, demand is 4000 units. For every $0.25 decrease in price, demand increases by 200 units. Find the price and quantity sold that maximize revenue.

22. At a price of $8 per ticket, a musical theater group can fill every seat in the theater, which has a capacity of 1500. For every additional dollar charged, the number of people buying tickets decreases by 75. What ticket price maximizes revenue?

23. A farmer uses x lb of fertilizer per acre at a cost of $2 per pound, leading to a revenue of $R = 700 - 400e^{-x/100}$ dollars per acre.

(a) How many pounds of fertilizer should be applied per acre to maximize profit?
(b) What is the maximum profit on a 200 acre farm?

24. The demand equation for a quantity q of a product at price p, in dollars, is $p = -5q + 4000$. Companies producing the product report the cost, C, in dollars, to produce a quantity q is $C = 6q + 5$ dollars.

(a) Express a company's profit, in dollars, as a function of q.
(b) What production level earns the company the largest profit?
(c) What is the largest profit possible?

25. **(a)** Production of an item has fixed costs of $10,000 and variable costs of $2 per item. Express the cost, C, of producing q items.
   (b) The relationship between price, p, and quantity, q, demanded is linear. Market research shows that 10,100 items are sold when the price is $5 and 12,872 items are sold when the price is $4.50. Express q as a function of price p.
   (c) Express the profit earned as a function of q.
   (d) How many items should the company produce to maximize profit? (Give your answer to the nearest integer.) What is the profit at that production level?

26. An online seller of knitted sweaters finds that it costs $35 to make her first sweater. Her cost for each additional sweater goes down until it reaches $25 for her 100th sweater, and after that it starts to rise again. If she can sell each sweater for $35, is the quantity sold that maximizes her profit less than 100? Greater than 100?

27. A landscape architect plans to enclose a 3000-square-foot rectangular region in a botanical garden. She will use shrubs costing $45 per foot along three sides and fencing costing $20 per foot along the fourth side. Find the minimum total cost.

28. You run a small furniture business. You sign a deal with a customer to deliver up to 400 chairs, the exact number to be determined by the customer later. The price will be $90 per chair up to 300 chairs, and above 300, the price will be reduced by $0.25 per chair (on the whole order) for every additional chair over 300 ordered. What are the largest and smallest revenues your company can make under this deal?

29. A warehouse selling cement has to decide how often and in what quantities to reorder. It is cheaper, on average, to place large orders, because this reduces the ordering cost per unit. On the other hand, larger orders mean higher storage costs. The warehouse always reorders cement in the same quantity, q. The total weekly cost, C, of ordering and storage is given by

$$C = \frac{a}{q} + bq, \quad \text{where } a, b \text{ are positive constants.}$$

   (a) Which of the terms, a/q and bq, represents the ordering cost and which represents the storage cost?
   (b) What value of q gives the minimum total cost?

30. A demand function is $p = 400 - 2q$, where q is the quantity of the good sold for price p.

   (a) Find an expression for the total revenue, R, in terms of q.
   (b) Differentiate R with respect to q to find the marginal revenue, MR, in terms of q. Calculate the marginal revenue when $q = 10$.
   (c) Calculate the change in total revenue when production increases from $q = 10$ to $q = 11$ units. Confirm that a one-unit increase in q gives a reasonable approximation to the exact value of MR obtained in part (b).

31. A business sells an item at a constant rate of r units per month. It reorders in batches of q units, at a cost of $a + bq$ dollars per order. Storage costs are k dollars per item per month, and, on average, $q/2$ items are in storage, waiting to be sold. [Assume r, a, b, k are positive constants.]

   (a) How often does the business reorder?
   (b) What is the average monthly cost of reordering?
   (c) What is the total monthly cost, C, of ordering and storage?
   (d) Obtain Wilson's lot size formula, the optimal batch size which minimizes cost.

32. **(a)** A cruise line offers a trip for $2000 per passenger. If at least 100 passengers sign up, the price is reduced for *all* the passengers by $10 for every additional passenger (beyond 100) who goes on the trip. The boat can accommodate 250 passengers. What number of passengers maximizes the cruise line's total revenue? What price does each passenger pay then?
   (b) The cost to the cruise line for n passengers is $80,000 + 400n$. What is the maximum profit that the cruise line can make on one trip? How many passengers must sign up for the maximum to be reached and what price will each pay?

33. The government imposes a tax of T dollars per item sold. The producers of the item react by raising the price to p dollars per item to consumers, which reduces the quantity of items sold to q thousand. We have

$$p = 50 + 0.2T$$
$$q = 10 - T.$$

   (a) What is the tax that maximizes the tax revenue collected by the government?
   (b) What is that maximum tax revenue?
   (c) With the tax that maximizes tax revenue, what is the producers' revenue after taxes?
   (d) What is the producers' revenue with no tax?

34. A company manufactures only one product. The quantity, q, of this product produced per month depends on the amount of capital, K, invested (i.e., the number of machines the company owns, the size of its building, and so on) and the amount of labor, L, available each month. We assume that q can be expressed as a *Cobb-Douglas production function*:

$$q = cK^\alpha L^\beta,$$

where c, α, β are positive constants, with $0 < \alpha < 1$ and $0 < \beta < 1$. In this problem we will see how the Russian government could use a Cobb-Douglas function to estimate how many people a newly privatized industry might employ. A company in such an industry has only a small amount of capital available to it and needs to use all of it, so K is fixed. Suppose L is measured in man-hours per month, and that each man-hour costs the company w rubles (a ruble is the unit of Russian currency). Suppose the company has no other costs besides labor, and that each unit of the good can be sold for a fixed price of p rubles. How many man-hours of labor per month should the company use in order to maximize its profit?

35. A company can produce and sell $f(L)$ tons of a product per month using L hours of labor per month. The wage of the workers is w dollars per hour, and the finished product sells for p dollars per ton.

    (a) The function $f(L)$ is the company's production function. Give the units of $f(L)$. What is the practical significance of $f(1000) = 400$?

(b) The derivative $f'(L)$ is the company's marginal product of labor. Give the units of $f'(L)$. What is the practical significance of $f'(1000) = 2$?

(c) The real wage of the workers is the quantity of product that can be bought with one hour's wages. Show that the real wage is w/p tons per hour.

(d) Show that the monthly profit of the company is

$$\pi(L) = pf(L) - wL.$$

(e) Show that when operating at maximum profit, the company's marginal product of labor equals the real wage:

$$f'(L) = \frac{w}{p}.$$

4.5   AVERAGE COST

To stay in business, a company needs to know whether it can turn a profit—which is possible if the price of its product can be set above the average cost of production.

In this section, we see how average cost can be calculated and visualized, and the relationship between average cost and marginal cost.

What Is Average Cost?

The average cost is the cost per unit of producing a certain quantity; it is the total cost divided by the number of units produced.

> If the cost of producing a quantity q is $C(q)$, then the **average cost**, $a(q)$, of producing a quantity q is given by
>
> $$a(q) = \frac{C(q)}{q}.$$

Although both are measured in the same units, for example, dollars per item, be careful not to confuse the average cost with the marginal cost (the cost of producing the next item).

Example 1   A salsa company has cost function $C(q) = 0.01q^3 - 0.6q^2 + 13q + 1000$ (in dollars), where q is the number of cases of salsa produced. If 100 cases are produced, find the average cost per case.

Solution   The total cost of producing the 100 cases is given by

$$C(100) = 0.01(100^3) - 0.6(100^2) + 13(100) + 1000 = \$6300.$$

We find the average cost per case by dividing by 100, the number of cases produced.

$$\text{Average cost} = \frac{6300}{100} = 63 \text{ dollars/case}.$$

If 100 cases of salsa are produced, the average cost is \$63 per case.

Visualizing Average Cost on the Total Cost Curve

We know that average cost is $a(q) = C(q)/q$. Since we can subtract zero from any number without changing it, we can write

$$a(q) = \frac{C(q)}{q} = \frac{C(q) - 0}{q - 0}.$$

This expression gives the slope of the line joining the points $(0, 0)$ and $(q, C(q))$ on the cost curve. See Figure 4.58.

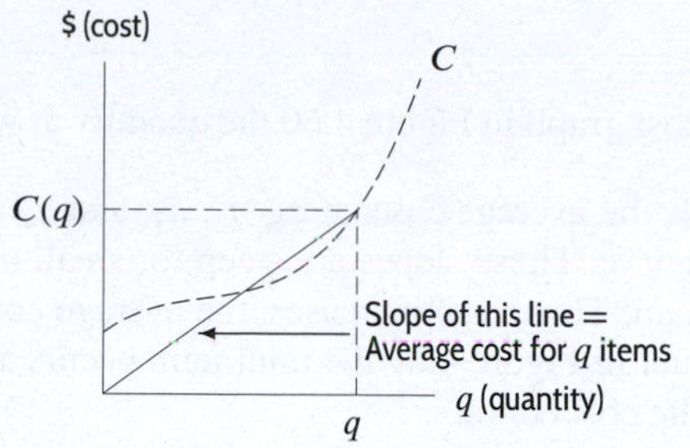

$ (cost)

$C(q)$

Slope of this line =
Average cost for q items

q (quantity)

Figure 4.58: Average cost is the slope of the line from the origin to a point on the cost curve

$$\text{Average cost to produce } q \text{ items} = \frac{C(q)}{q} = \text{Slope of the line from the origin to point } (q, C(q)) \text{ on cost curve.}$$

Minimizing Average Cost

We use the graphical representation of average cost to investigate the relationship between average and marginal cost, and to identify the production level which minimizes average cost.

Example 2   A cost function, in dollars, is $C(q) = 1000 + 20q$, where q is the number of units produced. Find and compare the marginal cost to produce the 100th unit and the average cost of producing 100 units. Illustrate your answer on a graph.

Solution   The cost function is linear with fixed costs of \$1000 and variable costs of \$20 per unit. Thus,

$$\text{Marginal cost} = C'(q) = 20 \text{ dollars per unit.}$$

This means that after 99 units have been produced, it costs an additional \$20 to produce the next unit. In contrast,

$$\text{Average cost of producing 100 units} = a(100) = \frac{C(100)}{100} = \frac{3000}{100} = 30 \text{ dollars/unit.}$$

Notice that the average cost includes the fixed costs of \$1000 spread over the entire production, whereas marginal cost does not. Thus, the average cost is greater than the marginal cost in this example. See Figure 4.59.

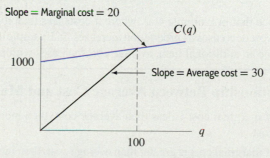

Figure 4.59: Average cost > Marginal cost

Example 3   Mark on the cost graph in Figure 4.60 the quantity at which the average cost is minimized.

Solution   In Figure 4.61, the average costs at q_1, q_2, q_3, and q_4 are given by the slopes of the lines from the origin to the curve. These slopes are steep for small q, become less steep as q increases, and then get steeper again. Thus, as q increases, the average cost decreases and then increases, so there is a minimum value. In Figure 4.61 the minimum occurs at the point q_0 where the line from the origin is tangent to the cost curve.

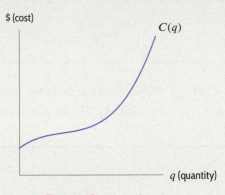

Figure 4.60: A cost function

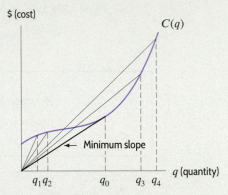

Figure 4.61: Minimum average cost occurs at q_0 where line is tangent to cost curve

In Figure 4.61, notice that average cost is a minimum (at q_0) when average cost equals marginal cost. The next example shows what happens when marginal cost and average cost are not equal.

Example 4   Suppose 100 items are produced at an average cost of \$2 per item. Find the average cost of producing 101 items if the marginal cost to produce the 101st item is: (a) \$1   (b) \$3.

Solution   If 100 items are produced at an average cost of \$2 per item, the total cost of producing the items is $100 \cdot \$2 = \$200$.

(a) Since the marginal, or additional, cost to produce the 101st item is \$1, the total cost of producing 101 items is $\$200 + \$1 = \$201$. The average cost to produce these items is 201/101, or \$1.99 per item. The average cost has gone down.

(b) In this case, the marginal cost to produce the 101st item is \$3. The total cost to produce 101 items is \$203 and the average cost is 203/101, or \$2.01 per item. The average cost has gone up.

Notice that in Example 4(a), where it costs less than the average to produce an additional item, average cost decreases as production increases. In Example 4(b), where it costs more than the average to produce an additional item, average cost increases with production. We summarize:

Relationship Between Average Cost and Marginal Cost

- If marginal cost is less than average cost, then increasing production decreases average cost.
- If marginal cost is greater than average cost, then increasing production increases average cost.
- Marginal cost equals average cost at critical points of average cost.

Example 5   Show analytically that critical points of average cost occur when marginal cost equals average cost.

Solution   Since $a(q) = C(q)/q = C(q)q^{-1}$, we use the product rule to find $a'(q)$:

$$a'(q) = C'(q)(q^{-1}) + C(q)(-q^{-2}) = \frac{C'(q)}{q} + \frac{-C(q)}{q^2} = \frac{qC'(q) - C(q)}{q^2}.$$

At critical points we have $a'(q) = 0$, so

$$\frac{qC'(q) - C(q)}{q^2} = 0$$

Therefore, we have

$$qC'(q) - C(q) = 0$$
$$qC'(q) = C(q)$$
$$C'(q) = \frac{C(q)}{q}.$$

In other words, at a critical point:

$$\text{Marginal cost} = \text{Average cost.}$$

Example 6   A total cost function, in thousands of dollars, is given by $C(q) = q^3 - 6q^2 + 15q$, where q is in thousands and $0 \leq q \leq 5$.

(a) Graph $C(q)$. Estimate visually the quantity at which average cost is minimized.
(b) Graph the average cost function. Use it to estimate the minimum average cost.
(c) Determine analytically the exact value of q at which average cost is minimized.
(d) Graph the marginal cost function on the same axes as the average cost.
(e) Show that at the minimum average cost, Marginal cost = Average cost. Explain how you can see this result on your graph of average and marginal costs.

Solution   (a) A graph of $C(q)$ is in Figure 4.62. Average cost is minimized at the point where a line from the origin to the point on the curve has minimum slope. This occurs where the line is tangent to the curve, which is at approximately $q = 3$, corresponding to a production of 3000 units.

(b) Since average cost is total cost divided by quantity, we have

$$a(q) = \frac{C(q)}{q} = \frac{q^3 - 6q^2 + 15q}{q} = q^2 - 6q + 15.$$

Figure 4.63 suggests that the minimum average cost occurs at $q = 3$.

(c) Average cost is minimized at a critical point of $a(q) = q^2 - 6q + 15$. Differentiating gives

$$a'(q) = 2q - 6 = 0$$
$$q = 3.$$

The minimum occurs at $q = 3$.

(d) See Figure 4.63. Marginal cost is the derivative of $C(q) = q^3 - 6q^2 + 15q$,

$$MC(q) = 3q^2 - 12q + 15.$$

(e) At $q = 3$, we have

$$\text{Marginal cost} = 3 \cdot 3^2 - 12 \cdot 3 + 15 = 6.$$
$$\text{Average cost} = 3^2 - 6 \cdot 3 + 15 = 6.$$

Thus, marginal and average cost are equal at $q = 3$. This result can be seen in Figure 4.63 since the marginal cost curve cuts the average cost curve at the minimum average cost.

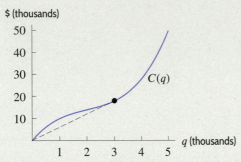

Figure 4.62: Cost function, showing the minimum average cost

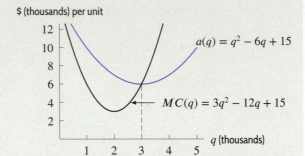

Figure 4.63: Average and marginal cost functions, showing minimum average cost

Problems for Section 4.5

1. For each cost function in Figure 4.64, is there a value of q at which average cost is minimized? If so, approximately where? Explain your answer.

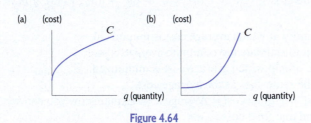

Figure 4.64

2. Figure 4.65 shows cost with $q = 10{,}000$ marked.

(a) Find the average cost when the production level is 10,000 units and interpret it.
(b) Represent your answer to part (a) graphically.
(c) At approximately what production level is average cost minimized?

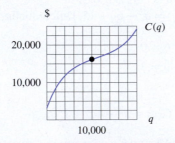

Figure 4.65

3. The graph of a cost function is given in Figure 4.66.

(a) At $q = 25$, estimate the following quantities and represent your answers graphically.

(i)   Average cost   (ii)   Marginal cost

(b) At approximately what value of q is average cost minimized?

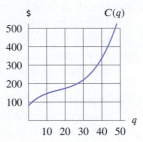

Figure 4.66

4. The cost of producing q items is $C(q) = 2500 + 12q$ dollars.

(a) What is the marginal cost of producing the 100th item? the 1000th item?
(b) What is the average cost of producing 100 items? 1000 items?

5. The cost function is $C(q) = 1000 + 20q$. Find the marginal cost to produce the 200th unit and the average cost of producing 200 units.

6. Graph the average cost function corresponding to the total cost function shown in Figure 4.67.

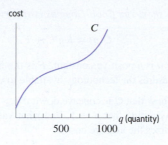

cost

Figure 4.67

7. The cost of producing q units of a good is $C(q) = 0.1q^2 + 1000$ million dollars.

   (a) What are the fixed costs? The marginal cost? The average cost?
   (b) At what production level does the average cost have a critical point?
   (c) Is the critical point a local maximum or a local minimum or neither?

8. The total cost of production, in thousands of dollars, is $C(q) = q^3 - 12q^2 + 60q$, where q is in thousands and $0 \le q \le 8$.

   (a) Graph $C(q)$. Estimate visually the quantity at which average cost is minimized.
   (b) Determine analytically the exact value of q at which average cost is minimized.

9. You are the manager of a firm that sells slippers for $20 a pair. You are producing 1200 pairs of slippers each month, at an average cost of $2 each. The marginal cost at a production level of 1200 is $3 per pair.

   (a) Are you making or losing money?
   (b) Will increasing production increase or decrease your average cost? Your profit?
   (c) Would you recommend that production be increased or decreased?

10. The average cost per item to produce q items is given by

$$a(q) = 0.01q^2 - 0.6q + 13, \quad \text{for} \quad q > 0.$$

   (a) What is the total cost, $C(q)$, of producing q goods?
   (b) What is the minimum marginal cost? What is the practical interpretation of this result?
   (c) At what production level is the average cost a minimum? What is the lowest average cost?
   (d) Compute the marginal cost at $q = 30$. How does this relate to your answer to part (c)? Explain this relationship both analytically and in words.

11. The marginal cost at a production level of 2000 units of an item is $10 per unit and the average cost of producing 2000 units is $15 per unit. If the production level were increased slightly above 2000, would the following quantities increase or decrease, or is it impossible to tell?

   (a) Average cost   (b) Profit

12. An agricultural worker in Uganda is planting clover to increase the number of bees making their home in the region. There are 100 bees in the region naturally, and for every acre put under clover, 20 more bees are found in the region.

   (a) Draw a graph of the total number, $N(x)$, of bees as a function of x, the number of acres devoted to clover.
   (b) Explain, both geometrically and algebraically, the shape of the graph of:
      (i) The marginal rate of increase of the number of bees with acres of clover, $N'(x)$.
      (ii) The average number of bees per acre of clover, $N(x)/x$.

13. (a) Find the average cost at each of the production levels q in the table.
    (b) At which q-value shown does the average cost have a critical point?
    (c) Is the critical value a local maximum, local minimum, or neither?

q	5	6	7	8	9	10
$C(q)$	89	100	113	128	145	164
$MC(q)$	10	12	14	16	18	20

14. A developer has purchased a laundromat and an adjacent factory. To keep smoke, which ruins the clothes, out of the dryers the developer can protect the laundromat or install filters on the factory's smokestacks. The daily cost of filters for the factory and the cost of protecting the laundromat against smoke depend on the number of filters used, as shown in the the table.

   (a) Make a table which shows, for each possible number of filters (0 through 7), the marginal cost of the filter, the average cost of the filters, and the marginal savings in protecting the laundromat from smoke.
   (b) What should the developer do to minimize total costs to the two businesses? Use the table from part (a) to explain your answer.
   (c) What should the developer do if, in addition to the cost of the filters, the filters must be mounted on a rack which costs $100?
   (d) What should the developer do if the rack costs $50?

Number of filters	Total cost of filters	Total cost of protecting laundromat from smoke
0	$0	$127
1	$5	$63
2	$11	$31
3	$18	$15
4	$26	$6
5	$35	$3
6	$45	$0
7	$56	$0

15. Show analytically that if marginal cost is less than average cost, then the derivative of average cost with respect to quantity satisfies $a'(q) < 0$.

16. Show analytically that if marginal cost is greater than average cost, then the derivative of average cost with respect to quantity satisfies $a'(q) > 0$.

17. A reasonably realistic model of a firm's costs is given

by the *short-run Cobb-Douglas cost curve*

$$C(q) = Kq^{1/a} + F,$$

where a is a positive constant, F is the fixed cost, and K measures the technology available to the firm.

(a) Show that C is concave down if $a > 1$.
(b) Assuming that $a < 1$, find what value of q minimizes the average cost.

4.6 ELASTICITY OF DEMAND

The sensitivity of demand to changes in price varies with the product. For example, a change in the price of light bulbs may not affect the demand for light bulbs much, because people need light bulbs no matter what their price. However, a change in the price of a particular make of car may have a significant effect on the demand for that car, because people can switch to another make.

Elasticity of Demand

We want to find a way to measure this sensitivity of demand to price changes. Our measure should work for products as diverse as light bulbs and cars. The prices of these two items are so different that it makes little sense to talk about absolute changes in price: Changing the price of light bulbs by \$1 is a substantial change, whereas changing the price of a car by \$1 is not. Instead, we use the percent change in price. How, for example, does a 1% increase in price affect the demand for the product?

Let Δp denote the change in the price p of a product and Δq denote the corresponding change in quantity q demanded. The percent change in price is $\Delta p/p$ and the percent change in quantity demanded is $\Delta q/q$. We assume in this book that Δp and Δq have opposite signs (because increasing the price usually decreases the quantity demanded). Then the effect of a price change on demand is measured by the absolute value of the ratio

$$\left| \frac{\text{Percent change in demand}}{\text{Percent change in price}} \right| = \left| \frac{\Delta q/q}{\Delta p/p} \right| = \left| \frac{\Delta q}{q} \cdot \frac{p}{\Delta p} \right| = \left| \frac{p}{q} \cdot \frac{\Delta q}{\Delta p} \right|$$

For small changes in p, we approximate $\Delta q/\Delta p$ by the derivative dq/dp. We define:

> The **elasticity of demand**[8] for a product, E, is given approximately by
>
> $$E \approx \left| \frac{\Delta q/q}{\Delta p/p} \right|, \quad \text{or exactly by} \quad E = \left| \frac{p}{q} \cdot \frac{dq}{dp} \right|.$$

Increasing the price of an item by 1% causes a drop of approximately $E\%$ in the quantity of goods demanded. For small changes, Δp, in price,

$$\frac{\Delta q}{q} \approx -E \frac{\Delta p}{p}.$$

If $E > 1$, a 1% increase in price causes demand to drop by more than 1%, and we say that demand is *elastic*. If $0 \leq E < 1$, a 1% increase in price causes demand to drop by less than 1%, and we say that demand is *inelastic*. In general, a larger elasticity causes a larger percent change in demand for a given percent change in price.

---

[8] When it is necessary to distinguish it from other elasticities, this quantity is called the elasticity of demand with respect to price, or the price elasticity of demand.

Example 1 Raising the price of hotel rooms from \$75 to \$80 per night reduces weekly sales from 100 rooms to 90 rooms.

(a) Approximate the elasticity of demand for rooms at a price of \$75.
(b) Should the owner raise the price?

Solution (a) The percent change in the price is

$$\frac{\Delta p}{p} = \frac{5}{75} = 0.067 = 6.7\%,$$

and the percent change in demand is

$$\frac{\Delta q}{q} = \frac{-10}{100} = -0.1 = -10\%.$$

The elasticity of demand is approximated by the ratio

$$E \approx \left| \frac{\Delta q/q}{\Delta p/p} \right| = \frac{0.10}{0.067} = 1.5.$$

The elasticity is greater than 1 because the percent change in the demand is greater than the percent change in the price.

(b) At a price of \$75 per room,

$$\text{Revenue} = (100 \text{ rooms})(\$75 \text{ per room}) = \$7500 \text{ per week.}$$

At a price of \$80 per room,

$$\text{Revenue} = (90 \text{ rooms})(\$80 \text{ per room}) = \$7200 \text{ per week.}$$

A price increase results in loss of revenue, so the price should not be raised.

Example 2 The demand curve for a product is given by $q = 1000 - 2p^2$, where p is the price. Find the elasticity at $p = 10$ and at $p = 15$. Interpret your answers.

Solution We first find the derivative $dq/dp = -4p$. At a price of $p = 10$, we have $dq/dp = -4 \cdot 10 = -40$, and the quantity demanded is $q = 1000 - 2 \cdot 10^2 = 800$. At this price, the elasticity is

$$E = \left| \frac{p}{q} \cdot \frac{dq}{dp} \right| = \left| \frac{10}{800}(-40) \right| = 0.5.$$

The demand is inelastic at a price of $p = 10$: a 1% increase in price results in approximately a 0.5% decrease in demand.

At a price of \$15, we have $q = 550$ and $dq/dp = -60$. The elasticity is

$$E = \left| \frac{p}{q} \cdot \frac{dq}{dp} \right| = \left| \frac{15}{550}(-60) \right| = 1.64.$$

The demand is elastic: a 1% increase in price results in approximately a 1.64% decrease in demand.

Revenue and Elasticity of Demand

Elasticity enables us to analyze the effect of a price change on revenue. An increase in price usually leads to a fall in demand. However, the revenue may increase or decrease. The revenue $R = pq$ is the product of two quantities, and as one increases, the other decreases. Elasticity measures the relative significance of these two competing changes.

Example 3   Three hundred units of an item are sold when the price of the item is $10. When the price of the item is raised by $1, what is the effect on revenue if the quantity sold drops by

(a)   10 units?   (b)   100 units?

Solution   Since Revenue = Price · Quantity, when the price is $10, we have

$$\text{Revenue} = 10 \cdot 300 = \$3000.$$

(a)   At a price of $11, the quantity sold is $300 - 10 = 290$, so

$$\text{Revenue} = 11 \cdot 290 = \$3190.$$

Thus, raising the price has increased revenue.

(b)   At a price of $11, the quantity sold is $300 - 100 = 200$, so

$$\text{Revenue} = 11 \cdot 200 = \$2200.$$

Thus, raising the price has decreased revenue.

Elasticity allows us to predict whether revenue increases or decreases with a price increase.

Example 4   The item in Example 3(a) is wool whose demand equation is $q = 400 - 10p$. The item in Example 3(b) is houseplants, whose demand equation is $q = 1300 - 100p$. Find the elasticity of wool and houseplants.

Solution   For wool, $q = 400 - 10p$, so $dq/dp = -10$. Thus,

$$E_{\text{Wool}} = \left| \frac{p \, dq}{q \, dp} \right| = \left| \frac{10}{300}(-10) \right| = \frac{1}{3}.$$

For houseplants, $q = 1300 - 100p$, so $dq/dp = -100$. Thus,

$$E_{\text{Houseplants}} = \left| \frac{p \, dq}{q \, dp} \right| = \left| \frac{10}{300}(-100) \right| = \frac{10}{3}.$$

Notice that $E_{\text{Wool}} < 1$ and revenue increases with an increase in price; $E_{\text{Houseplants}} > 1$ and revenue decreases with an increase in price. In the next example we see the relationship between elasticity and maximum revenue.

Example 5   Table 4.4 shows the demand, q, revenue, R, and elasticity, E, for the product in Example 2 at several prices. What price brings in the greatest revenue? What is the elasticity at that price?

Solution   Table 4.4 suggests that maximum revenue is achieved at a price of about $13, and at that price, E is about 1. At prices below $13, we have $E < 1$, so the reduction in demand caused by a price increase is small; thus, raising the price increases revenue. At prices above $13, we have $E > 1$, so the increase in demand caused by a price decrease is relatively large; thus lowering the price increases revenue.

Table 4.4   *Revenue and elasticity at different points*

Price p	10	11	12	13	14	15
Demand q	800	758	712	662	608	550
Revenue R	8000	8338	8544	8606	8512	8250
Elasticity E	0.5	0.64	0.81	1.02	1.29	1.64
	Inelastic	Inelastic	Inelastic	Elastic	Elastic	Elastic

Example 6 shows that revenue does have a local maximum when $E = 1$. We summarize as follows:

Relationship Between Elasticity and Revenue

- If $E < 1$, demand is inelastic and revenue is increased by raising the price.
- If $E > 1$, demand is elastic and revenue is increased by lowering the price.
- $E = 1$ occurs at critical points of the revenue function.

Example 6   Show analytically that critical points of the revenue function occur when $E = 1$.

Solution   We think of revenue as a function of price. Using the product rule to differentiate $R = pq$, we have

$$\frac{dR}{dp} = \frac{d}{dp}(pq) = p\frac{dq}{dp} + \frac{dp}{dp}q = p\frac{dq}{dp} + q.$$

At a critical point the derivative dR/dp equals zero, so we have

$$p\frac{dq}{dp} + q = 0$$

$$p\frac{dq}{dp} = -q$$

$$\frac{p}{q}\frac{dq}{dp} = -1$$

$$E = 1.$$

Elasticity of Demand for Different Products

Different products generally have different elasticities. See Table 4.5. If there are close substitutes for a product, or if the product is a luxury rather than a necessity, a change in price generally has a large effect on demand, and the demand for the product is elastic. On the other hand, if there are no close substitutes or if the product is a necessity, changes in price have a relatively small effect on demand, and the demand is inelastic. For example, demand for salt, penicillin, eyeglasses, and lightbulbs is inelastic over the usual range of prices for these products.

Table 4.5   *Elasticity of demand (with respect to price) for selected farm products*[9]

Cabbage	0.25	Oranges	0.62
Potatoes	0.27	Cream	0.69
Wool	0.33	Apples	1.27
Peanuts	0.38	Peaches	1.49
Eggs	0.43	Fresh tomatoes	2.22
Milk	0.49	Lettuce	2.58
Butter	0.62	Fresh peas	2.83

[9]Estimated by the US Department of Agriculture and reported in W. Adams & J. Brock, *The Structure of American Industry*, 10th ed (Englewood Cliffs: Prentice Hall, 2000).

Problems for Section 4.6

1. The elasticity of a good is $E = 0.5$. What is the effect on the quantity demanded of:

   (a) A 3% price increase? (b) A 3% price decrease?

2. The elasticity of a good is $E = 2$. What is the effect on the quantity demanded of:

   (a) A 3% price increase? (b) A 3% price decrease?

3. The elasticity[10] of bread is $E = 0.35$. If the price of bread increases from \$1.75 to \$2.00, does the demand for bread increase or decrease? Estimate the percent change in the quantity demanded.

4. The elasticity[11] of coffee is $E = 1.4$. If the price of coffee drops from \$12.00 to \$11.50, does the demand for coffee increase or decrease? Estimate the percent change in the quantity demanded.

5. What are the units of elasticity if:

   (a) Price p is in dollars and quantity q is in tons?
   (b) Price p is in yen and quantity q is in liters?
   (c) What can you conclude in general?

6. There are many brands of laundry detergent. Would you expect the elasticity of demand for any particular brand to be high or low? Explain.

7. Would you expect the demand for high-definition television sets to be elastic or inelastic? Explain.

8. There is only one company offering wireless service in a town. Would you expect the elasticity of demand for wireless service to be high or low? Explain.

9. What is the elasticity for peaches in Table 4.5? Explain what this number tells you about the effect of price increases on the demand for peaches. Is the demand for peaches elastic or inelastic? Is this what you expect? Explain.

10. What is the elasticity for potatoes in Table 4.5? Explain what this number tells you about the effect of price increases on the demand for potatoes. Is the demand for potatoes elastic or inelastic? Is this what you expect? Explain.

11. The demand for a product is given by $q = 200 - 2p^2$. Find the elasticity of demand when the price is \$5. Is the demand inelastic or elastic, or neither?

12. The demand for a product is given by $p = 90 - 10q$. Find the elasticity of demand when $p = 50$. If this price rises by 2%, calculate the corresponding percentage change in demand.

13. School organizations raise money by selling candy door to door. The table shows p, the price of the candy, and q, the quantity sold at that price.

(a) Estimate the elasticity of demand at a price of \$1.00. At this price, is the demand elastic or inelastic?
(b) Estimate the elasticity at each of the prices shown. What do you notice? Give an explanation for why this might be so.
(c) At approximately what price is elasticity equal to 1?
(d) Find the total revenue at each of the prices shown. Confirm that the total revenue appears to be maximized at approximately the price where $E = 1$.

p	\$1.00	\$1.25	\$1.50	\$1.75	\$2.00	\$2.25	\$2.50
q	2765	2440	1980	1660	1175	800	430

14. The demand for yams is given by $q = 5000 - 10p^2$, where q is in pounds of yams and p is the price of a pound of yams.

   (a) If the current price of yams is \$2 per pound, how many pounds will be sold?
   (b) Is the demand at \$2 elastic or inelastic? Is it more accurate to say "People want yams and will buy them no matter what the price" or "Yams are a luxury item and people will stop buying them if the price gets too high"?

15. The demand for yams is given in Problem 14.

   (a) At a price of \$2 per pound, what is the total revenue for the yam farmer?
   (b) Write revenue as a function of price, and then find the price that maximizes revenue.
   (c) What quantity is sold at the price you found in part (b), and what is the total revenue?
   (d) Show that $E = 1$ at the price you found in part (b).

16. In Kazakhstan, the demand curve[12] of dairy products is $q = kp^{-0.6}$ for some positive constant k. What is the elasticity of dairy products in Kazakhstan?

17. Find the exact price that maximizes revenue for sales of the product in Example 2.

18. If $E = 2$ for all prices p, how can you maximize revenue?

19. If $E = 0.5$ for all prices p, how can you maximize revenue?

20. (a) Let p be the price and q be the quantity sold of a good with a high elasticity of demand, E. Explain intuitively (without formulas) the effect of raising the price on the revenue, R.
   (b) Derive an expression for dR/dp in terms of q and E. Show all the steps and reasoning.
   (c) Explain how your answer to part (b) confirms your answer to part (a).

[10]www.ers.usda.gov/data-products/commodity-and-food-elasticities. Accessed December 2016.
[11]www.ers.usda.gov/data-products/commodity-and-food-elasticities. Accessed December 2016.
[12]Based on www.ers.usda.gov/data-products/commodity-and-food-elasticities. Accessed December 2016.

21. **(a)** If the demand equation is $pq = k$ for a positive constant k, compute the elasticity of demand.
    (b) Explain the answer to part (a) in terms of the revenue function.

22. Show that a demand equation $q = k/p^r$, where r is a positive constant, gives constant elasticity $E = r$.

23. If p is price and E is the elasticity of demand for a good, show analytically that

    $$\text{Marginal revenue} = p(1 - 1/E).$$

24. Suppose cost is proportional to quantity, $C(q) = kq$. Show that a firm earns maximum profit when

    $$\frac{\text{Profit}}{\text{Revenue}} = \frac{1}{E}.$$

    [Hint: Combine the result of Problem 23 with the fact that profit is maximized when $MR = MC$.]

25. A linear demand function is given in Figure 4.68. Economists compute elasticity of demand E for any quantity q_0 using the formula

    $$E = d_1/d_2,$$

    where d_1 and d_2 are the vertical distances shown in Figure 4.68.

    (a) Explain why this formula works.
    (b) Determine the prices, p, at which (i) $E > 1$
    (ii) $E < 1$ (iii) $E = 1$

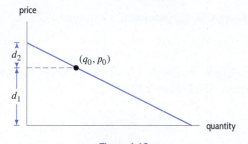

Figure 4.68

26. Show analytically that if elasticity of demand satisfies $E > 1$, then the derivative of revenue with respect to price satisfies $dR/dp < 0$.

27. Show analytically that if elasticity of demand satisfies $E < 1$, then the derivative of revenue with respect to price satisfies $dR/dp > 0$.

28. If q is the quantity of chicken demanded as a function of the price p of beef, the *cross-price* elasticity of demand for chicken with respect to the price of beef is defined as $E_{\text{cross}} = |p/q \cdot dq/dp|$. What does E_{cross} tell you about the sensitivity of the quantity of chicken bought to changes in the price of beef?

29. Dwell time, t, is the time in minutes that shoppers spend in a store. Sales, s, is the number of dollars they spend in the store. The elasticity of sales with respect to dwell time is 1.3. Explain what this means in simple language.

30. Elasticity of cost with respect to quantity is defined as $E_{C,q} = q/C \cdot dC/dq$.

    (a) What does this elasticity tell you about sensitivity of cost to quantity produced?
    (b) Show that $E_{C,q} = $ Marginal cost/Average cost.

31. The *income* elasticity of demand for a product is defined as $E_{\text{income}} = |I/q \cdot dq/dI|$ where q is the quantity demanded as a function of the income I of the consumer. What does E_{income} tell you about the sensitivity of the quantity of the product purchased to changes in the income of the consumer?

4.7 LOGISTIC GROWTH

In 1923, eighteen koalas were introduced to Kangaroo Island, off the coast of Australia.[13] The koalas thrived on the island and their population grew to about 5000 in 1997. Is it reasonable to expect the population to continue growing exponentially? Since there is only a finite amount of space on the island, the population cannot grow without bound forever. Instead we expect that there is a maximum population that the island can sustain. Population growth with an upper bound can be modeled with a *logistic* or *inhibited growth model*.

Modeling the US Population

Population projections first became important to political philosophers in the late eighteenth century. As concern for scarce resources has grown, so has the interest in accurate population projections. In the US, the population is recorded every ten years by a census. The first such census was in 1790. Table 4.6 contains the census data from 1790 to 2010.

[13] *Watertown Daily Times*, April 18, 1997.

Table 4.6   *US population,[14] in millions, 1790–2010*

Year	Population	Year	Population	Year	Population	Year	Population
1790	3.9	1850	23.2	1910	92.2	1970	203.2
1800	5.3	1860	31.4	1920	106.0	1980	226.5
1810	7.2	1870	38.6	1930	123.2	1990	248.7
1820	9.6	1880	50.2	1940	132.2	2000	281.4
1830	12.9	1890	63.0	1950	151.3	2010	308.7
1840	17.1	1900	76.2	1960	179.3		

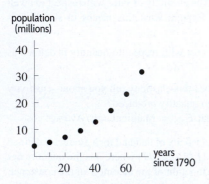

Figure 4.69: US population, 1790–1860

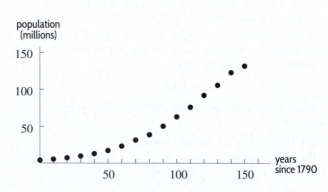

Figure 4.70: US population, 1790–1940

Figure 4.69 suggests that the population grew exponentially during the years 1790–1860. However, after 1860 the rate of growth began to decrease. See Figure 4.70.

The Years 1790–1860: An Exponential Model

We begin by modeling the US population for the years 1790–1860 using an exponential function. If t is the number of years since 1790 and P is the population in millions, regression gives the exponential function that fits the data as approximately[15]

$$P = 3.9(1.03)^t.$$

Thus, between 1790 and 1860, the US population was growing at an annual rate of about 3%.

The function $P = 3.9(1.03)^t$ is plotted in Figure 4.71 with the data; it fits the data remarkably well. Of course, since we used the data from throughout the 70-year period, we should expect good agreement throughout that period. What is surprising is that if we had used only the populations in 1790 and 1800 to create our exponential function, the predictions would still be very accurate. It is amazing that a person in 1800 could predict the population 60 years later so accurately, especially when one considers all the wars, recessions, epidemics, additions of new territory, and immigration that took place from 1800 to 1860.

---

[14] www.census.gov. Accessed February 12, 2012.

[15] See Appendix A: Fitting Formulas to Data. Different algorithms may give different formulas.

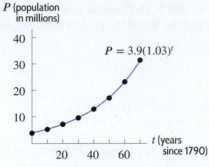

Figure 4.71: An exponential model for
the US population, 1790–1860

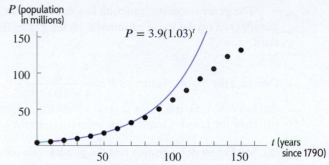

Figure 4.72: The exponential model and the US population,
1790–1940. Not a good fit beyond 1860

The Years 1790–1940: A Logistic Model

How well does the exponential function fit the US population beyond 1860? Figure 4.72 shows a graph of the US population from 1790 until 1940 with the exponential function $P = 3.9(1.03)^t$. The exponential function which fit the data so well for the years 1790–1860 does not fit very well beyond 1860. We must look for another way to model this data.

The graph of the function given by the data in Figure 4.70 is concave up for small values of t, but then appears to become concave down and to be leveling off. This kind of growth is modeled with a *logistic function*. If t is in years since 1790, the function

$$P = \frac{192}{1 + 48e^{-0.0317t}},$$

which is graphed in Figure 4.73, fits the data well up to 1940. Such a formula is found by logistic regression on a calculator or computer.[16]

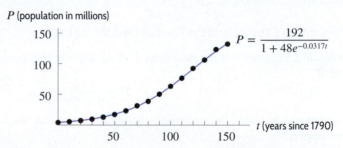

Figure 4.73: A logistic model for US population, 1790–1940

The Logistic Function

A logistic function, such as that used to model the US population, is everywhere increasing. Its graph is concave up at first, then becomes concave down, and levels off at a horizontal asymptote. As we saw in the US population model, a logistic function is approximately exponential for small[17] values of t. A logistic function can be used to model the sales of a new product and the spread of a virus.

> For positive constants L, C, and k, a **logistic function** has the form
>
> $$P = f(t) = \frac{L}{1 + Ce^{-kt}}.$$

---

[16]See Appendix A: Fitting Formulas to Data.
[17]Just how small is small enough depends on the values of the parameters C and k.

The general logistic function has three parameters: L, C, and k. In Example 1, we investigate the effect of two of these parameters on the graph; Problem 1 at the end of the section considers the third.

Example 1

Consider the logistic function $P = \dfrac{L}{1 + 100e^{-kt}}$.

(a) Let $k = 1$. Graph P for several values for L. Explain the effect of the parameter L.
(b) Now let $L = 1$. Graph P for several values for k. Explain the effect of the parameter k.

Solution

(a) See Figure 4.74. Notice that the graph levels off at the value L. The parameter L determines the horizontal asymptote and the upper bound for P.

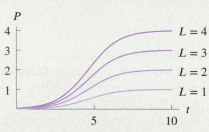

Figure 4.74: Graph of $P = \dfrac{L}{1 + 100e^{-t}}$ with various values of L

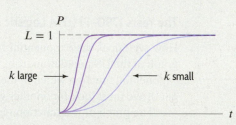

Figure 4.75: Graph of $P = \dfrac{1}{1 + 100e^{-kt}}$ with various values of k

(b) See Figure 4.75. Notice that as k increases, the curve approaches the asymptote more rapidly. The parameter k affects the steepness of the curve.

The Carrying Capacity and the Point of Diminishing Returns

Example 1 suggests that the parameter L of the logistic function is the value at which P levels off, where

$$P = \frac{L}{1 + Ce^{-kt}}.$$

This value L is called the *carrying capacity* and represents the largest population an environment can support.

One way to estimate the carrying capacity is to find the inflection point. The graph of a logistic curve is concave up at first and then concave down. At the inflection point, where the concavity changes, the slope is largest. To the left of this point, the graph is concave up and the rate of growth is increasing. To the right of this point, the graph is concave down and the rate of growth is diminishing. The inflection point is called the *point of diminishing returns*. Problem 24 shows that this point is at $P = L/2$. See Figure 4.76. Companies sometimes watch for this concavity change in the sales of a new product and use it to estimate the maximum potential sales.

Properties of the logistic function $P = \dfrac{L}{1 + Ce^{-kt}}$:

- The limiting value L represents the carrying capacity for P.

- The point of diminishing returns is the inflection point where P is growing the fastest. It occurs where $P = L/2$.

- The logistic function is approximately exponential for small values of t, with growth rate k.

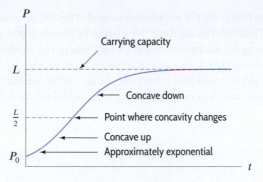

Figure 4.76: Logistic growth

The Years 1790–2010: Another Look at the US Population

We used a logistic function to model the US population between 1790 and 1940. How well does this model fit the US population since 1940? We now look at all the population data from 1790 to 2010.

Example 2   If t is in years since 1790 and P is in millions, we used the following logistic function to model the US population between 1790 and 1940:

$$P = \frac{192}{1 + 48e^{-0.0317t}}.$$

According to this function, what is the maximum US population? Is this prediction accurate? How well does this logistic model fit the growth of the US population since 1940?

Table 4.7   *Predicted versus actual US population, in millions, 1940–2010 (logistic model)*

Year	1940	1950	1960	1970	1980	1990	2000	2010
Actual	132.2	151.3	179.3	203.2	226.5	248.7	281.4	308.7
Predicted	136.0	147.7	157.6	165.7	172.2	177.2	181.0	183.9

Solution   Table 4.7 shows the actual US population between 1940 and 2010 and the predicted values using this logistic model. According to the formula for the logistic function, the upper bound for the population is $L = 192$ million. However, Table 4.7 shows that the actual US population was above this figure by 1970. The fit between the logistic function and the actual population is not a good one beyond 1940. See Figure 4.77.

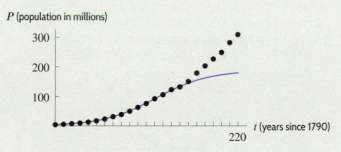

Figure 4.77: The logistic model and the US population, 1790–2010

Despite World War II, which depressed population growth between 1942 and 1945, in the last half of the 1940s the US population surged. The 1950s saw a population growth of 28 million, leaving our logistic model in the dust. This surge in population is referred to as the baby boom.

Once again we have reached a point where our model is no longer useful. This should not lead you to believe that a reasonable mathematical model cannot be found; rather, it points out that no model is perfect and that when one model fails, we seek a better one. Just as we abandoned the exponential model in favor of the logistic model for the US population, we could look further.

Sales Predictions

Total sales of a new product often follow a logistic model. For example, when a new application (app) for a mobile device comes on the market, sales first increase rapidly as word of the app spreads. Eventually, most of the people who want the app have already bought it and sales slow down. The graph of total sales against time is concave up at first and then concave down, with the upper bound L equal to the maximum potential sales.

Example 3   Table 4.8 shows the total sales (in thousands) of new app since it was introduced.

Table 4.8   *Total sales of a new app since its introduction*

t (months)	0	1	2	3	4	5	6	7
P (total sales in 1000s)	0.5	2	8	33	95	258	403	496

(a) Find the point where concavity changes in this function. Use it to estimate the maximum potential sales, L.

(b) Using logistic regression, fit a logistic function to this data. What maximum potential sales does this function predict?

Solution

(a) The rate of change of total sales increases until $t = 5$ and decreases after $t = 5$, so the inflection point is at approximately $t = 5$, when $P = 258$. So $L/2 = 258$ and $L = 516$. The maximum potential sales for this app are estimated to be 516,000.

(b) Logistic regression gives the following function:

$$P = \frac{532}{1 + 869e^{-1.33t}}.$$

Maximum potential sales predicted by this function are $L = 532$, or about 532,000. See Figure 4.78.

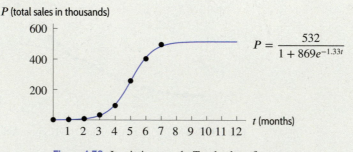

Figure 4.78: Logistic growth: Total sales of an app

Dose-Response Curves

A *dose-response curve* plots the intensity of physiological response to a drug as a function of the dose administered. As the dose increases, the intensity of the response increases, so a dose-response function is increasing. The intensity of the response is generally scaled as a percentage of the maximum response. The curve cannot go above the maximum response (or 100%), so the curve levels off at a horizontal asymptote. Dose-response curves are generally concave up for low doses and concave down for high doses. A dose-response curve can be modeled by a logistic function with the independent variable being the dose of the drug, not time.

A dose-response curve shows the amount of drug needed to produce the desired effect, as well as the maximum effect attainable and the dose required to obtain it. The slope of the dose-response curve gives information about the therapeutic safety margin of the drug.

Drugs need to be administered in a dose which is large enough to be effective but not so large as to be dangerous. Figure 4.79 shows two different dose-response curves: one with a small slope and one with a large slope. In Figure 4.79(a), there is a broad range of dosages at which the drug is both safe and effective. In Figure 4.79(b), where the slope of the curve is steep, the range of dosages at which the drug is both safe and effective is small. If the slope of the dose-response curve is steep, a small mistake in the dosage can have dangerous results. Administration of such a drug is difficult.

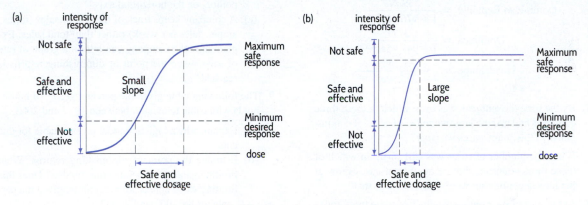

Figure 4.79: What does the slope of the dose-response curve tell us?

Example 4   Figure 4.80 shows dose-response curves for three different drugs used for the same purpose. Discuss the advantages and disadvantages of the three drugs.

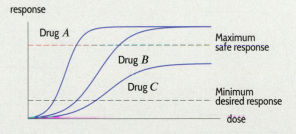

Figure 4.80: What are the advantages and disadvantages of each of these drugs?

Solution   Drugs *A* and *B* show the same maximum response, both above the maximum safe level. Drug *A* reaches this level more quickly. The maximum response of Drug *C* is significantly lower, while still reaching the minimum desired level. Thus, Drug *C* may be the preferred drug despite its lower maximum effect because it is the safest to administer.

Problems for Section 4.7

1. Investigate the effect of the parameter C on the logistic curve

$$P = \frac{10}{1 + Ce^{-t}}.$$

Substitute several values for C and explain, with a graph and with words, the effect of C on the graph.

■ Problems 2–5 are about chikungunya, a disease that arrived in the Americas in 2013 and spread rapidly in 2014. While seldom fatal, the disease causes debilitating joint pain and a high fever. In August 2014, a public challenge was issued to predict the number of cases in each of the affected countries. The winners, Joceline Lega and Heidi Brown, used a logistic model to make their predictions.[18] Let N be the total number of cases of chikungunya in a country by week t, where t is measured since the first cases were recorded in that country. The progress of the disease in three Caribbean countries is represented by the logistic functions:

$$\text{Dominican Republic} = \frac{539{,}227}{1 + 176.8e^{-0.35t}}$$

$$\text{Dominica} = \frac{3771}{1 + 941.75e^{-0.32t}}$$

$$\text{Guadeloupe} = \frac{81{,}780}{1 + 27{,}259e^{-0.32t}}$$

2. By the time the outbreak was over, which of these three countries had had the largest number of cases of chikungunya? How many cases was that?

3. When the number of cases was still small, in which of these three countries did the number of cases grow at the largest continuous rate? What rate was that?

4. At the start of their outbreak, which of these three countries had the smallest number of cases of chikungunya? How many cases was that?

5. Figure 4.81 is a graph of the three logistic functions given. Why does the graph seem to show only two? Which two are they?

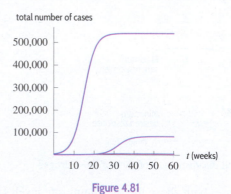

Figure 4.81

6. The following table shows the total sales, in thousands, since a new game was brought to market.

   (a) Plot this data and mark on your plot the point of diminishing returns.
   (b) Predict total possible sales of this game, using the point of diminishing returns.

Month	0	2	4	6	8	10	12	14
Sales	0	2.3	5.5	9.6	18.2	31.8	42.0	50.8

7. Write a paragraph explaining why sales of a new product often follow a logistic curve. Explain the benefit to the company of watching for the point of diminishing returns.

8. (a) Draw a logistic curve. Label the carrying capacity L and the point of diminishing returns t_0.
   (b) Draw the derivative of the logistic curve. Mark the point t_0 on the horizontal axis.
   (c) A company keeps track of the rate of sales (for example, sales per week) rather than total sales. Explain how the company can tell on a graph of rate of sales when the point of diminishing returns is reached.

9. The following table gives the percentage, P, of households with cable television between 1977 and 2003.[19]

   (a) Explain why a logistic model is reasonable for this data.
   (b) Estimate the point of diminishing returns. What limiting value L does this point predict? Does this limiting value appear to be accurate, given the percentages for 2002 and 2003?
   (c) If t is in years since 1977, the best fitting logistic function for this data turns out to be

$$P = \frac{68.8}{1 + 3.486e^{-0.237t}},$$

   What limiting value does this function predict?
   (d) Explain in terms of percentages of households what the limiting value is telling you. Do you think your answer to part (c) is an accurate prediction?

Year	1977	1978	1979	1980	1981	1982	1983
P	16.6	17.9	19.4	22.6	28.3	35.0	40.5
Year	1984	1985	1986	1987	1988	1989	1990
P	43.7	46.2	48.1	50.5	53.8	57.1	59.0
Year	1991	1992	1993	1994	1995	1996	1997
P	60.6	61.5	62.5	63.4	65.7	66.7	67.3
Year	1998	1999	2000	2001	2002	2003	
P	67.4	68.0	67.8	69.2	68.9	68.0	

---

[18] http://www.darpa.mil/news-events/2015-05-27. Based on J. Lega and H. Brown, "Data-Driven Outbreak Forecasting with a Simple Nonlinear Growth Model", *Epidemics,* Vol. 17, December 2016, pp. 19–26.
[19] *The World Almanac and Book of Facts 2005*, p. 310 (New York).

10. The Tojolobal Mayan Indian community in Southern Mexico has available a fixed amount of land.[20] The proportion, P, of land in use for farming t years after 1935 is modeled with the logistic function

$$P = \frac{1}{1 + 3e^{-0.0275t}}.$$

   (a) What proportion of the land was in use for farming in 1935?
   (b) What is the long-run prediction of this model?
   (c) When was half the land in use for farming?
   (d) When is the proportion of land used for farming increasing most rapidly?

11. In the spring of 2003, SARS (Severe Acute Respiratory Syndrome) spread rapidly in several Asian countries and Canada. Table 4.9 gives the total number, P, of SARS cases reported in Hong Kong[21] by day t, where $t = 0$ is March 17, 2003.

   (a) Find the average rate of change of P for each interval in Table 4.9.
   (b) In early April 2003, there was fear that the disease would spread at an ever-increasing rate for a long time. What is the earliest date by which epidemiologists had evidence to indicate that the rate of new cases had begun to slow?
   (c) Explain why an exponential model for P is not appropriate.
   (d) It turns out that a logistic model fits the data well. Estimate the value of t at the inflection point. What limiting value of P does this point predict?
   (e) The best-fitting logistic function for this data turns out to be

$$P = \frac{1760}{1 + 17.53e^{-0.1408t}}.$$

   What limiting value of P does this function predict?

Table 4.9   *Total number of SARS cases in Hong Kong by day t (where $t = 0$ is March 17, 2003)*

t	P	t	P	t	P	t	P
0	95	26	1108	54	1674	75	1739
5	222	33	1358	61	1710	81	1750
12	470	40	1527	68	1724	87	1755
19	800	47	1621				

12. Substitute $t = 0, 10, 20, \ldots, 70$ into the exponential function used in this section to model the US population 1790–1860. Compare the predicted values of the population with the actual values.

13. On page 215, a logistic function was used to model the US population. Use this function to predict the US population in each of the census years from 1790–1940. Compare the predicted and actual values.

14. A curve representing the total number of people, P, infected with a virus often has the shape of a logistic curve of the form

$$P = \frac{L}{1 + Ce^{-kt}},$$

with time t in weeks. Suppose that 10 people originally have the virus and that in the early stages the number of people infected is increasing approximately exponentially, with a continuous growth rate of 1.78. It is estimated that, in the long run, approximately 5000 people will become infected.

   (a) What should we use for the parameters k and L?
   (b) Use the fact that when $t = 0$, we have $P = 10$, to find C.
   (c) Now that you have estimated L, k, and C, what is the logistic function you are using to model the data? Graph this function.
   (d) Estimate the length of time until the rate at which people are becoming infected starts to decrease. What is the value of P at this point?

15. Find the point where the following curve is steepest:

$$y = \frac{50}{1 + 6e^{-2t}} \qquad \text{for } t \geq 0.$$

16. A dose-response curve is given by $R = f(x)$, where R is percent of maximum response and x is the dose of the drug in mg. The curve has the shape shown in Figure 4.79. The inflection point is at $(15, 50)$ and $f'(15) = 11$.

   (a) Explain what $f'(15)$ tells you in terms of dose and response for this drug.
   (b) Is $f'(10)$ greater than or less than 11? Is $f'(20)$ greater than or less than 11? Explain.

17. If R is percent of maximum response and x is dose in mg, the dose-response curve for a drug is given by

$$R = \frac{100}{1 + 100e^{-0.1x}}.$$

   (a) Graph this function.
   (b) What dose corresponds to a response of 50% of the maximum? This is the inflection point, at which the response is increasing the fastest.
   (c) For this drug, the minimum desired response is 20% and the maximum safe response is 70%. What range of doses is both safe and effective for this drug?

---

[20]Adapted from J. S. Thomas and M. C. Robbins, "The Limits to Growth in a Tojolobal Maya Ejido," *Geoscience and Man 26*, pp. 9–16 (Baton Rouge: Geoscience Publications, 1988).
[21]www.who.int/csr/country/en, accessed July 13, 2003.

18. Dose-response curves for three different products are given in Figure 4.82.

    (a) For the desired response, which drug requires the largest dose? The smallest dose?
    (b) Which drug has the largest maximum response? The smallest?
    (c) Which drug is the safest to administer? Explain.

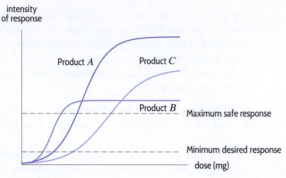

Figure 4.82

19. Explain why it is safer to use a drug for which the derivative of the dose-response curve is smaller.

■ There are two kinds of dose-response curves. One type, discussed in this section, plots the intensity of response against the dose of the drug. We now consider a dose-response curve in which the percentage of subjects showing a specific response is plotted against the dose of the drug. In Problems 20–21, the curve on the left shows the percentage of subjects exhibiting the desired response at the given dose, and the curve on the right shows the percentage of subjects for which the given dose is lethal.

20. In Figure 4.83, what range of doses appears to be both safe and effective for 99% of all patients?

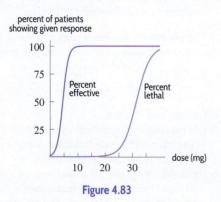

Figure 4.83

21. In Figure 4.84, discuss the possible outcomes and what percent of patients fall in each outcome when 50 mg of the drug is administered.

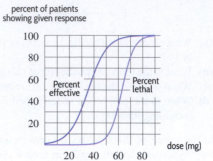

Figure 4.84

22. A population, P, growing logistically is given by

$$P = \frac{L}{1 + Ce^{-kt}}.$$

    (a) Show that

$$\frac{L - P}{P} = Ce^{-kt}.$$

    (b) Explain why part (a) shows that the ratio of the additional population the environment can support to the existing population decays exponentially.

23. Cell membranes contain ion channels. The fraction, f, of channels that are open is a function of the membrane potential V (the voltage inside the cell minus voltage outside), in millivolts (mV), given by

$$f(V) = \frac{1}{1 + e^{-(V+25)/2}}.$$

    (a) Find the values of L, k, and C in the logistic formula for f:

$$f(V) = \frac{L}{1 + Ce^{-kV}}.$$

    (b) At what voltages V are 10%, 50% and 90% of the channels open?

24. Consider a population P satisfying the *logistic equation*

$$\frac{dP}{dt} = kP\left(1 - \frac{P}{L}\right).$$

    (a) Use the chain rule to find d^2P/dt^2.
    (b) Show that the point of diminishing returns, where $d^2P/dt^2 = 0$, occurs where $P = L/2$.

4.8   THE SURGE FUNCTION AND DRUG CONCENTRATION

Nicotine in the Blood

When a person smokes a cigarette, the nicotine from the cigarette enters the body through the lungs, is absorbed into the blood, and spreads throughout the body. Most cigarettes contain between 0.5 and 2.0 mg of nicotine; approximately 20% (between 0.1 and 0.4 mg) is actually inhaled and absorbed into the person's bloodstream. As the nicotine leaves the blood, the smoker feels the need for another cigarette. The half-life of nicotine in the bloodstream is about two hours. The lethal dose is considered to be about 60 mg.

The nicotine level in the blood rises as a person smokes, and tapers off when smoking ceases. Table 4.10 shows blood nicotine concentration (in ng/ml) during and after the use of cigarettes. (Smoking occurred during the first ten minutes and the experimental data shown represent average values for ten people.[22])

The points in Table 4.10 are plotted in Figure 4.85. Functions with this behavior are called *surge functions*. They have equations of the form $y = ate^{-bt}$, where a and b are positive constants.

Table 4.10   *Blood nicotine concentrations during and after the use of cigarettes*

t (minutes)	0	5	10	15	20	25	30	45	60	75	90	105	120
C (ng/ml)	4	12	17	14	13	12	11	9	8	7.5	7	6.5	6

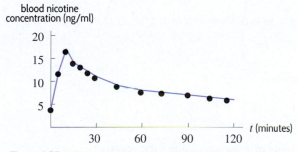

Figure 4.85: Blood nicotine concentrations during and after the use of cigarettes

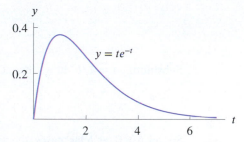

Figure 4.86: One member of the family $y = ate^{-bt}$, with $a = 1$ and $b = 1$

The Family of Functions $y = ate^{-bt}$

What effect do the positive parameters a and b have on the shape of the graph of $y = ate^{-bt}$? Start by looking at the graph with $a = 1$ and $b = 1$. See Figure 4.86. We consider the effect of the parameter b on the graph of $y = ate^{-bt}$ now; the parameter a is considered in Problem 2 of this section.

The Effect of the Parameter b on $y = te^{-bt}$

Graphs of $y = te^{-bt}$ for different positive values of b are shown in Figure 4.87. The general shape of the curve does not change as b changes, but as b decreases, the curve rises for a longer period of time and to a higher value.

[22]N. L. Benowitz, H. Porchet, L. Sheiner, P. Jacob III, "Nicotine Absorption and Cardiovascular Effects with Smokeless Tobacco Use: Comparison with Cigarettes and Nicotine Gum," *Clinical Pharmacology and Therapeutics*, 44 (1988), 24.

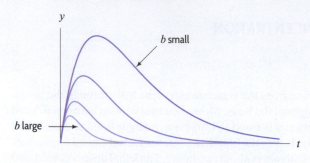

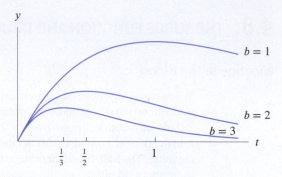

Figure 4.87: Graph of $y = te^{-bt}$, with b varying   **Figure 4.88:** How does the maximum depend on b?

We see in Figure 4.88 that, when $b = 1$, the maximum occurs at about $t = 1$. When $b = 2$, it occurs at about $t = \frac{1}{2}$, and when $b = 3$, it occurs at about $t = \frac{1}{3}$. The next example shows that the maximum of the function $y = te^{-bt}$ occurs at $t = 1/b$.

Example 1   For $b > 0$, show that the maximum value of $y = te^{-bt}$ occurs at $t = 1/b$ and increases as b decreases.

Solution   The maximum occurs at a critical point where $dy/dt = 0$. Differentiating gives

$$\frac{dy}{dt} = 1 \cdot e^{-bt} + t\left(-be^{-bt}\right) = e^{-bt} - bte^{-bt} = e^{-bt}(1 - bt).$$

So $dy/dt = 0$ where

$$1 - bt = 0$$
$$t = \frac{1}{b}.$$

Substituting $t = 1/b$ shows that at the maximum,

$$y = \frac{1}{b}e^{-b(1/b)} = \frac{e^{-1}}{b}.$$

So, for $b > 0$, as b increases, the maximum value of y decreases and vice versa.

The **surge function** $y = ate^{-bt}$, for positive constants a and b, increases rapidly and then decreases toward zero with a maximum at $t = 1/b$. See Figure 4.89.

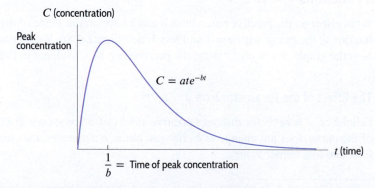

Figure 4.89: Curve showing drug concentration as a function of time

Drug Concentration Curves

As with nicotine, the graph of concentration against time is called the *drug concentration curve*. If t is the time since the drug was administered, the concentration, C, can be modeled by the surge function $C = ate^{-bt}$, where a and b are positive constants. (See Figure 4.89.)

Factors Affecting Drug Absorption

Drug interactions and the age of the patient can affect the drug concentration curve. In Problems 10 and 12, we see that food intake can also affect the rate of absorption of a drug, and (perhaps most surprising) that drug concentration curves can vary markedly between different commercial versions of the same drug.

Example 2 Figure 4.90 shows the drug concentration curves for paracetamol (acetaminophen) alone and for paracetamol taken in conjunction with propantheline. Figure 4.91 shows drug concentration curves for patients known to be slow absorbers of the drug, for paracetamol alone and for paracetamol in conjunction with metoclopramide. Discuss the effects of the additional drugs on peak concentration and the time to reach peak concentration.[23]

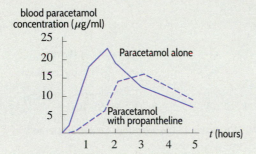

Figure 4.90: Drug concentration curves for paracetamol, normal patients

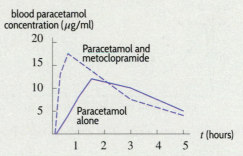

Figure 4.91: Drug concentration curves for paracetamol, patients with slow absorption

Solution Figure 4.90 shows it takes about 1.5 hours for the paracetamol to reach its peak concentration, and that the maximum concentration reached is about 23 μg of paracetamol per ml of blood. However, if propantheline is administered with the paracetamol, it takes much longer to reach the peak concentration (about three hours, or approximately double the time), and the peak concentration is much lower, at about 16 μg/ml.

Comparing the curves for paracetamol alone in Figures 4.90 and 4.91 shows that the time to reach peak concentration is the same (about 1.5 hours), but the maximum concentration is lower for patients with slow absorption. When metoclopramide is given with paracetamol in Figure 4.91, the peak concentration is reached faster and is higher.

Minimum Effective Concentration

The minimum effective concentration of a drug is the blood concentration necessary to achieve a pharmacological response. The time at which this concentration is reached is referred to as onset; termination occurs when the drug concentration falls below this level. See Figure 4.92.

[23]Graeme S. Avery, ed. *Drug Treatment: Principles and Practice of Clinical Pharmacology and Therapeutics* (Sydney: Adis Press, 1976).

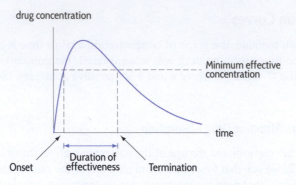

Figure 4.92: When is the drug effective?

Example 3   Depo-Provera was approved for use in the US in 1992 as a contraceptive. Figure 4.93 shows the drug concentration curve for a dose of 150 mg given intramuscularly.[24] The minimum effective concentration is about 4 ng/ml. How often should the drug be administered?

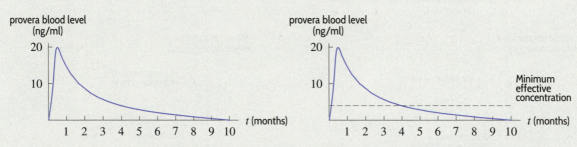

Figure 4.93: Drug concentration curve for Depo-Provera   Figure 4.94: When should the next dose be administered?

Solution   The minimum effective concentration on the drug concentration curve is plotted as a dotted horizontal line at 4 ng/ml. See Figure 4.94. We see that the drug becomes effective almost immediately and ceases to be effective after about four months. Doses should be given about every four months.

Although the dosage interval is four months, notice that it takes ten months after injections are discontinued for Depo-Provera to be entirely eliminated from the body. Fertility during that period is unpredictable.

Problems for Section 4.8

1. If time, t, is in hours and concentration, C, is in ng/ml, the drug concentration curve for a drug is given by

$$C = 12.4te^{-0.2t}.$$

   (a) Graph this curve.
   (b) How many hours does it take for the drug to reach its peak concentration? What is the concentration at that time?
   (c) If the minimum effective concentration is 10 ng/ml, during what time period is the drug effective?

   (d) Complications can arise whenever the level of the drug is above 4 ng/ml. How long must a patient wait before being safe from complications?

2. Let $b = 1$, and graph $C = ate^{-bt}$ using different values for a. Explain the effect of the parameter a.

3. Figure 4.95 shows drug concentration curves for anhydrous ampicillin for newborn babies and adults.[25] Discuss the differences between newborns and adults in the absorption of this drug.

---

[24]Robert M. Julien, *A Primer of Drug Action* (W. H. Freeman and Co, 1995).

[25]J. Silverio and J. Poole, "Serum Concentrations of Ampicillin in Newborn Infants after Oral Administration", *Pediatrics*, 1973, (51), p 578.

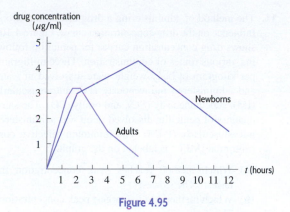

Figure 4.95

4. Absorption of different forms of the antibiotic erythromycin may be increased, decreased, delayed or not affected by food. Figure 4.96 shows the drug concentration levels of erythromycin in healthy, fasting human volunteers who received single oral doses of 500 mg erythromycin tablets, together with either large (250 ml) or small (20 ml) accompanying volumes of water.[26] Discuss the effect of the water on the concentration of erythromycin in the blood. How are the peak concentration and the time to reach peak concentration affected? When does the effect of the volume of water wear off?

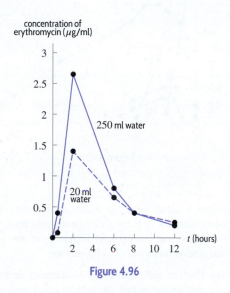

Figure 4.96

5. Hydrocodone bitartrate is a cough suppressant usually administered in a 10 mg oral dose. The peak concentration of the drug in the blood occurs 1.3 hours after consumption and the peak concentration is 23.6 ng/ml. Draw the drug concentration curve for hydrocodone bitartrate.

6. Figure 4.85 shows the concentration of nicotine in the blood during and after smoking a cigarette. Figure 4.97 shows the concentration of nicotine in the blood during and after using chewing tobacco or nicotine gum. (The chewing occurred during the first 30 minutes and the experimental data shown represent the average values for ten patients.)[27] Compare the three nicotine concentration curves (for cigarettes, chewing tobacco and nicotine gum) in terms of peak concentration, the time until peak concentration, and the rate at which the nicotine is eliminated from the bloodstream.

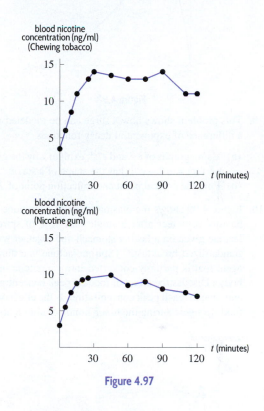

Figure 4.97

7. If t is in minutes since the drug was administered, the concentration, $C(t)$ in ng/ml, of a drug in a patient's bloodstream is given by

$$C(t) = 20te^{-0.03t}.$$

(a) How long does it take for the drug to reach peak concentration? What is the peak concentration?
(b) What is the concentration of the drug in the body after 15 minutes? After an hour?
(c) If the minimum effective concentration is 10 ng/ml, when should the next dose be administered?

---

[26] J. W. Bridges and L.F. Chasseaud, *Progress in Drug Metabolism* (New York: John Wiley and Sons, 1980).

[27] N. L. Benowitz, H. Porchet, L. Sheiner, P. Jacob III, "Nicotine Absorption and Cardiovascular Effects with Smokeless Tobacco Use: Comparison with Cigarettes and Nicotine Gum," *Clinical Pharmacology and Therapeutics*, 44 (1988), 24.

8. For time $t \geq 0$, the function $C = ate^{-bt}$ with positive constants a and b gives the concentration, C, of a drug in the body. Figure 4.98 shows the maximum concentration reached (in nanograms per milliliter, ng/ml) after a 10 mg dose of the cough medicine hydrocodone bitartrate. Find the values of a and b.

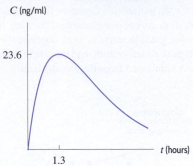

Figure 4.98

9. This problem shows how a surge can be modeled with a difference of exponential decay functions.

   (a) Using graphs of e^{-t} and e^{-2t}, explain why the graph of $f(t) = e^{-t} - e^{-2t}$ has the shape of a surge.
   (b) Find the critical point and inflection point of f.

10. Figure 4.99 shows the plasma levels of canrenone in a healthy volunteer after a single oral dose of spironolactone given on a fasting stomach and together with a standardized breakfast.[28] (Spironolactone is a diuretic agent that is partially converted into canrenone in the body.) Discuss the effect of food on peak concentration and time to reach peak concentration. Is the effect of the food strongest during the first 8 hours, or after 8 hours?

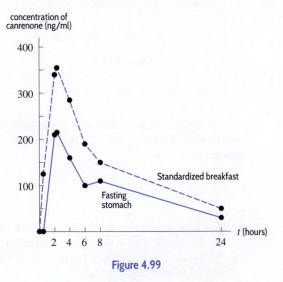

Figure 4.99

11. The method of administering a drug can have a strong influence on the drug concentration curve. Figure 4.100 shows drug concentration curves for penicillin following various routes of administration. Three milligrams per kilogram of body weight were dissolved in water and administered intravenously (IV), intramuscularly (IM), subcutaneously (SC), and orally (PO). The same quantity of penicillin dissolved in oil was administered intramuscularly (P-IM). The minimum effective concentration (MEC) is labeled on the graph.[29]

   (a) Which method reaches peak concentration the fastest? The slowest?
   (b) Which method has the largest peak concentration? The smallest?
   (c) Which method wears off the fastest? The slowest?
   (d) Which method has the longest effective duration? The shortest?
   (e) When penicillin is administered orally, for approximately what time interval is it effective?

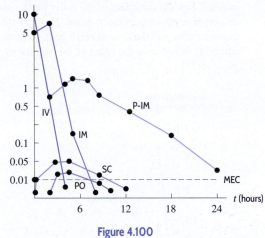

Figure 4.100

12. Figure 4.101 shows drug concentration curves after oral administration of 0.5 mg of four digoxin products. All the tablets met current USP standards of potency, disintegration time, and dissolution rate.[30]

   (a) Discuss differences and similarities in the peak concentration and the time to reach peak concentration.
   (b) Give possible values for minimum effective concentration and maximum safe concentration that would make Product C or Product D the preferred drug.

---

[28]P. Welling and F. Tse, *Pharmacokinetics of Cardiovascular, Central Nervous System, and Antimicrobial Drugs* (The Royal Society of Chemistry, 1985).

[29]J. W. Bridges and L. F. Chasseaud, *Progress in Drug Metabolism* (New York: John Wiley and Sons, 1980).

[30]Graeme S. Avery, ed. *Drug Treatment: Principles and Practice of Clinical Pharmacology and Therapeutics* (Sydney: Adis Press, 1976).

(c) Give possible values for minimum effective concentration and maximum safe concentration that would make Product A the preferred drug.

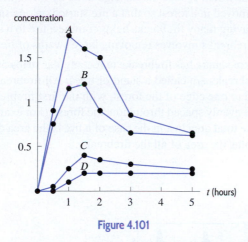

Figure 4.101

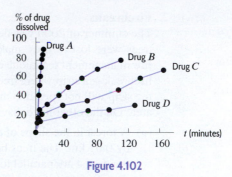

Figure 4.102

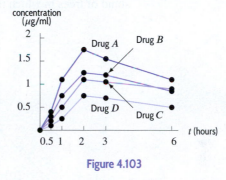

Figure 4.103

13. Figure 4.102 shows a graph of the percentage of drug dissolved against time for four tetracycline products A, B, C, and D. Figure 4.103 shows the drug concentration curves for the same four tetracycline products.[31] Discuss the effect of dissolution rate on peak concentration and time to reach peak concentration.

PROJECTS FOR CHAPTER FOUR

1. Average and Marginal Costs

The total cost of producing a quantity q is $C(q)$. The average cost $a(q)$ is given in Figure 4.104. The following rule is used by economists to determine the marginal cost $C'(q_0)$, for any q_0:

- Construct the tangent line t_1 to $a(q)$ at q_0.
- Let t_2 be the line with the same vertical intercept as t_1 but with twice the slope of t_1.

Then $C'(q_0)$ is the vertical distance shown in Figure 4.104. Explain why this rule works.

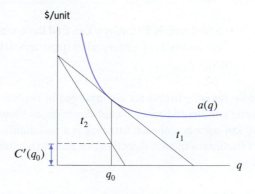

Figure 4.104

---

[31]J. W. Bridges and L.F. Chasseaud, *Progress in Drug Metabolism* (New York: John Wiley and Sons, 1980).

2. Firebreaks

The summer of 2000 was devastating for forests in the western US: over 3.5 million acres of trees were lost to fires, making this the worst fire season in 30 years. This project studies a fire management technique called *firebreaks*, which reduce the damage done by forest fires. A firebreak is a strip where trees have been removed in a forest so that a fire started on one side of the strip will not spread to the other side. Having many firebreaks helps confine a fire to a small area. On the other hand, having too many firebreaks involves removing large swaths of trees.[32]

(a) A forest in the shape of a 50 km by 50 km square has firebreaks in rectangular strips 50 km by 0.01 km. The trees between two firebreaks are called a stand of trees. All firebreaks in this forest are parallel to each other and to one edge of the forest, with the first firebreak at the edge of the forest. The firebreaks are evenly spaced throughout the forest. (For example, Figure 4.105 shows four firebreaks.) The total area lost in the case of a fire is the area of the stand of trees in which the fire started plus the area of all the firebreaks.

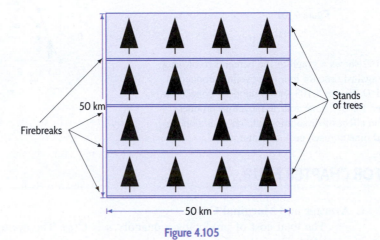

Figure 4.105

(i) Find the number of firebreaks that minimizes the total area lost to the forest in the case of a fire.

(ii) If a firebreak is 50 km by b km, find the optimal number of firebreaks as a function of b. If the width, b, of a firebreak is quadrupled, how does the optimal number of firebreaks change?

(b) Now suppose firebreaks are arranged in two equally spaced sets of parallel lines with the same number of breaks in each direction, as shown in Figure 4.106. The forest is a 50 km by 50 km square, and each firebreak is a rectangular strip 50 km by 0.01 km. Find the number of firebreaks in each direction that minimizes the total area lost to the forest in the case of a fire.

---

[32] Adapted from R. D. Harding and D. A. Quinney, *Calculus Connections* (New York: John Wiley & Sons, 1996).

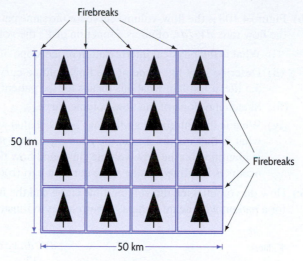

Firebreaks

Firebreaks

50 km

50 km

Figure 4.106

3. **Production and the Price of Raw Materials**

The production function $f(x)$ gives the number of units of an item that a manufacturing company can produce from x units of raw material. The company buys the raw material at price w dollars per unit and sells all it produces at a price of p dollars per unit. The quantity of raw material that maximizes profit is denoted by x^*.

(a) Do you expect the derivative $f'(x)$ to be positive or negative? Justify your answer.

(b) Explain why the formula $\pi(x) = pf(x) - wx$ gives the profit $\pi(x)$ that the company earns as a function of the quantity x of raw materials that it uses.

(c) Evaluate $f'(x^*)$.

(d) Assuming it is nonzero, is $f''(x^*)$ positive or negative?

(e) If the supplier of the raw materials is likely to change the price w, then it is appropriate to treat x^* as a function of w. Find a formula for the derivative dx^*/dw and decide whether it is positive or negative.

(f) If the price w goes up, should the manufacturing company buy more or less of the raw material?

4. **Medical Case Study: Impact of Asthma on Breathing[33]**

Asthma is a common breathing disease in which inflammation in the airways of the lungs causes episodes of shortness of breath, coughing, and chest tightness. Patients with asthma often have wheezing, an abnormal sound heard on exhalation due to turbulent airflow. Turbulent airflow is caused by swelling, mucus secretion, and constriction of muscle in the walls of the airways, shrinking the radius of the air passages leading to increased resistance to airflow and making it harder for patients to exhale.

An important breathing test for asthma is called *spirometry*. In this test, a patient takes in as deep a breath as he or she can, and then exhales as rapidly, forcefully, and for as long as possible through a tube connected to an analyzer. The analyzer measures a number of parameters and generates two graphs.

(a) Figure 4.107 is a volume-time curve for an asthma-free patient, showing the volume of air exhaled, V, as a function of time, t, since the test began.[34]

   (i) What is the physical interpretation of the slope of the volume-time curve?

   (ii) The volume VC shown on the volume-time graph is called the "(forced) vital capacity" (FVC or simply VC). Describe the physical meaning of VC.

[33]From David E. Sloane, M.D.
[34]Image based on www.aafp.org/afp/2004/0301/p1107.html, accessed July 9, 2011.

(b) Figure 4.108 is the flow-volume curve for the same patient.[35] The flow-volume curve shows the flow rate, dV/dt, of air as a function of V, the volume of air exhaled.

    (i) What is the physical interpretation of the slope of the flow-volume curve?

    (ii) Describe how the slope of the flow-volume curve changes as V increases from 0 to 5.5 liters. Explain what this means for the patient's breath.

    (iii) Sketch the slope of the flow-volume curve.

    (iv) What is the volume of air that has been exhaled when the flow rate is a maximum, and what is that maximal rate, the *peak expiratory flow*? Explain how this maximal rate is identified on the flow-volume curve and how the volume at which the maximal rate occurs is identified on the slope curve in part (b)(iii).

(c) How do you imagine the volume-time curve and the flow-volume curve would be different for a patient with acute asthma? Draw curves to illustrate your thinking.

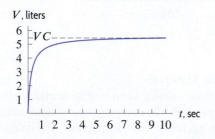

Figure 4.107: Volume-time curve

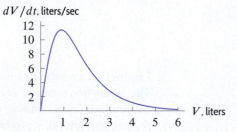

Figure 4.108: Flow-volume curve

---

[35]Image from http://www.aafp.org/afp/2004/0301/p1107.html, accessed July 9, 2011.

Chapter 5

ACCUMULATED CHANGE: THE DEFINITE INTEGRAL

CONTENTS

5.1   DISTANCE AND ACCUMULATED CHANGE

In Chapter 2, we used the derivative to find the rate of change of a function. Here we see how to go in the other direction. If we know the rate of change, can we find the original function? We start by finding the distance traveled from its rate of change, the velocity. For positive constant velocities, we can find the distance traveled using the formula

$$\text{Distance} = \text{Velocity} \times \text{Time}.$$

In this section we see how to estimate the distance when the velocity is not a constant.

A Thought Experiment: How Far Did the Car Go?

Suppose a car is moving with increasing velocity and that we measure the car's velocity every two seconds, obtaining the data in Table 5.1:

Table 5.1   *Velocity of car every two seconds*

Time (sec)	0	2	4	6	8	10
Velocity (ft/sec)	20	30	38	44	48	50

How far has the car traveled? Since we don't know how fast the car is moving at every moment, we can't calculate the distance exactly, but we can make an estimate. The velocity is increasing, so the car is going at least 20 ft/sec for the first two seconds. Since Distance = Velocity × Time, the car goes at least $20 \cdot 2 = 40$ feet during the first two seconds. Likewise, it goes at least $30 \cdot 2 = 60$ feet during the next two seconds, and so on. During the ten-second period it goes at least

$$20 \cdot 2 + 30 \cdot 2 + 38 \cdot 2 + 44 \cdot 2 + 48 \cdot 2 = 360 \text{ feet}.$$

Thus, 360 feet is an underestimate of the total distance traveled during the ten seconds.

To get an overestimate, we can reason in a similar way: During the first two seconds, the car's velocity is at most 30 ft/sec, so it moves at most $30 \cdot 2 = 60$ feet. In the next two seconds it moves at most $38 \cdot 2 = 76$ feet, and so on. Therefore, over the ten-second period it moves at most

$$30 \cdot 2 + 38 \cdot 2 + 44 \cdot 2 + 48 \cdot 2 + 50 \cdot 2 = 420 \text{ feet}.$$

Therefore,

$$360 \text{ feet} \leq \text{Total distance traveled} \leq 420 \text{ feet}.$$

There is a difference of 60 feet between the upper and lower estimates. A better estimate would be the average of the upper and lower estimates,

$$\frac{360 + 420}{2} = 390 \text{ feet}.$$

Visualizing Distance on the Velocity Graph

We can represent both upper and lower estimates on a graph of the velocity against time. The velocity can be graphed by plotting the data in Table 5.1 and drawing a curve through the data points. (See Figure 5.1.)

The area of the first dark rectangle is $20 \cdot 2 = 40$, the lower estimate of the distance moved during the first two seconds. The area of the second dark rectangle is $30 \cdot 2 = 60$, the lower estimate for the distance moved in the next two seconds. The total area of the dark rectangles represents the lower estimate for the total distance moved during the ten seconds.

If the dark and light rectangles are considered together, the first area is $30 \cdot 2 = 60$, the upper estimate for the distance moved in the first two seconds. The second area is $38 \cdot 2 = 76$, the upper estimate for the next two seconds. Continuing this calculation suggests that the upper estimate for the total distance is represented by the sum of the areas of the dark and light rectangles.

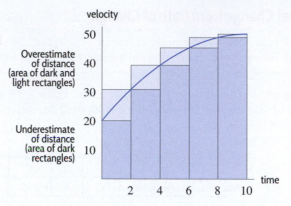

Figure 5.1: Shaded area estimates distance traveled. Velocity measured every 2 seconds

Visualizing Distance on the Velocity Graph: Area Under Curve

As we make more frequent velocity measurements, the rectangles used to estimate the distance traveled fit the curve more closely. See Figures 5.2 and 5.3. In the limit, as the number of subdivisions increases, we see that the distance traveled is given by the area between the velocity curve and the horizontal axis. See Figure 5.4. In general:

> If the velocity is positive, the total distance traveled is the area under the velocity curve.

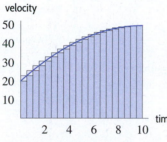

Figure 5.2: Velocity measured every 1/2 second

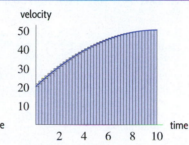

Figure 5.3: Velocity measured every 1/4 second

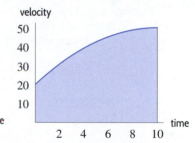

Figure 5.4: Distance traveled is area under curve

Example 1   With time t in seconds, the velocity of a bicycle, in feet per second, is given by $v(t) = 5t$. How far does the bicycle travel in the first 3 seconds after $t = 0$?

Solution   The velocity is linear. See Figure 5.5. The distance traveled is the area between the line $v(t) = 5t$ and the t-axis. Since this region is a triangle of height 15 and base 3,

$$\text{Distance traveled } = \text{ Area of triangle } = \frac{1}{2} \cdot 15 \cdot 3 = 22.5 \text{ feet.}$$

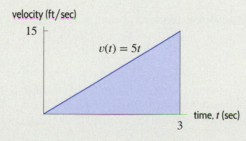

Figure 5.5: Shaded area represents distance traveled

Approximating Total Change from Rate of Change

We have seen how to use the rate of change of distance (the velocity) to calculate the total distance traveled. We can use the same method to find total change from the rate of change of other quantities.

Example 2

The rate of sales (in games per week) of a new video game is shown in Table 5.2. Assuming that the rate of sales increased throughout the 20-week period, estimate the total number of games sold during this period.

Table 5.2   *Weekly sales of a video game*

Time (weeks)	0	5	10	15	20
Rate of sales (games per week)	0	585	892	1875	2350

Solution

If the rate of sales is constant, we have

$$\text{Total sales} = \text{Rate of sales per week} \times \text{Number of weeks}.$$

How many games were sold during the first five weeks? During this time, sales went from 0 to 585 games per week. If we assume that 585 games were sold every week, we get an overestimate for the sales in the first five weeks of (585 games/week)(5 weeks) = 2925 games. Similar overestimates for each of the five-week periods give an overestimate for the entire 20-week period:

$$\text{Overestimate for total sales} = 585 \cdot 5 + 892 \cdot 5 + 1875 \cdot 5 + 2350 \cdot 5 = 28{,}510 \text{ games.}$$

We underestimate the total sales by taking the lower value for rate of sales during each of the five-week periods:

$$\text{Underestimate for total sales} = 0 \cdot 5 + 585 \cdot 5 + 892 \cdot 5 + 1875 \cdot 5 = 16{,}760 \text{ games.}$$

Thus, the total sales of the game during the 20-week period is between 16,760 and 28,510 games. A good single estimate of total sales is the average of these two numbers:

$$\text{Total sales} \approx \frac{16{,}760 + 28{,}510}{2} = 22{,}635 \text{ games.}$$

Improving the Approximation: The Role of n and Δt

In general, to approximate total change from a function $f(t)$ giving the rate of change at time t, we construct a sum. We use the notation Δt for the size of the t-intervals used. We use n to represent the number of subintervals of length Δt. In the following example, we see how decreasing Δt (and increasing n) improves the accuracy of the approximation.

Example 3

If t is in hours since the start of a 20-hour period, a bacteria population increases at a rate given by

$$f(t) = 3 + 0.1t^2 \text{ millions of bacteria per hour.}$$

Make an underestimate of the total change in the number of bacteria over this period using

(a) $\Delta t = 4$ hours   (b) $\Delta t = 2$ hours   (c) $\Delta t = 1$ hour

Solution

(a) The rate of change is $f(t) = 3 + 0.1t^2$. If we use $\Delta t = 4$, we measure the rate every 4 hours and $n = 20/4 = 5$. Using the formula to calculate $f(t)$, we obtain Table 5.3. An underestimate for the population change during the first 4 hours is (3.0 million/hour)(4 hours) = 12 million. Combining the contributions from all the subintervals gives the underestimate:

$$\text{Total change} \approx 3.0 \cdot 4 + 4.6 \cdot 4 + 9.4 \cdot 4 + 17.4 \cdot 4 + 28.6 \cdot 4 = 252.0 \text{ million bacteria.}$$

The rate of change is graphed in Figure 5.6(a); the area of the shaded rectangles represents this underestimate. Notice that $n = 5$ is the number of rectangles in the graph.

Table 5.3   *Rate of change with $\Delta t = 4$ using $f(t) = 3 + 0.1t^2$ million bacteria/hour*

t (hours)	0	4	8	12	16	20
$f(t)$	3.0	4.6	9.4	17.4	28.6	43.0

(b) If we use $\Delta t = 2$, we measure $f(t)$ every 2 hours and $n = 20/2 = 10$. See Table 5.4. The underestimate is

$$\text{Total change} \approx 3.0 \cdot 2 + 3.4 \cdot 2 + 4.6 \cdot 2 + \cdots + 35.4 \cdot 2 = 288.0 \text{ million bacteria.}$$

Figure 5.6(b) suggests that this estimate is more accurate than the estimate made in part (a).

Table 5.4   *Rate of change with $\Delta t = 2$ using $f(t) = 3 + 0.1t^2$ million bacteria/hour*

t (hours)	0	2	4	6	8	10	12	14	16	18	20
$f(t)$	3.0	3.4	4.6	6.6	9.4	13.0	17.4	22.6	28.6	35.4	43.0

(c) If we use $\Delta t = 1$, then $n = 20$ and a similar calculation shows that we have

$$\text{Total change} \approx 307.0 \text{ million bacteria.}$$

The shaded area in Figure 5.6(c) represents this estimate; it is the most accurate of the three.

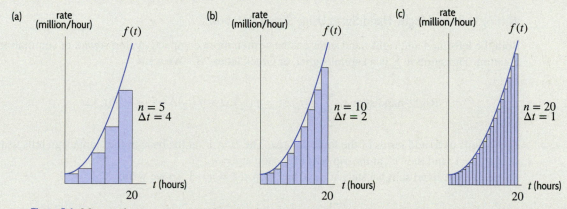

Figure 5.6: More and more accurate estimates of total change from rate of change. In each case, $f(t)$ is the rate of change, and the shaded area approximates total change. Largest n and smallest Δt give the best estimate.

Notice that as n gets larger, the estimate improves and the area of the shaded rectangles approaches the area under the curve.

Left- and Right-Hand Sums

Now we see how to express these sums for any function $f(t)$ that is continuous for $a \leq t \leq b$. We divide the interval from a to b into n equal subdivisions, each of width Δt, so

$$\Delta t = \frac{b - a}{n}.$$

We let $t_0, t_1, t_2, \ldots, t_n$ be endpoints of the subdivisions, as in Figures 5.7 and 5.8. We construct two sums, similar to the overestimates and underestimates earlier in this section. For a *left-hand sum*, we use the values of the function from the left end of the interval. For a *right-hand sum*, we use the values of the function from the right end of the interval. We have:

$$\text{Left-hand sum} = f(t_0)\Delta t + f(t_1)\Delta t + \cdots + f(t_{n-1})\Delta t$$

$$\text{Right-hand sum} = f(t_1)\Delta t + f(t_2)\Delta t + \cdots + f(t_n)\Delta t.$$

These sums represent the shaded areas in Figures 5.7 and 5.8, provided $f(t) \geq 0$. In Figure 5.7, the first rectangle has width Δt and height $f(t_0)$, since the top of its left edge just touches the curve, and hence it has area $f(t_0)\Delta t$. The second rectangle has width Δt and height $f(t_1)$, and hence has area $f(t_1)\Delta t$, and so on. The sum of all these areas is the left-hand sum. The right-hand sum, shown in Figure 5.8, is constructed in the same way, except that each rectangle touches the curve on its right edge instead of its left. We often use the shorthand "left sum" for a left-hand sum and "right sum" for a right-hand sum.

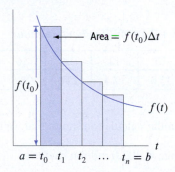

Figure 5.7: Left-hand sum: Area of rectangles

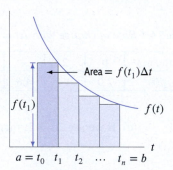

Figure 5.8: Right-hand sum: Area of rectangles

Writing Left- and Right-Hand Sums Using Sigma Notation

Both the left-hand and right-hand sums can be written more compactly using *sigma*, or summation, notation. The symbol $\sum$ is a capital sigma, or Greek letter "S." We write

$$\text{Right-hand sum} = \sum_{i=1}^{n} f(t_i)\Delta t = f(t_1)\Delta t + f(t_2)\Delta t + \cdots + f(t_n)\Delta t.$$

The $\sum$ tells us to add terms of the form $f(t_i)\Delta t$. The "$i = 1$" at the base of the sigma sign tells us to start at $i = 1$, and the "n" at the top tells us to stop at $i = n$.

In the left-hand sum we start at $i = 0$ and stop at $i = n - 1$, so we write

$$\text{Left-hand sum} = \sum_{i=0}^{n-1} f(t_i)\Delta t = f(t_0)\Delta t + f(t_1)\Delta t + \cdots + f(t_{n-1})\Delta t.$$

Problems for Section 5.1

1. You travel 30 miles/hour for 2 hours, then 40 miles/hour for 1/2 hour, then 20 miles/hour for 4 hours.

   (a) What is the total distance you traveled?
   (b) Sketch a graph of the velocity function for this trip.
   (c) Represent the total distance traveled on your graph in part (b).

2. Figure 5.9 shows the velocity of an object for $0 \leq t \leq 6$. Calculate the following estimates of the distance the object travels between $t = 0$ and $t = 6$, and indicate whether each result is an upper or lower estimate of the distance traveled.

   (a) A left sum with $n = 2$ subdivisions

(b) A right sum with $n = 2$ subdivisions

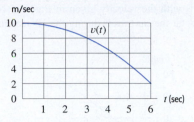

Figure 5.9

3. Figure 5.10 shows the velocity of an object for $0 \leq t \leq 8$. Calculate the following estimates of the distance the object travels between $t = 0$ and $t = 8$, and indicate whether each is an upper or lower estimate of the distance traveled.

(a) A left sum with $n = 2$ subdivisions
(b) A right sum with $n = 2$ subdivisions

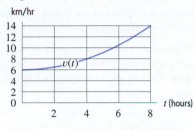

Figure 5.10

4. Figure 5.11 shows the velocity of a car for $0 \leq t \leq 12$ and the rectangles used to estimate the distance traveled.

(a) Do the rectangles represent a left or a right sum?
(b) Do the rectangles lead to an upper or a lower estimate?
(c) What is the value of n?
(d) What is the value of Δt?
(e) Give an approximate value for the estimate.

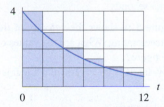

Figure 5.11

5. Figure 5.12 shows the velocity of a runner for $0 \leq t \leq 15$ and the rectangles used to estimate the distance traveled.

(a) Do the rectangles represent a left or a right sum?
(b) Do the rectangles lead to an upper or a lower estimate?
(c) What is the value of n?
(d) What is the value of Δt?
(e) Give an approximate value for the estimate.

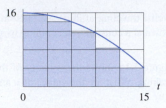

Figure 5.12

6. Figure 5.13 shows the velocity of a car for $0 \leq t \leq 24$ and the rectangles used to estimate the distance traveled.

(a) Do the rectangles represent a left or a right sum?
(b) Do the rectangles lead to an upper or a lower estimate?
(c) What is the value of n?
(d) What is the value of Δt?
(e) Estimate the distance traveled.

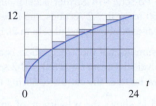

Figure 5.13

7. A car comes to a stop six seconds after the driver applies the brakes. While the brakes are on, the velocities recorded are in Table 5.5.

Table 5.5

Time since brakes applied (sec)	0	2	4	6
Velocity (ft/sec)	88	45	16	0

(a) Give lower and upper estimates for the distance the car traveled after the brakes were applied.
(b) On a sketch of velocity against time, show the lower and upper estimates of part (a).

8. A car starts moving at time $t = 0$ and goes faster and faster. Its velocity is shown in the following table. Estimate how far the car travels during the 12 seconds.

t (seconds)	0	3	6	9	12
Velocity (ft/sec)	0	10	25	45	75

9. Graph the rate of sales against time for the video game data in Example 2. Represent graphically the overestimate and the underestimate calculated in that example.

10. A bicyclist traveling at 20 ft/sec puts on the brakes to slow down at a constant rate, coming to a stop in 3 seconds.

   (a) Figure 5.14 shows the velocity of the bike during braking. What are the values of *a* and *b* in the figure?
   (b) How far does the bike travel while braking?

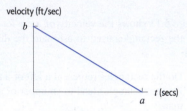

velocity (ft/sec)

Figure 5.14

11. A 2015 Porsche 918 Spyder accelerates from 0 to 88 ft/sec (60 mph) in 2.2 seconds, the fastest acceleration of any car available for retail sale in 2015.[1]

   (a) Assuming that the acceleration is constant, graph the velocity from $t = 0$ to $t = 2.2$ seconds.
   (b) How far does the car travel during this time?

12. Figure 5.15 shows the velocity, v, of an object (in meters/sec). Estimate the total distance the object traveled between $t = 0$ and $t = 6$.

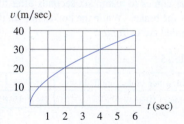

Figure 5.15

13. Roger runs a marathon. His friend Jeff rides behind him on a bicycle and clocks his speed every 15 minutes. Roger starts out strong, but after an hour and a half he is so exhausted that he has to stop. Jeff's data follow:

Time since start (min)	0	15	30	45	60	75	90
Speed (mph)	12	11	10	10	8	7	0

   (a) Assuming that Roger's speed is never increasing, give upper and lower estimates for the distance Roger ran during the first half hour.
   (b) Give upper and lower estimates for the distance Roger ran in total during the entire hour and a half.

14. A car accelerates smoothly from 0 to 60 mph in 10 seconds with the velocity given in Figure 5.16. Estimate how far the car travels during the 10-second period.

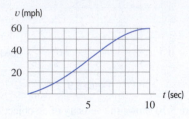

Figure 5.16

15. The velocity of a car is $f(t) = 5t$ meters/sec. Use a graph of $f(t)$ to find the exact distance traveled by the car, in meters, from $t = 0$ to $t = 10$ seconds.

16. A bicyclist accelerates at a constant rate, from 0 ft/sec to 15 ft/sec in 10 seconds.

   (a) Figure 5.17 shows the velocity of the bike while it is accelerating. What is the value of *b* in the figure?
   (b) How far does the bike travel while it is accelerating?

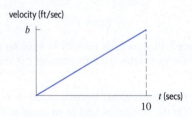

Figure 5.17

17. A car accelerates at a constant rate from 44 ft/sec to 88 ft/sec in 5 seconds.

   (a) Figure 5.18 shows the velocity of the car while it is accelerating. What are the values of *a*, *b* and *c* in the figure?
   (b) How far does the car travel while it is accelerating?

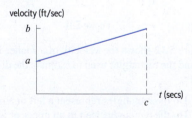

Figure 5.18

---

[1] K. C. Colwell, "First Test: 2015 Porsche 918 Spyder," *Car and Driver*, August 1, 2014.

18. A village wishes to measure the quantity of water that is piped to a factory during a typical morning. A gauge on the water line gives the flow rate (in cubic meters per hour) at any instant. The flow rate is about 100 m³/hr at 6 am and increases steadily to about 280 m³/hr at 9 am. Using only this information, give your best estimate of the total volume of water used by the factory between 6 am and 9 am.

19. Filters at a water treatment plant become less effective over time. The rate at which pollution passes through the filters into a nearby lake is given in the following table.

    (a) Estimate the total quantity of pollution entering the lake during the 30-day period.
    (b) Your answer to part (a) is only an estimate. Give bounds (lower and upper estimates) between which the true quantity of pollution must lie. (Assume the rate of pollution is continually increasing.)

Day	0	6	12	18	24	30
Rate (kg/day)	7	8	10	13	18	35

20. A car initially going 50 ft/sec brakes at a constant rate (constant negative acceleration), coming to a stop in 5 seconds.

    (a) Graph the velocity from $t = 0$ to $t = 5$.
    (b) How far does the car travel?
    (c) How far does the car travel if its initial velocity is doubled, but it brakes at the same constant rate?

21. The following table gives world oil consumption, in billions of barrels per year.[2] Estimate total oil consumption during this 25-year period.

Year	1985	1990	1995	2000	2005	2010
Oil (bn barrels/yr)	20.8	23.2	25.5	27.9	30.7	32.0

22. The following table gives the annual natural gas production, in trillions of cubic feet per year, in the US between 2002 and 2014.[3]

    (a) Estimate the total natural gas produced in the US between 2002 and 2014 using a

       (i) Left sum, $n = 6$ (ii) Right sum, $n = 6$

    (b) Can you determine whether the right sum is an over- or underestimate? Why or why not?

Year	2002	2004	2006	2008	2010	2012	2014
Gas (tr ft³/yr)	19.9	19.5	19.4	21.1	22.4	25.3	27.5

23. Figure 5.19 shows the rate of change of a fish population. Estimate the total change in the population during this 12-month period.

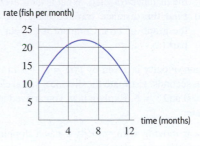

Figure 5.19

24. Two cars travel in the same direction along a straight road. Figure 5.20 shows the velocity, v, of each car at time t. Car B starts 2 hours after car A and car B reaches a maximum velocity of 50 km/hr.

    (a) For approximately how long does each car travel?
    (b) Estimate car A's maximum velocity.
    (c) Approximately how far does each car travel?

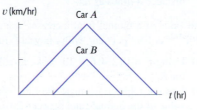

Figure 5.20

25. Two cars start at the same time and travel in the same direction along a straight road. Figure 5.21 gives the velocity, v, of each car as a function of time, t. Which car:

    (a) Attains the larger maximum velocity?
    (b) Stops first?
    (c) Travels farther?

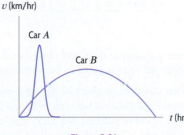

Figure 5.21

[2] http://www.indexmundi.com/energy.aspx, accessed January 2017.
[3] https://www.eia.gov, accessed January 2017.

26. The velocity of a car, in ft/sec, is $v(t) = 10t$ for t in seconds, $0 \leq t \leq 6$.

    (a) Use $\Delta t = 2$ to give upper and lower estimates for the distance traveled. What is their average?
    (b) Find the distance traveled using the area under the graph of $v(t)$. Compare it to your answer for part (a).

27. Your velocity is given by $v(t) = t^2 + 1$ in m/sec, with t in seconds. Estimate the distance, s, traveled between $t = 0$ and $t = 5$. Explain how you arrived at your estimate.

28. A car moving with velocity v has a stopping distance proportional to v^2.

    (a) If a car going 20 mi/hr has a stopping distance of 50 feet, what is its stopping distance going 40 mi/hr? What about 60 mi/hr?
    (b) After applying the brakes, a car going 30 ft/sec stops in 5 seconds and has $v = 30 - 6t$. Explain why the stopping distance is given by the area under the graph of v against t.
    (c) By looking at areas under graphs of v, explain why a car with the same deceleration as the car in part (b) but an initial speed of 60 ft/sec has a stopping distance 4 times as far.

29. The value of a mutual fund increases at a rate of $R = 500e^{0.04t}$ dollars per year, with t in years since 2010.

    (a) Using $t = 0, 2, 4, 6, 8, 10$, make a table of values for R.
    (b) Use the table to estimate the total change in the value of the mutual fund between 2010 and 2020.

30. An old rowboat has sprung a leak. Water is flowing into the boat at a rate, $r(t)$, given in the table.

    (a) Compute upper and lower estimates for the volume of water that has flowed into the boat during the 15 minutes.
    (b) Draw a graph to illustrate the lower estimate.

t minutes	0	5	10	15
$r(t)$ liters/min	12	20	24	16

31. The rate of change of the world's population, in millions of people per year, is given in the following table.

    (a) Use this data to estimate the total change in the world's population between 1950 and 2000.
    (b) The world population was 2555 million people in 1950 and 6085 million people in 2000. Calculate the true value of the total change in the population. How does this compare with your estimate in part (a)?

---

⁴www.motortrend.com/, accessed May 2011.

Year	1950	1960	1970	1980	1990	2000
Rate of change	37	41	78	77	86	79

32. A car speeds up at a constant rate from 10 to 70 mph over a period of half an hour. Its fuel efficiency (in miles per gallon) increases with speed; values are in the table. Make lower and upper estimates of the quantity of fuel used during the half hour.

Speed (mph)	10	20	30	40	50	60	70
Fuel efficiency (mpg)	15	18	21	23	24	25	26

■ Problems **33–36** concern hybrid cars such as the Toyota Prius that are powered by a gas-engine, electric-motor combination, but can also function in Electric-Vehicle (EV) only mode. Figure 5.22 shows the velocity, v, of a 2010 Prius Plug-in Hybrid Prototype operating in normal hybrid mode and EV-only mode, respectively, while accelerating from a stoplight.⁴

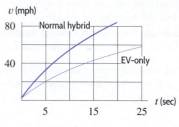

Figure 5.22

33. Could the car travel half a mile in EV-only mode during the first 25 seconds of movement?

34. About how far, in feet, does the 2010 Prius Prototype travel in EV-only mode during the first 15 seconds of movement?

35. Assume two identical cars, one running in normal hybrid mode and one running in EV-only mode, accelerate together in a straight path from a stoplight. Approximately how far apart are the cars after 5 seconds?

36. Assume two identical cars, one running in normal hybrid mode and one running in EV-only mode, accelerate together in a straight path from a stoplight. Approximately how far apart are the cars after 15 seconds?

37. A car is traveling at 80 feet per second (approximately 55 miles per hour) and slows down as it passes through a busy intersection. The car's velocity is shown in the following table.

Time since brakes applied (seconds)	0	3	6	9	12
velocity (ft/sec)	80	62	47	32	22

(a) Use a left sum to approximate the total distance the car has traveled over the 12 second interval.

(b) Is your answer in part (a) a lower or upper estimate? How can you tell?

(c) Use a right sum to approximate the total distance the car has traveled over the 12 second interval.

(d) Is your answer in part (c) a lower or upper estimate? How can you tell?

(e) Using your answers in parts (a) and (c), find a better estimate for the total distance the car has traveled.

38. Table 5.6 gives the ground speed of a small plane accelerating for takeoff. Find upper and lower estimates for the distance traveled by the plane during takeoff.

Table 5.6

Time (sec)	0	2	4	6	8	10
Speed (m/s)	2.7	2.7	4	6.3	8.5	11.6
Time (sec)	12	14	16	18	20	
Speed (m/s)	13.4	17.4	21.9	29.1	32.6	

39. The following table gives the total world emissions of CO_2 from fossil fuels, in billions of tons per year.[5]

(a) Use this data to estimate the total world CO_2 emissions between 1983 and 2013 using a left sum with $n = 6$.

(b) Is your answer in part (a) an upper or lower estimate? How can you tell?

(c) Use this data to estimate the total world CO_2 emissions between 1983 and 2013 using a right sum with $n = 3$.

(d) Is your answer in part (c) an upper or lower estimate? How can you tell?

Year	1983	1988	1993	1998	2003	2008	2013
CO_2 (bn ton/yr)	18.6	21.8	22.4	24.1	27.0	32.0	35.8

40. The rate of change of a quantity is given by $f(t) = t^2 + 1$. Make an underestimate and an overestimate of the total change in the quantity between $t = 0$ and $t = 8$ using

(a) $\Delta t = 4$ (b) $\Delta t = 2$ (c) $\Delta t = 1$

What is n in each case? Graph $f(t)$ and shade rectangles to represent each of your six answers.

41. Use the expressions for left and right sums on page 238 and Table 5.7.

(a) If $n = 4$, what is Δt? What are t_0, t_1, t_2, t_3, t_4? What are $f(t_0), f(t_1), f(t_2), f(t_3), f(t_4)$?

(b) Find the left and right sums using $n = 4$.

(c) If $n = 2$, what is Δt? What are t_0, t_1, t_2? What are $f(t_0), f(t_1), f(t_2)$?

(d) Find the left and right sums using $n = 2$.

Table 5.7

t	15	17	19	21	23
$f(t)$	10	13	18	20	30

5.2 THE DEFINITE INTEGRAL

In Section 5.1 we saw how to approximate total change given the rate of change. In this section we define a new quantity, the definite integral, which gives the total change exactly.

Taking the Limit to Obtain the Definite Integral

If f is a rate of change of some quantity, then the left-hand sum and the right-hand sum approximate the total change in the quantity. For most functions f, the approximation is improved by increasing the value of n. To find the total change exactly, we take larger and larger values of n and look at the values approached by the left and right sums. This is called taking the *limit* of these sums as n goes to infinity and is written $\lim_{n \to \infty}$. If f is continuous for $a \le t \le b$, the limits of the left- and right-hand sums exist and are equal. The *definite integral* is the common limit of these sums.

---

[5]http://cdiac.ornl.gov/ftp/ndp030/global.1751_2013.ems, accessed December 13, 2016.

Suppose f is continuous for $a \leq t \leq b$. The **definite integral** of f from a to b, written

$$\int_a^b f(t)\, dt,$$

is the limit of the left-hand or right-hand sums with n subdivisions of $a \leq t \leq b$ as n gets arbitrarily large. In other words, if $t_0, t_1, \ldots t_n$ are the endpoints of the subdivisions,

$$\int_a^b f(t)\, dt = \lim_{n \to \infty} (\text{Left-hand sum}) = \lim_{n \to \infty} \left(\sum_{i=0}^{n-1} f(t_i) \Delta t \right)$$

and

$$\int_a^b f(t)\, dt = \lim_{n \to \infty} (\text{Right-hand sum}) = \lim_{n \to \infty} \left(\sum_{i=1}^{n} f(t_i) \Delta t \right).$$

Each of these sums is called a *Riemann sum*, f is called the *integrand*, and a and b are called the *limits of integration*.

The "$\int$" notation comes from an old-fashioned "S," which stands for "sum" in the same way that $\sum$ does. The "dt" in the integral comes from the factor Δt. Notice that the limits on the $\sum$ symbol are 0 and $n - 1$ for the left-hand sum, and 1 and n for the right-hand sum, whereas the limits on the $\int$ sign are a and b.

When $f(t)$ is positive, the left- and right-hand sums are represented by the sums of areas of rectangles, so the definite integral is represented graphically by an area.

Example 1   Write the result of Example 3 in the previous section using integral notation.

Solution   Part (c) gave 307.0 million as the best estimate of the total change in the bacteria population. In integral notation,

$$\int_0^{20} f(t)\, dt = \int_0^{20} (3 + 0.1 t^2)\, dt \approx 307.0 \text{ million.}$$

Computing a Definite Integral

In practice, many calculators and computers calculate definite integrals by automatically computing sums for larger and larger values of n.

Example 2   Compute $\int_0^1 e^{-t^2}\, dt$ and represent this integral as an area.

Solution   Using a calculator or computer, we find

$$\int_0^1 e^{-t^2}\, dt = 0.747.$$

The integral represents the area between $t = 0$ and $t = 1$ under the curve $f(t) = e^{-t^2}$. See Figure 5.23.

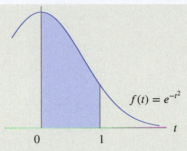

Figure 5.23: Shaded area $= \int_0^1 e^{-t^2}\, dt$

Estimating a Definite Integral from a Table or Graph

If we have a formula for the integrand, $f(x)$, we can calculate the integral $\int_a^b f(x)\, dx$ using a calculator or computer. If, however, we have only a table of values or a graph of $f(x)$, we can still estimate the integral. Note that we can use any letter (x, t, y, q, etc.) for the independent variable.

Example 3   Values for a function $f(t)$ are in the following table. Estimate $\int_{20}^{30} f(t)dt$.

t	20	22	24	26	28	30
$f(t)$	5	7	11	18	29	45

Solution   Since we have only a table of values, we use left- and right-hand sums to approximate the integral. The values of $f(t)$ are spaced 2 units apart, so $\Delta t = 2$ and $n = (30 - 20)/2 = 5$. Calculating the left-hand and right-hand sums gives

$$\text{Left-hand sum} = f(20) \cdot 2 + f(22) \cdot 2 + f(24) \cdot 2 + f(26) \cdot 2 + f(28) \cdot 2$$
$$= 5 \cdot 2 + 7 \cdot 2 + 11 \cdot 2 + 18 \cdot 2 + 29 \cdot 2$$
$$= 140.$$

$$\text{Right-hand sum} = f(22) \cdot 2 + f(24) \cdot 2 + f(26) \cdot 2 + f(28) \cdot 2 + f(30) \cdot 2$$
$$= 7 \cdot 2 + 11 \cdot 2 + 18 \cdot 2 + 29 \cdot 2 + 45 \cdot 2$$
$$= 220.$$

Both left- and right-hand sums approximate the integral. We generally get a better estimate by averaging the two:

$$\int_{20}^{30} f(t)dt \approx \frac{140 + 220}{2} = 180.$$

Example 4   The function $f(x)$ is graphed in Figure 5.24. Estimate $\int_0^6 f(x)\, dx$.

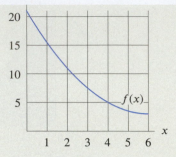

Figure 5.24: Estimate $\int_0^6 f(x)\,dx$

Solution   We can approximate the integral using left- and right-hand sums with $n = 3$, so $\Delta x = 2$. Figures 5.25 and 5.26 give

$$\text{Left-hand sum} = f(0) \cdot 2 + f(2) \cdot 2 + f(4) \cdot 2 = 21 \cdot 2 + 11 \cdot 2 + 5 \cdot 2 = 74,$$
$$\text{Right-hand sum} = f(2) \cdot 2 + f(4) \cdot 2 + f(6) \cdot 2 = 11 \cdot 2 + 5 \cdot 2 + 3 \cdot 2 = 38.$$

We estimate the integral by taking the average:

$$\int_0^6 f(x)\,dx \approx \frac{74 + 38}{2} = 56.$$

Figure 5.25: Area of shaded region is left-hand sum with $n = 3$

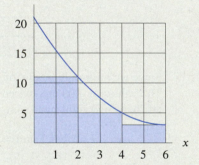

Figure 5.26: Area of shaded region is right-hand sum with $n = 3$

Alternatively, since the integral equals the area under the curve between $x = 0$ and $x = 6$, we can estimate it by counting grid boxes. Each grid box has area $5 \cdot 1 = 5$, and the region under $f(x)$ includes about 10.5 grid boxes, so the area is about $10.5 \cdot 5 = 52.5$.

Rough Estimates of a Definite Integral

When calculating an integral using a calculator or computer, it is useful to have a rough idea of the value you expect. This helps detect errors in entering the integral.

Example 5   Figure 5.27 shows the graph of a function $f(t)$. Which of the numbers 0.023, 1.099, 1.526, and 11.984 could be the value of $\int_1^3 f(t)\,dt$? Explain your reasoning.

Solution   Figure 5.27 shows left- and right-hand approximations to $\int_1^3 f(t)\,dt$ with $n = 1$. We see that the right-hand sum $2(1/3)$ is an underestimate of $\int_1^3 f(t)\,dt$. Since 0.023 is less than 2/3, this value must be wrong. Similarly, we see that the left-hand sum $2(1)$ is an overestimate of the integral. Since 11.984 is larger than 2, this value is wrong. Since the graph is concave up, the value of the integral is closer to the smaller value of the two sums, 2/3, than to the larger value, 2. Thus, the integral is less than the average of the two sums, 4/3. Since $1.526 > 4/3$, this value is wrong too. Since $2/3 < 1.099 < 4/3$, the value 1.099 could be correct.

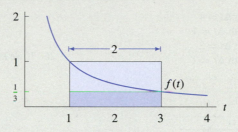

Figure 5.27: Left- and right-hand sums with $n = 1$

Problems for Section 5.2

■ In Problems 1–2, rectangles have been drawn to approximate $\int_0^6 g(x)\,dx$.

(a) Do the rectangles represent a left or a right sum?

(b) Do the rectangles lead to an upper or a lower estimate?

(c) What is the value of n?

(d) What is the value of Δx?

1. **2.**

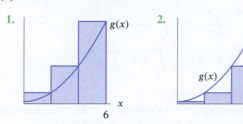

3. Figure 5.28 shows a Riemann sum approximation with n subdivisions to $\int_a^b f(x)\,dx$.

(a) Is it a left- or right-hand approximation? Would the other one be larger or smaller?

(b) What are a, b, n, and Δx?

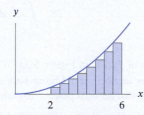

Figure 5.28

4. Estimate $\int_0^6 2^x\,dx$ using a left-hand sum with $n = 2$.

5. Estimate $\int_0^{12} \dfrac{1}{x+1}\,dx$ using a left-hand sum with $n = 3$.

6. Estimate $\int_0^1 e^{-x^2}\,dx$ using $n = 5$ rectangles to form a

(a) Left-hand sum   (b) Right-hand sum

■ In Problems 7–14, estimate the integral using a left-hand sum and a right-hand sum with the given value of n.

7. $\displaystyle\int_0^{12} x^2\,dx$, $n = 4$ **8.** $\displaystyle\int_{-2}^{8} \frac{1}{4}x^4\,dx$, $n = 5$

9. $\displaystyle\int_{-1}^{8} 2^x\,dx$, $n = 3$ **10.** $\displaystyle\int_{-4}^{4} \frac{1}{2^x} + 1\,dx$, $n = 4$

11. $\displaystyle\int_1^3 \frac{3}{x}\,dx$, $n = 4$ **12.** $\displaystyle\int_{5.5}^{7} (x^2 + 3x)\,dx$, $n = 3$

13. $\displaystyle\int_1^4 \sqrt{x}\,dx$, $n = 3$ **14.** $\displaystyle\int_0^{\pi} \sin x\,dx$, $n = 4$

15. Use the following table to estimate $\int_{10}^{26} f(x)\,dx$.

x	10	14	18	22	26
$f(x)$	100	88	72	50	28

16. Use the table to estimate $\int_0^{40} f(x)\,dx$. What values of n and Δx did you use?

x	0	10	20	30	40
$f(x)$	350	410	435	450	460

17. Use the following table to estimate $\int_0^{15} f(x)\,dx$.

x	0	3	6	9	12	15
$f(x)$	50	48	44	36	24	8

18. Use the following table to estimate $\int_3^4 W(t)\,dt$. What are n and Δt?

t	3.0	3.2	3.4	3.6	3.8	4.0
$W(t)$	25	23	20	15	9	2

19. Using Figure 5.29, draw rectangles representing each of the following Riemann sums for the function f on the interval $0 \le t \le 8$. Calculate the value of each sum.

    (a) Left-hand sum with $\Delta t = 4$
    (b) Right-hand sum with $\Delta t = 4$
    (c) Left-hand sum with $\Delta t = 2$
    (d) Right-hand sum with $\Delta t = 2$

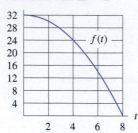

Figure 5.29

20. Use Figure 5.30 to estimate $\int_0^{20} f(x)\,dx$.

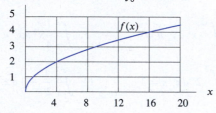

Figure 5.30

21. Use Figure 5.31 to estimate $\int_{-10}^{15} f(x)\,dx$.

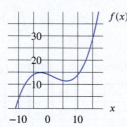

Figure 5.31

■ Use the graphs in Problems **22–23** to estimate $\int_0^3 f(x)\,dx$.

22. 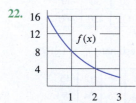   **23.**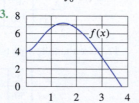

24. Using Figure 5.32, find the value of $\int_1^6 f(x)\,dx$.

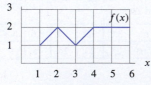

Figure 5.32

25. Without calculation, what can you say about the relationship between the values of the two integrals:

$$\int_0^2 e^{x^2}\,dx \quad \text{and} \quad \int_0^2 e^{t^2}\,dt?$$

26. If we know $\int_2^5 f(x)\,dx = 4$, what is the value of

$$3\left(\int_2^5 f(x)\,dx\right) + 1?$$

■ In Problems **27–34**, use a calculator or computer to evaluate the integral.

27. $\displaystyle\int_1^4 \frac{1}{\sqrt{1+x^2}}\,dx$ **28.** $\displaystyle\int_{-1}^1 \frac{1}{e^t}\,dt$

29. $\displaystyle\int_{-1}^1 \frac{x^2+1}{x^2-4}\,dx$ **30.** $\displaystyle\int_{1.1}^{1.7} e^t \ln t\,dt$

31. $\displaystyle\int_0^3 \ln(y^2+1)\,dy$ **32.** $\displaystyle\int_3^4 \sqrt{e^z+z}\,dz$

33. $\displaystyle\int_0^2 \frac{1}{1+x^2}\,dx + \int_0^2 \frac{1}{1+y^2}\,dy$

34. $\displaystyle\int_0^2 \sqrt{4+t^2}\,dt - \int_0^2 \sqrt{4+x^2}\,dx$

■ For Problems **35–38**:

    (a) Use a graph of the integrand to make a rough estimate of the integral. Explain your reasoning.

    (b) Use a computer or calculator to find the value of the definite integral.

35. $\displaystyle\int_0^1 x^3\,dx$ **36.** $\displaystyle\int_0^3 \sqrt{x}\,dx$

37. $\displaystyle\int_0^1 3^t\,dt$ **38.** $\displaystyle\int_1^2 x^x\,dx$

39. The graph of $f(t)$ is in Figure 5.33. Which of the following four numbers could be an estimate of $\int_0^1 f(t)\,dt$ accurate to two decimal places? Explain your choice.

    I. -98.35 II. 71.84 III. 100.12 IV. 93.47

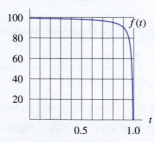

Figure 5.33

40. (a) Use a calculator or computer to find $\int_0^6 (x^2 + 1)\,dx$. Represent this value as the area under a curve.

(b) Estimate $\int_0^6 (x^2 + 1)\,dx$ using a left-hand sum with $n = 3$. Represent this sum graphically on a sketch of $f(x) = x^2 + 1$. Is this sum an overestimate or underestimate of the true value found in part (a)?

(c) Estimate $\int_0^6 (x^2 + 1)\,dx$ using a right-hand sum with $n = 3$. Represent this sum on your sketch. Is this sum an overestimate or underestimate?

41. Use Table 5.8 to evaluate the Riemann sums:

(a) $\displaystyle\sum_{i=0}^{n-1} f\left(t_i\right) \Delta t$ where $t_0 = 3, t_n = 15, n = 4$

(b) $\displaystyle\sum_{i=1}^{n} f\left(t_i\right) \Delta t$ where $t_0 = 3, t_n = 15, n = 3$

(c) $\displaystyle\sum_{i=1}^{n} f\left(t_i\right) \Delta t$ where $t_0 = 5, t_n = 13, n = 4$

Table 5.8

t	3	4	5	6	7	8	9
$f(t)$	−40	−17	4	23	40	55	68
t	10	11	12	13	14	15	
$f(t)$	79	88	95	100	103	104	

42. For $g(x) = 4x - 1$, evaluate the Riemann sums:

(a) $\displaystyle\sum_{i=0}^{3} g\left(x_i\right) \Delta x$ where $\Delta x = 3, x_0 = 2$

(b) $\displaystyle\sum_{i=1}^{4} g\left(x_i\right) \Delta x$ where $\Delta x = 2, x_0 = 4$

(c) $\displaystyle\sum_{i=2}^{5} g\left(x_i\right) \Delta x$ where $\Delta x = 3, x_0 = 1$

43. For $h(x) = \dfrac{1}{2}x + 5$ evaluate the Riemann sums:

(a) $\displaystyle\sum_{i=0}^{4} h\left(x_i\right) \Delta x$ where $\Delta x = 2, x_0 = 2$

(b) $\displaystyle\sum_{i=2}^{5} h\left(x_i\right) \Delta x$ where $\Delta x = 3, x_0 = 0$

(c) $\displaystyle\sum_{i=4}^{7} h\left(x_i\right) \Delta x$ where $\Delta x = 2, x_0 = 1$

5.3 THE DEFINITE INTEGRAL AS AREA

Relationship Between Definite Integral and Area: When $f(x)$ is positive

If $f(x)$ is continuous and positive, each term $f(x_0)\Delta x, f(x_1)\Delta x, \ldots$ in a left- or right-hand Riemann sum represents the area of a rectangle. See Figure 5.34. As the width Δx of the rectangles approaches zero, the rectangles fit the curve of the graph more exactly, and the sum of their areas gets closer to the area under the curve shaded in Figure 5.35. In other words:

> When $f(x)$ is positive and $a < b$:
>
> $$\text{Area under graph of } f \text{ between } a \text{ and } b = \int_a^b f(x)\,dx.$$

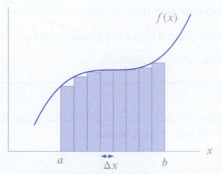

Figure 5.34: Area of rectangles approximating the area under the curve

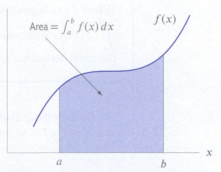

Figure 5.35: Shaded area is the definite integral $\int_a^b f(x)\,dx$

Relationship Between Definite Integral and Area: When $f(x)$ is Not Positive

We assumed in drawing Figure 5.35 that the graph of $f(x)$ lies above the x-axis. If the graph lies below the x-axis, then each value of $f(x)$ is negative, so each $f(x) \, \Delta x$ is negative, and the area is counted negatively. In that case, the definite integral is the negative of the area between the graph of f and the horizontal axis.

Example 1   What is the relation between the area between the parabola $y = x^2 - 1$ and the x-axis and these definite integrals?

(a) $\displaystyle \int_1^2 (x^2 - 1) \, dx$

(b) $\displaystyle \int_{-1}^1 (x^2 - 1) \, dx$

Solution   (a) The parabola lies above the x-axis between $x = 1$ and $x = 2$. (See Figure 5.36.) So,

$$\int_1^2 (x^2 - 1) \, dx = \text{Area} = 1.33.$$

(b) The parabola lies below the x-axis between $x = -1$ and $x = 1$. (See Figure 5.37.) So,

$$\int_{-1}^1 (x^2 - 1) \, dx = -\text{Area} = -1.33.$$

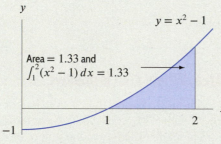

Figure 5.36: Integral $\int_1^2 (x^2 - 1) \, dx$ is positive, equal to shaded area

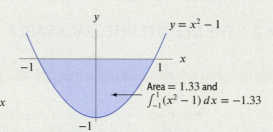

Figure 5.37: Integral $\int_{-1}^1 (x^2 - 1) \, dx$ is negative of shaded area

Summarizing, assuming $f(x)$ is continuous, we have:

When $f(x)$ is positive for some x-values and negative for others, and $a < b$:

$\displaystyle \int_a^b f(x) \, dx$ is the sum of the areas between a and b above the x-axis, counted positively, and the areas between a and b below the x-axis, counted negatively.

In the following example, we break up the integral. The properties that allow us to do this are discussed in the Focus on Theory section at the end of this chapter.

Example 2   Interpret the definite integral $\displaystyle \int_0^4 (x^3 - 7x^2 + 11x) \, dx$ in terms of areas.

Solution   Figure 5.38 shows the graph of $f(x) = x^3 - 7x^2 + 11x$ crossing below the x-axis at about $x = 2.38$. The integral is the area above the x-axis, A_1, minus the area below the x-axis, A_2. Computing the integral with a calculator or computer shows that

$$\int_0^4 (x^3 - 7x^2 + 11x)\, dx = 2.67.$$

Breaking the integral into two parts and calculating each one separately gives

$$\int_0^{2.38} (x^3 - 7x^2 + 11x)\, dx = 7.72 \quad \text{and} \quad \int_{2.38}^4 (x^3 - 7x^2 + 11x)\, dx = -5.05,$$

so $A_1 = 7.72$ and $A_2 = 5.05$. Then, as we would expect,

$$\int_0^4 (x^3 - 7x^2 + 11x)\, dx = A_1 - A_2 = 7.72 - 5.05 = 2.67.$$

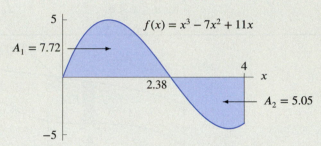

Figure 5.38: Integral $\displaystyle\int_0^4 (x^3 - 7x^2 + 11x)\, dx = A_1 - A_2$

Example 3   Find the total area of the shaded regions in Figure 5.38.

Solution   We saw in Example 2 that $A_1 = 7.72$ and $A_2 = 5.05$. Thus we have

$$\text{Total shaded area} = A_1 + A_2 = 7.72 + 5.05 = 12.77.$$

Example 4   For each of the functions graphed in Figure 5.39, decide whether $\int_0^5 f(x)\, dx$ is positive, negative or approximately zero.

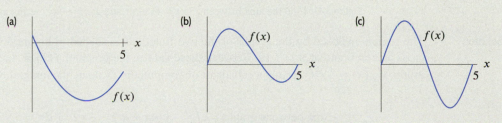

Figure 5.39: Is $\int_0^5 f(x)dx$ positive, negative or zero?

Solution   (a) The graph lies almost entirely below the x-axis, so the integral is negative.

(b) The graph lies partly below the x-axis and partly above the x-axis. However, the area above the x-axis is larger than the area below the x-axis, so the integral is positive.

(c) The graph lies partly below the x-axis and partly above the x-axis. Since the areas above and below the x-axis appear to be approximately equal in size, the integral is approximately zero.

Area Between Two Curves

We can use rectangles to approximate the area between two curves. If $g(x) \leq f(x)$, as in Figure 5.40, the height of a rectangle is $f(x) - g(x)$. The area of the rectangle is $(f(x) - g(x))\Delta x$, and we have the following result:

> If f and g are continuous functions, and $g(x) \leq f(x)$ for $a \leq x \leq b$:
>
> $$\text{Area between graphs of } f(x) \text{ and } g(x) \text{ for } a \leq x \leq b = \int_a^b (f(x) - g(x))\, dx.$$

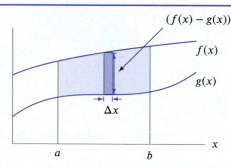

Figure 5.40: Area between two curves $= \int_a^b (f(x) - g(x))\, dx$

Example 5   Graphs of $f(x) = 4x - x^2$ and $g(x) = \frac{1}{2}x^{3/2}$ for $x \geq 0$ are shown in Figure 5.41. Use a definite integral to estimate the area enclosed by the graphs of these two functions.

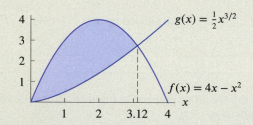

Figure 5.41: Find the area between $f(x) = 4x - x^2$ and $g(x) = \frac{1}{2}x^{3/2}$ using an integral

Solution   The region enclosed by the graphs of the two functions is shaded in Figure 5.41. The two graphs cross at $x = 0$ and at $x \approx 3.12$. Between these values, the graph of $f(x) = 4x - x^2$ lies above the graph of $g(x) = \frac{1}{2}x^{3/2}$. Using a calculator or computer to evaluate the integral, we get

$$\text{Area between graphs} = \int_0^{3.12} \left((4x - x^2) - \frac{1}{2}x^{3/2} \right) dx = 5.906.$$

Problems for Section 5.3

1. Find the area under the graph of $f(x) = x^2 + 2$ between $x = 0$ and $x = 6$.

2. Find the area under $P = 100(0.6)^t$ between $t = 0$ and $t = 8$.

3. Find the total area between $y = 4 - x^2$ and the x-axis for $0 \leq x \leq 3$.

4. Find the area between $y = x + 5$ and $y = 2x + 1$ between $x = 0$ and $x = 2$.

5. Find the area enclosed by $y = 3x$ and $y = x^2$.

6. (a) What is the total area between the graph of $f(x)$ in Figure 5.42 and the x-axis, between $x = 0$ and $x = 5$?
   (b) What is $\int_0^5 f(x)\, dx$?

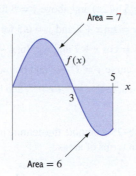

Figure 5.42

7. Using Figure 5.43, decide whether each of the following definite integrals is positive or negative.

   (a) $\int_{-5}^{-4} f(x)\, dx$ (b) $\int_{-4}^{1} f(x)\, dx$
   (c) $\int_{1}^{3} f(x)\, dx$ (d) $\int_{-5}^{3} f(x)\, dx$

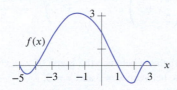

Figure 5.43

8. Using Figure 5.43, arrange the following definite integrals in ascending order:
   $\int_{-5}^{-3} f(x)\, dx$, $\int_{-5}^{-1} f(x)\, dx$, $\int_{-5}^{1} f(x)\, dx$, $\int_{-5}^{3} f(x)\, dx$.

9. (a) Estimate (by counting the squares) the total area shaded in Figure 5.44.
   (b) Using Figure 5.44, estimate $\int_0^8 f(x)\, dx$.

(c) Why are your answers to parts (a) and (b) different?

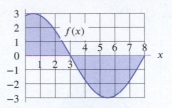

Figure 5.44

10. Using Figure 5.45, estimate $\int_{-3}^{5} f(x)\, dx$.

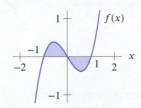

Figure 5.45

11. Given $\int_{-1}^{0} f(x)\, dx = 0.25$ and Figure 5.46, estimate:
    (a) $\int_0^1 f(x)\, dx$ (b) $\int_{-1}^{1} f(x)\, dx$
    (c) The total shaded area.

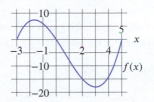

Figure 5.46

12. Using Figure 5.47, list the following integrals in increasing order (from smallest to largest). Which integrals are negative, which are positive? Give reasons.

    I. $\int_a^b f(x)\, dx$ II. $\int_a^c f(x)\, dx$ III. $\int_a^e f(x)\, dx$
    IV. $\int_b^e f(x)\, dx$ V. $\int_b^c f(x)\, dx$

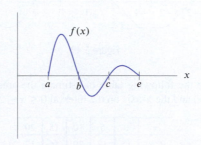

Figure 5.47

13. Use Figure 5.48 to find the values of

    (a) $\int_a^b f(x)\,dx$ **(b)** $\int_b^c f(x)\,dx$

    (c) $\int_a^c f(x)\,dx$ **(d)** $\int_a^c |f(x)|\,dx$

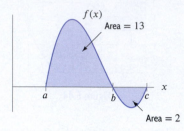

Figure 5.48

■ In Problems 14–17, match the graph with one of the following possible values for the integral $\int_0^5 f(x)\,dx$:

    I. -10.4 II. -2.1 III. 5.2 IV. 10.4

14.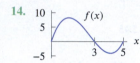

15.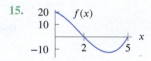

16.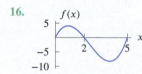

17.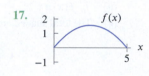

18. **(a)** Graph $f(x) = x(x+2)(x-1)$.

    (b) Find the total area between the graph and the x-axis between $x = -2$ and $x = 1$.

    (c) Find $\int_{-2}^1 f(x)\,dx$ and interpret it in terms of areas.

19. **(a)** Using Figure 5.49, find $\int_{-3}^0 f(x)\,dx$.

    (b) If the area of the shaded region is A, estimate $\int_{-3}^4 f(x)\,dx$.

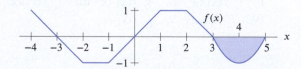

Figure 5.49

20. Use the following table to estimate the area between $f(x)$ and the x-axis on the interval $0 \le x \le 20$.

x	0	5	10	15	20
$f(x)$	15	18	20	16	12

21. Use Figure 5.50 to find the values of

    (a) $\int_0^2 f(x)\,dx$ **(b)** $\int_3^7 f(x)\,dx$

    (c) $\int_2^7 f(x)\,dx$ **(d)** $\int_5^8 f(x)\,dx$

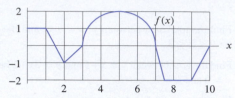

Figure 5.50: Graph consists of a semicircle and line segments

■ In Problems 22–27, use an integral to find the specified area.

22. Under $y = 6x^3 - 2$ for $5 \le x \le 10$.

23. Under $y = 5\ln(2x)$ and above $y = 3$ for $3 \le x \le 5$.

24. Between $y = \sin x + 2$ and $y = 0.5$ for $6 \le x \le 10$.

25. Between $y = \cos x + 7$ and $y = \ln(x-3)$, $5 \le x \le 7$.

26. Above the curve $y = x^4 - 8$ and below the x-axis.

27. Above the curve $y = -e^x + e^{2(x-1)}$ and below the x-axis, for $x \ge 0$.

■ For Problems 28–29, compute the definite integral and interpret the result in terms of areas.

28. $\displaystyle\int_1^4 \frac{x^2 - 3}{x}\,dx$.   **29.** $\displaystyle\int_1^4 (x - 3\ln x)\,dx$.

■ In Problems 30–33, find the integral by finding the area of the region between the curve and the horizontal axis.

30. $\displaystyle\int_0^{10} (x - 5)\,dx$ **31.** $\displaystyle\int_0^8 (6 - 2x)\,dx$

32. $\displaystyle\int_{-8}^6 \left(\frac{1}{2}x + 3\right)\,dx$ **33.** $\displaystyle\int_{-10}^1 \frac{-4x - 16}{3}\,dx$

34. Find the area between the graph of $y = x^2 - 2$ and the x-axis, between $x = 0$ and $x = 3$.

35. **(a)** Find the total area between $f(x) = x^3 - x$ and the x-axis for $0 \le x \le 3$.

    (b) Find $\displaystyle\int_0^3 f(x)\,dx$.

    (c) Are the answers to parts (a) and (b) the same? Explain.

36. Compute the definite integral $\int_0^4 \cos\sqrt{x}\,dx$ and interpret the result in terms of areas.

■ In Problems 37–40, use Figure 5.51 to find limits a and b in the interval $[0, 5]$ with $a < b$ satisfying the given condition.

37. $\displaystyle\int_0^b f(x)\,dx$ is largest

38. $\displaystyle\int_a^4 f(x)\,dx$ is smallest

39. $\displaystyle\int_a^b f(x)\,dx$ is largest

40. $\displaystyle\int_a^b f(x)\,dx$ is smallest

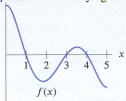

Figure 5.51

5.4   INTERPRETATIONS OF THE DEFINITE INTEGRAL

This section focuses on interpretations of the definite integral. It includes when the integrand is a rate of change with time—as it was in Section 5.1 when we integrated velocity. The next section focuses on the Fundamental Theorem of Calculus, where we integrate more general rates of change.

The Notation and Units for the Definite Integral

Just as the Leibniz notation dy/dx for the derivative reminds us that the derivative is the limit of a quotient of differences, the notation for the definite integral,

$$\int_a^b f(x)\,dx,$$

reminds us that an integral is a limit of a sum. Since dx can be thought of as a small difference in x, the terms being added are products of the form "$f(x)$ times a difference in x." We have the following result:

> The unit of measurement for $\displaystyle\int_a^b f(x)\,dx$ is the product of the units for $f(x)$ and the units for x.

Example: If x and $f(x)$ have the same units, then the integral $\displaystyle\int_a^b f(x)\,dx$ is measured in square units, say cm $\times$ cm $=$ cm^2. This is as we expect, since the integral represents an area.

Example: If $v = f(t)$ is velocity in meters/second and t is time in seconds, then the integral $\displaystyle\int_a^b f(t)\,dt$ has units of (meters/sec) $\times$ (sec) $=$ meters, as we expect, since the integral represents change in position.

Example 1   Let $C(t)$ represent the cost per day to heat your home in dollars per day, where t is time measured in days and $t = 0$ corresponds to January 1, 2014. Interpret $\displaystyle\int_0^{90} C(t)\,dt$.

Solution   The units for the integral $\int_0^{90} C(t)\,dt$ are (dollars/day) $\times$ (days) $=$ dollars. The integral represents the cost in dollars to heat your house for the first 90 days of 2014, namely the months of January, February, and March.

In Section 5.1, we saw that the integral of a rate of change gives total change. This is a particular case of the Fundamental Theorem of Calculus, which is considered in more depth in Section 5.5.

If $f(t)$ is a rate of change of a quantity, then

$$\int_a^b f(t)\,dt = \text{Total change in quantity between } t = a \text{ and } t = b.$$

The units correspond: If $f(t)$ is a rate of change, with units of quantity/time, then $f(t)\Delta t$ and the definite integral have units of (quantity/time) $\times$ (time) = quantity.

Example 2   A bacteria colony initially has a population of 14 million bacteria. Suppose that t hours later the population is growing at a rate of $f(t) = 2^t$ million bacteria per hour.

(a) Give a definite integral that represents the total change in the bacteria population during the time from $t = 0$ to $t = 2$.

(b) Find the population at time $t = 2$.

Solution   (a) Since $f(t) = 2^t$ gives the rate of change of population, we have

$$\text{Change in population between } t = 0 \text{ and } t = 2 = \int_0^2 2^t\,dt.$$

(b) Using a calculator, we find $\int_0^2 2^t\,dt = 4.328$. The bacteria population was 14 million at time $t = 0$ and increased 4.328 million between $t = 0$ and $t = 2$. Therefore, at time $t = 2$,

$$\text{Population} = 14 + 4.328 = 18.328 \text{ million bacteria.}$$

Example 3   A man starts 50 miles away from his home and takes a trip in his car. He moves on a straight line, and his home lies on this line. His velocity is given in Figure 5.52.

(a) When is the man closest to his home? Approximately how far away is he then?

(b) When is the man farthest from his home? How far away is he then?

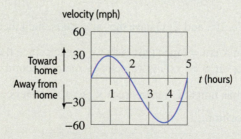

Figure 5.52: Velocity of trip starting 50 miles from home

Solution   What happens on this trip? The velocity function is positive the first two hours and negative between $t = 2$ and $t = 5$. So the man moves toward his home during the first two hours, then turns around at $t = 2$ and moves away from his home. The distance he travels is represented by the area between the graph of velocity and the t-axis; since the area below the axis is greater than the area above the axis, we see that he ends up farther away from home than when he started. Thus he is closest to home at $t = 2$ and farthest from home at $t = 5$. We can estimate how far he went in each direction by estimating areas.

(a) The man starts out 50 miles from home. The distance the man travels during the first two hours is the area under the curve between $t = 0$ and $t = 2$. This area corresponds to about one grid box. Since each grid box has area (30 miles/hour)(1 hour) = 30 miles, the man travels about 30 miles toward home. He is closest to home after 2 hours, and he is about 20 miles away at that time.

(b) Between $t = 2$ and $t = 5$, the man moves away from his home. Since this area is equal to about 3.5 grid boxes, which is (3.5)(30) =105 miles, he has moved 105 miles farther from home. He was already 20 miles from home at $t = 2$, so at $t = 5$ he is about 125 miles from home. He is farthest from home at $t = 5$.

Notice that the man has covered a total distance of $30 + 105 = 135$ miles. However, he went toward his home for 30 miles and away from his home for 105 miles. His *net* change in position is 75 miles.

Example 4   The rates of growth of the populations of two species of plants (measured in new plants per year) are shown in Figure 5.53. Assume that the populations of the two species are equal at time $t = 0$.

(a) Which population is larger after one year? After two years?
(b) How much does the population of species 1 increase during the first two years?

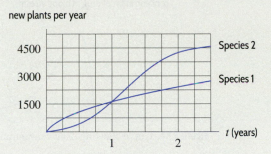

new plants per year

Figure 5.53: Population growth rates for two species of plants

Solution   (a) The rate of growth of the population of species 1 is higher than that of species 2 throughout the first year, so the population of species 1 is larger after one year. After two years, the situation is less clear, since the population of species 1 increased faster for the first year and that of species 2 for the second. However, if $r(t)$ is the rate of growth of a population, we have

$$\text{Total change in population during first two years} = \int_0^2 r(t)\,dt.$$

This integral is the area under the graph of $r(t)$. For $t = 0$ to $t = 2$, the area under the species 1 graph in Figure 5.53 is smaller than the area under the species 2 graph, so the population of species 2 is larger after two years.

(b) The population change for species 1 is the area of the region under the graph of $r(t)$ between $t = 0$ and $t = 2$ in Figure 5.53. The region consists of about 16.5 grid boxes, each of area (750 plants/year)(0.25 year) = 187.5 plants, giving a total of (16.5)(187.5) = 3093.75 plants. The population of species 1 increases by about 3100 plants during the two years.

Bioavailability of Drugs

In pharmacology, the definite integral is used to measure the overall exposure of a person to a drug during the course of treatment; this determines the *bioavailability* of the drug under different administration regimes.

The concentration of a drug in the blood is generally not constant, but increases as the drug is absorbed and then decreases as the drug is broken down and excreted.[6] (See Figure 5.54.) To calculate the exposure to a drug in the blood with concentration $C(t)$ $\mu g/cm^3$ at time t over a small interval Δt, we estimate

$$\text{Exposure} \approx \text{Concentration} \times \text{Time} = C(t)\,\Delta t.$$

Summing over all subintervals gives

$$\text{Total exposure} \approx \sum C(t)\,\Delta t.$$

In the limit as $n \to \infty$, where n is the number of intervals of width Δt, the sum becomes an integral. So for $0 \leq t \leq T$, we have

$$\text{Exposure} = \int_0^T C(t)\,dt.$$

That is, the total exposure of a drug is equal to the area under the drug concentration curve.

concentration of drug in blood stream

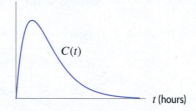

Figure 5.54: Curve showing drug concentration as a function of time

Example 5   Concentration curves for a drug administered by two methods are shown in Figure 5.55. Describe the differences and similarities between the two administration methods in terms of peak concentration, speed of absorption, and total exposure.

concentration of drug in blood stream

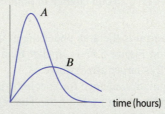

Figure 5.55: Concentration curves of two drug administration methods

Solution   Method A has a peak concentration more than twice as high as that of Method B. Because Method A achieves peak concentration sooner than Method B, the drug is absorbed more rapidly using Method A than Method B. Finally, Method A provides greater total exposure to the drug, since the area under the graph of the concentration for Method A is greater than the area under the graph for Method B.

---

[6]en.wikipedia.org/wiki/Bioavailability. Accessed January 2017.

Problems for Section 5.4

■ In Problems 1–4, explain in words what the integral represents and give units.

1. $\int_1^3 v(t)\, dt$, where $v(t)$ is velocity in meters/sec and t is time in seconds.

2. $\int_0^6 a(t)\, dt$, where $a(t)$ is acceleration in km/hr² and t is time in hours.

3. $\int_{2005}^{2011} f(t)\, dt$, where $f(t)$ is the rate at which world population is growing in year t, in billion people per year.

4. $\int_0^5 s(x)\, dx$, where $s(x)$ is rate of change of salinity (salt concentration) in gm/liter per cm in sea water, and where x is depth below the surface of the water in cm.

5. Your velocity is $v(t) = \ln(t^2 + 1)$ ft/sec for t in seconds, $0 \le t \le 3$. Find the distance traveled during this time.

6. Oil leaks out of a tanker at a rate of $r = f(t)$ gallons per minute, where t is in minutes. Write a definite integral expressing the total quantity of oil which leaks out of the tanker in the first hour.

7. Pollution is removed from a lake at a rate of $f(t)$ kg/day on day t.

   (a) Explain the meaning of the statement $f(12) = 500$.
   (b) If $\int_5^{15} f(t)\, dt = 4000$, give the units of the 5, the 15, and the 4000.
   (c) Give the meaning of $\int_5^{15} f(t)\, dt = 4000$.

8. Annual coal production in the US (in billion tons per year) is given in the table.[7] Estimate the total amount of coal produced in the US between 1997 and 2009. If $r = f(t)$ is the rate of coal production t years since 1997, write an integral to represent the 1997–2009 coal production.

Year	1997	1999	2001	2003	2005	2007	2009
Rate	1.090	1.094	1.121	1.072	1.132	1.147	1.073

9. The table gives annual US emissions, $H(t)$, of hydrofluorocarbons, or "super greenhouse gases," in millions of metric tons of carbon-dioxide equivalent. Let t be in years since 2000.[8]

   (a) What are the units and meaning of $\int_0^{12} H(t)\, dt$?
   (b) Estimate $\int_0^{12} H(t)\, dt$.

Year	2000	2002	2004	2006	2008	2010	2012
$H(t)$	158.9	154.4	152.1	153.9	162.7	167.0	173.6

10. Table 5.9 shows the upward vertical velocity $v(t)$, in ft/min, of a small plane at time t seconds during a short flight.

    (a) When is the plane going up? Going down?
    (b) If the airport is located 110 ft above sea level, estimate the maximum altitude the plane reaches during the flight.

Table 5.9

t	$v(t)$	t	$v(t)$
0	10	100	−140
10	20	110	−180
20	60	120	0
30	490	130	−820
40	890	140	−1270
50	980	150	−780
60	830	160	−940
70	970	170	−540
80	300	180	−230
90	10	190	0

11. World annual natural gas[9] production, N, in billion cubic meters, is approximated by $N = 2711 + 77t$, where t is in years since 2004.

    (a) How much natural gas was produced in 2004? In 2015?
    (b) Estimate the total amount of natural gas produced during the 10-year period from 2004 to 2014.

12. Solar photovoltaic (PV) cells are the world's fastest-growing energy source. In year t since 2007, PV cells were manufactured worldwide at a rate of $S = 3.7e^{0.61t}$ gigawatts per year.[10] Estimate the total solar energy-generating capacity of the PV cells manufactured between 2007 and 2011.

■ Problems 13–16 show the velocity, in cm/sec, of a particle moving along a number line. (Positive velocities represent movement to the right; negative velocities to the left.) Find the change in position and total distance traveled between times $t = 0$ and $t = 5$ seconds.

---

[7] http://www.eia.doe.gov/cneaf/coal/page/special/tbl1.html. Accessed May 2011.
[8] www.epa.gov/climatechange/ghgemissions/inventoryexplorer/#allsectors/allgas/gas/all. Accessed April 2015.
[9] *BP Statistical Review of World Energy 2015*, www.bp.com/, accessed December 15, 2016.
[10] www.earth-policy.org/indicators/Temp/solar_power_2011. Accessed February 2013. A gigawatt measures power, the rate of energy generation. In this case, we are considering gigawatts of capacity manufactured per year.

13. **14.**

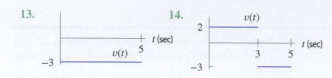

15. **16.**

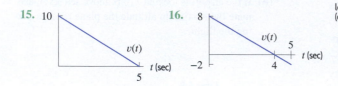

17. A forest fire covers 2000 acres at time $t = 0$. The fire is growing at a rate of $8\sqrt{t}$ acres per hour, where t is in hours. How many acres are covered 24 hours later?

18. Water is pumped out of a holding tank at a rate of $5 - 5e^{-0.12t}$ liters/minute, where t is in minutes since the pump is started. If the holding tank contains 1000 liters of water when the pump is started, how much water does it hold one hour later?

19. With t in seconds, the velocity of an object is $v(t) = 10 + 8t - t^2$ m/sec.

(a) Represent the distance traveled during the first 5 seconds as a definite integral and as an area.

(b) Estimate the distance traveled by the object during the first 5 seconds by estimating the area.

(c) Calculate the distance traveled.

20. A bungee jumper leaps off the starting platform at time $t = 0$ and rebounds once during the first 5 seconds. With velocity measured downward, for t in seconds and $0 \leq t \leq 5$, the jumper's velocity is approximated[11] by $v(t) = -4t^2 + 16t$ meters/sec.

(a) How many meters does the jumper travel during the first five seconds?

(b) Where is the jumper relative to the starting position at the end of the five seconds?

(c) What does $\int_0^5 v(t)\,dt$ represent in terms of the jump?

21. After a foreign substance is introduced into the blood, the rate at which antibodies are made is given by

$$r(t) = \frac{t}{t^2 + 1} \text{ thousands of antibodies per minute,}$$

where time, t, is in minutes. Assuming there are no antibodies present at time $t = 0$, find the total quantity of antibodies in the blood at the end of 4 minutes.

22. Figure 5.56 shows the length growth rate of a human fetus.

(a) What feature of a graph of length as a function of age corresponds to the maximum in Figure 5.56?

(b) Estimate the length of a baby born in week 40.

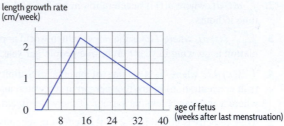

Figure 5.56

23. At 11:57 pm on March 12, 1928, the two-year-old St Francis dam on the outskirts of Los Angeles failed catastrophically and the dam emptied. The resulting flood was one of the worst US civil engineering disasters of the 20th century, claiming over 400 lives.[12] The volume of water discharging from the dam, in acre-feet per minute, with time in minutes after midnight on March 13, is in Figure 5.57.

(a) How long did it take to empty the dam?

(b) What was the maximum discharge rate and when did it occur?

(c) How much water was discharged in total?

(d) How long did it take for the dam to be half-empty?

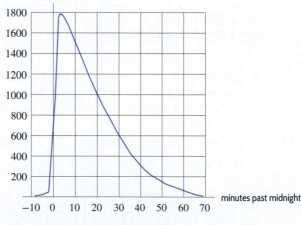

Figure 5.57

[11] Based on www.itforus.oeiizk.waw.pl/tresc/activ//modules/bj.pdf. Accessed February 12, 2012.

[12] web.mst.edu/rogersda/st_francis_dam/Mapping%20the%20St%20Francis%20Dam%20Outburst%20Flood%20with%20GIS.pdf, math.ucsd.edu/~ashenk/Math_20B_Summer_2008/Rog_Sec_5_1-5.5_Review.pdf, and en.wikipedia.org/wiki/St._Francis_Dam. Accessed January 14, 2017.

24. Figure 5.58 gives your velocity during a trip starting from home. Positive velocities take you away from home and negative velocities take you toward home. Where are you at the end of the 5 hours? When are you farthest from home? How far away are you at that time?

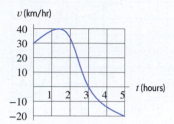

Figure 5.58

25. A bicyclist pedals along a straight road with velocity, v, given in Figure 5.59. She starts 5 miles from a lake; positive velocities take her away from the lake and negative velocities take her toward the lake. When is the cyclist farthest from the lake, and how far away is she then?

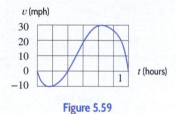

Figure 5.59

26. Figure 5.60 shows the rate of growth of two trees. If the two trees are the same height at time $t = 0$, which tree is taller after 5 years? After 10 years?

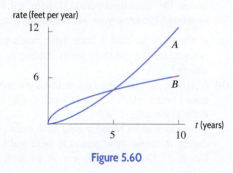

Figure 5.60

■ Problems **27–29** concern the future of the US Social Security Trust Fund, out of which pensions are paid. Figure 5.61 shows the rates (billions of dollars per year) at which income,

$I(t)$, from taxes and interest is projected to flow into the fund and at which expenditures, $E(t)$, flow out of the fund. Figure 5.62 shows the value of the fund as a function of time.[13]

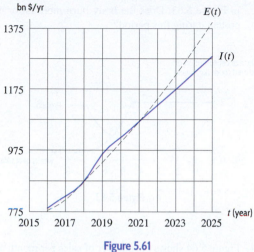

Figure 5.61

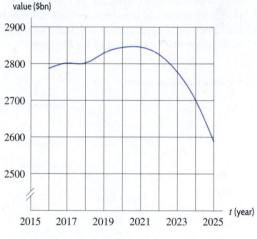

Figure 5.62

27. (a) Write each of the following areas in Figure 5.61 from 2016 to 2020 as an integral and explain its significance for the fund.

    (i) Under the income curve

    (ii) Under the expenditure curve

    (iii) Between the income and expenditure curves

(b) Use Figure 5.62 to estimate the area between the two curves from 2016 to 2020.

28. Decide when the value of the fund is projected to be a maximum using

    (a) Figure 5.61   **(b)** Figure 5.62

[13]The 2016 OASDI Trustees Report, www.ssa.gov/OACT/TR/2016, accessed September 1, 2016.

29. Express the projected change in value of the fund from 2016 to 2025 as an integral.

30. The rates of consumption of stores of protein and fat in the human body during 8 weeks of starvation are shown in Figure 5.63. Does the body burn more fat or more protein during this period?

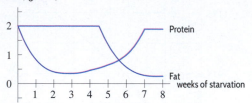

Figure 5.63

31. Figure 5.64 shows the number of sales per month made by two salespeople. Which person has the most total sales after 6 months? After the first year? At approximately what times (if any) have they sold roughly equal total amounts? Approximately how many total sales has each person made at the end of the first year?

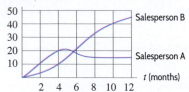

Figure 5.64

32. Height velocity graphs are used by endocrinologists to follow the progress of children with growth deficiencies. Figure 5.65 shows the height velocity curves of an average boy and an average girl between ages 3 and 18.

   (a) Which curve is for girls and which is for boys? Explain how you can tell.
   (b) About how much does the average boy grow between ages 3 and 10?
   (c) The growth spurt associated with adolescence and the onset of puberty occurs between ages 12 and 15 for the average boy and between ages 10 and 12.5 for the average girl. Estimate the height gained by each average child during this growth spurt.
   (d) When fully grown, about how much taller is the average man than the average woman? (The average boy and girl are about the same height at age 3.)

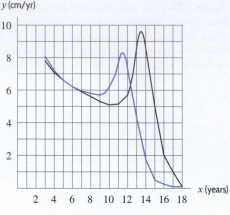

Figure 5.65

■ A healthy human heart pumps about 5 liters of blood per minute. Problems **33–34** refer to Figure 5.66, which shows the response of the heart to bleeding. The pumping rate drops and then returns to normal if the person recovers fully, or drops to zero if the person dies.

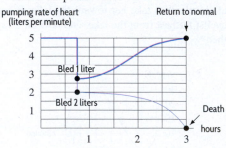

Figure 5.66

33. (a) If the body is bled 2 liters, how much blood is pumped during the three hours leading to death?
   (b) If $f(t)$ is the pumping rate in liters per minute at time t hours, express your answer to part (a) as a definite integral.
   (c) How much more blood would have been pumped during the same time period if there had been no bleeding? Illustrate your answer on the graph.

34. (a) If the body is bled 1 liter, how much blood is pumped during the three hours leading to full recovery?
   (b) If $g(t)$ is the pumping rate in liters per minute at time t hours, express your answer to part (a) as a definite integral.
   (c) How much more blood would have been pumped during the same time period if there had been no bleeding? Show your answer as an area on the graph.

35. The amount of waste a company produces, W, in tons per week, is approximated by $W = 3.75e^{-0.008t}$, where t is in weeks since January 1, 2016. Waste removal for the company costs \$150/ton. How much did the company pay for waste removal during the year 2016?

36. Figure 5.67 shows plasma concentration curves for two products containing a drug used to slow a rapid heart rate. Compare the two products in terms of level of peak concentration, time until peak concentration, and overall exposure. Which product should be used to achieve the fastest response?

concentration of drug in plasma

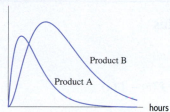

Product B

Product A

hours

Figure 5.67

37. Figure 5.68 compares the concentration in blood plasma for two pain relievers. Compare the two products in terms of level of peak concentration, time until peak concentration, and overall exposure.

concentration of drug in plasma

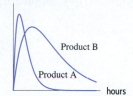

Product B

Product A

hours

Figure 5.68

38. Draw plasma concentration curves for two products A and B if product A has the highest peak concentration, but product B is absorbed more quickly and provides greater overall exposure.

39. A two-day environmental cleanup started at 9 am on the first day. The number of workers fluctuated as shown in Figure 5.69. If the workers were paid $10 per hour, how much was the total personnel cost of the cleanup?

workers

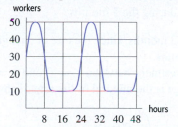

hours

Figure 5.69

40. In Problem 39, suppose workers were paid $10 per hour for work between 9 am and 5 pm and $15 per hour for work during the rest of the day. What would the total personnel costs have been under these conditions?

41. At the site of a spill of radioactive iodine, radiation levels were four times the maximum acceptable limit, so an evacuation was ordered. If R_0 is the initial radiation level (at $t = 0$) and t is the time in hours, the radiation level $R(t)$, in millirems/hour, is given by

$$R(t) = R_0(0.996)^t.$$

(a) How long does it take for the site to reach the acceptable level of radiation of 0.6 millirems/hour?

(b) How much total radiation (in millirems) has been emitted by that time?

42. If you jump out of an airplane and your parachute fails to open, your downward velocity (in meters per second) t seconds after the jump is approximated by

$$v(t) = 49(1 - (0.8187)^t).$$

(a) Write an expression for the distance you fall in T seconds.

(b) If you jump from 5000 meters above the ground, estimate, using trial and error, how many seconds you fall before hitting the ground.

43. The Montgolfier brothers (Joseph and Etienne) were eighteenth-century pioneers of hot-air ballooning. Had they had the appropriate instruments, they might have left us a record, like that shown in Figure 5.70, of one of their early experiments. The graph shows their vertical velocity, v, with upward as positive.

(a) Over what intervals was the acceleration positive? Negative?

(b) What was the greatest altitude achieved, and at what time?

(c) This particular flight ended on top of a hill. How do you know that it did, and what was the height of the hill above the starting point?

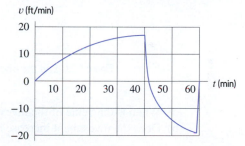

Figure 5.70

44. Figure 5.71 shows solar radiation, in watts per square meter (w/m²), in Santa Rosa, California, throughout a typical January day.[14] Estimate the daily energy produced, in kwh, by a 20-square-meter solar array located in Santa Rosa if it converts 18% of solar radiation into energy.

solar radiation (w/m²)

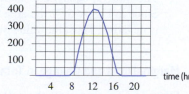

time (hrs past midnight)

Figure 5.71

[14]http://cdec.water.ca.gov/, accessed January 8, 2015.

5.5 TOTAL CHANGE AND THE FUNDAMENTAL THEOREM OF CALCULUS

In Section 5.4, we saw that the total change of a quantity can be obtained by integrating its rate of change. Since the change in $F(x)$ between $x = a$ and $x = b$ is $F(b) - F(a)$ and the rate of change is $F'(x)$, we have the following result:

The Fundamental Theorem of Calculus

If $F'(x)$ is continuous for $a \leq x \leq b$, then

$$\int_a^b F'(x)\,dx = F(b) - F(a).$$

In words:
The definite integral of the derivative of a function gives the total change in the function.

In Section 6.3 we see how to use the Fundamental Theorem to compute a definite integral. The Focus on Theory Section gives another version of the Fundamental Theorem.

Example 1   Figure 5.72 shows $F'(t)$, the rate of change of the value, $F(t)$, of an investment over a 5-month period.

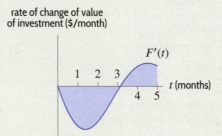

rate of change of value
of investment (\$/month)

Figure 5.72: Did the investment increase or decrease in value over these 5 months?

(a) When is the value of the investment increasing in value and when is it decreasing?
(b) Does the investment increase or decrease in value during the 5 months?

Solution   (a) The investment decreased in value during the first 3 months, since the rate of change of value is negative then. The value rose during the last 2 months.
(b) We want the total change in the value of the investment between $t = 0$ and $t = 5$. By the Fundamental Theorem of Calculus, the total change is the integral of the rate of change, $F'(t)$:

$$\text{Total change in value} = \int_0^5 F'(t)\,dt.$$

The integral equals the shaded area above the t-axis minus the shaded area below the t-axis. Since in Figure 5.72 the area below the axis is greater than the area above the axis, the integral is negative. Thus, the value of the investment during this time period has decreased.

Marginal Cost and Change in Total Cost

Suppose $C(q)$ represents the cost of producing q items. The derivative, $C'(q)$, is the marginal cost. Since marginal cost $C'(q)$ is the rate of change of the cost with respect to quantity, q, by the Fundamental Theorem, the integral represents the total change in the cost between $q = a$ and $q = b$:

$$\int_a^b C'(q)\,dq = C(b) - c(a).$$

In other words, the integral gives the amount it costs to increase production from a units to b units.

The cost of producing 0 units is the fixed cost $C(0)$. The area under the marginal cost curve between $q = 0$ and $q = b$ is the total increase in cost between a production of 0 and a production of b. This is called the *total variable cost*. Adding this to the fixed cost gives the total cost to produce b units. In summary:

If $C'(q)$ is a marginal cost function and $C(0)$ is the fixed cost,

$$\text{Cost to increase production from } a \text{ units to } b \text{ units } = C(b) - C(a) = \int_a^b C'(q)\,dq$$

$$\text{Total variable cost to produce } b \text{ units } = \int_0^b C'(q)\,dq$$

$$\text{Total cost of producing } b \text{ units } = \text{Fixed cost } + \text{ Total variable cost}$$

$$= C(0) + \int_0^b C'(q)\,dq$$

Example 2 A marginal cost curve is given in Figure 5.73. If the fixed cost is \$1000, estimate the total cost of producing 250 items.

Solution The total cost of production is Fixed cost + Total variable cost. The total variable cost of producing 250 items is represented by the area under the marginal cost curve. The area in Figure 5.73 between $q = 0$ and $q = 250$ is about 20 grid boxes. Each grid box has area (2 dollars/item)(50 items) = 100 dollars, so

$$\text{Total variable cost } = \int_0^{250} C'(q)\,dq \approx 20(100) = 2000.$$

The total cost to produce 250 items is given by :

$$\text{Total cost } = \text{Fixed cost } + \text{ Total variable cost}$$
$$\approx \$1000 + \$2000 = \$3000.$$

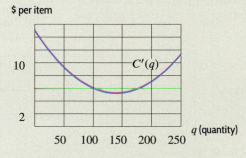

Figure 5.73: A marginal cost curve

Problems for Section 5.5

1. A cup of coffee at 90°C is put into a 20°C room when $t = 0$. The coffee's temperature is changing at a rate of $r(t) = -7(0.9^t)$ °C per minute, with t in minutes. Estimate the coffee's temperature when $t = 10$.

2. If the marginal cost function $C'(q)$ is measured in dollars per ton, and q gives the quantity in tons, what are the units of measurement for $\int_{800}^{900} C'(q)\, dq$? What does this integral represent?

3. The marginal cost of drilling an oil well depends on the depth at which you are drilling; drilling becomes more expensive, per meter, as you dig deeper into the earth. The fixed costs are 1,000,000 riyals (the riyal is the unit of currency of Saudi Arabia), and, if x is the depth in meters, the marginal costs are

$$C'(x) = 4000 + 10x \quad \text{riyals/meter.}$$

   Find the total cost of drilling a 500-meter well.

4. The population of Tokyo grew at the rate shown in Figure 5.74. Estimate the change in population between 1970 and 1990.

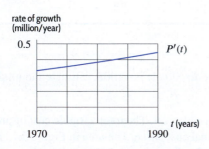

Figure 5.74

5. A marginal cost function $C'(q)$ is given in Figure 5.75. If the fixed costs are $10,000, estimate:

   (a) The total cost to produce 30 units.
   (b) The additional cost if the company increases production from 30 units to 40 units.
   (c) The value of $C'(25)$. Interpret your answer in terms of costs of production.

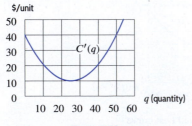

Figure 5.75

6. Figure 5.76 shows the rate of change of the quantity of water in a water tower, in liters per day, during the month of April. If the tower had 12,000 liters of water in it on April 1, estimate the quantity of water in the tower on April 30.

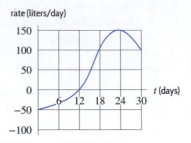

Figure 5.76

7. The concentration of a medication in the plasma changes at a rate of $h(t)$ mg/ml per hour, t hours after the delivery of the drug.

   (a) Explain the meaning of the statement $h(1) = 50$.
   (b) There is 250 mg/ml of the medication present at time $t = 0$ and $\int_0^3 h(t)\, dt = 480$. What is the plasma concentration of the medication present three hours after the drug is administered?

8. The total cost in dollars to produce q units of a product is $C(q)$. Fixed costs are $20,000. The marginal cost is

$$C'(q) = 0.005q^2 - q + 56.$$

   (a) On a graph of $C'(q)$, illustrate graphically the total variable cost of producing 150 units.
   (b) Estimate $C(150)$, the total cost to produce 150 units.
   (c) Find the value of $C'(150)$ and interpret your answer in terms of costs of production.
   (d) Use parts (b) and (c) to estimate $C(151)$.

9. The marginal cost $C'(q)$ (in dollars per unit) of producing q units is given in the following table.

   (a) If fixed cost is $10,000, estimate the total cost of producing 400 units.
   (b) How much would the total cost increase if production were increased one unit, to 401 units?

q	0	100	200	300	400	500	600
$C'(q)$	25	20	18	22	28	35	45

10. The marginal cost function for a company is given by

$$C'(q) = q^2 - 16q + 70 \quad \text{dollars/unit,}$$

   where q is the quantity produced. If $C(0) = 500$, find the total cost of producing 20 units. What is the fixed cost and what is the total variable cost for this quantity?

11. The marginal cost function of producing q mountain bikes is

$$C'(q) = \frac{600}{0.3q + 5}.$$

(a) If the fixed cost in producing the bicycles is $2000, find the total cost to produce 30 bicycles.

(b) If the bikes are sold for $200 each, what is the profit (or loss) on the first 30 bicycles?

(c) Find the marginal profit on the 31st bicycle.

12. The marginal revenue function on sales of q units of a product is $R'(q) = 200 - 12\sqrt{q}$ dollars per unit.

(a) Graph $R'(q)$.

(b) Estimate the total revenue if sales are 100 units.

(c) What is the marginal revenue at 100 units? Use this value and your answer to part (b) to estimate the total revenue if sales are 101 units.

13. Figure 5.77 shows $P'(t)$, the rate of change of the price of stock in a certain company at time t.

(a) At what time during this five-week period was the stock at its highest value? At its lowest value?

(b) If $P(t)$ represents the price of the stock, arrange the following quantities in increasing order:

$$P(0), \ P(1), \ P(2), \ P(3), \ P(4), \ P(5).$$

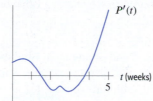

Figure 5.77

14. The net worth, $f(t)$, of a company is growing at a rate of $f'(t) = 2000 - 12t^2$ dollars per year, where t is in years since 2005. How is the net worth of the company expected to change between 2005 and 2015? If the company is worth $40,000 in 2005, what is it worth in 2015?

15. The graph of a derivative $f'(x)$ is shown in Figure 5.78. Fill in the table of values for $f(x)$ given that $f(0) = 2$.

x	0	1	2	3	4	5	6
$f(x)$	2						

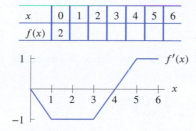

Figure 5.78: Graph of f', not f

16. The derivative $f'(x)$ is graphed in Figure 5.79. Fill in the table of values for $f(x)$ given that $f(0) = -10$.

x	0	1	2	3	4	5	6
$f(x)$	-10						

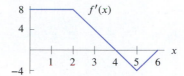

Figure 5.79: Graph of f', not f

17. Figure 5.80 shows the rate of change in the average plasma concentration of the drug Omeprazole (in ng/ml per hour) for six hours after the first dose is administered using two different capsules: immediate-release and delayed-release.[15]

(a) Which graph corresponds to which capsule?

(b) Do the two capsules provide the same maximum concentration? If not, which provides the larger maximum concentration?

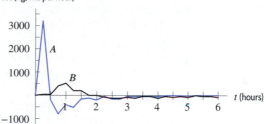

Figure 5.80

■ In Problems **18–19**, oil is pumped from a well at a rate of $r(t)$ barrels per day, with t in days. Assume $r'(t) < 0$ and $t_0 > 0$.

18. What does the value of $\int_0^{t_0} r(t)\, dt$ tell us about the oil well?

19. Rank in order from least to greatest:

$$\int_0^{2t_0} r(t)\, dt, \qquad \int_{t_0}^{2t_0} r(t)\, dt, \qquad \int_{2t_0}^{3t_0} r(t)\, dt.$$

---

[15]Data adapted from C. W. Howden, "Review article: "Immediate-Release Proton-Pump Inhibitor Therapy—Potential Advantages", *Alimententary Pharmacology and Therapeutics,* Vol 22, Issue s3 (2005).

■ In Problems **20–22**, let $C(n)$ be a city's cost, in millions of dollars, for plowing the roads when n inches of snow have fallen. Let $c(n) = C'(n)$. Evaluate the expressions and interpret your answers in terms of the cost of plowing snow, given

$$c'(n) < 0, \qquad \int_0^{15} c(n)\, dn = 7.5, \qquad c(15) = 0.7,$$

$$c(24) = 0.4, \qquad C(15) = 8, \qquad C(24) = 13.$$

20. $\displaystyle \int_{15}^{24} c(n)\, dn$ **21.** $C(0)$

22. $\displaystyle c(15) + \int_{15}^{24} c'(n)\, dn$

■ Problems **23–25** refer to a May 2, 2010, article:[16]

"The crisis began around 10 am yesterday when a 10-foot wide pipe in Weston sprang a leak, which worsened throughout the afternoon and eventually cut off Greater Boston from the Quabbin Reservoir, where most of its water supply is stored…Before water was shut off to the ruptured pipe [at 6:40 pm], brown water had been roaring from a massive crater [at a rate of] 8 million gallons an hour rushing into the nearby Charles River."

Let $r(t)$ be the rate in gallons/hr that water flowed from the pipe t hours after it sprang its leak.

23. Which is larger: $\displaystyle \int_0^2 r(t)\, dt$ or $\displaystyle \int_2^4 r(t)\, dt$?

24. Which is larger: $\displaystyle \int_0^4 r(t)\, dt$ or $4r(4)$?

25. Give a reasonable overestimate of $\displaystyle \int_0^8 r(t)\, dt$.

■ In Problems **26–27**, list the expressions (I)–(III) in order from smallest to largest, where $r(t)$ is the hourly rate that an animal burns calories and $R(t)$ is the total number of calories burned since time $t = 0$. Assume $r(t) > 0$ and $r'(t) < 0$ for $0 \le t \le 12$.

26. I. $R(10)$ II. $R(12)$ III. $R(10)+r(10)\cdot 2$

27. I. $\displaystyle \int_5^8 r(t)\, dt$ II. $\displaystyle \int_8^{11} r(t)\, dt$ III. $R(12)-R(9)$

5.6   AVERAGE VALUE

In this section we show how to interpret the definite integral as the average value of a function.

The Definite Integral as an Average

We know how to find the average of n numbers: Add them and divide by n. But how do we find the average value of a continuously varying function? Let us consider an example. Suppose $f(t)$ is the temperature at time t, measured in hours since midnight, and that we want to calculate the average temperature over a 24-hour period. One way to start would be to average the temperatures at n equally spaced times, $t_1, t_2, \ldots, t_n$, during the day.

$$\text{Average temperature} \approx \frac{f(t_1) + f(t_2) + \cdots + f(t_n)}{n}.$$

The larger we make n, the better the approximation. We can rewrite this expression as a Riemann sum over the interval $0 \le t \le 24$ if we use the fact that $\Delta t = 24/n$, so $n = 24/\Delta t$:

$$\text{Average temperature} \approx \frac{f(t_1) + f(t_2) + \cdots + f(t_n)}{24/\Delta t}$$

$$= \frac{f(t_1)\Delta t + f(t_2)\Delta t + \cdots + f(t_n)\Delta t}{24}$$

$$= \frac{1}{24} \sum_{i=1}^{n} f(t_i)\Delta t.$$

As $n \to \infty$, the Riemann sum tends toward an integral, and the approximation gets better. We expect that

$$\text{Average temperature} = \lim_{n \to \infty} \frac{1}{24} \sum_{i=1}^{n} f(t_i)\Delta t$$

$$= \frac{1}{24} \int_0^{24} f(t)\, dt.$$

[16] "A catastrophic rupture hits region's water system," *The Boston Globe*, May 2, 2010.

Generalizing for any function f, if $a < b$, we have

$$\text{Average value of } f \text{ on the interval from } a \text{ to } b = \frac{1}{b-a}\int_a^b f(x)\,dx.$$

The units of the average value of $f(x)$ are the same as the units of $f(x)$.

How to Visualize the Average on a Graph

The definition of average value tells us that

$$(\text{Average value of } f)\cdot(b-a) = \int_a^b f(x)\,dx.$$

Let's interpret the integral as the area under the graph of f. If $f(x)$ is positive, then the average value of f is the height of a rectangle whose base is $(b-a)$ and whose area is the same as the area between the graph of f and the x-axis. (See Figure 5.81.)

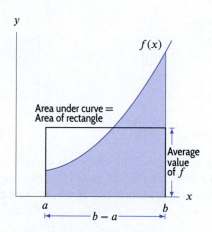

Figure 5.81: Area and average value

Example 1

Suppose that $C(t)$ represents the daily cost of heating your house, in dollars per day, where t is time in days and $t = 0$ corresponds to January 1, 2010. Interpret $\dfrac{1}{90-0}\displaystyle\int_0^{90} C(t)\,dt$.

Solution

The units for the integral $\int_0^{90} C(t)\,dt$ are (dollars/day)$\times$(days) = dollars. The integral represents the total cost in dollars to heat your house for the first 90 days of 2010, namely the months of January, February, and March. The expression $\frac{1}{90-0}\int_0^{90} C(t)\,dt$ represents the average cost per day to heat your house during the first 90 days of 2010. It is measured in (1/days)$\times$(dollars) = dollars/day, the same units as $C(t)$.

Example 2   The population of McAllen, Texas can be modeled by the function

$$P = f(t) = 570(1.037)^t,$$

where P is in thousands of people and t is in years since 2000. Use this function to predict the average population of McAllen between the years 2020 and 2040.

Solution   We want the average value of $f(t)$ between $t = 20$ and $t = 40$. Using a calculator to evaluate the integral, we get

$$\text{Average population} = \frac{1}{40 - 20} \int_{20}^{40} f(t)\, dt = \frac{1}{20}(34{,}656.2) = 1732.81.$$

The average population of McAllen between 2020 and 2040 is predicted to be about 1733 thousand people.

Example 3   (a) For the function $f(x)$ graphed in Figure 5.82, evaluate $\int_0^5 f(x)\, dx$.

(b) Find the average value of $f(x)$ on the interval $x = 0$ to $x = 5$. Check your answer graphically.

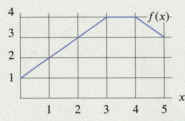

Figure 5.82: Estimate $\int_0^5 f(x)dx$

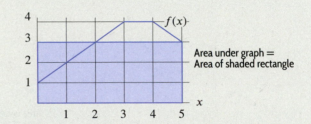

Figure 5.83: Average value of $f(x)$ is 3

Solution   (a) Since $f(x) \geq 0$, the definite integral is the area of the region under the graph of $f(x)$ between $x = 0$ and $x = 5$. Figure 5.82 shows that this region consists of 13 full grid squares and 4 half grid squares, each grid square of area 1, for a total area of 15, so

$$\int_0^5 f(x)\, dx = 15.$$

(b) The average value of $f(x)$ on the interval from 0 to 5 is given by

$$\text{Average value} = \frac{1}{5 - 0} \int_0^5 f(x)\, dx = \frac{1}{5}(15) = 3.$$

To check the answer graphically, draw a horizontal line at $y = 3$ on the graph of $f(x)$. (See Figure 5.83.) Then observe that, between $x = 0$ and $x = 5$, the area under the graph of $f(x)$ is equal to the area of the rectangle with height 3.

Problems for Section 5.6

1. (a) Use Figure 5.84 to find $\int_0^6 f(x)\,dx$.
(b) What is the average value of f on the interval $x = 0$ to $x = 6$?

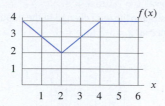

Figure 5.84

2. Use Figure 5.85 to estimate the following:

(a) The integral $\int_0^5 f(x)\,dx$.
(b) The average value of f between $x = 0$ and $x = 5$ by estimating visually the average height.
(c) The average value of f between $x = 0$ and $x = 5$ by using your answer to part (a).

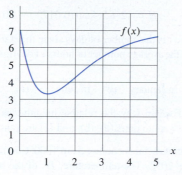

Figure 5.85

■ In Problems **3–4**, find the average value of the function over the given interval.

3. $g(t) = 1 + t$ for $0 \le t \le 2$

4. $g(t) = e^t$ for $0 \le t \le 10$

5. (a) What is the average value of $f(x) = \sqrt{1 - x^2}$ over the interval $0 \le x \le 1$?
(b) How can you tell whether this average value is more or less than 0.5 without doing any calculations?

■ In Problems **6–7**, estimate the average value of the function between $x = 0$ and $x = 7$.

6.

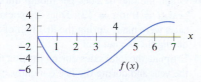

[17] www.eia.gov. Accessed on December 21, 2016.

7.

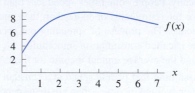

■ In Problems **8–9**, estimate the average value of $f(x)$ from $x = a$ to $x = b$.

8.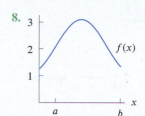

9.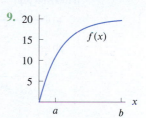

10. The value, V, of a Tiffany lamp, worth \$225 in 1975, increases at 15% per year. Its value in dollars t years after 1975 is given by

$$V = 225(1.15)^t.$$

Find the average value of the lamp over the period 1975–2010.

11. If t is measured in days since June 1, the inventory $I(t)$ for an item in a warehouse is given by

$$I(t) = 5000(0.9)^t.$$

(a) Find the average inventory in the warehouse during the 90 days after June 1.
(b) Graph $I(t)$ and illustrate the average graphically.

12. The population of the world t years after 2010 is predicted to be $P = 6.9e^{0.012t}$ billion.

(a) What population is predicted in 2020?
(b) What is the predicted average population between 2010 and 2020?

13. Crude oil production in the US increased approximately linearly from 1210 barrels per day in 1920 to 9637 barrels per day in 1970.[17]

(a) Find the average daily oil production between 1920 and 1970 using an integral.
(b) Find the average daily oil production by averaging the 1920 and 1970 values. Explain why it is the same as your answer to part (a).

■ In Problems 14–15 annual income for ages 25 to 85 is given graphically. People sometimes spend less than their income (to save for retirement) or more than their income (taking out a loan). The process of spreading out spending over a lifetime is called consumption smoothing.

(a) Find the average annual income for these years.

(b) Assuming that people spend at a constant rate equal to their average income, when are they spending less than they earn, and when are they spending more?

14.

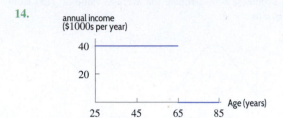

15.

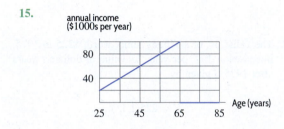

■ Problems 16–17 refer to Figure 5.86, which shows human arterial blood pressure during the course of one heartbeat.

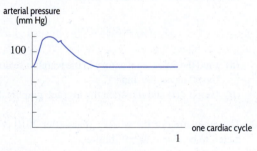

Figure 5.86

16. (a) Estimate the maximum blood pressure, called the systolic pressure.
    (b) Estimate the minimum blood pressure, called the diastolic pressure.
    (c) Calculate the average of the systolic and diastolic pressures.
    (d) Is the average arterial pressure over the entire cycle greater than, less than, or equal to the answer for part (c)?

17. Estimate the average arterial blood pressure over one cardiac cycle.

18. Figure 5.87 shows the rate, $f(x)$, in thousands of algae per hour, at which a population of algae is growing, where x is in hours.

(a) Estimate the average value of the rate over the interval $x = -1$ to $x = 3$.

(b) Estimate the total change in the population over the interval $x = -3$ to $x = 3$.

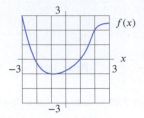

Figure 5.87

19. The number of hours, H, of daylight in Madrid as a function of date is approximated by the formula

$$H = 12 + 2.4 \sin\left(0.0172(t - 80)\right),$$

where t is the number of days since the start of the year. Find the average number of hours of daylight in Madrid:

(a) in January   (b) in June   (c) over a year

(d) Explain why the relative magnitudes of your answers to parts (a), (b), and (c) are reasonable.

20. A bar of metal is cooling from 1000°C to room temperature, 20°C. The temperature, H, of the bar t minutes after it starts cooling is given, in °C, by

$$H = 20 + 980e^{-0.1t}.$$

(a) Find the temperature of the bar at the end of one hour.

(b) Find the average value of the temperature over the first hour.

(c) Is your answer to part (b) greater or smaller than the average of the temperatures at the beginning and the end of the hour? Explain this in terms of the concavity of the graph of H.

21. The rate of sales (in sales per month) of a company is given, for t in months since January 1, by

$$r(t) = t^4 - 20t^3 + 118t^2 - 180t + 200.$$

(a) Graph the rate of sales per month during the first year ($t = 0$ to $t = 12$). Does it appear that more sales were made during the first half of the year, or during the second half?

(b) Estimate the total sales during the first 6 months of the year and during the last 6 months of the year.

(c) What are the total sales for the entire year?

(d) Find the average sales per month during the year.

22. Throughout much of the 20th century, the yearly consumption of electricity in the US increased exponentially at a continuous rate of 7% per year. Assume this trend continues and that the electrical energy consumed in 1900 was 1.4 million megawatt-hours.

(a) Write an expression for yearly electricity consumption as a function of time, t, in years since 1900.

(b) Find the average yearly electrical consumption throughout the 20th century.

(c) During what year was electrical consumption closest to the average for the century?

(d) Without doing the calculation for part (c), how could you have predicted which half of the century the answer would be in?

23. Using Figure 5.88, list the following numbers from least to greatest:

(a) $f'(1)$

(b) The average value of f on $0 \leq x \leq 4$

(c) $\int_0^1 f(x)dx$

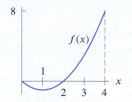

Figure 5.88

24. Using Figure 5.89, list from least to greatest,

(a) $f'(1)$.

(b) The average value of $f(x)$ on $0 \leq x \leq a$.

(c) The average value of the rate of change of $f(x)$, for $0 \leq x \leq a$.

(d) $\int_0^a f(x)\,dx$.

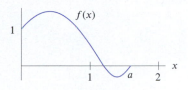

Figure 5.89

PROJECTS FOR CHAPTER FIVE

1. **Carbon Dioxide in Pond Water** Biological activity in a pond is reflected in the rate at which carbon dioxide, CO_2, is added to or withdrawn from the water. Plants take CO_2 out of the water during the day for photosynthesis and put CO_2 into the water at night. Animals put CO_2 into the water all the time as they breathe. Biologists are interested in how the net rate at which CO_2 enters a pond varies during the day. Figure 5.90 shows this rate as a function of time of day.[18] The rate is measured in millimoles (mmol) of CO_2 per liter of water per hour; time is measured in hours past dawn. At dawn, there were 2.600 mmol of CO_2 per liter of water.

(a) What can be concluded from the fact that the rate is negative during the day and positive at night?

(b) Some scientists have suggested that plants respire (breathe) at a constant rate at night, and that they photosynthesize at a constant rate during the day. Does Figure 5.90 support this view?

(c) When was the CO_2 content of the water at its lowest? How low did it go?

(d) How much CO_2 was released into the water during the 12 hours of darkness? Compare this quantity with the amount of CO_2 withdrawn from the water during the 12 hours of daylight. How can you tell by looking at the graph whether the CO_2 in the pond is in equilibrium?

(e) Estimate the CO_2 content of the water at three-hour intervals throughout the day. Use your estimates to plot a graph of CO_2 content throughout the day.

---

[18]Data from R. J. Beyers, *The Pattern of Photosynthesis and Respiration in Laboratory Microsystems* (Mem. 1st Ital. Idrobiol., 1965).

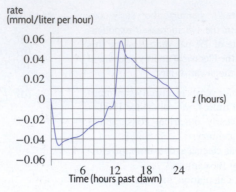

Figure 5.90: Rate at which CO_2 is entering the pond

2. **Flooding in the Grand Canyon**

The Glen Canyon Dam at the top of the Grand Canyon prevents natural flooding. In 1996, scientists decided an artificial flood was necessary to restore the environmental balance. Water was released through the dam at a controlled rate[19] shown in Figure 5.91. The figure also shows the rate of flow of the last natural flood in 1957.

(a) At what rate was water passing through the dam in 1996 before the artificial flood?

(b) At what rate was water passing down the river in the pre-flood season in 1957?

(c) Estimate the maximum rates of discharge for the 1996 and 1957 floods.

(d) Approximately how long did the 1996 flood last? How long did the 1957 flood last?

(e) Estimate how much additional water passed down the river in 1996 as a result of the artificial flood.

(f) Estimate how much additional water passed down the river in 1957 as a result of the flood.

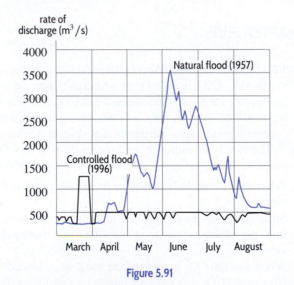

Figure 5.91

3. **Medical Case Study: Flux of Fluid from a Capillary**[20]

The heart pumps blood throughout the body, the arteries are the blood vessels carrying blood away from the heart, and the veins return blood to the heart. Close to the heart, arteries

---

[19] Adapted from M. Collier, R. Webb, E. Andrews, "Experimental Flooding in Grand Canyon," *Scientific American* (January 1997).

[20] From David E. Sloane, M.D.

are very large. As they progress toward tissues (such as the brain), the arteries branch repeatedly, getting smaller as they do. The smallest blood vessels are capillaries—microscopic, living tubes that link the smallest arteries to the smallest veins. The capillary is where nutrients and fluids move out of the blood into the adjacent tissues and waste products from the tissues move into the blood. The key to this process is the capillary wall, which is only one cell thick, allowing the unfettered passage of small molecules, such as water, ions, oxygen, glucose, and amino acids, while preventing the passage of large components of the blood (such as large proteins and blood cells).

Precise measurements demonstrate that the flux (rate of flow) of fluid through the capillary wall is not constant over the length of the capillary. Fluids in and around the capillary are subjected to two forces. The *hydrostatic pressure*, resulting from the heart's pumping, pushes fluid out of the capillary into the surrounding tissue. The *oncotic pressure* drives absorption in the other direction. At the start of a capillary, where the capillary branches off the small artery, the hydrostatic pressure is high while the oncotic pressure is low. Along the length of the capillary, the hydrostatic pressure decreases while the oncotic pressure is approximately constant. See Figure 5.92.

For most capillaries there is a net positive value for flow: more fluid flows from the capillaries into the surrounding tissue than the other way around. This presents a major problem for maintaining fluid balance in the body. How is the fluid left in the tissues to get back into circulation? If it cannot, the tissues progressively swell (a condition called âĂIJedemaâĂÍ). Evolution's solution is to provide humans and other mammals with a second set of vessels, the lymphatics, that absorb extra tissue fluid and provide one-way routes back to the bloodstream.

Along a cylindrical capillary of length $L = 0.1$ cm and radius $r = 0.0004$ cm, the hydrostatic pressure, p_h, varies from 35 mm Hg at the artery end to 15 mm Hg at the vein end. (mm Hg, millimeters of mercury, is a unit of pressure.) The oncotic pressure, p_o, is approximately 23 mm Hg throughout the length of the capillary.

(a) Find a formula for p_h as a function of x, the distance in centimeters from the artery end of the capillary, assuming that p_h is a linear function of x.

(b) Find a formula for p, the net outward pressure, as a function of x.

(c) Write and evaluate an integral for the average outward net pressure in the capillary.

(d) The rate of movement, j, of fluid volume per capillary wall area across the capillary wall is proportional to the net pressure. We have $j = k \cdot p$ where k, the hydraulic conductivity, has value

$$k = 10^{-7} \frac{\text{cm}}{\text{sec} \cdot \text{mm Hg}}.$$

Check that j has units of volume per time per area.

(e) Estimate the net volume flow rate (volume per unit time) through the wall of the entire capillary using the average pressure.

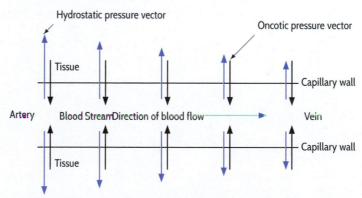

Figure 5.92: Vectors representing pressure in and out of capillary

4. **Medical Case Study: Testing for Kidney Disease**[21]

Patients with kidney disease often have protein in their urine. While small amounts of protein are not very worrisome, more than 1 gram of protein excreted in 24 hours warrants active treatment. The most accurate method for measuring urine protein is to have the patient collect all his or her urine in a container for a full 24-hour period. The total mass of protein can then be found by measuring the volume and protein concentration of the urine.

However, this process is not as straightforward as it sounds. Since the urine is collected intermittently throughout the 24-hour period, the first urine voided sits in the container longer than the last urine voided. During this time, the proteins slowly fall to the bottom of the container. Thus, at the end of a 24-hour collection period, there is a higher concentration of protein on the bottom of the container than at the top.

One could try to mix the urine so that the protein concentration is more uniform, but this forms bubbles that trap the protein, leading to an underestimate of the total amount excreted. A better way to determine the total protein is to measure the concentration at the top and at the bottom, and then calculate the average protein concentration.

Suppose a patient voids 2 litres (2000 ml) of urine in 24 hours and collects it in a cylindrical container of diameter 10 cm (note that 1 cm^3 = 1 ml). A technician determines that the protein concentration at the top is 0.14 mg/ml, and at the bottom is 0.96 mg/ml. Assume that the concentration of protein varies linearly from the top to the bottom.

(a) Find a formula for c, the protein concentration in mg/ml, as a function of y, the distance in centimeters from the base of the cylinder.

(b) Write an integral that gives the average protein concentration in the urine sample.

(c) Estimate the quantity of protein in the sample by multiplying the average protein concentration by the volume of the urine collected. Does this patient require active treatment?

---

[21] From David E. Sloane, M.D.

FOCUS ON THEORY

THEOREMS ABOUT DEFINITE INTEGRALS

The Second Fundamental Theorem of Calculus

The Fundamental Theorem of Calculus tells us that if we have a function F whose derivative is a continuous function f, then the definite integral of f is given by

$$\int_a^b f(t)\,dt = F(b) - F(a).$$

We now take a different point of view. If a is fixed and the upper limit is x, then the value of the integral is a function of x. We define a new function G on the interval by

$$G(x) = \int_a^x f(t)\,dt.$$

To visualize G, suppose that f is positive and $x > a$. Then $G(x)$ is the area under the graph of f in Figure 5.93. If f is continuous on an interval containing a, then it can be shown that G is defined for all x on that interval.

We now consider the derivative of G. Using the definition of the derivative, we have

$$G'(x) = \lim_{h \to 0} \frac{G(x+h) - G(x)}{h}.$$

Suppose f and h are positive. Then we can visualize

$$G(x) = \int_a^x f(t)\,dt$$

and

$$G(x+h) = \int_a^{x+h} f(t)\,dt$$

as areas, which leads to representing

$$G(x+h) - G(x) = \int_x^{x+h} f(t)\,dt$$

as a difference of two areas.

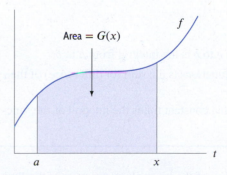

Figure 5.93: Representing $G(x)$ as an area

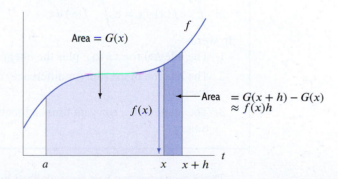

Figure 5.94: $G(x+h) - G(x)$ is the area of a roughly rectangular region

From Figure 5.94, we see that, if h is small, $G(x + h) - G(x)$ is roughly the area of a rectangle of height $f(x)$ and width h (shaded darker in Figure 5.94), so we have

$$G(x + h) - G(x) \approx f(x)h;$$

hence

$$\frac{G(x + h) - G(x)}{h} \approx f(x).$$

The same result holds when h is negative, suggesting that

$$G'(x) = \lim_{h \to 0} \frac{G(x + h) - G(x)}{h} = f(x).$$

This result is another form of the Fundamental Theorem of Calculus. It is usually stated as follows:

Second Fundamental Theorem of Calculus

If f is a continuous function on an interval, and if a is any number in that interval, then the function G defined on the interval by

$$G(x) = \int_a^x f(t)\, dt$$

has derivative f; that is, $G'(x) = f(x)$.

Properties of the Definite Integral

In this chapter, we have used the following properties to break up definite integrals.

Sums and Multiples of Definite Integrals

If a, b, and c are any numbers and f and g are continuous functions, then

1. $\displaystyle \int_a^c f(x)\, dx + \int_c^b f(x)\, dx = \int_a^b f(x)\, dx.$

2. $\displaystyle \int_a^b (f(x) \pm g(x))\, dx = \int_a^b f(x)\, dx \pm \int_a^b g(x)\, dx.$

3. $\displaystyle \int_a^b cf(x)\, dx = c \int_a^b f(x)\, dx.$

In words:

1. The integral from a to c plus the integral from c to b is the integral from a to b.

2. The integral of the sum (or difference) of two functions is the sum (or difference) of their integrals.

3. The integral of a constant times a function is that constant times the integral of the function.

These properties can best be visualized by thinking of the integrals as areas or as the limit of the sum of areas of rectangles.

Problems on the Second Fundamental Theorem of Calculus

■ For Problems 1–4, find $G'(x)$.

1. $G(x) = \int_a^x t^3 \, dt$

2. $G(x) = \int_a^x 3^t \, dt$

3. $G(x) = \int_a^x te^t \, dt$

4. $G(x) = \int_a^x \ln y \, dy$

5. Let $F(b) = \int_0^b 2^x \, dx$.

  (a) What is $F(0)$?

  (b) Does the value of F increase or decrease as b increases? (Assume $b \geq 0$.)

  (c) Estimate $F(1)$, $F(2)$, and $F(3)$.

6. For $x = 0, 0.5, 1.0, 1.5,$ and 2.0, make a table of values for $I(x) = \int_0^x \sqrt{t^4 + 1} \, dt$.

7. Assume that $F'(t) = \sin t \cos t$ and $F(0) = 1$. Find $F(b)$ for $b = 0, \ 0.5, \ 1, \ 1.5, \ 2, \ 2.5,$ and 3.

■ In Problems 8–11, find the integral, given that $\int_a^b f(x) \, dx = 8$, $\int_a^b (f(x))^2 \, dx = 12$, $\int_a^b g(t) \, dt = 2$, and $\int_a^b (g(t))^2 \, dt = 3$.

8. $\int_a^b (f(x) + g(x)) \, dx$

9. $\int_a^b \left((f(x))^2 - (g(x))^2\right) dx$

10. $\int_a^b (f(x))^2 \, dx - \left(\int_a^b f(x) \, dx\right)^2$

11. $\int_a^b c f(z) \, dz$

Chapter 6

ANTIDERIVATIVES AND APPLICATIONS

CONTENTS

6.1   ANALYZING ANTIDERIVATIVES GRAPHICALLY AND NUMERICALLY

What Is an Antiderivative?

If the derivative of $F(x)$ is $f(x)$, that is, if $F'(x) = f(x)$, then we call $F(x)$ an *antiderivative* of $f(x)$. For example, the derivative of x^2 is $2x$, so we say that

$$x^2 \text{ is an antiderivative of } 2x.$$

In this section, we see how values of an antiderivative, F, are computed using the Fundamental Theorem of Calculus when the derivative, F', and one value of the function, $F(a)$, are known.

Example 1   Suppose $F'(t) = (1.8)^t$ and $F(0) = 2$. Find the value of $F(b)$ for $b = 0, 0.1, 0.2, \ldots, 1.0$.

Solution   We apply the Fundamental Theorem with $F'(t) = (1.8)^t$ and $a = 0$ to get values for $F(b)$:

$$F(b) - F(0) = \int_0^b F'(t)\, dt = \int_0^b (1.8)^t\, dt.$$

Since $F(0) = 2$, we have

$$F(b) = 2 + \int_0^b (1.8)^t\, dt.$$

We use a calculator or computer to estimate the definite integral $\int_0^b (1.8)^t\, dt$ for each value of b. For example, when $b = 0.1$, we find that $\int_0^b (1.8)^t\, dt = 0.103$. Thus, $F(0.1) = 2.103$. Similar calculations give the values in Table 6.1.

Table 6.1   *Approximate values of F*

b	0	0.1	0.2	0.3	0.4	0.5	0.6	0.7	0.8	0.9	1.0
$F(b)$	2	2.103	2.212	2.328	2.451	2.581	2.719	2.866	3.021	3.186	3.361

Notice from the table that values of $F(b)$ increase between $b = 0$ and $b = 1$. This is because the derivative $F'(t) = (1.8)^t$ is positive for t between 0 and 1.

Graphing an Antiderivative

To graph an antiderivative of F from its derivative $F' = f$ we can calculate values of F, as in Example 1, or we can use the graph of f and the following results from Section 2.2, page 99.

> Suppose F is an antiderivative of f:
> If $f > 0$ on an interval, then F is *increasing* on that interval.
> If $f < 0$ on an interval, then F is *decreasing* on that interval.
> If $f = 0$ on an interval, then F is *constant* on that interval.

Example 2   Figure 6.1 shows the derivative $f'(x)$ of a function $f(x)$ and the values of some areas. If $f(0) = 10$, sketch a graph of the function $f(x)$. Give the coordinates of the local maxima and minima.

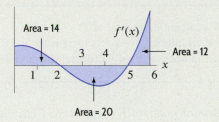

Figure 6.1: The graph of a derivative f'

Solution   Figure 6.1 shows that the derivative f' is positive between 0 and 2, negative between 2 and 5, and positive between 5 and 6. Therefore, the function f is increasing between 0 and 2, decreasing between 2 and 5, and increasing between 5 and 6. There is a local maximum at $x = 2$ and a local minimum at $x = 5$.

Notice that we can sketch the general shape of the graph of f in Figure 6.2 without knowing any of the areas in Figure 6.1. The areas are used to make the graph more precise. We are told that $f(0) = 10$, so we plot the point $(0, 10)$ on the graph of f in Figure 6.3. The Fundamental Theorem and Figure 6.1 show that

$$f(2) - f(0) = \int_0^2 f'(x)\,dx = 14.$$

Therefore, the total change in f between $x = 0$ and $x = 2$ is 14. Since $f(0) = 10$, we have

$$f(2) = 10 + 14 = 24.$$

The point $(2, 24)$ is on the graph of f. See Figure 6.3.

Figure 6.1 shows that the area between $x = 2$ and $x = 5$ is 20. Since this area lies entirely below the x-axis, the Fundamental Theorem gives

$$f(5) - f(2) = \int_2^5 f'(x)\,dx = -20.$$

The total change in f is -20 between $x = 2$ and $x = 5$. Since $f(2) = 24$, we have

$$f(5) = 24 - 20 = 4.$$

Thus, the point $(5, 4)$ lies on the graph of f. Finally,

$$f(6) = f(5) + \int_5^6 f'(x)\,dx = 4 + 12 = 16,$$

so the point $(6, 16)$ is on the graph of f.

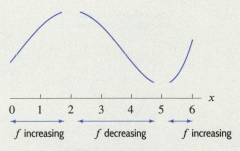

Figure 6.2: The shape of f

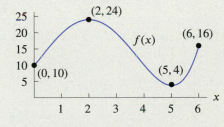

Figure 6.3: The graph of f

Concavity

If F is an antiderivative of f, then $F'' = f'$. Thus, we can determine the concavity of the graph of F from the properties of f' (see Section 2.4, on page 114).

> Suppose F is an antiderivative of f:
> If $f' > 0$ on an interval, then f is increasing and the graph of F is concave up on that interval.
> If $f' < 0$ on an interval, then f is decreasing and the graph of F is concave down on that interval.

Thus, inflection points of F occur where f changes from increasing to decreasing or vice versa.

Problems for Section 6.1

1. Suppose $F'(x) = 2x^2 + 5$ and $F(0) = 3$. Find the value of $F(b)$ for $b = 0, 0.1, 0.2, 0.5$, and 1.0.

2. Suppose $G'(t) = (1.12)^t$ and $G(5) = 1$. Find the value of $G(b)$ for $b = 5, 5.1, 5.2, 5.5$, and 6.0.

3. Suppose $f'(t) = (0.82)^t$ and $f(2) = 9$. Find the value of $f(b)$ for $b = 2, 4, 6, 10$, and 20.

4. If $F(0) = 5$ and $F(x)$ is an antiderivative of $f(x) = 3e^{-x^2}$, use a calculator to find $F(2)$.

5. If $G(1) = 50$ and $G(x)$ is an antiderivative of $g(x) = \ln x$, use a calculator to find $G(4)$.

6. Let $F(x)$ be an antiderivative of $f(x)$, with $F(0) = 50$ and $\int_0^5 f(x)\,dx = 12$. What is $F(5)$?

7. Let $F(x)$ be an antiderivative of $f(x)$, with $F(1) = 20$ and $\int_1^4 f(x)\,dx = -7$. What is $F(4)$?

8. (a) Using Figure 6.4, estimate $\int_0^7 f(x)\,dx$.
(b) If F is an antiderivative of the same function f and $F(0) = 25$, estimate $F(7)$.

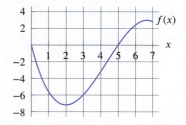

Figure 6.4

9. Figure 6.5 shows f. If $F' = f$ and $F(0) = 0$, find $F(b)$ for $b = 1, 2, 3, 4, 5, 6$.

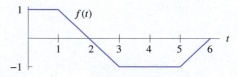

Figure 6.5

10. Let $G'(t) = g(t)$ and $G(0) = 4$. Use Figure 6.6 to find the values of $G(t)$ at $t = 5, 10, 20, 25$.

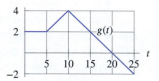

Figure 6.6

11. Figure 6.7 shows the derivative g'. If $g(0) = 0$, graph g. Give (x, y)-coordinates of all local maxima and minima.

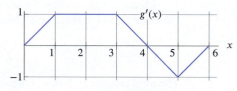

Figure 6.7

12. Figure 6.8 shows the rate of change of the concentration of adrenaline, in micrograms per milliliter per minute, in a person's body. Sketch a graph of the concentration of adrenaline, in micrograms per milliliter, in the body as a function of time, in minutes.

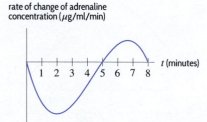

rate of change of adrenaline concentration (μg/ml/min)

Figure 6.8

■ Problems **13–14** show the derivative f' of f.
  (a) Where is f increasing and where is f decreasing? What are the x-coordinates of the local maxima and minima of f?
  (b) Sketch a possible graph for f. (You don't need a scale on the vertical axis.)

13. 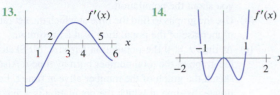   **14.**

15. The derivative $F'(t)$ is graphed in Figure 6.9. Given that $F(0) = 5$, calculate $F(t)$ for $t = 1, 2, 3, 4, 5$.

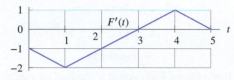

Figure 6.9

■ In Problems **16–21**, sketch two functions F such that $F' = f$. In one case let $F(0) = 0$ and in the other, let $F(0) = 1$.

16. 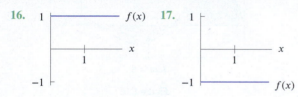   **17.**

18. 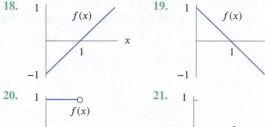   **19.**

20. 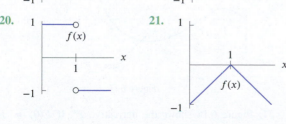   **21.**

22. (a) Estimate $\int_0^4 f(x)\,dx$ for $f(x)$ in Figure 6.10.
  (b) Let $F(x)$ be an antiderivative of $f(x)$. Is $F(x)$ increasing or decreasing on the interval $0 \le x \le 4$?
  (c) If $F(0) = 20$, what is $F(4)$?

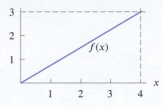

Figure 6.10

23. Figure 6.11 shows the derivative F' of a function F. If $F(20) = 150$, estimate the maximum value attained by F.

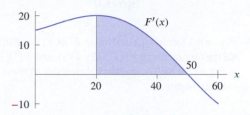

Figure 6.11

24. Table 6.2 shows the monthly change in water stored in Lake Sonoma, California, from March through November 2014. The change is measured in acre-feet per month.[1] On March 1, the water stored was 182,566 acre-feet. Let $S(t)$ be the total water, in acre-feet, stored in month t, where $t = 0$ is March.

  (a) Find and interpret $S(0)$ and $S(3)$.
  (b) Approximately when do maximum and minimum values of $S(t)$ occur?
  (c) Does $S(t)$ appear to have inflection points? If so, approximately when?

Table 6.2 *Change in water in acre-feet per month*

Month	Mar	Apr	May	June	July
Change in water	3003	−5631	−8168	−8620	−8270

Month	Aug	Sept	Oct	Nov	
Change in water	−7489	−6245	−4593	54,743	

25. Urologists are physicians who specialize in the health of the bladder. In a common diagnostic test, urologists monitor the emptying of the bladder using a device that produces two graphs. In one of the graphs the flow rate (in milliliters per second) is measured as a function of time (in seconds). In the other graph, the volume emptied from the bladder is measured (in milliliters) as a function of time (in seconds). See Figure 6.12.

  (a) Which graph is the flow rate and which is the volume?
  (b) Which one of these graphs is an antiderivative of the other?

---

[1] Date from http://cdec.water.ca.gov/cgi-progs/stationInfo?station_id=WRS, accessed June, 2015. An acre-foot is the amount of water it takes to cover one acre of area with 1 foot of water.

(I)

(II)

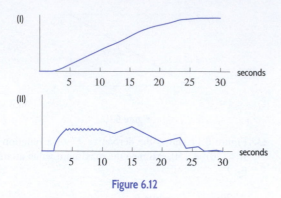

Figure 6.12

26. Figure 6.13 shows the derivative F' of F. Let $F(0) = 0$. Of the four numbers $F(1)$, $F(2)$, $F(3)$, and $F(4)$, which is largest? Which is smallest? How many of these numbers are negative?

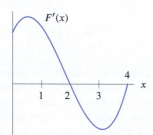

Figure 6.13

27. During photosynthesis, plants absorb sunlight and release oxygen. The rate at which a leaf releases oxygen, the rate of photosynthesis, is a function of the age of the leaf. Figure 6.14 shows the rate of photosynthesis for a soybean leaf.[2]

   (a) Sketch a graph of the amount of oxygen, $E(t)$, generated by this leaf during the first 20 days of life.
   (b) Give the t-coordinate of the inflection point in your sketch and interpret what this point tells you about soybean leaves and oxygen.
   (c) Give a symbolic expression that represents the total amount of oxygen released by this leaf in its first 10 days of life.
   (d) Does the soybean leaf release more oxygen during the first 10 days of its life or during the second 10 days of its life?

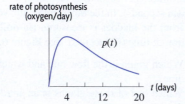

Figure 6.14

28. The birth rate, B, in births per hour, of a bacteria population is given in Figure 6.15. The curve marked D gives the death rate, in deaths per hour, of the same population.

   (a) Explain what the shape of each of these graphs tells you about the population.
   (b) Use the graphs to find the time at which the net rate of increase of the population is at a maximum.
   (c) At time $t = 0$ the population has size N. Sketch the graph of the total number born by time t. Also sketch the graph of the number alive at time t. Estimate the time at which the population is a maximum.

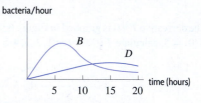

Figure 6.15

29. Using Figure 6.16, sketch a graph of an antiderivative $G(t)$ of $g(t)$ satisfying $G(0) = 5$. Label each critical point of $G(t)$ with its coordinates.

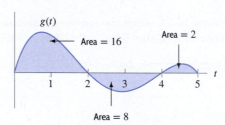

Figure 6.16

30. Use Figure 6.17 and the fact that $F(2) = 3$ to sketch the graph of $F(x)$. Label the values of at least four points.

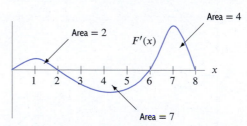

Figure 6.17

31. Figure 6.18 shows the derivative F'. If $F(0) = 14$, graph F. Give (x, y)-coordinates of all local maxima and minima.

[2]Based on: P. Reich et al., Response of Soybean to Low Concentrations of Ozone: I. Reductions in Leaf and Whole Plant Net Photosynthesis and Leaf Chlorophyll Content, *Journal of Environmental Quality,* 1986, 15(1).

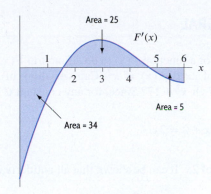

Area = 25
F′(x)
1
5
6
x
2
3
4
Area = 5
Area = 34

Figure 6.18

32. Figure 6.19 shows the derivative $F'(t)$. If $F(0) = 3$, find the values of $F(2)$, $F(5)$, $F(6)$. Graph $F(t)$.

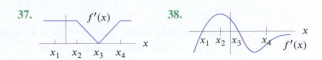

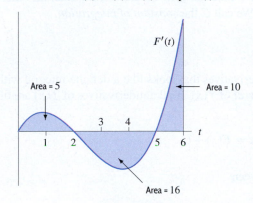

F′(t)
Area = 5
Area = 10
3
4
t
1
2
5
6
Area = 16

Figure 6.19

33. **(a)** Estimate $\int_0^4 f(x)\,dx$ for $f(x)$ in Figure 6.20.
(b) Let $F(x)$ be an antiderivative of $f(x)$. If $F(0) = 100$, what is $F(4)$?

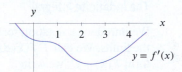

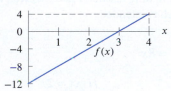

4
0
−4
−8
−12
1
2
3
4
x
f(x)

Figure 6.20

■ In Problems 34–35, a graph of f is given. Let $F'(x) = f(x)$.
(a) What are the x-coordinates of the critical points of $F(x)$?
(b) Which critical points are local maxima, which are local minima, and which are neither?
(c) Sketch a possible graph of $F(x)$.

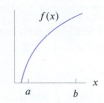

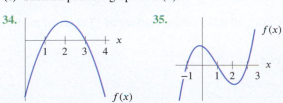

34.

1
2
3
4
x
f(x)

35.

f(x)
x
−1
1
2
3

36. Let $F(x)$ be an antiderivative of $f(x) = 1 - x^2$.
(a) On what intervals is $F(x)$ increasing?
(b) On what intervals is the graph of $F(x)$ concave up?

■ Problems 37–38 give a graph of $f'(x)$. Graph $f(x)$. Mark the points $x_1, \ldots, x_4$ on your graph and label local maxima, local minima and points of inflection.

37.

f′(x)
x
x1
x2
x3
x4

38.

x
x1
x2
x3
x4
f′(x)

■ Problems 39–40 concern the graph of f' in Figure 6.21.

y
x
1
2
3
4
y = f′(x)

Figure 6.21: Note: Graph of f', not f

39. Which is greater, $f(0)$ or $f(1)$?

40. List the following in increasing order:
$$\frac{f(4) - f(2)}{2}, \quad f(3) - f(2), \quad f(4) - f(3).$$

■ For Problems 41–44, show the following quantities on Figure 6.22.

f(x)
x
a
b

Figure 6.22

41. A length representing $f(b) - f(a)$.

42. A slope representing $\dfrac{f(b) - f(a)}{b - a}$.

43. An area representing $F(b) - F(a)$, where $F' = f$.

44. A length roughly approximating
$$\frac{F(b) - F(a)}{b - a}, \text{ where } F' = f.$$

6.2 ANTIDERIVATIVES AND THE INDEFINITE INTEGRAL

In the previous section we defined an antiderivative of a function $f(x)$ to be a function $F(x)$ whose derivative is $f(x)$. For example, an antiderivative of $2x$ is $F(x) = x^2$. Can you think of another function whose derivative is $2x$? How about $x^2 + 1$? Or $x^2 + 17$? Since, for any constant C,

$$\frac{d}{dx}(x^2 + C) = 2x + 0 = 2x,$$

any function of the form $x^2 + C$ is an antiderivative of $2x$. It can be shown that all antiderivatives of $2x$ are of this form, so we say that

$$x^2 + C \text{ is the family of antiderivatives of } 2x.$$

Once we know one antiderivative $F(x)$ for a function $f(x)$ on an interval, then all other antiderivatives of $f(x)$ are of the form $F(x) + C$. We call C the *constant of integration*.

The Indefinite Integral

We introduce a notation for the family of antiderivatives that looks like a definite integral without the limits. We call $\int f(x)\,dx$ the *indefinite integral* of $f(x)$. If all antiderivatives of $f(x)$ are of the form $F(x) + C$, we write

$$\int f(x)\,dx = F(x) + C.$$

It is important to understand the difference between

$$\int_a^b f(x)\,dx \qquad \text{and} \qquad \int f(x)\,dx.$$

The first is a number and the second is a family of functions. Because the notation is similar, the word "integration" is frequently used for the process of finding antiderivatives as well as of finding definite integrals. The context usually makes clear which is intended.

Example 1

(a) Decide which functions are antiderivatives of $f(x) = 8x^3$:

$$F(x) = 24x^2, \qquad G(x) = 8x^4, \qquad H(x) = 2x^4.$$

(b) Find $\int 8x^3\,dx$.

Solution

(a) An antiderivative of $8x^3$ is any function whose derivative is $8x^3$. We have:

$$F'(x) = 48x, \qquad G'(x) = 32x^3, \qquad H'(x) = 8x^3.$$

Since $H'(x) = 8x^3$, the function $H(x)$ is an antiderivative of $8x^3$. The two functions $F(x)$ and $G(x)$ are not antiderivatives of $8x^3$.

(b) Since $H(x) = 2x^4$ is one antiderivative of $f(x)$, all other antiderivatives of $f(x)$ are of the form $H(x) + C$. Thus,

$$\int 8x^3\,dx = 2x^4 + C.$$

Finding Formulas for Antiderivatives

Finding antiderivatives of functions is like taking square roots of numbers: if we pick a number at random, such as 7 or 493, we may have trouble figuring out its square root without a calculator. But if we happen to pick a number such as 25 or 64, which we know is a perfect square, then we can easily guess its square root. Similarly, if we pick a function which we recognize as a derivative, then we can find its antiderivative easily.

For example, noticing that $2x$ is the derivative of x^2 tells us that x^2 is an antiderivative of $2x$. If we divide by 2, then we guess that

$$\text{An antiderivative of } x \text{ is } \frac{x^2}{2}.$$

To check this statement, take the derivative of $x^2/2$:

$$\frac{d}{dx}\left(\frac{x^2}{2}\right) = \frac{1}{2} \cdot \frac{d}{dx}x^2 = \frac{1}{2} \cdot 2x = x.$$

What about an antiderivative of x^2? The derivative of x^3 is $3x^2$, so the derivative of $x^3/3$ is $3x^2/3 = x^2$. Thus,

$$\text{An antiderivative of } x^2 \text{ is } \frac{x^3}{3}.$$

Can you see the pattern? It looks like

$$\text{An antiderivative of } x^n \text{ is } \frac{x^{n+1}}{n+1}.$$

(We assume $n \neq -1$, or we would have $x^0/0$, which does not make sense.) It is easy to check this formula by differentiation:

$$\frac{d}{dx}\left(\frac{x^{n+1}}{n+1}\right) = \frac{(n+1)x^n}{n+1} = x^n.$$

Thus, in indefinite integral notation, we see that

$$\int x^n \, dx = \frac{x^{n+1}}{n+1} + C, \quad n \neq -1.$$

Can you think of an antiderivative of the function $f(x) = 5$? We know that the derivative of $5x$ is 5, so $F(x) = 5x$ is an antiderivative of $f(x) = 5$. In general, if k is a constant, the derivative of kx is k, so we have the result:

If k is constant,

$$\int k \, dx = kx + C.$$

Example 2   Find $\int (3x + x^2) \, dx$.

Solution   We know that $x^2/2$ is an antiderivative of x and that $x^3/3$ is an antiderivative of x^2, so we expect

$$\int (3x + x^2) \, dx = 3\left(\frac{x^2}{2}\right) + \frac{x^3}{3} + C.$$

Again, we can check by differentiation:

$$\frac{d}{dx}\left(\frac{3}{2}x^2 + \frac{x^3}{3} + C\right) = \frac{3}{2} \cdot 2x + \frac{3x^2}{3} = 3x + x^2.$$

The preceding example illustrates that the sum and constant multiplication rules of differentiation work in reverse:

Properties of Antiderivatives: Sums and Constant Multiples

In indefinite integral notation,

1. $\displaystyle\int (f(x) \pm g(x))\, dx = \int f(x)\, dx \pm \int g(x)\, dx$

2. $\displaystyle\int c f(x)\, dx = c \int f(x)\, dx.$

In words,

1. An antiderivative of the sum (or difference) of two functions is the sum (or difference) of their antiderivatives.

2. An antiderivative of a constant times a function is the constant times an antiderivative of the function.

Example 3   Find all antiderivatives of each of the following: (a) x^5 (b) t^8 (c) $12x^3$ (d) $q^3 - 6q^2$

Solution

(a) $\displaystyle\int x^5\, dx = \frac{x^6}{6} + C.$

(b) $\displaystyle\int t^8\, dt = \frac{t^9}{9} + C.$

(c) $\displaystyle\int 12x^3\, dx = 12\left(\frac{x^4}{4}\right) + C = 3x^4 + C.$

(d) $\displaystyle\int (q^3 - 6q^2)\, dq = \frac{q^4}{4} - 6\left(\frac{q^3}{3}\right) + C = \frac{q^4}{4} - 2q^3 + C.$

To check, differentiate the antiderivative; you should get the original function.

What is an antiderivative of x^n when $n = -1$? In other words, what is an antiderivative of $1/x$? Fortunately, we know a function whose derivative is $1/x$, namely, the natural logarithm. Thus, since

$$\frac{d}{dx}(\ln x) = \frac{1}{x},$$

we know that

$$\int \frac{1}{x}\, dx = \ln x + C, \quad \text{for } x > 0.$$

If $x < 0$, then $\ln x$ is not defined, so it can't be an antiderivative of $1/x$. In this case, we can try $\ln(-x)$:

$$\frac{d}{dx}\ln(-x) = (-1)\frac{1}{-x} = \frac{1}{x}$$

so

$$\int \frac{1}{x}\, dx = \ln(-x) + C, \quad \text{for } x < 0.$$

This means $\ln x$ is an antiderivative of $1/x$ if $x > 0$, and $\ln(-x)$ is an antiderivative of $1/x$ if $x < 0$. Since $|x| = x$ when $x > 0$ and $|x| = -x$ when $x < 0$, we can collapse these two formulas into one. On any interval that does not contain 0,

$$\int \frac{1}{x}\, dx = \ln|x| + C.$$

Since the exponential function e^x is its own derivative, it is also one of its own antiderivatives; thus

$$\int e^x \, dx = e^x + C.$$

What about e^{kx}? We know that the derivative of e^{kx} is ke^{kx}, so, for $k \neq 0$, we have

$$\int e^{kx} \, dx = \frac{1}{k} e^{kx} + C.$$

Example 4 Find all antiderivatives of each of the following: (a) $8x^3 + \dfrac{1}{x}$ (b) $12e^{0.2t}$

Solution (a) $\displaystyle\int \left(8x^3 + \frac{1}{x} \right) dx = 8 \left(\frac{x^4}{4} \right) + \ln |x| + C = 2x^4 + \ln |x| + C.$

(b) $\displaystyle\int 12e^{0.2t} \, dt = 12 \left(\frac{1}{0.2} e^{0.2t} \right) + C = 60e^{0.2t} + C.$

Differentiate your answers to check them.

Antiderivatives of Periodic Functions

Antiderivatives of the sine and cosine are easy to guess. Since

$$\frac{d}{dx} \sin x = \cos x \qquad \text{and} \qquad \frac{d}{dx} \cos x = -\sin x,$$

we get

$$\int \cos x \, dx = \sin x + C \quad \text{and} \quad \int \sin x \, dx = -\cos x + C.$$

Since the derivative of $\sin(kx)$ is $k \cos(kx)$ and the derivative of $\cos(kx)$ is $-k \sin(kx)$, we have, for $k \neq 0$,

$$\int \cos(kx) \, dx = \frac{1}{k} \sin(kx) + C \quad \text{and} \quad \int \sin(kx) \, dx = -\frac{1}{k} \cos(kx) + C.$$

Example 5 Find $\displaystyle\int (\sin x + 3 \cos(5x)) \, dx.$

Solution We break the antiderivative into two terms:

$$\int (\sin x + 3 \cos(5x)) \, dx = \int \sin x \, dx + 3 \int \cos(5x) \, dx = -\cos x + \frac{3}{5} \sin(5x) + C.$$

Check by differentiating:

$$\frac{d}{dx} \left(-\cos x + \frac{3}{5} \sin(5x) + C \right) = \sin x + 3 \cos(5x).$$

Problems for Section 6.2

■ In Problems 1–6, decide if the function is an antiderivative of $f(x) = 2e^{2x}$.

1. $F(x) = e^{2x} + 5$

2. $F(x) = e^{2x}$

3. $F(x) = 5e^{2x}$

4. $F(x) = xe^{2x}$

5. $F(x) = 2e^{2x}$

6. $F(x) = e^{2x} + \int_0^1 e^{2t}\, dt$

7. Which of (I)–(V) are antiderivatives of $f(x) = e^{x/2}$?

  I. $e^{x/2}$ II. $2e^{x/2}$ III. $2e^{(1+x)/2}$

  IV. $2e^{x/2} + 1$ V. $e^{x^2/4}$

8. Which of (I)–(V) are antiderivatives of $f(x) = 1/x$?

  I. $\ln x$ II. $-1/x^2$ III. $\ln x + \ln 3$

  IV. $\ln(2x)$ V. $\ln(x+1)$

9. Which of (I)–(V) are antiderivatives of

$$f(x) = 2 \sin x \cos x?$$

  I. $-2\sin x \cos x$ II. $2\cos^2 x - 2\sin^2 x$

  III. $\sin^2 x$ IV. $-\cos^2 x$

  V. $2\sin^2 x + \cos^2 x$

■ In Problems 10–16, decide whether the expression is a number or a family of functions. (Assume $f(x)$ is a function.)

10. $5 + \displaystyle\int 2f(x)\, dx$

11. $\displaystyle\int_3^5 2f(x)\, dx$

12. $\displaystyle\int 5\, dx$

13. $\displaystyle\int f(5)\, dx$

14. $\displaystyle\int \frac{1}{x+1}\, dx$

15. $\displaystyle\int_1^3 \left(\frac{x}{3} + \frac{3}{x}\right) dx$

16. $\displaystyle\int_2^3 f(x)\, dx + \int f(x)\, dx$

■ In Problems 17–42, find an antiderivative.

17. $f(t) = 5t$

18. $f(x) = x^2$

19. $g(t) = t^2 + t$

20. $f(x) = 5$

21. $f(x) = x^4$

22. $g(t) = t^7 + t^3$

23. $f(q) = 5q^2$

24. $g(x) = 6x^3 + 4$

25. $h(y) = 3y^2 - y^3$

26. $k(x) = 10 + 8x^3$

27. $p(r) = 2\pi r$

28. $f(x) = x + x^5 + x^{-5}$

29. $g(z) = \sqrt{z}$

30. $p(x) = x^2 - 6x + 17$

31. $f(x) = 5x - \sqrt{x}$

32. $p(t) = t^3 - \dfrac{t^2}{2} - t$

33. $r(t) = \dfrac{1}{t^2}$

34. $g(z) = \dfrac{1}{z^3}$

35. $f(z) = e^z + 3$

36. $f(x) = x^6 - \dfrac{1}{7x^6}$

37. $g(x) = \dfrac{1}{x} + \dfrac{1}{x^2} + \dfrac{1}{x^3}$

38. $p(z) = (\sqrt{z})^3$

39. $g(t) = e^{-3t}$

40. $h(t) = \cos t$

41. $g(t) = 5 + \cos t$

42. $g(\theta) = \sin\theta - 2\cos\theta$

■ In Problems 43–46, decide which function is an antiderivative of the other.

43. $f(x) = \dfrac{1}{\sqrt{x}}; g(x) = 2\sqrt{x}$

44. $f(x) = -\sin x - \cos x; g(x) = \cos x - \sin x$

45. $f(x) = \dfrac{2}{3}e^{3x}; g(x) = 2e^{3x}$

46. $f(x) = 1 - \dfrac{1}{x^2}; g(x) = \dfrac{1}{x} + x$

■ In Problems 47–52, find an antiderivative $F(x)$ with $F'(x) = f(x)$ and $F(0) = 0$. Is there only one possible solution?

47. $f(x) = 3$

48. $f(x) = 2 + 4x + 5x^2$

49. $f(x) = \dfrac{1}{4}x$

50. $f(x) = \sqrt{x}$

51. $f(x) = x^2$

52. $f(x) = e^x$

■ In Problems 53–82, find the indefinite integrals.

53. $\displaystyle\int (5x + 7)\, dx$

54. $\displaystyle\int 9x^2\, dx$

55. $\displaystyle\int e^{-0.05t}\, dt$

56. $\displaystyle\int \left(1 + \frac{1}{p}\right) dp$

57. $\displaystyle\int t^{12}\, dt$

58. $\displaystyle\int \left(x^2 + \frac{1}{x^2}\right) dx$

59. $\displaystyle\int (t^2 + 5t + 1)\, dt$

60. $\displaystyle\int 5e^z\, dz$

61. $\displaystyle\int \left(\frac{3}{t} - \frac{2}{t^2}\right) dt$

62. $\displaystyle\int (t^3 + 6t^2)\, dt$

63. $\displaystyle\int 3\sqrt{w}\, dw$

64. $\displaystyle\int (x^2 + 4x - 5)\, dx$

65. $\displaystyle\int e^{2t}\,dt$

66. $\displaystyle\int\left(x+\frac{1}{\sqrt{x}}\right)dx$

67. $\displaystyle\int (x^3+5x^2+6)dx$

68. $\displaystyle\int (e^x+5)\,dx$

69. $\displaystyle\int\left(x^2+\frac{1}{x}\right)dx$

70. $\displaystyle\int (x^5-12x^3)dx$

71. $\displaystyle\int e^{3r}\,dr$

72. $\displaystyle\int \sin t\,dt$

73. $\displaystyle\int 25e^{-0.04q}\,dq$

74. $\displaystyle\int 100e^{4x}dx$

75. $\displaystyle\int \cos\theta\,d\theta$

76. $\displaystyle\int\left(\frac{2}{x}+\pi\sin x\right)dx$

77. $\displaystyle\int \sin(3x)dx$

78. $\displaystyle\int (3\cos x-7\sin x)\,dx$

79. $\displaystyle\int 6\cos(3x)dx$

80. $\displaystyle\int (10+8\sin(2x))dx$

81. $\displaystyle\int (2e^x-8\cos x)\,dx$

82. $\displaystyle\int (12\sin(2x)+15\cos(5x))dx$

■ In Problems 83–87, find an antiderivative and use differentiation to check your answer.

83. $f(x)=\dfrac{2}{x}+\dfrac{x}{2}$

84. $g(x)=x\sqrt{x}$

85. $h(x)=\dfrac{x}{\sqrt{x}}+\dfrac{\sqrt{x}}{x}$

86. $p(x)=e^{2x}-e^{-2x}$

87. $q(x)=7\sin x-\sin(7x)$

88. The marginal revenue function of a monopolistic producer is $MR=20-4q$.

   (a) Find the total revenue function.
   (b) Find the corresponding demand curve.

89. A firm's marginal cost function is $MC=3q^2+4q+6$. Find the total cost function if the fixed costs are 200.

■ For Problems 90–93, find an antiderivative $F(x)$ with $F'(x)=f(x)$ and $F(0)=5$.

90. $f(x)=6x-5$

91. $f(x)=x^2+1$

92. $f(x)=8\sin(2x)$

93. $f(x)=6e^{3x}$

94. In drilling an oil well, the total cost, C, consists of fixed costs (independent of the depth of the well) and marginal costs, which depend on depth; drilling becomes more expensive, per meter, deeper into the earth. Suppose the fixed costs are 1,000,000 riyals (the riyal is the unit of currency of Saudi Arabia), and the marginal costs are

$$C'(x)=4000+10x\ \text{riyals/meter},$$

where x is the depth in meters. Find the total cost of drilling a well x meters deep.

95. Over the past fifty years the carbon dioxide level in the atmosphere has increased. Carbon dioxide is believed to drive temperature, so predictions of future carbon dioxide levels are important. If $C(t)$ is carbon dioxide level in parts per million (ppm) and t is time in years since 1950, three possible models are:[3]

  I $C'(t)=1.3$
  II $C'(t)=0.5+0.03t$
  III $C'(t)=0.5e^{0.02t}$

   (a) Given that the carbon dioxide level was 311 ppm in 1950, find $C(t)$ for each model.
   (b) Find the carbon dioxide level in 2020 predicted by each model.

6.3 USING THE FUNDAMENTAL THEOREM TO FIND DEFINITE INTEGRALS

In the previous section we calculated antiderivatives. In this section we see how antiderivatives are used to calculate definite integrals exactly. If f is continuous, then it has an antiderivative F, so we can use the Fundamental Theorem:

> **Fundamental Theorem of Calculus**
> If f is continuous on the interval $[a,b]$ and $F(x)$ is an antiderivative of f, that is $F'(x)=f(x)$, then
> $$\int_a^b f(x)\,dx=F(b)-F(a).$$

So far we have approximated definite integrals using a graph or left- and right-hand sums. The Fundamental Theorem gives us another method of calculating definite integrals. To find $\int_a^b f(x)\,dx$,

---

[3] Based on data from www.esrl.noaa.gov/gmd/ccgg. Accessed March 13, 2013.

we first try to find an antiderivative, F, of $f(x)$ and then calculate $F(b) - F(a)$. This method of computing definite integrals has an important advantage: it gives an exact answer. However, the method works only when we can find a formula for an antiderivative $F(x)$. We must approximate definite integrals of functions such as $f(x) = e^{-x^2}$ where we can not find a formula for an antiderivative.

Example 1   Compute $\displaystyle\int_1^3 2x\,dx$ numerically and using the Fundamental Theorem.

Solution   Using a calculator, we obtain

$$\int_1^3 2x\,dx = 8.000.$$

The Fundamental Theorem allows us to compute the integral exactly. An antiderivative of $f(x) = 2x$ is $F(x) = x^2$ and we obtain

$$\int_1^3 2x\,dx = F(3) - F(1) = 3^2 - 1^2 = 8.$$

In Example 1 we used the antiderivative $F(x) = x^2$, but $F(x) = x^2 + C$ works just as well for any constant C, because the constant cancels out when we subtract $F(a)$ from $F(b)$:

$$\int_1^3 2x\,dx = F(3) - F(1) = (3^2 + C) - (1^2 + C) = 8.$$

It is helpful to introduce a shorthand notation for $F(b) - F(a)$: we write

$$F(x)\Big|_a^b = F(b) - F(a).$$

For example:

$$\int_1^3 2x\,dx = x^2\Big|_1^3 = 3^2 - 1^2 = 8.$$

Example 2   Use the Fundamental Theorem to compute the following definite integrals:

(a) $\displaystyle\int_0^2 6x^2\,dx$ (b) $\displaystyle\int_0^2 t^3\,dt$ (c) $\displaystyle\int_1^2 (8x + 5)\,dx$ (d) $\displaystyle\int_0^1 8e^{2t}\,dt$

Solution   (a) Since $f(x) = 6x^2$, we take $F(x) = 6(x^3/3) = 2x^3$. So

$$\int_0^2 6x^2 dx = 6\left(\frac{x^3}{3}\right)\Big|_0^2 = 2x^3\Big|_0^2 = 2\cdot 2^3 - 2\cdot 0^3 = 16.$$

(b) Since $f(t) = t^3$, we take $F(t) = t^4/4$, so

$$\int_0^2 t^3\,dt = F(t)\Big|_0^2 = F(2) - F(0) = \frac{2^4}{4} - \frac{0^4}{4} = \frac{16}{4} - 0 = 4.$$

(c) Since $f(x) = 8x + 5$, we take $F(x) = 4x^2 + 5x$, giving

$$\int_1^2 (8x + 5)dx = (4x^2 + 5x)\Big|_1^2 = (4\cdot 2^2 + 5\cdot 2) - (4\cdot 1^2 + 5\cdot 1)$$

$$= 26 - 9 = 17.$$

(d) Since $f(t) = 8e^{2t}$, we take $F(t) = (8e^{2t})/2 = 4e^{2t}$, so

$$\int_0^1 8e^{2t}\,dt = 8\left(\frac{1}{2}e^{2t}\right)\Big|_0^1 = 4e^{2t}\Big|_0^1 = 4e^2 - 4e^0 = 25.556.$$

Example 3   Write a definite integral to represent the area under the graph of $f(t) = e^{0.5t}$ between $t = 0$ and $t = 4$. Use the Fundamental Theorem to calculate the area.

Solution   The function is graphed in Figure 6.23. We have

$$\text{Area} = \int_0^4 e^{0.5t}\,dt = 2e^{0.5t}\Big|_0^4 = 2e^{0.5(4)} - 2e^{0.5(0)} = 2e^2 - 2 = 12.778.$$

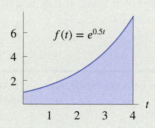

Figure 6.23: Shaded area $= \int_0^4 e^{0.5t}\,dt$

Improper Integrals

So far, in our discussion of the definite integral $\int_a^b f(x)\,dx$, we have assumed that the interval $a \leq x \leq b$ is of finite length and the integrand f is continuous. An *improper integral* is a definite integral in which one (or both) of the limits of integration is infinite or the integrand is unbounded. An example of an improper integral is

$$\int_1^\infty \frac{1}{x^2}\,dx.$$

Example 4   Interpret and estimate the value of

$$\int_1^\infty \frac{1}{x^2}\,dx.$$

Solution   This integral represents the area under the graph of $\frac{1}{x^2}$ from $x = 1$ infinitely far to the right. (See Figure 6.24.)

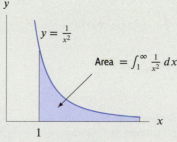

Figure 6.24: Area representation of improper integral

To estimate the value of the integral, we find the area under the curve between $x = 1$ and $x = b$, where b is large. The larger the value of b, the better the estimate. For example for $b = 10, 100, 1000$:

$$\int_1^{10} \frac{1}{x^2}\,dx = 0.9, \qquad \int_1^{100} \frac{1}{x^2}\,dx = 0.99, \qquad \int_1^{1000} \frac{1}{x^2}\,dx = 0.999,$$

These calculations suggest that as the upper limit of integration tends to infinity, the estimates tend to 1.

We say that the improper integral $\int_1^{\infty} \frac{1}{x^2}\,dx$ *converges* to 1. To show that the integral converges to exactly 1 (and not to 1.0001, say), we need to use the Fundamental Theorem of Calculus. (See Problem 35.) It may seem surprising that the region shaded in Figure 6.24 (which has infinite length) can have *finite* area. The area is finite because the values of the function $1/x^2$ shrink to zero very fast as $x \to \infty$. In other examples (where the integrand does not shrink to zero so fast), the area represented by an improper integral may not be finite. In that case, we say the improper integral *diverges*. (See Problem 41.)

Problems for Section 6.3

■ In Problems **1–20**, use the Fundamental Theorem to evaluate the definite integral exactly.

1. $\int_0^4 6x\,dx$

2. $\int_1^3 5\,dx$

3. $\int_0^3 t^3\,dt$

4. $\int_0^2 (12x^2 + 1)\,dx$

5. $\int_0^2 (3t^2 + 4t + 3)\,dt$

6. $\int_1^2 \frac{1}{x}\,dx$

7. $\int_1^4 \frac{1}{\sqrt{x}}\,dx$

8. $\int_0^1 (6q^2 + 4)\,dq$

9. $\int_0^5 3x^2\,dx$

10. $\int_0^1 2e^x\,dx$

11. $\int_1^2 5t^3\,dt$

12. $\int_1^3 6x^2\,dx$

13. $\int_0^1 (y^2 + y^4)\,dy$

14. $\int_1^2 \frac{1}{x^2}\,dx$

15. $\int_2^5 (x^3 - \pi x^2)\,dx$

16. $\int_0^1 e^{-0.2t}\,dt$

17. $\int_{-1}^1 \cos t\,dt$

18. $\int_0^{\pi/4} (\sin t + \cos t)\,dt$

19. $\int_0^3 e^{0.05t}\,dt$

20. $\int_4^9 \sqrt{x}\,dx$

■ In Problems **21–24**, find the area under the graph of $f(t)$ for $0 \le t \le 5$ using the Fundamental Theorem of Calculus.

Compare your answer with what you get using areas of triangles.

21. $f(t) = 6t$

22. $f(t) = 10 - 2t$

23. $f(t) = t$

24. $f(t) = -2t$

25. Find the exact area between $y = x^2$ and $y = x$.

26. Use the Fundamental Theorem to find the area between $f(x) = 2x$ and $g(x) = 6 - x$ from the vertical axis to their point of intersection.

27. Find the exact area of the region bounded by the x-axis and the graph of $y = x^3 - x$.

28. Use the Fundamental Theorem of Calculus to find the average value of $f(x) = e^{0.5x}$ between $x = 0$ and $x = 3$. Show the average value on a graph of $f(x)$.

29. Use the Fundamental Theorem to determine the value of b if the area under the graph of $f(x) = x^2$ between $x = 0$ and $x = b$ is equal to 100. Assume $b > 0$.

30. Use the Fundamental Theorem to determine the value of b if the area under the graph of $f(x) = 4x$ between $x = 1$ and $x = b$ is equal to 240. Assume $b > 1$.

31. If t is in years, and $t = 0$ is January 1, 2005, worldwide energy consumption, r, in quadrillion (10^{15}) BTUs per year, is modeled by

$$r = 462e^{0.019t}.$$

   (a) Write a definite integral for the total energy use between the start of 2005 and the start of 2010.

   (b) Use the Fundamental Theorem of Calculus to evaluate the integral. Give units with your answer.

32. Oil is leaking out of a ruptured tanker at the rate of $r(t) = 50e^{-0.02t}$ thousand liters per minute.

  (a) At what rate, in liters per minute, is oil leaking out at $t = 0$? At $t = 60$?

  (b) How many liters leak out during the first hour?

33. (a) Between 2005 and 2015, ACME Widgets sold widgets at a continuous rate of $R = R_0 e^{0.125t}$ widgets per year, where t is time in years since January 1, 2005. Suppose they were selling widgets at a rate of 1000 per year on January 1, 2005. How many widgets did they sell between 2005 and 2015? How many did they sell if the rate on January 1, 2005 was 1,000,000 widgets per year?

  (b) In the first case (1000 widgets per year on January 1, 2005), how long did it take for half the widgets in the ten-year period to be sold? In the second case (1,000,000 widgets per year on January 1, 2005), when had half the widgets in the ten-year period been sold?

  (c) In 2015, ACME advertised that half the widgets it had sold in the previous ten years were still in use. Based on your answer to part (b), how long must a widget last in order to justify this claim?

34. Decide if the improper integral $\int_0^\infty e^{-2t}\, dt$ converges, and if so, to what value, by the following method.

  (a) Use a computer or calculator to find $\int_0^b e^{-2t}\, dt$ for $b = 3, 5, 7, 10$. What do you observe? Make a guess about the convergence of the improper integral.

  (b) Find $\int_0^b e^{-2t}\, dt$ using the Fundamental Theorem. Your answer will contain b.

  (c) Take the limit $b \to \infty$ of your answer to part (b). Does this limit confirm your guess from part (a)?

35. In this problem, you will show that the following improper integral converges to 1:

$$\int_1^\infty \frac{1}{x^2}\, dx.$$

  (a) Use the Fundamental Theorem to find $\int_1^b 1/x^2\, dx$. Your answer will contain b.

  (b) Now take the limit as $b \to \infty$. What does this tell you about the improper integral?

36. (a) Graph $f(x) = e^{-x^2}$ and shade the area represented by the improper integral $\int_{-\infty}^\infty e^{-x^2}\, dx$.

  (b) Use a calculator or computer to find $\int_{-a}^a e^{-x^2}\, dx$ for $a = 1, a = 2, a = 3, a = 5$.

  (c) The improper integral $\int_{-\infty}^\infty e^{-x^2}\, dx$ converges to a finite value. Use your answers from part (b) to estimate that value.

37. Graph $y = 1/x^2$ and $y = 1/x^3$ on the same axes. Which do you think is larger: $\int_1^\infty 1/x^2\, dx$ or $\int_1^\infty 1/x^3\, dx$? Why?

38. At a time t hours after taking a tablet, the rate at which a drug is being eliminated is

$$r(t) = 50 \left(e^{-0.1t} - e^{-0.2t} \right) \text{ mg/hr.}$$

Assuming that all the drug is eventually eliminated, calculate the original dose.

39. The rate, r, at which people get sick during an epidemic of the flu can be approximated by

$$r = 1000te^{-0.5t},$$

where r is measured in people/day and t is measured in days since the start of the epidemic.

  (a) Write an improper integral representing the total number of people that get sick.

  (b) Use a graph of r to represent the improper integral from part (a) as an area.

40. An island has a carrying capacity of 1 million rabbits. (That is, no more than 1 million rabbits can be supported by the island.) The rabbit population is two at time $t = 1$ day and grows at a rate of $r(t)$ thousand rabbits/day until the carrying capacity is reached. For each of the following formulas for $r(t)$, is the carrying capacity ever reached? Explain your answer.

  (a) $r(t) = 1/t^2$ **(b)** $r(t) = t$

  (c) $r(t) = 1/\sqrt{t}$

41. Consider the improper integral

$$\int_1^\infty \frac{1}{\sqrt{x}}\, dx.$$

  (a) Use a calculator or computer to find $\int_1^b 1/(\sqrt{x})\, dx$ for $b = 100, 1000, 10{,}000$. What do you notice?

  (b) Find $\int_1^b 1/(\sqrt{x})\, dx$ using the Fundamental Theorem of Calculus. Your answer will contain b.

  (c) Now take the limit as $b \to \infty$. What does this tell you about the improper integral?

6.4 APPLICATION: CONSUMER AND PRODUCER SURPLUS

Supply and Demand Curves

As we saw in Chapter 1, the quantity of a certain item produced and sold can be described by the supply and demand curves of the item. The *supply curve* shows what quantity, q, of the item the producers supply at different prices, p. The consumers' behavior is reflected in the *demand curve*, which shows what quantity of goods are bought at various prices. See Figure 6.25.

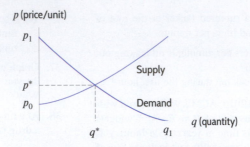

Figure 6.25: Supply and demand curves

It is assumed that the market settles at the *equilibrium price* p^* and *equilibrium quantity* q^* where the graphs cross. At equilibrium, a quantity q^* of an item is produced and sold for a price of p^* each.

Consumer and Producer Surplus

Notice that at equilibrium, a number of consumers have bought the item at a lower price than they would have been willing to pay. (For example, there are some consumers who would have been willing to pay prices up to p_1.) Similarly, there are some suppliers who would have been willing to produce the item at a lower price (down to p_0, in fact). We define the following terms:

> • The **consumer surplus** measures the consumers' gain from trade. It is the total amount gained by consumers by buying the item at the current price, rather than at the price they would have been willing to pay.
>
> • The **producer surplus** measures the suppliers' gain from trade. It is the total amount gained by producers by selling at the current price, rather than at the price they would have been willing to accept.
>
> In the absence of price controls, the current price is assumed to be the equilibrium price.

Both consumers and producers are richer for having traded. The consumer and producer surplus measure how much richer they are.

Suppose that all consumers buy the good at the maximum price they are willing to pay. Subdivide the interval from 0 to q^* into intervals of length Δq. Figure 6.26 shows that a quantity Δq of items are sold at a price of about p_1, another Δq are sold for a slightly lower price of about p_2, the next Δq for a price of about p_3, and so on. Thus, the consumers' total expenditure is about

$$p_1 \Delta q + p_2 \Delta q + p_3 \Delta q + \cdots = \sum p_i \Delta q.$$

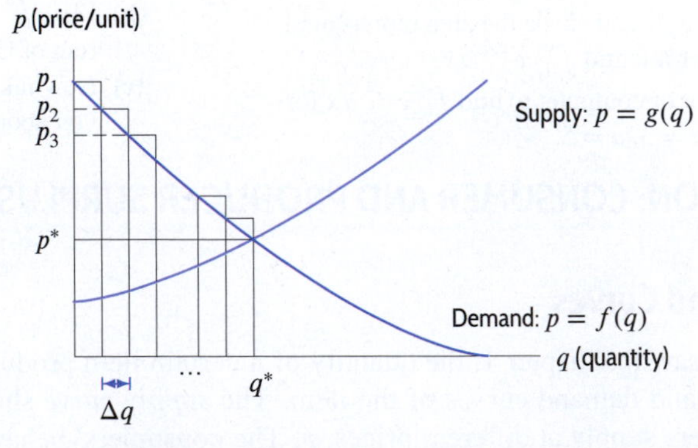

Figure 6.26: Calculation of consumer surplus

If the demand curve has equation[4] $p = f(q)$, and if all consumers who were willing to pay more than p^* paid as much as they were willing, then as $\Delta q \to 0$, we would have

$$\begin{matrix} \text{Consumer} \\ \text{expenditure} \end{matrix} = \int_0^{q^*} f(q)\, dq = \begin{matrix} \text{Area under demand} \\ \text{curve from 0 to } q^*. \end{matrix}$$

If all goods are sold at the equilibrium price, the consumers' actual expenditure is only p^*q^*, which is the area of the rectangle between the q-axis and the line $p = p^*$ from $q = 0$ to $q = q^*$. The consumer surplus is the difference between the total consumer expenditure if all consumers pay the maximum they are willing to pay and the actual consumer expenditure if all consumers pay the current price. The consumer surplus is represented by the area in Figure 6.27. Similarly, the producer surplus is represented by the area in Figure 6.28. (See Problems 17 and 18.) Thus:

$$\begin{matrix} \text{Consumer surplus} \\ \text{at price } p^* \end{matrix} = \begin{matrix} \text{Area between demand curve} \\ \text{and horizontal line at } p^*. \end{matrix}$$

$$\begin{matrix} \text{Producer surplus} \\ \text{at price } p^* \end{matrix} = \begin{matrix} \text{Area between supply curve} \\ \text{and horizontal line at } p^*. \end{matrix}$$

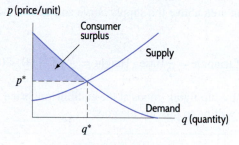

Figure 6.27: Consumer surplus

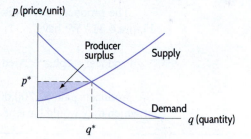

Figure 6.28: Producer surplus

Example 1   The supply and demand curves for a product are given in Figure 6.29.

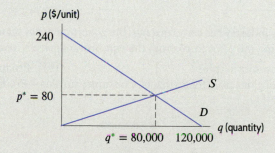

Figure 6.29: Supply and demand curves for a product

(a) What are the equilibrium price and quantity?

(b) At the equilibrium price, calculate and interpret the consumer and producer surplus.

---

[4]Note that here p is written as a function of q.

Solution   (a) The equilibrium price is $p^* = \$80$ per unit and the equilibrium quantity is $q^* = 80{,}000$ units.

(b) The consumer surplus is the area under the demand curve and above the line $p = 80$. (See Figure 6.30.) We have

$$\text{Consumer surplus} = \text{Area of triangle} = \frac{1}{2}\text{Base}\cdot\text{Height} = \frac{1}{2}80{,}000\cdot 160 = \$6{,}400{,}000.$$

This tells us that consumers gain $6,400,000 in buying goods at the equilibrium price instead of at the price they would have been willing to pay.

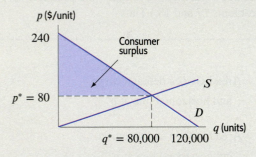

Figure 6.30: Consumer surplus

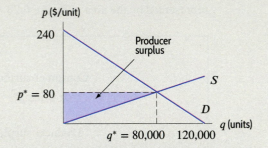

Figure 6.31: Producer surplus

The producer surplus is the area above the supply curve and below the line $p = 80$. (See Figure 6.31.) We have

$$\text{Producer surplus} = \text{Area of triangle} = \frac{1}{2}\text{Base}\cdot\text{Height} = \frac{1}{2}\cdot 80{,}000\cdot 80 = \$3{,}200{,}000.$$

So, producers gain $3,200,000 by supplying goods at the equilibrium price instead of the price at which they would have been willing to provide the goods.

Wage and Price Controls

In a free market, the price of a product generally moves to the equilibrium price, unless outside forces keep the price artificially high or artificially low. Rent control, for example, keeps prices below market value, whereas cartel pricing or the minimum-wage law raise prices above market value. What happens to consumer and producer surplus at non-equilibrium prices?

Example 2   The dairy industry has cartel pricing: the government has set milk prices artificially high. What effect does raising the price to p^+ from the equilibrium price have on:

(a) Consumer surplus?   (b) Producer surplus?

(c) Total gains from trade (that is, Consumer surplus + Producer surplus)?

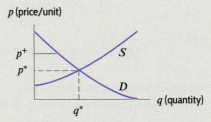

Figure 6.32: What is the effect of the artificially high price, p^+, on consumer and producer surplus?
(q^* and p^* are equilibrium values)

Solution   (a) A graph of possible supply and demand curves for the milk industry is given in Figure 6.32. Suppose that the price is fixed at p^+, above the equilibrium price. Consumer surplus is the difference between the amount the consumers paid (p^+) and the amount they would have been willing to pay (given on the demand curve). This is the area shaded in Figure 6.33. This consumer surplus is less than the consumer surplus at the equilibrium price, shown in Figure 6.34.

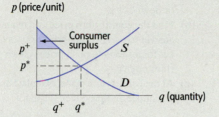

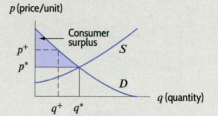

Figure 6.33: Consumer surplus: Artificial price

Figure 6.34: Consumer surplus: Equilibrium price

(b) At a price of p^+, the quantity sold, q^+, is less than it would have been at the equilibrium price. The producer surplus is represented by the area between p^+ and the supply curve at this reduced demand. This area is shaded in Figure 6.35. Compare this producer surplus (at the artificially high price) to the producer surplus in Figure 6.36 (at the equilibrium price). In this case, producer surplus appears to be greater at the artificial price than at the equilibrium price. (However, different supply and demand curves might lead to a different answer.)

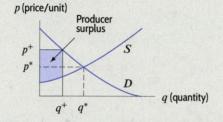

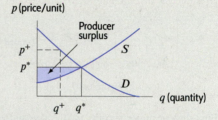

Figure 6.35: Producer surplus: Artificial price

Figure 6.36: Producer surplus: Equilibrium price

(c) The total gains from trade (Consumer surplus + Producer surplus) at the price of p^+ is represented by the area shaded in Figure 6.37. The total gains from trade at the equilibrium price of p^* is represented by the area shaded in Figure 6.38. Under artificial price conditions, the total gains from trade decreases. The total financial effect of the artificially high price on all producers and consumers combined is negative.

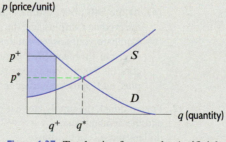

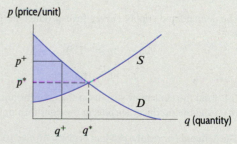

Figure 6.37: Total gains from trade: Artificial price

Figure 6.38: Total gains from trade: Equilibrium price

Problems for Section 6.4

1. (a) What are the equilibrium price and quantity for the supply and demand curves in Figure 6.39?

(b) Shade the areas representing the consumer and producer surplus and estimate them.

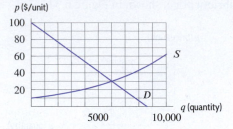

Figure 6.39

2. The supply and demand curves for a product are given in Figure 6.40. Estimate the equilibrium price and quantity and the consumer and producer surplus. Shade areas representing the consumer surplus and the producer surplus.

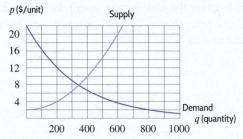

Figure 6.40

3. Find the consumer surplus for the demand curve $p = 100 - 3q^2$ when $q^* = 5$ units are sold at the equilibrium price.

4. Given the demand curve $p = 35 - q^2$ and the supply curve $p = 3 + q^2$, find the producer surplus when the market is in equilibrium.

5. Find the consumer surplus for the demand curve $p = 100 - 4q$ when $q^* = 10$ items are sold at the equilibrium price.

6. The demand and supply curves for a product are given as

$$2q - 15p = -120$$

$$3q + 6p = 105.$$

(a) Find the consumer surplus at the equilibrium.

(b) Find the producer surplus at the equilibrium.

7. The demand curve for a product is given by $q = 100 - 2p$ and the supply curve is given by $q = 3p - 50$.

(a) Find the consumer surplus at the equilibrium.

(b) Find the producer surplus at the equilibrium.

8. For a product, the demand curve is $p = 100e^{-0.008q}$ and the supply curve is $p = 4\sqrt{q} + 10$ for $0 \le q \le 500$, where q is quantity and p is price in dollars per unit.

(a) At a price of \$50, what quantity are consumers willing to buy and what quantity are producers willing to supply? Will the market push prices up or down?

(b) Find the equilibrium price and quantity. Does your answer to part (a) support the observation that market forces tend to push prices closer to the equilibrium price?

(c) At the equilibrium price, calculate and interpret the consumer and producer surplus.

9. Sketch possible supply and demand curves where the consumer surplus at the equilibrium price is

(a) Greater than the producer surplus.

(b) Less than the producer surplus.

10. Supply and demand curves for a product are in Figure 6.41.

(a) Estimate the equilibrium price and quantity.

(b) Estimate the consumer and producer surplus. Shade them.

(c) What are the total gains from trade for this product?

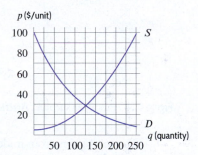

Figure 6.41

11. Supply and demand curves are in Figure 6.41. A price of \$40 is artificially imposed.

(a) At the \$40 price, estimate the consumer surplus, the producer surplus, and the total gains from trade.

(b) Compare your answers in this problem to your answers in Problem 10. Discuss the effect of price controls on the consumer surplus, producer surplus, and gains from trade in this case.

12. For a product, the supply curve is given by $q = 10p - 30$ and the demand curve by $q = -2p + 30$, where p is in dollars and q is in millions.

(a) What is the equilibrium price and quantity?

(b) Find the consumer and producer surplus using the Fundamental Theorem of Calculus.

13. (a) Estimate the equilibrium price and quantity for the supply and demand curves in Figure 6.42.
    (b) Estimate the consumer and producer surplus.
    (c) The price is set artificially low at $p^- = 4$ dollars per unit. Estimate the consumer and producer surplus at this price. Compare your answers to the consumer and producer surplus at the equilibrium price.

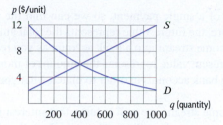

Figure 6.42

14. Rent controls on apartments are an example of price controls on a commodity. They keep the price artificially low (below the equilibrium price). Sketch a graph of supply and demand curves, and label on it a price p^- below the equilibrium price. What effect does forcing the price down to p^- have on:

    (a) The producer surplus?
    (b) The consumer surplus?
    (c) The total gains from trade (Consumer surplus + Producer surplus)?

15. Supply and demand data are in Tables 6.3 and 6.4.

    (a) Which table shows supply and which shows demand?
    (b) Estimate the equilibrium price and quantity.
    (c) Estimate the consumer and producer surplus.

Table 6.3

q (quantity)	0	100	200	300	400	500	600
p ($/unit)	60	50	41	32	25	20	17

Table 6.4

q (quantity)	0	100	200	300	400	500	600
p ($/unit)	10	14	18	22	25	28	34

16. The total gains from trade (consumer surplus + producer surplus) is largest at the equilibrium price. What about the consumer surplus and producer surplus separately?

    (a) Suppose a price is artificially high. Can the consumer surplus at the artificial price be larger than the consumer surplus at the equilibrium price? What about the producer surplus? Sketch possible supply and demand curves to illustrate your answers.
    (b) Suppose a price is artificially low. Can the consumer surplus at the artificial price be larger than the consumer surplus at the equilibrium price? What about the producer surplus? Sketch possible supply and demand curves to illustrate your answers.

■ In Problems 17–19, the supply and demand curves have equations $p = S(q)$ and $p = D(q)$, respectively, with equilibrium at (q^*, p^*).

17. Using Riemann sums, explain the economic significance of $\int_0^{q^*} S(q)\,dq$ to the producers.

18. Using Riemann sums, give an interpretation of producer surplus, $\int_0^{q^*} (p^* - S(q))\,dq$, analogous to the interpretation of consumer surplus.

19. Referring to Figures 6.27 and 6.28 on page 299, mark the regions representing the following quantities and explain their economic meaning:

    (a) $p^* q^*$

    (b) $\int_0^{q^*} D(q)\,dq$

    (c) $\int_0^{q^*} S(q)\,dq$

    (d) $\int_0^{q^*} D(q)\,dq - p^* q^*$

    (e) $p^* q^* - \int_0^{q^*} S(q)\,dq$

    (f) $\int_0^{q^*} (D(q) - S(q))\,dq$

6.5 APPLICATION: PRESENT AND FUTURE VALUE

In Section 1.7 on page 60, we introduced the present and future value of a single payment. In this section we see how to calculate the present and future value of a continuous stream of payments.

Income Stream

When we consider payments made to or by an individual, we usually think of *discrete* payments, that is, payments made at specific moments in time. However, we may think of payments made by a company as being *continuous*. The revenues earned by a huge corporation, for example, come in essentially all the time, and therefore they can be represented by a continuous *income stream*. Since the rate at which revenue is earned may vary from time to time, the income stream is described by

$$S(t) \text{ dollars/year.}$$

Notice that $S(t)$ is a *rate* at which payments are made (its units are dollars per year, for example) and that the rate depends on the time, t, usually measured in years from the present.

Present and Future Values of an Income Stream

Just as we can find the present and future values of a single payment, so we can find the present and future values of a stream of payments. As before, the future value represents the total amount of money that you would have if you deposited an income stream into a bank account as you receive it and let it earn interest until that future date. The present value represents the amount of money you would have to deposit today (in an interest-bearing bank account) in order to match what you would get from the income stream by that future date.

When we are working with a continuous income stream, we will assume that interest is compounded continuously. If the interest rate is r, the present value, P, of a deposit, B, made t years in the future is

$$P = Be^{-rt}.$$

Suppose that we want to calculate the present value of the income stream described by a rate of $S(t)$ dollars per year, and that we are interested in the period from now until M years in the future. In order to use what we know about single deposits to calculate the present value of an income stream, we divide the stream into many small deposits, and imagine each deposited at one instant. Dividing the interval $0 \leq t \leq M$ into subintervals of length Δt gives:

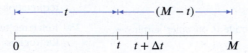

Assuming Δt is small, the rate, $S(t)$, at which deposits are being made does not vary much within one subinterval. Thus, between t and $t + \Delta t$:

$$\text{Amount paid} \approx \text{Rate of deposits} \times \text{Time}$$
$$\approx (S(t) \text{ dollars/year})(\Delta t \text{ years})$$
$$= S(t)\Delta t \text{ dollars.}$$

The deposit of $S(t)\Delta t$ is made t years in the future. Thus, assuming a continuous interest rate r,

$$\begin{array}{l}\text{Present value of money} \\ \text{deposited in interval } t \text{ to } t + \Delta t\end{array} \approx S(t)\Delta t e^{-rt}.$$

Summing over all subintervals gives

$$\text{Total present value} \approx \sum S(t)e^{-rt}\Delta t \text{ dollars.}$$

In the limit as $\Delta t \to 0$, we get the following integral:

$$\boxed{\text{Present value} = \int_0^M S(t)e^{-rt}dt.}$$

As in Section 1.7, the value M years in the future is given by

$$\boxed{\text{Future value} = \text{Present value} \cdot e^{rM}.}$$

Example 1   Find the present and future values of a constant income stream of $1000 per year over a period of 20 years, assuming an interest rate of 6% compounded continuously.

Solution   Using $S(t) = 1000$ and $r = 0.06$ and a calculator or computer to evaluate the integral, we have

$$\text{Present value} = \int_0^{20} 1000e^{-0.06t} dt = \$11{,}647.$$

Alternately, using the Fundamental Theorem and the fact that an antiderivative of $1000e^{-0.06t}$ is $\dfrac{1000}{-0.06}e^{-0.06t}$, we have

$$\int_0^{20} 1000e^{-0.06t} dt = \frac{1000}{-0.06}e^{-0.06t}\Big|_0^{20} = \frac{1000}{-0.06}e^{1.2} - \frac{1000}{-0.06}e^0 = \$11{,}647.$$

We can get the future value, B, from the present value, P, using $B = Pe^{rt}$, so

$$\text{Future value} = 11{,}647e^{0.06(20)} = \$38{,}669.$$

Notice that since money was deposited at a rate of $1000 a year for 20 years, the total amount deposited was $20,000. The future value is $38,669, so the money has almost doubled because of the interest.

Example 2   Suppose you want to have $50,000 in 8 years' time in a bank account earning 2% interest, compounded continuously.

(a) If you make one lump sum deposit now, how much should you deposit?
(b) If you deposit money continuously throughout the 8-year period, at what rate should you deposit it?

Solution   (a) If you deposit a lump sum of $\$P$, then $\$P$ is the present value of $50,000. So, using $B = Pe^{rt}$ with $B = 50{,}000$ and $r = 0.02$ and $t = 8$:

$$50000 = Pe^{0.02(8)}$$
$$P = \frac{50000}{e^{0.02(8)}} = 42{,}607.$$

If you deposit $42,607 into the account now, you will have $50,000 in 8 years' time.

(b) Suppose you deposit money at a constant rate of $\$S$ per year. Then

$$\text{Present value of deposits} = \int_0^8 Se^{-0.02t} \, dt$$

Since S is a constant, we can take it out in front of the integral sign and use a calculator to evaluate the integral:

$$\text{Present value} = S \int_0^8 e^{-0.02t} \, dt = S(7.3928).$$

Alternately, using the Fundamental Theorem and the fact that an antiderivative of $e^{-0.02t}$ is $\dfrac{e^{-0.02t}}{-0.02}$, we have

$$S \int_0^8 e^{-0.02t} dt = S\frac{e^{-0.02t}}{-0.02}\Big|_0^8 = S\left(\frac{e^{-0.16}}{-0.02} - \frac{e^0}{-0.02}\right) = S(7.3928).$$

But the present value of the continuous deposit must be the same as the present value of the lump sum deposit; that is, $42,607. So

$$42{,}607 = S(7.3928)$$
$$S = \$5763.$$

To meet your goal of $50,000, you need to deposit money at a continuous rate of $5763 per year, or about $480 per month.

Problems for Section 6.5

1. Draw a graph, with time in years on the horizontal axis, of what an income stream might look like for a company that sells sunscreen in the northeast United States.

2. Find the present and future values of an income stream of $12,000 a year for 20 years. The interest rate is 6%, compounded continuously.

3. Find the present and future values of an income stream of $2000 per year for 15 years, assuming a 5% interest rate compounded continuously.

4. **(a)** Find the present and future value of an income stream of $5000 per year for a period of 8 years if the interest rate, compounded continuously, is 2%.
   (b) Explain, in plain language, what the present and future values mean in terms of the income stream.

5. Assuming an interest rate of 3% compounded continuously,

   (a) Find the future value in 10 years of a payment of $10,000 made today.
   (b) Find the future value of an income stream of $1000 per year over 10 years.
   (c) Which is larger, the future value from the lump sum in part (a) or from the income stream in part (b)? Explain why this makes sense financially.

6. Assuming an interest rate of 5% compounded continuously,

   (a) Find the future value in 6 years of a payment of $12,000 made today.
   (b) Find the future value of an income stream of $2000 per year over 6 years.
   (c) Which is larger, the future value from the lump sum in part (a) or from the income stream in part (b)? Explain why this makes sense financially.

7. A person deposits money into an account, which pays 5% interest compounded continuously, at a rate of $1000 per year for 15 years. Calculate:

   (a) The balance in the account at the end of the 15 years.

   (b) The amount of money actually deposited into the account.
   (c) The interest earned during the 15 years.

8. A person deposits money into an account, which pays 6% interest compounded continuously, at a rate of $1000 per year for 30 years. Calculate:

   (a) The balance in the account at the end of the 30 years.
   (b) The amount of money actually deposited into the account.
   (c) The interest earned during the 30 years.

9. **(a)** Find the present and future value of an income stream of $6000 per year for a period of 10 years if the interest rate, compounded continuously, is 5%.
   (b) How much of the future value is from the income stream? How much is from interest?

10. A small business expects an income stream of $5000 per year for a four-year period.

    (a) Find the present value of the business if the annual interest rate, compounded continuously, is

      (i)   3%   (ii)   10%

    (b) In each case, find the value of the business at the end of the four-year period.

11. A small business, Company A, believes it can generate an income stream of $120,000 per year for the next 20 years. A larger firm, Company B, offers to purchase Company A for a lump sum of $2 million today. With a continuous interest rate of 2% per year, is this a reasonable deal for Company A? For Company B? Explain in terms of present value.

12. The Hershey Company is the largest US producer of chocolate. Between 2011 and 2014, Hershey generated net sales at a rate approximated by $5.3 + 0.42t$ billion dollars per year, where t is the time in years since January 1, 2011.[5] Assume this rate continues through the year 2016 and that the interest rate is 2% per year compounded continuously.

---

[5] www.statista.com/statistics, accessed August 24, 2016.

(a) Find the value, on January 1, 2011, of Hershey's net sales during the five-year period from from January 1, 2011 to January 1, 2016.

(b) Find the value, on January 1, 2016, of Hershey's net sales during the same five-year period.

13. German TV and radio users pay an annual license fee of 200 euros. In 2008, the German poet Friedrich Schiller, who died in 1805, was sent reminders to pay his license fee.[6] (The reminders were sent to an elementary school named after Schiller, despite a teacher pointing out that "the addressee is no longer in a position to listen to the radio or watch television".) Assume the license fee had been charged at a continuous rate of 200 euros per year since 1805 and that the continous interest rate was 3% per year over this period. If Schiller were charged the license fee, with interest, since his death, how much would he have owed by 2008?

14. A recently installed machine earns the company revenue at a continuous rate of $60{,}000t + 45{,}000$ dollars per year during the first six months of operation and at the continuous rate of 75,000 dollars per year after the first six months. The cost of the machine is $150,000, the interest rate is 7% per year, compounded continuously, and t is time in years since the machine was installed.

(a) Find the present value of the revenue earned by the machine during the first year of operation.

(b) Find how long it will take for the machine to pay for itself; that is, how long it will take for the present value of the revenue to equal the cost of the machine.

15. **(a)** An investment account earns 10% interest compounded continuously. At what constant, continuous rate must a parent deposit money into such an account in order to save $100,000 in 10 years for a child's college expenses?

(b) If the parent decides instead to deposit a lump sum now in order to attain the goal of $100,000 in 10 years, how much must be deposited now?

16. Your company needs $500,000 in two years' time for renovations and can earn 9% interest on investments.

(a) What is the present value of the renovations?

(b) If your company deposits money continuously at a constant rate throughout the two-year period, at what rate should the money be deposited so that you have the $500,000 when you need it?

17. A company is expected to earn $50,000 a year, at a continuous rate, for 8 years. You can invest the earnings at an interest rate of 7%, compounded continuously. You have the chance to buy the rights to the earnings of the company now for $350,000. Should you buy? Explain.

18. Sales of Version 6.0 of a computer software package start out high and decrease exponentially. At time t, in years, the sales are $s(t) = 50{,}000e^{-t}$ dollars per year. After two years, Version 7.0 of the software is released and replaces Version 6.0. You can invest earnings at an interest rate of 6%, compounded continuously. Calculate the total present value of sales of Version 6.0 over the two-year period.

19. Intel Corporation is a leading manufacturer of integrated circuits. In 2011, Intel generated profits at a continuous rate of 34.6 billion dollars per year based on a total revenue of 54 billion dollars.[7] Assume the interest rate was 6% per year compounded continuously.

(a) What was the present value of Intel's profits over the 2011 one-year time period?

(b) What was the value at the end of the year of Intel's profits over the 2011 one-year time period?

20. Harley-Davidson Inc. manufactures motorcycles. During the years following 2013 (the company's 110th anniversary year), the company's net revenue can be approximated[8] by $5.9 + 0.3t$ billion dollars per year, where t is time in years since January 1, 2013. Assume this rate holds through January 1, 2016, and assume a continuous interest rate of 2.0% per year.

(a) What was the net revenue of the Harley-Davidson Company in 2013? What is the projected net revenue in 2016?

(b) What was the present value, on January 1, 2013, of Harley-Davidson's net revenue for the three years from January 1, 2013 to January 1, 2016?

(c) What is the future value, on January 1, 2016, of net revenue for the preceding 3 years?

21. McDonald's Corporation licenses and operates over 30,000 fast-food restaurants throughout the world. Between 2005 and 2008, McDonald's generated revenue at continuous rates between 17.9 and 22.8 billion dollars per year.[9] Suppose that McDonald's rate of revenue stays within this range. Use an interest rate of 4.5% per year compounded continuously. Fill in the blanks:

(a) The present value of McDonald's revenue over a five-year time period is between _____ and _____ billion dollars.

(b) The present value of McDonald's revenue over a twenty-five-year time period is between _____ and _____ billion dollars.

---

[6]http://news.bbc.co.uk/2/hi/europe/7648021.stm, accessed January 2013.
[7]http://www.intc.com/financials.cfm, accessed May 2012.
[8]www.marketwatch.com, accessed August 25, 2016.
[9]McDonald's Annual Report 2007, www.mcdonalds.com, accessed March 13, 2013.

22. Your company is considering buying new production machinery. You want to know how long it will take for the machinery to pay for itself; that is, you want to find the length of time over which the present value of the profit generated by the new machinery equals the cost of the machinery. The new machinery costs \$130,000 and earns profit at the continuous rate of \$80,000 per year. Use an interest rate of 8.5% per year compounded continuously.

23. An oil company discovered an oil reserve of 100 million barrels. For time $t > 0$, in years, the company's extraction plan is a linear declining function of time as follows:

$$q(t) = a - bt,$$

where $q(t)$ is the rate of extraction of oil in millions of barrels per year at time t and $b = 0.1$ and $a = 10$.

(a) How long does it take to exhaust the entire reserve?

(b) The oil price is a constant \$20 per barrel, the extraction cost per barrel is a constant \$10, and the market interest rate is 10% per year, compounded continuously. What is the present value of the company's profit?

24. The value of some good wine increases with age. Thus, if you are a wine dealer, you have the problem of deciding whether to sell your wine now, at a price of \$P a bottle, or to sell it later at a higher price. Suppose you know that the amount a wine-drinker is willing to pay for a bottle of this wine t years from now is $\$P(1 + 20\sqrt{t})$. Assuming continuous compounding and a prevailing interest rate of 5% per year, when is the best time to sell your wine?

6.6 INTEGRATION BY SUBSTITUTION

In Chapter 3, we learned rules to differentiate any function obtained by combining constants, powers of x, $\sin x$, $\cos x$, e^x, and $\ln x$, using addition, multiplication, division, or composition of functions. Such functions are called *elementary*.

There is a great difference between looking for derivatives and looking for antiderivatives. Every elementary function has elementary derivatives, but most elementary functions—such as $\sqrt{x^3 + 1}$ and e^{-x^2}—do not have elementary antiderivatives.

In this section, we introduce integration by substitution, which reverses the chain rule. According to the chain rule,

$$\frac{d}{dx}(f(g(x))) = \underbrace{f'}_{\text{Derivative of outside}} \overbrace{(\ g(x)\ )}^{\text{Inside}} \cdot \underbrace{g'(x)}_{\text{Derivative of inside}}.$$

Thus, any function which is the result of differentiating with the chain rule is the product of two factors: the "derivative of the outside" and the "derivative of the inside." If a function has this form, its antiderivative is $f(g(x)) + C$.

Example 1 Use the chain rule to find $f'(x)$ and then write the corresponding antidifferentiation formula.

(a) $f(x) = e^{x^2}$ **(b)** $f(x) = \frac{1}{6}(x^2 + 1)^6$ **(c)** $f(x) = \ln(x^2 + 4)$

Solution **(a)** Using the chain rule, we see

$$\frac{d}{dx}\left(e^{x^2}\right) = e^{x^2} \cdot 2x \quad \text{so} \quad \int e^{x^2} \cdot 2x\, dx = e^{x^2} + C.$$

(b) Using the chain rule, we see

$$\frac{d}{dx}\left(\frac{1}{6}(x^2 + 1)^6\right) = (x^2 + 1)^5 \cdot 2x \quad \text{so} \quad \int (x^2 + 1)^5 \cdot 2x\, dx = \frac{1}{6}(x^2 + 1)^6 + C.$$

(c) Using the chain rule, we see

$$\frac{d}{dx}(\ln(x^2 + 4)) = \frac{1}{x^2 + 4} \cdot 2x \quad \text{so} \quad \int \frac{1}{x^2 + 4} \cdot 2x\, dx = \ln(x^2 + 4) + C.$$

In Example 1, the derivative of each inside function is $2x$. Notice that the derivative of the inside function is a factor in the integrand in each antidifferentiation formula.

Finding an inside function whose derivative appears as a factor is key to the method of substitution. We formalize this method as follows:

> ## To Make a Substitution in an Integral
>
> Let w be the "inside function" and $dw = w'(x)\,dx = \dfrac{dw}{dx}\,dx$. Then express the integrand in terms of w.

Example 2 Make a substitution to find each of the following integrals:

(a) $\displaystyle \int e^{x^2} \cdot 2x\,dx$ (b) $\displaystyle \int (x^2 + 1)^5 \cdot 2x\,dx$ (c) $\displaystyle \int \frac{1}{x^2 + 4} \cdot 2x\,dx$

Solution

(a) We look for an inside function whose derivative appears as a factor. In this case, the inside function is x^2, with derivative $2x$. We let $w = x^2$. Then $dw = w'(x)\,dx = 2x\,dx$. The original integrand can now be rewritten in terms of w:

$$\int e^{x^2} \cdot 2x\,dx = \int e^w\,dw = e^w + C = e^{x^2} + C.$$

By changing the variable to w, we simplified the integrand. The final step, after antidifferentiating, is to convert back to the original variable, x.

(b) Here, the inside function is $x^2 + 1$, with derivative $2x$. We let $w = x^2 + 1$. Then $dw = w'(x)\,dx = 2x\,dx$. Rewriting the original integral in terms of w, we have

$$\int (x^2 + 1)^5 \cdot 2x\,dx = \int w^5\,dw = \frac{1}{6}w^6 + C = \frac{1}{6}(x^2 + 1)^6 + C.$$

Again, by changing the variable to w, we simplified the integrand.

(c) The inside function is $x^2 + 4$, so we let $w = x^2 + 4$. Then $dw = w'(x)\,dx = 2x\,dx$. Substituting, we have

$$\int \frac{1}{x^2 + 4} \cdot 2x\,dx = \int \frac{1}{w}\,dw = \ln|w| + C = \ln(x^2 + 4) + C, \text{ since } w = x^2 + 4 > 0.$$

Notice that the derivative of the inside function must be present in the integral for this method to work. The method works, however, even when the derivative is missing a constant factor, as in the next two examples.

Example 3 Find $\displaystyle \int t e^{(t^2 + 1)}\,dt$.

Solution

Here the inside function is $t^2 + 1$, with derivative $2t$. Since there is a factor of t in the integrand, we try $w = t^2 + 1$. Then $dw = w'(t)\,dt = 2t\,dt$. Notice, however, that the original integrand has only $t\,dt$, not $2t\,dt$. We therefore write

$$\frac{1}{2}\,dw = t\,dt$$

and then substitute:

$$\int t e^{(t^2 + 1)}\,dt = \int \overbrace{e^{(t^2 + 1)}}^{e^w} \cdot \underbrace{t\,dt}_{\frac{1}{2}dw} = \int e^w \frac{1}{2}\,dw = \frac{1}{2}\int e^w\,dw = \frac{1}{2}e^w + C = \frac{1}{2}e^{(t^2 + 1)} + C.$$

Why didn't we put $\frac{1}{2} \int e^w \, dw = \frac{1}{2} e^w + \frac{1}{2} C$ in the preceding example? Since the constant C is arbitrary, it does not really matter whether we add C or $\frac{1}{2} C$. The convention is always to add C to whatever antiderivative we have calculated.

Example 4   Find $\displaystyle\int x^3 \sqrt{x^4 + 5} \, dx$.

Solution   The inside function is $x^4 + 5$, with derivative $4x^3$. The integrand has a factor of x^3, and since the only thing missing is a constant factor, we try

$$w = x^4 + 5.$$

Then

$$dw = w'(x) \, dx = 4x^3 \, dx,$$

giving

$$\frac{1}{4} dw = x^3 \, dx.$$

Thus,

$$\int x^3 \sqrt{x^4 + 5} \, dx = \int \sqrt{w} \, \frac{1}{4} dw = \frac{1}{4} \int w^{1/2} \, dw = \frac{1}{4} \cdot \frac{w^{3/2}}{3/2} + C = \frac{1}{6}(x^4 + 5)^{3/2} + C.$$

Warning

We saw in the preceding examples that we can apply the substitution method when a *constant* factor is missing from the derivative of the inside function. However, we may not be able to use substitution if anything other than a constant factor is missing. For example, setting $w = x^4 + 5$ to find

$$\int x^2 \sqrt{x^4 + 5} \, dx$$

does us no good because $x^2 \, dx$ is not a constant multiple of $dw = 4x^3 \, dx$. In order to use substitution, it helps if the integrand contains the derivative of the inside function, *to within a constant factor*.

Example 5   Find $\displaystyle\int \frac{t^2}{1 + t^3} \, dt$.

Solution   Observing that the derivative of $1 + t^3$ is $3t^2$, we take $w = 1 + t^3$, $dw = 3t^2 \, dt$, so $\frac{1}{3} dw = t^2 \, dt$. Thus,

$$\int \frac{t^2}{1 + t^3} \, dt = \int \frac{\frac{1}{3} dw}{w} = \frac{1}{3} \ln |w| + C = \frac{1}{3} \ln |1 + t^3| + C.$$

Since the numerator is $t^2 \, dt$, we might have tried $w = t^3$. This substitution leads to the integral $\frac{1}{3} \int 1/(1 + w) \, dw$. To evaluate this integral we would have to make a second substitution $u = 1 + w$. There is often more than one way to do an integral by substitution.

Using Substitution with Periodic Functions

The method of substitution can be used for integrals involving periodic functions.

Example 6   Find $\int 3x^2 \cos(x^3)\,dx$.

Solution   We look for an inside function whose derivative appears—in this case x^3. We let $w = x^3$. Then $dw = w'(x)\,dx = 3x^2\,dx$. The original integrand can now be completely rewritten in terms of the new variable w:

$$\int 3x^2 \cos(x^3)\,dx = \int \cos\underbrace{(x^3)}_{w} \cdot \underbrace{3x^2\,dx}_{dw} = \int \cos w\,dw = \sin w + C = \sin(x^3) + C.$$

By changing the variable to w, we have simplified the integrand to $\cos w$, which can be antidifferentiated more easily. The final step, after antidifferentiating, is to convert back to the original variable, x.

Example 7   Find $\int e^{\cos\theta} \sin\theta\,d\theta$.

Solution   We let $w = \cos\theta$ since its derivative is $-\sin\theta$ and there is a factor of $\sin\theta$ in the integrand. This gives

$$dw = w'(\theta)\,d\theta = -\sin\theta\,d\theta,$$

so

$$-dw = \sin\theta\,d\theta.$$

Thus,

$$\int e^{\cos\theta} \sin\theta\,d\theta = \int e^w(-dw) = (-1)\int e^w\,dw = -e^w + C = -e^{\cos\theta} + C.$$

Definite Integrals by Substitution

The following example shows two ways of computing a definite integral by substitution.

Example 8   Compute $\int_0^2 xe^{x^2}\,dx$.

Solution   To evaluate this definite integral using the Fundamental Theorem of Calculus, we first need to find an antiderivative of $f(x) = xe^{x^2}$. The inside function is x^2, so we let $w = x^2$. Then $dw = 2x\,dx$, so $\frac{1}{2}\,dw = x\,dx$. Thus,

$$\int xe^{x^2}\,dx = \int e^w \frac{1}{2}\,dw = \frac{1}{2}e^w + C = \frac{1}{2}e^{x^2} + C.$$

Now we find the definite integral:

$$\int_0^2 xe^{x^2}\,dx = \frac{1}{2}e^{x^2}\Big|_0^2 = \frac{1}{2}(e^4 - e^0) = \frac{1}{2}(e^4 - 1).$$

There is another way to look at the same problem. After we established that

$$\int xe^{x^2}\, dx = \frac{1}{2}e^w + C,$$

our next two steps were to replace w by x^2, and then x by 2 and 0. We could have directly replaced the original limits of integration, $x = 0$ and $x = 2$, by the corresponding w limits. Since $w = x^2$, the w limits are $w = 0^2 = 0$ (when $x = 0$) and $w = 2^2 = 4$ (when $x = 2$), so we get

$$\int_{x=0}^{x=2} xe^{x^2}\, dx = \frac{1}{2}\int_{w=0}^{w=4} e^w\, dw = \frac{1}{2}e^w\Big|_0^4 = \frac{1}{2}\left(e^4 - e^0\right) = \frac{1}{2}(e^4 - 1).$$

As we would expect, both methods give the same answer.

To Use Substitution to Find Definite Integrals

Either
- Compute the indefinite integral, expressing an antiderivative in terms of the original variable, and then evaluate the result at the original limits,

or
- Convert the original limits to new limits in terms of the new variable and do not convert the antiderivative back to the original variable.

Problems for Section 6.6

1. (a) Find the derivatives of $\sin(x^2 + 1)$ and $\sin(x^3 + 1)$.
  (b) Use your answer to part (a) to find antiderivatives of:
    (i) $x\cos(x^2 + 1)$ (ii) $x^2\cos(x^3 + 1)$
  (c) Find the general antiderivatives of:
    (i) $x\sin(x^2 + 1)$ (ii) $x^2\sin(x^3 + 1)$

■ In Problems **2–5**, explain how you can tell if substitution can be used to find an antiderivative.

2. $\displaystyle\int x(1 - 5x^2)^5\, dx$ **3.** $\displaystyle\int \frac{\sqrt{\ln x}}{x}\, dx$

4. $\displaystyle\int \frac{x}{\sqrt{\ln x}}\, dx$ **5.** $\displaystyle\int \sin^9 t\, \cos t\, dt$

■ Find the integrals in Problems **6–48**. Check your answers by differentiation.

6. $\displaystyle\int 2x(x^2 + 1)^5 dx$ **7.** $\displaystyle\int \frac{x}{\sqrt{x^2 + 4}}\, dx$

8. $\displaystyle\int (5x - 7)^{10} dx$ **9.** $\displaystyle\int x\sqrt{x^2 + 1}\, dx$

10. $\displaystyle\int 2qe^{q^2 + 1}\, dq$ **11.** $\displaystyle\int 5e^{5t + 2}\, dt$

12. $\displaystyle\int xe^{-x^2}\, dx$ **13.** $\displaystyle\int e^{-0.1t + 4}\, dt$

14. $\displaystyle\int 100e^{-0.2t}\, dt$ **15.** $\displaystyle\int t^2(t^3 - 3)^{10}\, dt$

16. $\displaystyle\int x\sin(x^2)\, dx$ **17.** $\displaystyle\int x(x^2 - 4)^{7/2}\, dx$

18. $\displaystyle\int x(x^2 + 3)^2\, dx$ **19.** $\displaystyle\int \frac{1}{(3x + 1)^2}\, dx$

20. $\displaystyle\int \frac{4x^3}{x^4 + 1}\, dx$ **21.** $\displaystyle\int 12x^2\cos(x^3)\, dx$

22. $\displaystyle\int (2t - 7)^{73}\, dt$ **23.** $\displaystyle\int (x^2 + 3)^2\, dx$

24. $\displaystyle\int y^2(1 + y)^2\, dy$ **25.** $\displaystyle\int \sin\theta(\cos\theta + 5)^7\, d\theta$

26. $\displaystyle\int \sin^6\theta\cos\theta\, d\theta$ **27.** $\displaystyle\int \sqrt{\cos 3t}\,\sin 3t\, dt$

28. $\int \sin(3-t)\,dt$

29. $\int \dfrac{t}{1+3t^2}\,dt$

30. $\int x^2 e^{x^3+1}\,dx$

31. $\int \sin^3\alpha\cos\alpha\,d\alpha$

32. $\int x\sin(4x^2)\,dx$

33. $\int \sin^2 x\cos x\,dx$

34. $\int e^{3x-4}\,dx$

35. $\int \dfrac{1+e^x}{\sqrt{x+e^x}}\,dx$

36. $\int xe^{3x^2}\,dx$

37. $\int x\sqrt{3x^2+4}\,dx$

38. $\int \dfrac{e^x-e^{-x}}{e^x+e^{-x}}\,dx$

39. $\int \dfrac{(\ln z)^2}{z}\,dz$

40. $\int \dfrac{y}{y^2+4}\,dy$

41. $\int \dfrac{q}{5q^2+8}\,dq$

42. $\int \dfrac{e^t+1}{e^t+t}\,dt$

43. $\int \dfrac{e^{\sqrt{y}}}{\sqrt{y}}\,dy$

44. $\int \dfrac{\cos\sqrt{x}}{\sqrt{x}}\,dx$

45. $\int \dfrac{x+1}{x^2+2x+19}\,dx$

46. $\int \dfrac{e^t}{e^t+1}\,dt$

47. $\int \sin^6(5\theta)\cos(5\theta)\,d\theta$

48. $\int \dfrac{x\cos(x^2)}{\sqrt{\sin(x^2)}}\,dx$

49. If appropriate, evaluate the following integrals by substitution. If substitution is not appropriate, say so, and do not evaluate.

(a) $\int x\sin(x^2)\,dx$ (b) $\int x^2\sin x\,dx$

(c) $\int \dfrac{x^2}{1+x^2}\,dx$ (d) $\int \dfrac{x}{(1+x^2)^2}\,dx$

(e) $\int x^3 e^{x^2}\,dx$ (f) $\int \dfrac{\sin x}{2+\cos x}\,dx$

50. Use substitution to express each of the following integrals as a multiple of $\int_a^b (1/w)\,dw$ for some a and b. Then evaluate the integrals.

(a) $\int_0^1 \dfrac{x}{1+x^2}\,dx$ (b) $\int_0^{\pi/4} \dfrac{\sin x}{\cos x}\,dx$

■ Use integration by substitution and the Fundamental Theorem to evaluate the definite integrals in Problems 51–60.

51. $\int_0^2 x(x^2+1)^2\,dx$

52. $\int_0^3 \dfrac{2x}{x^2+1}\,dx$

53. $\int_0^{\pi/2} e^{-\cos\theta}\sin\theta\,d\theta$

54. $\int_1^4 \dfrac{e^{\sqrt{x}}}{\sqrt{x}}\,dx$

55. $\int_0^3 \dfrac{1}{\sqrt{t+1}}\,dt$

56. $\int_{-1}^{e-2} \dfrac{1}{t+2}\,dt$

57. $\int_0^2 \dfrac{x}{(1+x^2)^2}\,dx$

58. $\int_0^1 2te^{-t^2}\,dt$

59. $\int_{-1}^2 \sqrt{x+2}\,dx$

60. $\int_1^3 \dfrac{dt}{(t+7)^2}$

■ In Problems 61–62, find the exact area.

61. Under $f(x)=xe^{x^2}$ between $x=0$ and $x=2$.

62. Under $f(x)=1/(x+1)$ between $x=0$ and $x=2$.

63. Find the exact average value of $f(x)=1/(x+1)$ on the interval $x=0$ to $x=2$. Sketch a graph showing the function and the average value.

64. Suppose $\int_0^2 g(t)\,dt=5$. Calculate the following:

(a) $\int_0^4 g(t/2)\,dt$ (b) $\int_0^2 g(2-t)\,dt$

65. (a) Find $\int (x+5)^2\,dx$ in two ways:
   (i) By multiplying out
   (ii) By substituting $w=x+5$

(b) Are the results the same? Explain.

66. Find $\int 4x(x^2+1)\,dx$ using two methods:

(a) Do the multiplication first, and then antidifferentiate.

(b) Use the substitution $w=x^2+1$.

(c) Explain how the expressions from parts (a) and (b) are different. Are they both correct?

67. (a) Find $\int \sin\theta\cos\theta\,d\theta$.

(b) You probably solved part (a) by making the substitution $w=\sin\theta$ or $w=\cos\theta$. (If not, go back and do it that way.) Now find $\int \sin\theta\cos\theta\,d\theta$ by making the *other* substitution.

(c) There is yet another way of finding this integral which involves the trigonometric identities

$$\sin(2\theta)=2\sin\theta\cos\theta$$
$$\cos(2\theta)=\cos^2\theta-\sin^2\theta.$$

Find $\int \sin\theta\cos\theta\,d\theta$ using one of these identities and then the substitution $w=2\theta$.

(d) You should now have three different expressions for the indefinite integral $\int \sin\theta\cos\theta\,d\theta$. Are they really different? Are they all correct? Explain.

68. At the start of 2014, the world's known copper reserves were 690 million tons. With t in years since the start of 2014, copper has been mined at a rate given by $17.9e^{0.025t}$ million tons per year.[10]

(a) Assuming that copper continues to be extracted at the same rate, write an expression for the total quantity mined in the first T years after the start of 2014.

(b) Under these assumptions, when is the world predicted to run out of copper?

6.7 INTEGRATION BY PARTS

Now we introduce *integration by parts*, which is a technique for finding integrals based on the product rule. We begin with the product rule:

$$\frac{d}{dx}(uv) = u'v + uv'$$

where u and v are functions of x with derivatives u' and v', respectively. We rewrite this as:

$$uv' = \frac{d}{dx}(uv) - u'v$$

and then integrate both sides:

$$\int uv' \, dx = \int \frac{d}{dx}(uv) \, dx - \int u'v \, dx.$$

Since an antiderivative of $\frac{d}{dx}(uv)$ is just uv, we get the following formula:

Integration by Parts

$$\int uv' \, dx = uv - \int u'v \, dx.$$

This technique is useful when the integrand can be viewed as a product and when the integral on the right-hand side is simpler than that on the left.

Example 1   Use integration by parts to find $\displaystyle\int xe^x \, dx$.

Solution   We let $xe^x = (x) \cdot (e^x) = uv'$, and choose $u = x$ and $v' = e^x$. Thus, $u' = 1$ and $v = e^x$, so

$$\int \underbrace{(x)}_{u} \underbrace{(e^x)}_{v'} \, dx = \underbrace{(x)}_{u} \underbrace{(e^x)}_{v} - \int \underbrace{(1)}_{u'} \underbrace{(e^x)}_{v} \, dx = xe^x - e^x + C.$$

Let's look again at Example 1. Notice what would have happened if we took $v = e^x + C_1$. Then

$$\int xe^x \, dx = x(e^x + C_1) - \int (e^x + C_1) \, dx$$

$$= xe^x + C_1 x - e^x - C_1 x + C$$

$$= xe^x - e^x + C,$$

---

[10]Data from http://minerals.usgs.gov/minerals/pubs/commodity/ Accessed February 8, 2015.

as before. Thus, it is not necessary to include an arbitrary constant in the antiderivative for v; any antiderivative will do.

What would have happened if we had picked u and v' the other way around in Example 1? If $u = e^x$ and $v' = x$, then $u' = e^x$ and $v = x^2/2$. The formula for integration by parts then gives

$$\int xe^x \, dx = \frac{x^2}{2}e^x - \int \frac{x^2}{2} \cdot e^x \, dx,$$

which is true but not helpful, since the new integral on the right seems harder than the original one on the left. To use this method, we must choose u and v' to make the integral on the right no harder to find than the integral on the left.

> ### How to Choose u and v'
>
> - Whatever you let v' be, you need to be able to find v.
> - It helps if $u'v$ is simpler (or at least no more complicated) than uv'.

Example 2   Find $\displaystyle\int xe^{3x} \, dx$.

Solution   We let $xe^{3x} = (x) \cdot (e^{3x}) = uv'$, and choose $u = x$ and $v' = e^{3x}$. Thus, $u' = 1$ and $v = \frac{1}{3}e^{3x}$. We have

$$\int \underbrace{(x)}_{u}\, \underbrace{(e^{3x})}_{v'} \, dx = \underbrace{(x)}_{u}\, \underbrace{\left(\frac{1}{3}e^{3x}\right)}_{v} - \int \underbrace{(1)}_{u'}\, \underbrace{\left(\frac{1}{3}e^{3x}\right)}_{v} \, dx = \frac{1}{3}xe^{3x} - \frac{1}{9}e^{3x} + C.$$

There are some examples which don't look like good candidates for integration by parts because they don't appear to involve products, but for which the method works well. Such examples often involve $\ln x$. Here is one:

Example 3   Find $\displaystyle\int_2^3 \ln x \, dx$.

Solution   The integrand does not look like a product unless we write $\ln x = (1)(\ln x)$. We might say $u = 1$ so $u' = 0$, which certainly makes things simpler. But if $v' = \ln x$, what is v? If we knew, we would not need integration by parts. Let's try the other way: if $u = \ln x$, $u' = 1/x$ and if $v' = 1$, $v = x$, so

$$\int_2^3 \underbrace{(\ln x)}_{u}\, \underbrace{(1)}_{v'} \, dx = \underbrace{(\ln x)}_{u}\, \underbrace{(x)}_{v}\, \Big|_2^3 - \int_2^3 \underbrace{\left(\frac{1}{x}\right)}_{u'} \cdot \underbrace{(x)}_{v} \, dx$$

$$= x \ln x \Big|_2^3 - \int_2^3 1 \, dx = (x \ln x - x)\Big|_2^3$$

$$= 3 \ln 3 - 3 - 2 \ln 2 + 2 = 3 \ln 3 - 2 \ln 2 - 1.$$

Notice that when doing a definite integral by parts, we must remember to put the limits of integration (here 2 and 3) on the uv term (in this case $x \ln x$) as well as on the integral $\int u'v \, dx$.

Example 4   Find $\int x^6 \ln x \, dx$.

Solution   We can write $x^6 \ln x$ as uv' where $u = \ln x$ and $v' = x^6$. Then $v = \frac{1}{7}x^7$ and $u' = 1/x$, so integration by parts gives us:

$$\int x^6 \ln x \, dx = \int (\ln x)x^6 \, dx = (\ln x)\left(\frac{1}{7}x^7\right) - \int \frac{1}{7}x^7 \cdot \frac{1}{x}\, dx$$

$$= \frac{1}{7}x^7 \ln x - \frac{1}{7}\int x^6 \, dx$$

$$= \frac{1}{7}x^7 \ln x - \frac{1}{49}x^7 + C.$$

In Example 4 we did not choose $v' = \ln x$, because it is not immediately clear what v would be. In fact, we used integration by parts in Example 3 to find the antiderivative of $\ln x$. Also, using $u = \ln x$, as we have done, gives $u'v = x^6/7$, which is simpler to integrate than $uv' = x^6 \ln x$. This example shows that u does not have to be the first factor in the integrand (here x^6).

Problems for Section 6.7

■ Find the integrals in Problems 1–16.

1. $\int te^{5t} \, dt$

2. $\int pe^{-0.1p} \, dp$

3. $\int y \ln y \, dy$

4. $\int (z + 1)e^{2z} \, dz$

5. $\int q^5 \ln 5q \, dq$

6. $\int y\sqrt{y+3}\, dy$

7. $\int x^3 \ln x \, dx$

8. $\int (t+2)\sqrt{2+3t}\, dt$

9. $\int \frac{y}{\sqrt{5-y}}\, dy$

10. $\int \frac{z}{e^z}\, dz$

11. $\int \frac{t+7}{\sqrt{5-t}}\, dt$

12. $\int \frac{\ln x}{x^2}\, dx$

13. $\int t \sin t \, dt$

14. $\int (\theta+1)\sin(\theta+1)\, d\theta$

15. $\int \sqrt{x} \ln x \, dx$

16. $\int y\sqrt{1-y}\, dy$

17. Find $\int_1^2 \ln x \, dx$ numerically. Find $\int_1^2 \ln x \, dx$ using antiderivatives. Check that your answers agree.

■ Evaluate the integrals in Problems 18–22 both exactly [e.g. $\ln(3\pi)$] and numerically [e.g. $\ln(3\pi) \approx 2.243$].

18. $\int_0^{10} ze^{-z}\, dz$

19. $\int_1^5 \ln t \, dt$

20. $\int_0^5 \ln(1+t)\, dt$

21. $\int_1^3 t \ln t \, dt$

22. $\int_3^5 x \cos x \, dx$

■ In Problems 23–24, use integration by parts twice to evaluate the integral.

23. $\int (\ln t)^2 \, dt$

24. $\int t^2 e^{5t} \, dt$

25. For each of the following integrals, indicate whether integration by substitution or integration by parts is more appropriate. Do not evaluate the integrals.

(a) $\int \frac{x^2}{1+x^3}\, dx$ **(b)** $\int xe^{x^2}\, dx$

(c) $\int x^2 \ln(x^3 + 1)\, dx$ **(d)** $\int \frac{1}{\sqrt{3x+1}}\, dx$

(e) $\int x^2 \ln x \, dx$ **(f)** $\int \ln x \, dx$

■ In Problems 26–27, find the exact area.

26. Under $y = te^{-t}$ for $0 \le t \le 2$.

27. Between $y = \ln x$ and $y = \ln(x^2)$ for $1 \le x \le 2$.

28. The concentration, C, in ng/ml, of a drug in the blood as a function of the time, t, in hours since the drug was administered is given by $C = 15te^{-0.2t}$. The area under the concentration curve is a measure of the overall exposure of a person to the drug. Find the total exposure provided by the drug between $t = 0$ and $t = 3$.

29. During a surge in the demand for electricity, the rate, r, at which energy is used can be approximated by

$$r = te^{-at},$$

where t is the time in hours and a is a positive constant.

(a) Find the total energy, E, used in the first T hours. Give your answer as a function of a.

(b) What happens to E as $T \to \infty$?

PROJECTS FOR CHAPTER SIX

1. Quabbin Reservoir The Quabbin Reservoir in the western part of Massachusetts provides most of Boston's water. The graph in Figure 6.43 represents the flow of water in and out of the Quabbin Reservoir throughout 2016.

(a) Sketch a graph of the quantity of water in the reservoir, as a function of time.

(b) When, in the course of 2016, was the quantity of water in the reservoir largest? Smallest? Mark and label these points on the graph you drew in part (a).

(c) When was the quantity of water increasing most rapidly? Decreasing most rapidly? Mark and label these times on both graphs.

(d) By July 2017 the quantity of water in the reservoir was about the same as in January 2016. Draw plausible graphs for the flow into and the flow out of the reservoir for the first half of 2017.

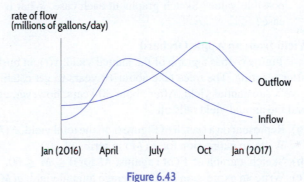

Figure 6.43

2. Distribution of Resources

Whether a resource is distributed evenly among members of a population is often an important political or economic question. How can we measure this? How can we decide if the distribution of wealth in this country is becoming more or less equitable over time? How can we measure which country has the most equitable income distribution? This problem describes a way of making such measurements. Suppose the resource is distributed evenly. Then any 20% of the population will have 20% of the resource. Similarly, any 30% will have 30% of the resource and so on. If, however, the resource is not distributed evenly, the poorest $p\%$ of the population (in terms of this resource) will not have $p\%$ of the goods. Suppose $F(x)$ represents the fraction of the resource owned by the poorest fraction x of the population. Thus $F(0.4) = 0.1$ means that the poorest 40% of the population owns 10% of the resource.

(a) What would F be if the resource were distributed evenly?

(b) What must be true of any such F? What must $F(0)$ and $F(1)$ equal? Is F increasing or decreasing? Is the graph of F concave up or concave down?

(c) Gini's index of inequality, G, is one way to measure how evenly the resource is distributed. It is defined by

$$G = 2 \int_0^1 (x - F(x))\, dx.$$

Show graphically what G represents.

(d) Graphical representations of Gini's index for two countries are given in Figures 6.44 and 6.45. Which country has the more equitable distribution of wealth? Discuss the distribution of wealth in each of the two countries.

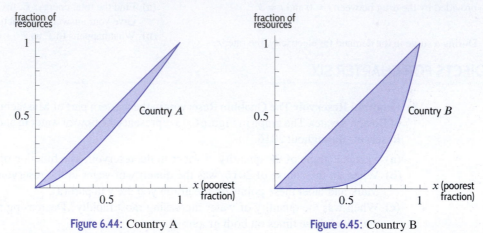

Figure 6.44: Country A   Figure 6.45: Country B

(e) What is the maximum possible value of Gini's index of inequality, G? What is the minimum possible value? Sketch graphs in each case. What is the distribution of resources in each case?

3. Yield from an Apple Orchard

Figure 6.46 is a graph of the annual yield, $y(t)$, in bushels per year, from an orchard t years after planting. The trees take about 10 years to get established, but for the next 20 years they give a substantial yield. After about 30 years, however, age and disease start to take their toll, and the annual yield falls off.[11]

(a) Represent on a sketch of Figure 6.46 the total yield, $F(M)$, up to M years, with $0 \le M \le 60$. Write an expression for $F(M)$ in terms of $y(t)$.

(b) Sketch a graph of $F(M)$ against M for $0 \le M \le 60$.

(c) Write an expression for the average annual yield, $a(M)$, up to M years.

(d) When should the orchard be cut down and replanted? Assume that we want to maximize average revenue per year, and that fruit prices remain constant, so that this is achieved by maximizing average annual yield. Use the graph of $y(t)$ to estimate the time at which the average annual yield is a maximum. Explain your answer geometrically and symbolically.

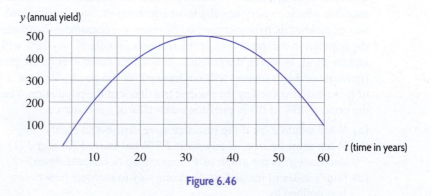

Figure 6.46

[11] From Peter D. Taylor, *Calculus: The Analysis of Functions* (Toronto: Wall & Emerson, Inc., 1992).

FOCUS ON PRACTICE

■ For Problems 1–47, evaluate the integrals. Assume a, b, A, B, P_0, h, and k are constants.

1. $\int y(y^2 + 5)^8 \, dy$

2. $\int (q^3 + 8q + 15) \, dq$

3. $\int (u^4 + 5) \, du$

4. $\int (x^2 + 1) \, dx$

5. $\int e^{-3t} \, dt$

6. $\int (6\sqrt{x}) \, dx$

7. $\int (ax^2 + b) \, dx$

8. $\int \frac{dq}{\sqrt{q}}$

9. $\int (x^3 + 4x + 8) \, dx$

10. $\int 100e^{-0.5t} \, dt$

11. $\int \left(q + \frac{1}{q^3}\right) dq$

12. $\int (w^4 - 12w^3 + 6w^2 - 10) \, dw$

13. $\int (q^2 + 5q + 2) \, dq$

14. $\int (Ax^3 + Bx) \, dx$

15. $\int \left(\frac{4}{x} + \frac{5}{x^2}\right) dx$

16. $\int \left(\frac{6}{\sqrt{x}} + 8\sqrt{x}\right) dx$

17. $\int 3\sin\theta \, d\theta$

18. $\int 30e^{-0.2t} \, dt$

19. $\int (p^2 + \frac{5}{p}) \, dp$

20. $\int 1000e^{0.075t} \, dt$

21. $\int (5\sin x + 3\cos x) \, dx$

22. $\int (10 + 5\sin x) \, dx$

23. $\int (Aq + B) \, dq$

24. $\int \frac{5}{w} \, dw$

25. $\int \pi r^2 h \, dr$

26. $\int xe^x \, dx$

27. $\int 15p^2 q^4 \, dp$

28. $\int 15p^2 q^4 \, dq$

29. $\int (3x^2 + 6e^{2x}) \, dx$

30. $\int 5e^{2q} \, dq$

31. $\int \left(p^3 + \frac{1}{p}\right) dp$

32. $\int 12\cos(4x) \, dx$

33. $\int \frac{1}{y+2} \, dy$

34. $\int (6\sqrt{x} + 15) \, dx$

35. $\int (x^2 + 8 + e^x) \, dx$

36. $\int (t^2 - 6t + 5) \, dt$

37. $\int \left(\frac{a}{x} + \frac{b}{x^2}\right) dx$

38. $\int_0^{10} ze^{-z} \, dz$

39. $\int (e^{2t} + 5) \, dt$

40. $\int \cos(4x) \, dx$

41. $\int P_0 e^{kt} \, dt$

42. $\int \sin(3x) \, dx$

43. $\int A\sin(Bt) \, dt$

44. $\int \sqrt{3x+1} \, dx$

45. $\int \frac{e^x}{2 + e^x} \, dx$

46. $\int \frac{\cos x}{\sqrt{1 + \sin x}} \, dx$

47. $\int x\ln x \, dx$

Chapter 7

PROBABILITY

CONTENTS

7.1   DENSITY FUNCTIONS

Understanding the distribution of various quantities through the population can be important to decision makers. For example, the income distribution gives useful information about the economic structure of a society. In this section we look at the distribution of ages in the US. To allocate funding for education, health care, and social security, the government needs to know how many people are in each age group. We see how to represent such information by a density function.

US Age Distribution

Table 7.1 *Distribution of ages in the US in 2012*

Age group	Fraction of total population
0–20	27% = 0.27
20–40	27% = 0.27
40–60	28% = 0.28
60–80	15% = 0.15
80–100	3% = 0.03

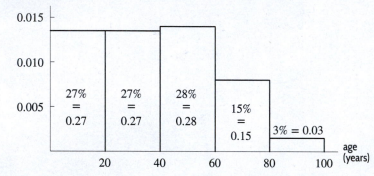

Figure 7.1: How ages were distributed in the US in 2012

Suppose we have the data in Table 7.1 showing how the ages of the US population were distributed in 2012.[1] To represent this information graphically we use a type of *histogram*[2] putting a vertical bar above each age group in such a way that the *area* of each bar represents the percentage in that age group. The total area of all the rectangles is 100% = 1. We only consider people who are less than 100 years old.[3] For the 0–20 age group, the base of the rectangle is 20, and we want the area to be 27%, so the height must be 27%/20 = 1.35%. We treat ages as though they were continuously distributed. The category 0–20, for example, contains people who are just one day short of their twentieth birthday. Notice that the vertical axis is measured in percent/year. (See Figure 7.1.)

Example 1   In 2012, estimate what percentage of the US population was:

(a)   Between 20 and 60 years old?   (b)   Less than 10 years old?

(c)   Between 75 and 80 years old?   (d)   Between 80 and 85 years old?

Solution   (a)   We add the percentages, so 27% + 28% = 55%.

(b)   To find the percentage less than 10 years old, we could assume, for example, that the population was distributed evenly over the 0–20 group. (This means we are assuming that babies were born at a fairly constant rate over the last 20 years, which is probably reasonable.) If we make this assumption, then we can say that the population less than 10 years old was about half that in the 0–20 group, that is, 13.5%. Notice that we get the same result by computing the area of the rectangle from 0 to 10. (See Figure 7.2.)

(c)   To find the population between 75 and 80 years old, since 15% of Americans in 2012 were in the 60-80 group, we might apply the same reasoning and say that $\frac{1}{4}(15\%) = 3.75\%$ of the population was in this age group. This result is represented as an area in Figure 7.2. The assumption that the population was evenly distributed is not a good one here; certainly there were more people between the ages of 60 and 65 than between 75 and 80. Thus, the estimate of 3.75% is certainly too high.

---

[1] www.census.gov, accessed April 22, 2015.

[2] There are other types of histograms which have frequency on the vertical axis.

[3] In fact, 0.02% of the population is over 100, but this is too small to be visible on the histogram.

(d) Again using the (faulty) assumption that ages in each group were distributed uniformly, we would find that the percentage between 80 and 85 was $\frac{1}{4}(3\%) = 0.75\%$. (See Figure 7.2.) This estimate is also poor—there were certainly more people in the 80–85 group than, say, the 95–100 group, and so the 0.75% estimate is too low.

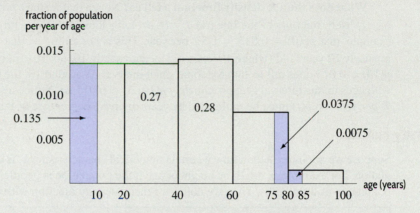

Figure 7.2: Ages in the US in 2012—various subgroups (for Example 1)

Smoothing Out the Histogram

We could get better estimates if we had smaller age groups (each age group in Figure 7.1 is 20 years, which is quite large) or if the histogram were smoother. Suppose we have the more detailed data in Table 7.2, which leads to the new histogram in Figure 7.3.

As we get more detailed information, the upper silhouette of the histogram becomes smoother, but the area of any of the bars still represents the percentage of the population in that age group. Imagine, in the limit, replacing the upper silhouette of the histogram by a smooth curve in such a way that area under the curve above one age group is the same as the area in the corresponding rectangle. The total area under the whole curve is again $100\% = 1$. (See Figure 7.3.)

The Age Density Function

If t is age in years, we define $p(t)$, the age *density function*, to be a function which "smooths out" the age histogram. This function has the property that

$$\text{Fraction of population between ages } a \text{ and } b = \text{Area under graph of } p \text{ between } a \text{ and } b = \int_a^b p(t)\,dt.$$

Table 7.2 *Ages in the US in 2012 (more detailed)*

Age group	Fraction of total population
0–10	13% = 0.13
10–20	14% = 0.14
20–30	14% = 0.14
30–40	13% = 0.13
40–50	14% = 0.14
50–60	14% = 0.14
60–70	10% = 0.10
70–80	5% = 0.05
80–90	2% = 0.02
90–100	1% = 0.01

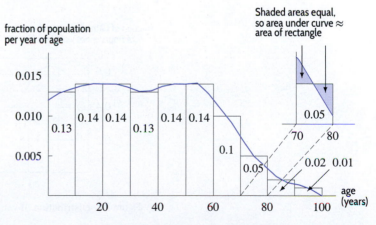

Figure 7.3: Smoothing out the age histogram

If a and b are the smallest and largest possible ages (say, $a = 0$ and $b = 100$), so that the ages of all of the population are between a and b, then

$$\int_a^b p(t)dt = \int_0^{100} p(t)dt = 1.$$

What does the age density function p tell us? Notice that we have not talked about the meaning of $p(t)$ itself, but *only* of the integral $\int_a^b p(t)\,dt$. Let's look at this in a bit more detail. Suppose, for example, that $p(10) = 0.015 = 1.5\%$ per year. This is *not* telling us that 1.5% of the population is precisely 10 years old (where 10 years old means exactly 10, not $10\frac{1}{2}$, not $10\frac{1}{4}$, not 10.1). However, $p(10) = 0.015$ does tell us that for some small interval Δt around 10, the fraction of the population with ages in this interval is approximately $p(10)\,\Delta t = 0.015\,\Delta t$. Notice also that the units of $p(t)$ are *% per year*, so $p(t)$ must be multiplied by years to give a percentage of the population.

The Density Function

Suppose we are interested in how a certain numerical characteristic, x, is distributed through a population. For example, x might be height or age if the population is people, or might be wattage for a population of light bulbs. Then we define a general density function with the following properties:

> The function, $p(x)$, is a **density function** if
>
> $$\begin{array}{ccccc} \text{Fraction of population} & & \text{Area under} & & \\ \text{for which } x \text{ is} & = & \text{graph of } p & = & \int_a^b p(x)dx. \\ \text{between } a \text{ and } b & & \text{between } a \text{ and } b & & \end{array}$$
>
> $$\int_{-\infty}^{\infty} p(x)\,dx = 1 \quad \text{and} \quad p(x) \geq 0 \quad \text{for all } x.$$

The density function must be nonnegative if its integral always gives a fraction of the population. The fraction of the population with x between $-\infty$ and ∞ is 1 because the entire population has the characteristic x between $-\infty$ and ∞. The function p that was used to smooth out the age histogram satisfies this definition of a density function. Notice that we do not assign a meaning to the value $p(x)$ directly, but rather interpret $p(x)\,\Delta x$ as the fraction of the population with the characteristic in a short interval of length Δx around x.

Example 2 Figure 7.4 gives the density function for the amount of time spent waiting at a doctor's office.

(a) What is the longest time anyone has to wait?
(b) Approximately what fraction of patients wait between 1 and 2 hours?
(c) Approximately what fraction of patients wait less than an hour?

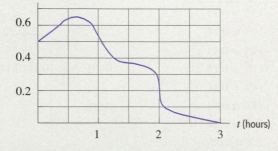

fraction of patients per number
of hours spent waiting

Figure 7.4: Distribution of waiting time at a doctor's office

Solution   (a) The density function is zero for all $t > 3$, so no one waits more than 3 hours. The longest time anyone has to wait is 3 hours.

(b) The fraction of patients who wait between 1 and 2 hours is equal to the area under the density curve between $t = 1$ and $t = 2$. We can estimate this area by counting squares: There are about 7.5 squares in this region, each of area $(0.5)(0.1) = 0.05$. The area is approximately $(7.5)(0.05) = 0.375$. Thus about 37.5% of patients wait between 1 and 2 hours.

(c) This fraction is equal to the area under the density function for $t < 1$. There are about 12 squares in this area, and each has area 0.05 as in part (b), so our estimate for the area is $(12)(0.05) = 0.60$. Therefore, about 60% of patients see the doctor in less than an hour.

Problems for Section 7.1

■ In Problems 1–4, the distribution of the heights, x, in meters, of trees is represented by the density function $p(x)$. In each case, calculate the fraction of trees which are:

(a) Less than 5 meters high

(b) More than 6 meters high

(c) Between 2 and 5 meters high

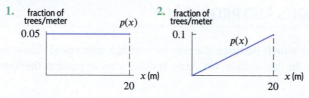

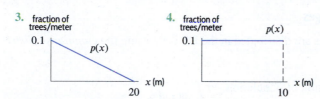

5. The density function $p(t)$ for the length of the larval stage, in days, for a breed of insect is given in Figure 7.5. What fraction of these insects are in the larval stage for between 10 and 12 days? For less than 8 days? For more than 12 days? In which one-day interval is the length of a larval stage most likely to fall?

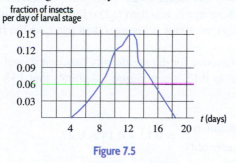

Figure 7.5

6. Figure 7.6[4] shows the distribution of elevation, in miles, across the earth's surface. Positive elevation denotes land above sea level; negative elevation shows land below sea level (that is, the ocean floor).

(a) Describe in words the elevation of most of the earth's surface.

(b) Approximately what fraction of the earth's surface is below sea level?

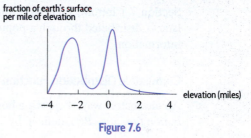

Figure 7.6

7. Let $p(x)$ be the density function for annual family income, where x is in thousands of dollars. What is the meaning of the statement $p(70) = 0.05$?

8. The density function for heights of American men, in inches is $p(x)$. What is the meaning of the statement $p(68) = 0.2$?

■ In Problems 9–12, calculate the value of c if p is a density function.

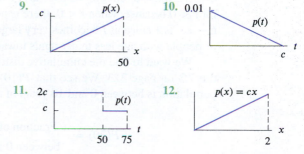

[4] Adapted from D. Freedman, R. Pisani, R. Purves, and A. Adikhari, *Statistics*, 2nd *Edition* (New York: Norton, 1991).

13. A machine lasts up to 10 years. Figure 7.7 shows the density function, $p(t)$, for the length of time it lasts.

    (a) What is the value of C?
    (b) Is a machine more likely to break in its first year or in its tenth year? In its first or second year?
    (c) What fraction of the machines lasts 2 years or less? Between 5 and 7 years? Between 3 and 6 years?

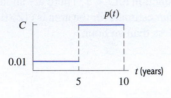

Figure 7.7

14. Find a density function $p(x)$ such that $p(x) = 0$ when $x \geq 5$ and when $x < 0$, and is decreasing when $0 \leq x \leq 5$.

■ In Problems **15–17**, graph a possible density function representing crop yield (in kilograms) from a field under the given circumstance.

15. All yields from 0 to 100 kg are equally likely; the field never yields more than 100 kg.

16. High yields are more likely than low. The maximum yield is 200 kg.

17. A drought makes low yields most common, and there is no yield greater than 30 kg.

18. Which of the following functions makes the most sense as a model for the probability density representing the time (in minutes, starting from $t = 0$) that the next customer walks into a store?

    (a) $p(t) = \begin{cases} \cos t & 0 \leq t \leq 2\pi \\ e^{t-2\pi} & t \geq 2\pi \end{cases}$
    (b) $p(t) = 3e^{-3t}$ for $t \geq 0$
    (c) $p(t) = e^{-3t}$ for $t \geq 0$
    (d) $p(t) = 1/4$ for $0 \leq t \leq 4$

7.2   CUMULATIVE DISTRIBUTION FUNCTIONS AND PROBABILITY

Section 7.1 introduced density functions which describe the way in which a numerical characteristic is distributed through a population. In this section we study another way to present the same information.

Cumulative Distribution Function for Ages

An alternative way of showing how ages are distributed in the US is by using the *cumulative distribution function* $P(t)$, defined by

$$P(t) = \frac{\text{Fraction of population}}{\text{of age less than } t} = \int_0^t p(x)\,dx.$$

Thus, P is the antiderivative of p with $P(0) = 0$, so $P(t)$ is the area under the density curve between 0 and t. See the left-hand part of Figure 7.8.

Notice that the cumulative distribution function is nonnegative and increasing (or at least nondecreasing), since the number of people younger than age t increases as t increases. Another way of seeing this is to notice that $P' = p$, and p is positive (or nonnegative). Thus the cumulative age distribution is a function which starts with $P(0) = 0$ and increases as t increases. We have $P(t) = 0$ for $t < 0$ because, when $t < 0$, there is no one whose age is less than t. The limiting value of P, as $t \to \infty$, is 1 since as t becomes very large (100 say), everyone is younger than age t, so the fraction of people with age less than t tends toward 1.

We want to find the cumulative distribution function for the age density function shown in Figure 7.3 on page 323. We see that $P(10)$ is equal to 0.13, since Figure 7.3 shows that 13% of the population is between 0 and 10 years of age. Also,

$$P(20) = \frac{\text{Fraction of the population}}{\text{between 0 and 20 years old}} = 0.13 + 0.14 = 0.27$$

and similarly

$$P(30) = 0.13 + 0.14 + 0.14 = 0.41.$$

Continuing in this way gives the values for $P(t)$ in Table 7.3. These values were used to graph $P(t)$ in the right-hand part of Figure 7.8.

Table 7.3 *Cumulative distribution function, $P(t)$, giving fraction of US population of age less than t years*

t	0	10	20	30	40	50	60	70	80	90	100
$P(t)$	0	0.13	0.27	0.41	0.54	0.68	0.82	0.92	0.97	0.99	1.00

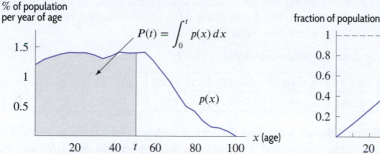

Figure 7.8: Graph of $p(x)$, the age density function, and its relation to $P(t)$, the cumulative age distribution function

Cumulative Distribution Function

A **cumulative distribution function**, $P(t)$, of a density function p, is defined by

$$P(t) = \int_{-\infty}^{t} p(x)\,dx = \begin{array}{l}\text{Fraction of population having}\\ \text{values of } x \text{ below } t.\end{array}$$

Thus, P is an antiderivative of p, that is, $P' = p$.
Any cumulative distribution function has the following properties:

- P is increasing (or nondecreasing).
- $\lim_{t \to \infty} P(t) = 1$ and $\lim_{t \to -\infty} P(t) = 0$.

- $\begin{array}{l}\text{Fraction of population}\\ \text{having values of } x\\ \text{between } a \text{ and } b\end{array} = \int_{a}^{b} p(x)\,dx = P(b) - P(a).$

Example 1   The time to conduct a routine maintenance check on a machine has a cumulative distribution function $P(t)$, which gives the fraction of maintenance checks completed in time less than or equal to t minutes. Values of $P(t)$ are given in Table 7.4.

Table 7.4   *Cumulative distribution function for time to conduct maintenance checks*

t (minutes)	0	5	10	15	20	25	30
$P(t)$ (fraction completed)	0	0.03	0.08	0.21	0.38	0.80	0.98

(a) What fraction of maintenance checks are completed in 15 minutes or less?
(b) What fraction of maintenance checks take longer than 30 minutes?
(c) What fraction take between 10 and 15 minutes?
(d) Draw a histogram showing how times for maintenance checks are distributed.
(e) In which of the given 5-minute intervals is the length of a maintenance check most likely to fall?
(f) Give a rough sketch of the density function.
(g) Sketch a graph of the cumulative distribution function.

Solution   (a) The fraction of maintenance checks completed in 15 minutes is $P(15) = 0.21$, or 21%.

(b) Since $P(30) = 0.98$, we see that 98% of maintenance checks take 30 minutes or less. Therefore, only 2% take more than 30 minutes.

(c) Since 8% take 10 minutes or less and 21% take 15 minutes or less, the fraction taking between 10 and 15 minutes is $0.21 - 0.08 = 0.13$, or 13%.

(d) We begin by making a table showing how the times are distributed. Table 7.4 shows that the fraction of checks completed between 0 and 5 minutes is 0.03, and the fraction completed between 5 and 10 minutes is 0.05, and so on. See Table 7.5.

Table 7.5   *Distribution of time to conduct maintenance checks*

t (minutes)	0 – 5	5 – 10	10 – 15	15 – 20	20 – 25	25 – 30	> 30
Fraction completed	0.03	0.05	0.13	0.17	0.42	0.18	0.02

The histogram in Figure 7.9 is drawn so that the area of each bar is the fraction of checks completed in the corresponding time period. For instance, the first bar has area 0.03 and width 5 minutes, so its height is $0.03/5 = 0.006$.

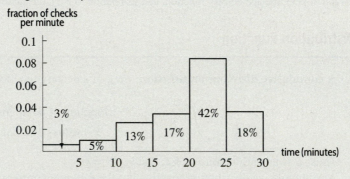

Figure 7.9: Histogram of times for maintenance checks

(e) From Figure 7.9, we see that more of the checks take between 20 and 25 minutes to complete, so this is the most likely length of time.

(f) The density function, $p(t)$, is a smoothed version of the histogram in Figure 7.9. A reasonable sketch is given in Figure 7.10.

(g) A graph of $P(t)$ is given in Figure 7.11. Since $P(t)$ is a cumulative distribution function, $P(t)$ is approaching 1 as t gets large, but is never larger than 1.

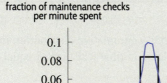
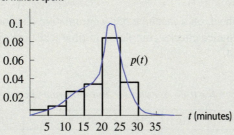

Figure 7.10: Density function for time to conduct maintenance checks

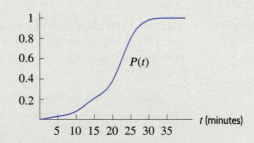

Figure 7.11: Cumulative distribution function for time to conduct maintenance checks

Probability

Suppose we pick a member of the US population at random. What is the probability that we pick a person who is between, say, the ages of 70 and 80? We saw in Table 7.2 on page 323 that 5% of the population is in this age group. We say that the probability, or chance, that the person is between 70 and 80 is 0.05. Using any age density function $p(t)$, we define probabilities as follows:

$$\begin{array}{ccc}
\text{Probability that a person is} & & \text{Fraction of population} \\
\text{between ages } a \text{ and } b & = & \text{between ages } a \text{ and } b
\end{array} = \int_a^b p(t)\, dt.$$

Since the cumulative distribution gives the fraction of the population younger than age t, the cumulative distribution function can also be used to calculate the probability that a randomly selected person is in a given age group.

$$\begin{array}{ccc}
\text{Probability that a person is} & & \text{Fraction of population} \\
\text{younger than age } t & = & \text{younger than age } t
\end{array} = P(t) = \int_0^t p(x)\, dx.$$

In the next example, both a density function and a cumulative distribution function are used to describe the same situation.

Example 2 Suppose you want to analyze the fishing industry in a small town. Each day, the boats bring back at least 2 tons of fish, but never more than 8 tons.

(a) Using the density function describing the daily catch in Figure 7.12, find and graph the corresponding cumulative distribution function and explain its meaning.

(b) What is the probability that the catch is between 5 and 7 tons?

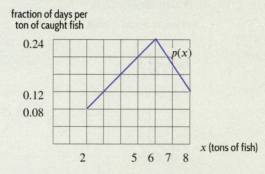

fraction of days per
ton of caught fish

Figure 7.12: Density function of daily catch

Solution (a) The cumulative distribution function $P(t)$ is equal to the fraction of days on which the catch is less than t tons of fish. Since the catch is never less than 2 tons, we have $P(t) = 0$ for $t \leq 2$. Since the catch is always less than 8 tons, we have $P(t) = 1$ for $t \geq 8$. For t in the range $2 < t < 8$ we must evaluate the integral

$$P(t) = \int_{-\infty}^t p(x)\,dx = \int_2^t p(x)\,dx.$$

This integral equals the area under the graph of $p(x)$ between $x = 2$ and $x = t$. It can be computed by counting grid squares in Figure 7.12; each square has area 0.04. For example,

$$P(3) = \int_2^3 p(x)\,dx \approx \text{Area of 2.5 squares} = 2.5(0.04) = 0.10.$$

Table 7.6 contains values of $P(t)$; the graph is shown in Figure 7.13.

Table 7.6   *Estimates for P(t) of daily catch*

t (tons of fish)	P(t) (fraction of fishing days)
2	0
3	0.10
4	0.24
5	0.42
6	0.64
7	0.85
8	1

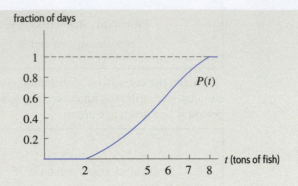

Figure 7.13: Cumulative distribution, $P(t)$, of daily catch

(b) The probability that the catch is between 5 and 7 tons can be found using either the density function p or the cumulative distribution function P. When we use the density function, this probability is represented by the shaded area in Figure 7.14, which is about 10.75 squares, so

$$\begin{array}{l}\text{Probability catch is} \\ \text{between 5 and 7 tons}\end{array} = \int_5^7 p(x)\,dx \approx \text{Area of 10.75 squares} = 10.75(0.04) = 0.43.$$

The probability can be found from the cumulative distribution function as follows:

$$\begin{array}{l}\text{Probability catch is} \\ \text{between 5 and 7 tons}\end{array} = P(7) - P(5) = 0.85 - 0.42 = 0.43.$$

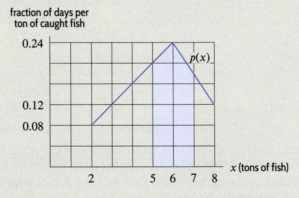

Figure 7.14: Shaded area represents the probability that the catch is between 5 and 7 tons

Problems for Section 7.2

1. (a) Using the density function in Example 2 on page 324, fill in values for the cumulative distribution function $P(t)$ for the length of time people wait in the doctor's office.

t (hours)	0	1	2	3	4
P(t) (fraction of people waiting)					

   (b) Graph $P(t)$.

2. Show that the area under the fishing density function in Figure 7.12 on page 329 is 1. Why is this to be expected?

3. Figure 7.15 shows a density function and the corresponding cumulative distribution function.[5]

   (a) Which curve represents the density function and which represents the cumulative distribution function? Give a reason for your choice.

---

[5] Adapted from David A. Smith and Lawrence C. Moore. *Calculus* (Lexington: D.C. Heath, 1994).

(b) Put reasonable values on the tick marks on each of the axes.

Figure 7.15

4. In an agricultural experiment, the quantity of grain from a given size field is measured. The yield can be anything from 0 kg to 50 kg. For each of the following situations, pick the graph that best represents the:
(i) Probability density function
(ii) Cumulative distribution function.

   (a) Low yields are more likely than high yields.
   (b) All yields are equally likely.
   (c) High yields are more likely than low yields.

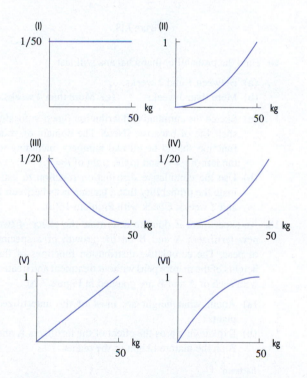

■ Decide if the function graphed in Problems **5–10** is a probability density function (pdf) or a cumulative distribution function (cdf). Give reasons. Find the value of c. Sketch and label the other function. (That is, sketch and label the cdf if the problem shows a pdf, and the pdf if the problem shows a cdf.)

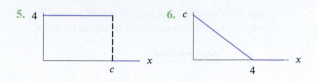

5.

6.

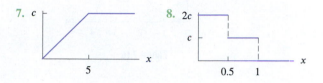

7.

8.

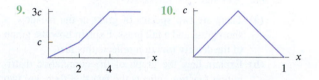

9.

10.

11. A person who travels regularly on the 9:00 am bus from Oakland to San Francisco reports that the bus is almost always a few minutes late but rarely more than five minutes late. The bus is never more than two minutes early, although it is on very rare occasions a little early.

   (a) Sketch a density function, $p(t)$, where t is the number of minutes that the bus is late. Shade the region under the graph between $t = 2$ minutes and $t = 4$ minutes. Explain what this region represents.
   (b) Now sketch the cumulative distribution function $P(t)$. What measurement(s) on this graph correspond to the area shaded? What do the inflection point(s) on your graph of P correspond to on the graph of p? Interpret the inflection points on the graph of P without referring to the graph of p.

12. The cumulative distribution function for heights (in meters) of trees in a forest is $F(x)$.

   (a) Explain in terms of trees the meaning of the statement $F(7) = 0.6$.
   (b) Which is greater, $F(6)$ or $F(7)$? Justify your answer in terms of trees.

13. Students at the University of California were surveyed and asked their grade point average. (The GPA ranges from 0 to 4, where 2 is just passing.) The distribution of GPAs is shown in Figure 7.16.[6]

   (a) Roughly what fraction of students are passing?
   (b) Roughly what fraction of the students have honor grades (GPAs above 3)?

[6]Adapted from D. Freedman, R. Pisani, R. Purves, and A. Adikhari, *Statistics*, 2nd *Edition* (New York: Norton, 1991).

(c) Why do you think there is a peak around 2?

(d) Sketch the cumulative distribution function.

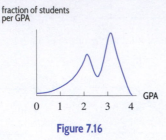

fraction of students
per GPA

Figure 7.16

14. The density function and cumulative distribution function of heights of grass plants in a meadow are in Figures 7.17 and 7.18, respectively.

(a) There are two species of grass in the meadow, a short grass and a tall grass. Explain how the graph of the density function reflects this fact.

(b) Explain how the graph of the cumulative distribution function reflects the fact that there are two species of grass in the meadow.

(c) About what percentage of the grasses in the meadow belong to the short grass species?

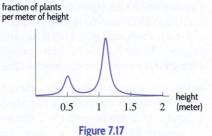

fraction of plants
per meter of height

Figure 7.17

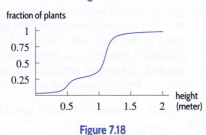

fraction of plants

Figure 7.18

15. A congressional committee is investigating a defense contractor whose projects often incur cost overruns. The data in Table 7.7 show y, the fraction of the projects with an overrun of at most $C\%$.

(a) Plot the data with C on the horizontal axis. Is this a density function or a cumulative distribution function? Sketch a curve through these points.

(b) If you think you drew a density function in part (a), sketch the corresponding cumulative distribution function on another set of axes. If you think you drew a cumulative distribution function in part (a), sketch the corresponding density function.

(c) Based on the table, what is the probability that there will be a cost overrun of 50% or more? Between 20% and 50%? Near what percent is the cost overrun most likely to be?

Table 7.7   *Fraction, y, of overruns that are at most C%*

C	−20%	−10%	0%	10%	20%	30%	40%	50%
y	0.01	0.08	0.19	0.32	0.50	0.80	0.94	0.99

■ For Problems 16–17, let $p(t) = -0.0375t^2 + 0.225t$ be the density function for the shelf life of a brand of banana, with t in weeks and $0 \leq t \leq 4$. See Figure 7.19.

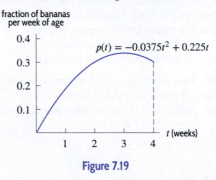

fraction of bananas
per week of age

$p(t) = -0.0375t^2 + 0.225t$

Figure 7.19

16. Find the probability that a banana will last

(a) Between 1 and 2 weeks.

(b) More than 3 weeks.   **(c)** More than 4 weeks.

17. **(a)** Sketch the cumulative distribution function for the shelf life of bananas. [Note: The domain of your function should be all real numbers, including to the left of $t = 0$ and to the right of $t = 4$.]

(b) Use the cumulative distribution function to estimate the probability that a banana lasts between 1 and 2 weeks. Check with Problem 16(a).

18. An experiment is done to determine the effect of two new fertilizers A and B on the growth of a species of peas. The cumulative distribution functions of the heights of the mature peas without treatment and treated with each of A and B are graphed in Figure 7.20.

(a) About what height are most of the unfertilized plants?

(b) Explain in words the effect of the fertilizers A and B on the mature height of the plants.

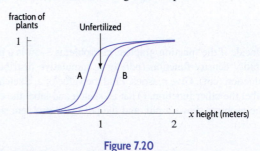

fraction of
plants

Unfertilized

Figure 7.20

19. A group of people have received treatment for cancer. Let t be the survival time, the number of years a person lives after the treatment. The density function giving the distribution of t is $p(t) = Ce^{-Ct}$ for some positive constant C. What is the practical meaning of the cumulative distribution function $P(t) = \int_0^t p(x)\, dx$?

20. The probability of a transistor failing between $t = a$ months and $t = b$ months is given by $c \int_a^b e^{-ct} dt$, for some constant c.

    (a) If the probability of failure within the first six months is 10%, what is c?
    (b) Given the value of c in part (a), what is the probability the transistor fails within the second six months?

21. While taking a walk along the road where you live, you accidentally drop your glove, but you don't know where. The probability density $p(x)$ for having dropped the glove x kilometers from home (along the road) is

$$p(x) = 2e^{-2x} \quad \text{for } x \geq 0.$$

    (a) What is the probability that you dropped it within 1 kilometer of home?
    (b) At what distance y from home is the probability that you dropped it within y km of home equal to 0.95?

7.3 THE MEDIAN AND THE MEAN

It is often useful to be able to give an "average" value for a distribution. Two measures that are in common use are the *median* and the *mean*.

The Median

> A **median** of a quantity x distributed through a population is a value T such that half the population has values of x less than (or equal to) T, and half the population has values of x greater than (or equal to) T. Thus, if p is the density function, a median T satisfies
>
> $$\int_{-\infty}^{T} p(x)\, dx = 0.5.$$
>
> In other words, half the area under the graph of p lies to the left of T. Equivalently, if P is the cumulative distribution function,
>
> $$P(T) = 0.5.$$

Example 1 Let t days be the length of time a pair of jeans remains in a shop before it is sold. The density function of t is graphed in Figure 7.21 and given by

$$p(t) = 0.04 - 0.0008t.$$

(a) What is the longest time a pair of jeans remains unsold?
(b) Would you expect the median time till sale to be less than, equal to, or greater than 25 days?
(c) Find the median time required to sell a pair of jeans.

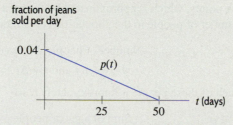

Figure 7.21: Density function for time till sale of a pair of jeans

Solution   (a) The density function is 0 for all times $t > 50$, so all jeans are sold within 50 days.
(b) The area under the graph of the density function in the interval $0 \leq t \leq 25$ is greater than the area under the graph in the interval $25 \leq t \leq 50$. So more than half the jeans are sold before their 25th day in the shop. The median time till sale is less than 25 days.
(c) Let P be the cumulative distribution function. We want to find the value of T such that

$$P(T) = \int_{-\infty}^{T} p(t)\, dt = \int_{0}^{T} p(t)\, dt = 0.5.$$

Using a calculator to evaluate the integrals, we obtain the values for P in Table 7.8.

Table 7.8   *Cumulative distribution for selling time*

T (days)	0	5	10	15	20	25
$P(T)$ (fraction of jeans sold by day T)	0	0.19	0.36	0.51	0.64	0.75

Since about half the jeans are sold within 15 days, the median time to sale is about 15 days. See Figures 7.22 and 7.23. We could also use the Fundamental Theorem of Calculus to find the median exactly. See Problem 12.

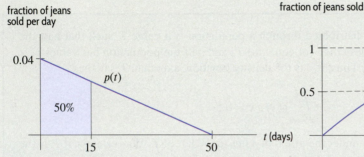

Figure 7.22: Median and density function

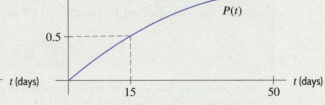

Figure 7.23: Median and cumulative distribution function

The Mean

Another commonly used average value is the *mean*. To find the mean of N numbers, we add the numbers and divide the sum by N. For example, the mean of the numbers 1, 2, 7, and 10 is $(1 + 2 + 7 + 10)/4 = 5$. The mean age of the entire US population is therefore defined as

$$\text{Mean age} = \frac{\sum \text{Ages of all people in the US}}{\text{Total number of people in the US}}.$$

Calculating the sum of all the ages directly would be an enormous task; we approximate the sum by an integral. We consider the people whose age is between t and $t + \Delta t$. How many are there?

The fraction of the population with age between t and $t + \Delta t$ is the area under the graph of p between these points, which is approximated by the area of the rectangle, $p(t)\Delta t$. (See Figure 7.24.) If the total number of people in the population is N, then

$$\begin{array}{c} \text{Number of people with age} \\ \text{between } t \text{ and } t + \Delta t \end{array} \approx p(t)\Delta t N.$$

The age of each of these people is approximately t, so

$$\begin{array}{c} \text{Sum of ages of people} \\ \text{between age } t \text{ and } t + \Delta t \end{array} \approx t p(t)\Delta t N.$$

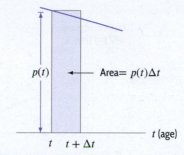

Figure 7.24: Shaded area is percentage of population with age between t and $t + \Delta t$

Therefore, adding and factoring out an N gives us

$$\text{Sum of ages of all people} \approx \left(\sum tp(t)\Delta t \right) N.$$

In the limit, as Δt shrinks to 0, the sum becomes an integral. Assuming no one is over 100 years old, we have

$$\text{Sum of ages of all people} = \left(\int_0^{100} tp(t)dt \right) N.$$

Since N is the total number of people in the US,

$$\text{Mean age} = \frac{\text{Sum of ages of all people in US}}{N} = \int_0^{100} tp(t)dt.$$

We can give the same argument for any[7] density function $p(x)$.

> If a quantity has density function $p(x)$,
>
> $$\textbf{Mean value of the quantity} = \int_{-\infty}^{\infty} xp(x)\,dx.$$

It can be shown that the mean is the point on the horizontal axis where the region under the graph of the density function, if it were made out of cardboard, would balance.

Example 2   Find the mean time for jeans sales, using the density function of Example 1.

Solution   The formula for p is $p(t) = 0.04 - 0.0008t$. We compute

$$\text{Mean time} = \int_0^{50} tp(t)\,dt = \int_0^{50} t(0.04 - 0.0008t)\,dt = 16.67 \text{ days.}$$

The mean is represented by the balance point in Figure 7.25. Notice that the mean is different from the median computed in Example 1.

---

[7]Provided all the relevant improper integrals converge.

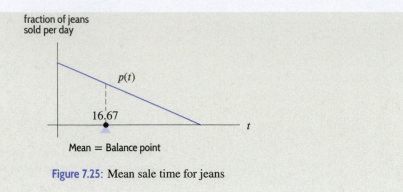

Figure 7.25: Mean sale time for jeans

Normal Distributions

How much rain do you expect to fall in your home town this year? If you live in Anchorage, Alaska, the answer is something close to 15 inches (including the snow). Of course, you don't expect exactly 15 inches. Some years there are more than 15 inches, and some years there are less. Most years, however, the amount of rainfall is close to 15 inches; only rarely is it well above or well below 15 inches. What does the density function for the rainfall look like? To answer this question, we look at rainfall data over many years. Records show that the distribution of rainfall is well approximated by a *normal distribution*. The graph of its density function is a bell-shaped curve which peaks at 15 inches and slopes downward approximately symmetrically on either side.

Normal distributions are frequently used to model real phenomena, from grades on an exam to the number of airline passengers on a particular flight. A normal distribution is characterized by its *mean*, μ, and its *standard deviation*, σ. The mean tells us the location of the central peak. The standard deviation tells us how closely the data is clustered around the mean. A small value of σ tells us that the data is close to the mean; a large σ tells us the data is spread out. In the following formula for a normal distribution, the factor of $1/(\sigma\sqrt{2\pi})$ makes the area under the graph equal to 1.

A **normal distribution** has a density function of the form

$$p(x) = \frac{1}{\sigma\sqrt{2\pi}}e^{-(x-\mu)^2/(2\sigma^2)},$$

where μ is the mean of the distribution and σ is the standard deviation, with $\sigma > 0$.

To model the rainfall in Anchorage, we use a normal distribution with $\mu = 15$ and $\sigma = 1$. (See Figure 7.26.)

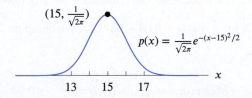

Figure 7.26: Normal distribution with $\mu = 15$ and $\sigma = 1$

Example 3   For Anchorage's rainfall, use the normal distribution with the density function with $\mu = 15$ and $\sigma = 1$ to compute the fraction of the years with rainfall between
(a) 14 and 16 inches,   (b) 13 and 17 inches,   (c) 12 and 18 inches.

Solution

(a) The fraction of the years with annual rainfall between 14 and 16 inches is $\int_{14}^{16} \frac{1}{\sqrt{2\pi}} e^{-(x-15)^2/2} \, dx$.

Since there is no elementary antiderivative for $e^{-(x-15)^2/2}$, we find the integral numerically. Its value is about 0.68.

$$\begin{array}{l} \text{Fraction of years with rainfall} \\ \text{between 14 and 16 inches} \end{array} = \int_{14}^{16} \frac{1}{\sqrt{2\pi}} e^{-(x-15)^2/2} \, dx \approx 0.68.$$

(b) Finding the integral numerically again:

$$\begin{array}{l} \text{Fraction of years with rainfall} \\ \text{between 13 and 17 inches} \end{array} = \int_{13}^{17} \frac{1}{\sqrt{2\pi}} e^{-(x-15)^2/2} \, dx \approx 0.95.$$

(c)

$$\begin{array}{l} \text{Fraction of years with rainfall} \\ \text{between 12 and 18 inches} \end{array} = \int_{12}^{18} \frac{1}{\sqrt{2\pi}} e^{-(x-15)^2/2} \, dx \approx 0.997.$$

Since 0.95 is so close to 1, we expect that most of the time the rainfall will be between 13 and 17 inches a year.

Among the normal distributions, the one having $\mu = 0$, $\sigma = 1$ is called the *standard normal distribution*. Values of the corresponding cumulative distribution function are published in tables.

Problems for Section 7.3

1. Estimate the median daily catch for the fishing data given in Example 2 of Section 7.2.

2. (a) Use the cumulative distribution function in Figure 7.27 to estimate the median.
   (b) Describe the density function: For what values is it positive? Increasing? Decreasing? Identify all local maximum and minimum values.

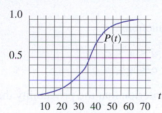

Figure 7.27

3. A quantity x has density function $p(x) = 0.5(2 - x)$ for $0 \le x \le 2$ and $p(x) = 0$ otherwise. Find the mean and median of x.

4. A quantity x has cumulative distribution function $P(x) = x - x^2/4$ for $0 \le x \le 2$ and $P(x) = 0$ for $x < 0$ and $P(x) = 1$ for $x > 2$. Find the mean and median of x.

■ For Problems 5–6, let $p(t) = -0.0375t^2 + 0.225t$ be the density function for the shelf life of a brand of banana which lasts up to 4 weeks. Time, t, is measured in weeks and $0 \le t \le 4$.

5. Find the median shelf life of a banana using $p(t)$. Plot the median on a graph of $p(t)$. Does it look like half the area is to the right of the median and half the area is to the left?

6. Find the mean shelf life of a banana using $p(t)$. Plot the mean on a graph of $p(t)$. Does it look like the mean is the place where the density function balances?

7. Suppose that x measures the time (in hours) it takes for a student to complete an exam. All students are done within two hours and the density function for x is

$$p(x) = \begin{cases} x^3/4 & \text{if } 0 < x < 2 \\ 0 & \text{otherwise.} \end{cases}$$

   (a) What proportion of students take between 1.5 and 2.0 hours to finish the exam?
   (b) What is the mean time for students to complete the exam?
   (c) Compute the median of this distribution.

8. Let $p(t) = 0.1e^{-0.1t}$ be the density function for the waiting time at a subway stop, with t in minutes, $0 \le t \le 60$.

   (a) Graph $p(t)$. Use the graph to estimate visually the median and the mean.
   (b) Calculate the median and the mean. Plot both on the graph of $p(t)$.
   (c) Interpret the median and mean in terms of waiting time.

9. The speeds of cars on a road are approximately normally distributed with a mean $\mu = 58$ km/hr and standard deviation $\sigma = 4$ km/hr.

   (a) What is the probability that a randomly selected car is going between 60 and 65 km/hr?
   (b) What fraction of all cars are going slower than 52 km/hr?

10. The distribution of IQ scores can be modeled by a normal distribution with mean 100 and standard deviation 15.

    (a) Write the formula for the density function of IQ scores.
    (b) Estimate the fraction of the population with IQ between 115 and 120.

11. Let $P(x)$ be the cumulative distribution function for the household income distribution in the US in 2009.[8] Values of $P(x)$ are in the following table:

Income x (thousand \$)	20	40	60	75	100
$P(x)$ (%)	29.5	50.1	66.8	76.2	87.1

   (a) What percent of the households made between \$40,000 and \$60,000? More than \$100,000?
   (b) Approximately what was the median income?
   (c) Is the statement "More than one-third of households made between \$40,000 and \$75,000" true or false?

12. Find the median of the density function given by $p(t) = 0.04 - 0.0008t$ for $0 \le t \le 50$ using the Fundamental Theorem of Calculus.

PROJECTS FOR CHAPTER SEVEN

1. **Triangular Probability Distribution**

   Triangular probability distributions, such as the one with density function graphed in Figure 7.28, are used in business to model uncertainty. Such a distribution can be used to model a variable where only three pieces of information are available: a lower bound ($x = a$), a most likely value ($x = c$), and an upper bound ($x = b$).

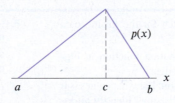

   Figure 7.28

   Thus, we can write the function $p(x)$ as two linear functions:

   $$p(x) = \begin{cases} m_1 x + b_1 & a \le x \le c \\ m_2 x + b_2 & c < x \le b. \end{cases}$$

   (a) Find the value of $p(c)$ geometrically, using the criterion that the probability that x takes on some value between a and b is 1.

   Suppose a new product costs between \$6 and \$10 per unit to produce, with a most likely cost of \$9.

   (b) Find $p(9)$.
   (c) Use the fact that $p(6) = p(10) = 0$ and the value of $p(9)$ you found in part (b) to find m_1, m_2, b_1, and b_2.
   (d) What is the probability that the production cost per unit will be less than \$8?
   (e) What is the median cost?
   (f) Write a formula for the cumulative probability distribution function $P(x)$ for

   (i) $6 \le x \le 9$,   (ii) $9 < x \le 10$.

   Sketch the graph of $P(x)$.

---

[8]http://www.census.gov/hhes/www/income/income.html, accessed January 7, 2012.

Chapter 8

FUNCTIONS OF SEVERAL VARIABLES

CONTENTS

8.1 UNDERSTANDING FUNCTIONS OF TWO VARIABLES

In business, science, and politics, outcomes are rarely determined by a single variable. For example, consider an airline's ticket pricing. To avoid flying planes with many empty seats, it sells some tickets at full price and some at a discount. For a particular route, the airline's revenue, R, earned in a given time period is determined by the number of full-price tickets, x, and the number of discount tickets, y, sold. We say that R is a function of x and y, and we write

$$R = f(x, y).$$

This is just like the function notation of one-variable calculus. The variable R is the dependent variable and the variables x and y are the independent variables. The letter f stands for the *function* or rule that gives the value, or output, of R corresponding to given values of x and y. The collection of all possible inputs, (x, y), is called the *domain* of f. We say a function is an *increasing* (*decreasing*) function of one of its variables if it increases (decreases) as that variable increases while the other independent variables are held constant.

A function of two variables can be represented numerically by a table of values, algebraically by a formula, or pictorially by a contour diagram. In this section we give numerical and algebraic examples; contour diagrams are introduced in Section 8.2.

Functions Given Numerically

The revenue, R (in dollars), from a particular airline route is shown in Table 8.1 as a function of the number of full-price tickets and the number of discount tickets sold.

Table 8.1   *Revenue from ticket sales as a function of x and y*

		Number of full-price tickets, x			
		100	200	300	400
	200	75,000	110,000	145,000	180,000
	400	115,000	150,000	185,000	220,000
Number of discount tickets, y	600	155,000	190,000	225,000	260,000
	800	195,000	230,000	265,000	300,000
	1000	235,000	270,000	305,000	340,000

Values of x are shown across the top, values of y are down the left side, and the corresponding values of $f(x, y)$ are in the table. For example, to find the value of $f(300, 600)$, we look in the column corresponding to $x = 300$ at the row $y = 600$, where we find the number 225,000. Thus,

$$f(300, 600) = 225,000.$$

This means that the revenue from 300 full-price tickets and 600 discount tickets is \$225,000. We see in Table 8.1 that f is an increasing function of x and an increasing function of y.

Notice how this differs from the table of values of a one-variable function, where one row or one column is enough to list the values of the function. Here many rows and columns are needed because the function has an output for every pair of values of the independent variables.

Functions Given Algebraically

The function given in Table 8.1 can be represented by a formula. Looking across the rows, we see that each additional 100 full-price tickets sold raises the revenue by \$35,000, so each full-price ticket must cost \$350. Similarly, looking down a column shows that an additional 200 discount tickets sold increases the revenue by \$40,000, so each discount ticket must cost \$200. Thus, the revenue function is given by the formula

$$R = 350x + 200y.$$

Example 1   Give a formula for the function $M = f(B, t)$ where M is the amount of money in a bank account t years after an initial investment of B thousand dollars, if interest accrues at a rate of 5% per year compounded (a) Annually   (b) Continuously.

Solution   (a) Annual compounding means that M increases by a factor of 1.05 every year, so

$$M = f(B, t) = B(1.05)^t.$$

(b) Continuous compounding means that M grows according to the function e^{kt}, with $k = 0.05$, so

$$M = f(B, t) = Be^{0.05t}.$$

Example 2   A car rental company charges $40 a day and 15 cents a mile for its cars.

(a) Write a formula for the cost, C, of renting a car as a function of the number of days, d, and the number of miles driven, m.

(b) If $C = f(d, m)$, find $f(5, 300)$ and interpret it.

Solution   (a) The total cost in dollars of renting a car is 40 times the numbers of days plus 0.15 times the number of miles, so

$$C = 40d + 0.15m.$$

(b) We have

$$f(5, 300) = 40(5) + 0.15(300)$$
$$= 200 + 45$$
$$= 245.$$

We see that $f(5, 300) = 245$. This tells us that if we rent a car for 5 days and drive it 300 miles, it costs us $245.

Strategy to Investigate Functions of Two Variables: Vary One Variable at a Time

We can learn a great deal about a function of two variables by letting one variable vary while holding the other fixed. This gives a function of one variable, called a *cross-section* of the original function.

Concentration of a Drug in the Blood

When a drug is injected into muscle tissue, it diffuses into the bloodstream. The concentration of the drug in the blood increases until it reaches a maximum, and then decreases. The concentration, C (in mg per liter), of the drug in the blood is a function of two variables: x, the amount (in mg) of the drug given in the injection, and t, the time (in hours) since the injection was administered. We are told that

$$C = f(x, t) = xte^{-t} \qquad \text{for } 0 \le x \le 6 \text{ and } t \ge 0.$$

Example 3   In terms of the drug concentration in the blood, explain the significance of the cross-sections:

(a) $f(4, t)$ (b) $f(x, 1)$

Solution   (a) Holding x fixed at 4 means that we are considering an injection of 4 mg of the drug; letting t vary means we are watching the effect of this dose as time passes. Thus the function $f(4, t)$ describes the concentration of the drug in the blood resulting from a 4-mg injection as a function of time. Figure 8.1 shows $f(4, t) = 4te^{-t}$. Notice that the concentration in the blood is at a maximum 1 hour after the injection, and that the concentration in the blood eventually approaches zero.

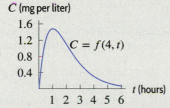

Figure 8.1: The function $f(4, t)$ shows the concentration in the blood resulting from a 4-mg injection

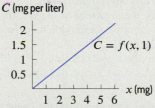

Figure 8.2: The function $f(x, 1)$ shows the concentration in the blood 1 hour after the injection

(b) Holding t fixed at 1 means that we are focusing on the blood 1 hour after the injection; letting x vary means we are considering the effect of different doses at that instant. Thus, the function $f(x, 1)$ gives the concentration of the drug in the blood 1 hour after injection as a function of the amount injected. Figure 8.2 shows the graph of $f(x, 1) = xe^{-1}$. Notice that $f(x, 1)$ is an increasing function of x. This makes sense: If we administer more of the drug, the concentration in the bloodstream is higher.

Example 4   Continue with $C = f(x, t) = xte^{-t}$. Graph the cross-sections of $f(a, t)$ for $a = 1, 2, 3$, and 4 on the same axes. Describe how the graph changes for larger values of a and explain what this means in terms of drug concentration in the blood.

Solution   The one-variable function $f(a, t)$ represents the effect of an injection of a mg at time t. Figure 8.3 shows the graphs of the four functions $f(1, t) = te^{-t}$, $f(2, t) = 2te^{-t}$, $f(3, t) = 3te^{-t}$, and $f(4, t) = 4te^{-t}$, corresponding to injections of 1, 2, 3, and 4 mg of the drug. The general shape of the graph is the same in every case: The concentration in the blood is zero at the time of injection $t = 0$, then increases to a maximum value, and then decreases toward zero again. We see that if a larger dose of the drug is administered, the peak of the graph is higher. This makes sense, since a larger dose will produce a higher concentration.

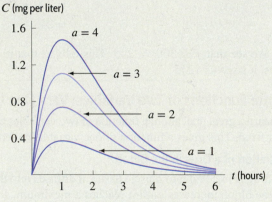

Figure 8.3: Concentration $C = f(a, t)$ of the drug resulting from an a-mg injection

Problems for Section 8.1

■ Problems 1–2 concern the cost, C, of renting a car from a company which charges $40 a day and 15 cents a mile, so $C = f(d, m) = 40d + 0.15m$, where d is the number of days, and m is the number of miles.

1. Make a table of values for C, using $d = 1, 2, 3, 4$ and $m = 100, 200, 300, 400$. You should have 16 values in your table.

2. (a) Find $f(3, 200)$ and interpret it.
   (b) Explain the significance of $f(3, m)$ in terms of rental car costs. Graph this function, with C as a function of m.
   (c) Explain the significance of $f(d, 100)$ in terms of rental car costs. Graph this function, with C as a function of d.

3. A cable company charges $100 for a monthly subscription to its services and $5 for each special feature movie that a subscriber chooses to watch.

   (a) Write a formula for the monthly revenue, R in dollars, earned by the cable company as a function of s, the number of monthly subscribers it serves, and m, the total number of special feature movies that its subscribers view.
   (b) If $R = f(s, m)$, find $f(1000, 5000)$ and interpret it in terms of revenue.

4. The number, n, of new cars sold in a year is a function of the price of new cars, c, and the average price of gas, g.

   (a) If c is held constant, is n an increasing or decreasing function of g? Why?
   (b) If g is held constant, is n an increasing or decreasing function of c? Why?

■ For Problems 5–7, refer to Table 8.2 which gives a person's body mass index, BMI, in terms of their weight w (in lbs) and height h (in inches).

Table 8.2   *Body mass index (BMI)*

	Weight w (lbs)				
	120	140	160	180	200
Height h 60	23.4	27.3	31.2	35.2	39.1
(inches) 63	21.3	24.8	28.3	31.9	35.4
66	19.4	22.6	25.8	29.0	32.3
69	17.7	20.7	23.6	26.6	29.5
72	16.3	19.0	21.7	24.4	27.1
75	15.0	17.5	20.0	22.5	25.0

5. Compute a table of values of BMI, with h fixed at 60 inches and w between 120 and 200 lbs at intervals of 20.

6. Medical evidence suggests that BMI values between 18.5 and 24.9 are healthy values.[1] Estimate the range of weights that are considered healthy for a woman who is 6 feet tall.

7. Estimate the BMI of a man who weighs 90 kilograms and is 1.9 meters tall.

■ Problems 8–12 refer to Table 8.3, which shows[2] the weekly beef consumption, C, (in lbs) of an average household as a function of p, the price of beef (in \$/lb) and I, annual household income (in \$1000s).

Table 8.3   *Quantity of beef bought (lbs/household/week)*

			p	
	3.00	3.50	4.00	4.50
I 20	2.65	2.59	2.51	2.43
40	4.14	4.05	3.94	3.88
60	5.11	5.00	4.97	4.84
80	5.35	5.29	5.19	5.07
100	5.79	5.77	5.60	5.53

8. Give tables for beef consumption as a function of p, with I fixed at $I = 20$ and $I = 100$. Give tables for beef consumption as a function of I, with p fixed at $p = 3.00$ and $p = 4.00$. Comment on what you see in the tables.

9. How does beef consumption vary as a function of household income if the price of beef is held constant?

10. Make a table showing the amount of money, M, that the average household spends on beef (in dollars per household per week) as a function of the price of beef and household income.

11. Make a table of the proportion, P, of household income spent on beef per week as a function of price and income. (Note that P is the fraction of income spent on beef.)

12. Express P, the proportion of household income spent on beef per week, in terms of the original function $f(I, p)$ which gave consumption as a function of p and I.

13. The total sales of a product, S, can be expressed as a function of the price p charged for the product and the amount, a, spent on advertising, so $S = f(p, a)$. Do you expect f to be an increasing or decreasing function of p? Do you expect f to be an increasing or decreasing function of a? Why?

14. Graph the bank-account function f in Example 1(a) on page 340, holding B fixed at $B = 10, 20, 30$ and letting t vary. Then graph f, holding t fixed at $t = 0, 5, 10$ and letting B vary. Explain what you see.

15. The heat index is a temperature which tells you how hot it feels as a result of the combination of temperature and humidity. See Table 8.4. Heat exhaustion is likely to occur when the heat index reaches 105°F.

 (a) If the temperature is 80°F and the humidity is 50%, how hot does it feel?
 (b) At what humidity does 90°F feel like 90°F?
 (c) Make a table showing the approximate temperature at which heat exhaustion becomes a danger, as a function of humidity.
 (d) Explain why the heat index is sometimes above the actual temperature and sometimes below it.

Table 8.4   *Heat index (°F) as a function of humidity ($H\%$) and temperature ($T°F$)*

					T					
	70	75	80	85	90	95	100	105	110	115
H 0	64	69	73	78	83	87	91	95	99	103
10	65	70	75	80	85	90	95	100	105	111
20	66	72	77	82	87	93	99	105	112	120
30	67	73	78	84	90	96	104	113	123	135
40	68	74	79	86	93	101	110	123	137	151
50	69	75	81	88	96	107	120	135	150	
60	70	76	82	90	100	114	132	149		

16. Using Table 8.4, graph heat index as a function of humidity with temperature fixed at 70°F and at 100°F. Explain the features of each graph and the difference between them in common-sense terms.

---

[1] http://www.cdc.gov. Accessed January 10, 2016.
[2] From Richard G. Lipsey, *An Introduction to Positive Economics*, 3rd ed. (London: Weidenfeld and Nicolson, 1971).

17. A person's basal metabolic rate (BMR) is the minimal number of daily calories needed to keep their body functioning at rest. The BMR (in kcal/day) of a man of mass m (in kg), height h (in cm) and age a (in years) can be approximated by[3]

$$P = f(m, h, a) = 14m + 5h - 7a + 66$$

and for women by

$$P = g(m, h, a) = 10m + 2h - 5a + 655.$$

(a) What is the BMR of a 28-year-old man 180 cm tall weighing 59 kg?

(b) What is the BMR of a 43-year-old woman 162 cm tall weighing 52 kg?

(c) If a 40-year-old man 175 cm tall weighing 77 kg restricts himself to a diet with a daily caloric intake of 1600 kcal, should he expect to lose weight?

18. The monthly cost, in dollars, of a cell phone bill is

$$P = f(t, m, d) = 0.25t + 0.2m + 0.01d$$

where t is the number of minutes talked, m is the number of messages sent and d is the number of kilobytes of data used that month.

(a) Find $f(250, 200, 100)$ and interpret it.

(b) Find a formula for monthly cost if your data is disabled.

(c) You are considering switching to an unlimited talk, text and data plan at \$50 per month. If your average usage is $t = 120$, $m = 100$ and $d = 250$, does it make sense to switch?

19. The monthly payments, P dollars, on a mortgage in which A dollars were borrowed at an annual interest rate of $r\%$ for t years is given by $P = f(A, r, t)$. Is f an increasing or decreasing function of A? Of r? Of t?

20. An airport can be cleared of fog by heating the air. The amount of heat required, $H(T, w)$ (in calories per cubic meter of fog), depends on the temperature of the air, T (in °C), and the wetness of the fog, w (in grams per cubic meter of fog). Figure 8.4 shows several graphs of H against T with w fixed.

(a) Estimate $H(20, 0.3)$ and explain what information it gives us.

(b) Make a table of values for $H(T, w)$. Use $T = 0$, 10, 20, 30, 40, and $w = 0.1, 0.2, 0.3, 0.4$.

[3]www.wikipedia.org, accessed May 11, 2016.

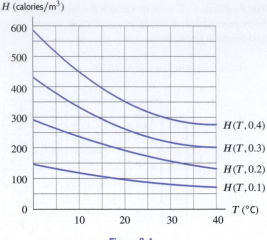

Figure 8.4

21. Figure 8.5 shows the annual energy production $E(d, w)$ (in kilowatt-hours) of a wind turbine as a function of its diameter, d, (in feet) and three different wind speeds, w, (in mph).

(a) Estimate $E(15, 12)$ and explain what information it gives us.

(b) For a wind turbine of diameter 10 feet, how much does annual energy production increase when the wind speed increases from 12 mph to 15 mph?

(c) If the wind is blowing at 12 mph, how much more energy does a 15-ft-diameter turbine generate than a 10-ft-diameter turbine?

(d) You have a 10-ft-diameter wind turbine in a location where the wind is 12 mph. Which would produce the biggest increase in energy production?

- Replacing your 10-ft turbine with a 15-ft turbine in the same location.
- Moving your 10-ft turbine to a new location where the wind is 15 mph.

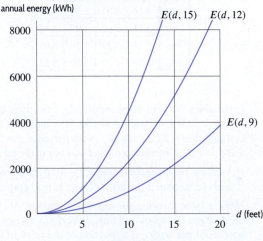

Figure 8.5

■ For Problems 22–25, a person's body mass index (BMI) is a function of their weight W (in kg) and height H (in m) given by $B(W, H) = W/H^2$.

22. What is the BMI of a 1.72 m tall man weighing 72 kg?

23. A 1.58 m tall woman has a BMI of 23.2. What is her weight?

24. With a BMI less than 18.5, a person is considered underweight. What is the possible range of weights for an underweight person 1.58 m tall?

25. For weight w in lbs and height h in inches, a persons BMI is approximated using the formula $f(w, h) = 703w/h^2$. Check this approximation by converting the formula $B(W, H)$.

■ In Problems 26–27, the fallout, V (in kilograms per square kilometer), from a volcanic explosion depends on the distance, d, from the volcano and the time, t, since the explosion:

$$V = f(d, t) = \left(\sqrt{t}\right)e^{-d}.$$

26. On the same axes, graph cross-sections of f with $t = 1$ and $t = 2$. As distance from the volcano increases, how does the fallout change? Look at the relationship between the graphs: how does the fallout change as time passes? Explain your answers in terms of volcanoes.

27. On the same axes, graph cross-sections of f with $d = 0$, $d = 1$, and $d = 2$. As time passes since the explosion, how does the fallout change? Look at the relationship between the graphs: how does fallout change as a function of distance? Explain your answers in terms of volcanoes.

■ In Problems 28–29, the atmospheric pressure, $P = f(y, t) = (950 + 2t)e^{-y/7}$, in millibars, on a weather balloon, is a func-

tion of its height $y \geq 0$, in km above sea level after t hours with $0 \leq t \leq 48$.

28. Find $f(2, 12)$. Give units and interpret this quantity in the context of atmospheric pressure.

29. Graph the following single-variable functions and explain the significance of the shape of the graph in terms of atmospheric pressure.

(a) $f(3, t)$ (b) $f(y, 24)$

30. The pressure of a fixed amount of compressed nitrogen gas in a cylinder is given, in atmospheres, by

$$P = f(T, V) = \frac{10T}{V},$$

where T is the temperature of the gas, in Kelvin, and V is the volume of the cylinder, in liters. Figures 8.6 and 8.7 give cross-sections of the function f.

(a) Which figure shows cross-sections of f with T fixed? What does the shape of the cross-sections tell you about the pressure?

(b) Which figure shows cross-sections of f with V fixed? What does the shape of the cross-sections tell you about the pressure?

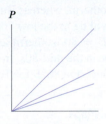 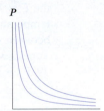

Figure 8.6   Figure 8.7

8.2 CONTOUR DIAGRAMS

How can we visualize a function of two variables? Just as a function of one variable can be represented by a graph, a function of two variables can be represented by a surface in space or by a *contour diagram* in the plane. Numerical information is more easily obtained from contour diagrams, so we concentrate on their use.

Weather Maps

Figure 8.8 shows a weather map. This contour diagram shows the predicted high temperature, T, in degrees Fahrenheit (°F), throughout the US on that day. The curves on the map, called *isotherms*, separate the country into zones, according to whether T is in the 60s, 70s, 80s, 90s, or 100s. (*Iso* means same and *therm* means heat.) Notice that the isotherm separating the 80s and 90s zones connects all the points where the temperature is predicted to be exactly 90°F.

If the function $T = f(x, y)$ gives the predicted high temperature (in °F) on this particular day as a function of latitude x and longitude y, then the isotherms are graphs of the equations

$$f(x, y) = c$$

where c is a constant. In general, such curves are called *contours*, and a graph showing selected contours of a function is called a contour diagram.

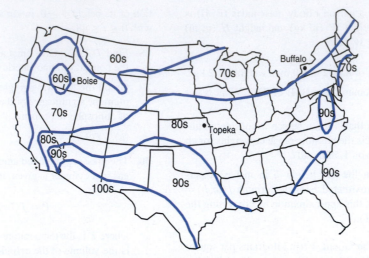

Figure 8.8: Weather map showing predicted high temperatures, T, on a summer day

Example 1   Estimate the predicted value of T in Boise, Idaho; Topeka, Kansas; and Buffalo, New York.

Solution   Boise and Buffalo are in the 70s region, and Topeka is in the 80s region. Thus, the predicted temperature in Boise and Buffalo is between 70 and 80 while the predicted temperature in Topeka is between 80 and 90.

   In fact, we can say more. Although both Boise and Buffalo are in the 70s, Boise is quite close to the $T = 70$ isotherm, whereas Buffalo is quite close to the $T = 80$ isotherm. So we estimate that the temperature will be in the low 70s in Boise and in the high 70s in Buffalo. Topeka is about halfway between the $T = 80$ isotherm and the $T = 90$ isotherm. Thus, we guess that the temperature in Topeka will be in the mid-80s. In fact, the actual high temperatures for that day were 71°F for Boise, 79°F for Buffalo, and 86°F for Topeka.

Topographical Maps

Another common example of a contour diagram is a topographical map like that shown in Figure 8.9. Here, the contours separate regions of lower elevation from regions of higher elevation, and give an overall picture of the nature of the terrain. Such topographical maps are frequently colored green at the lower elevations and brown, red, or even white at the higher elevations.

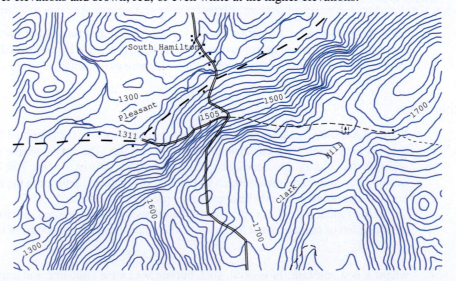

Figure 8.9: A topographical map showing the region around South Hamilton, NY

Example 2   Explain why the topographical map shown in Figure 8.10 corresponds to the surface of the terrain shown in Figure 8.11.

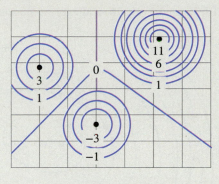

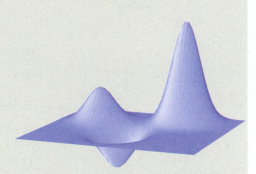

Figure 8.11: Terrain corresponding to the topographical map in Figure 8.10

Figure 8.10: A topographical map

Solution   We see from the topographical map in Figure 8.10 that there are two hills, one with height about 12 and the other with height about 4. Most of the terrain is around height 0, and there is one valley with height about −4. This matches the surface of the terrain in Figure 8.11 since there are two hills (one taller than the other) and one valley.

The contours on a topographical map outline the contour or shape of the land. Because every point along the same contour has the same elevation, contours are also called *level curves* or *level sets*. We usually draw contours for equally spaced values of the function. The more closely spaced the contours, the steeper the terrain; the more widely spaced the contours, the flatter the terrain (provided, of course, that the elevation between contours varies by a constant amount). Certain features have distinctive characteristics. A mountain peak is typically surrounded by contours like those in Figure 8.12. A pass in a range of mountains may have contours that look like Figure 8.13. A long valley has parallel contours indicating the rising elevations on both sides of the valley (see Figure 8.14); a long ridge of mountains has the same type of contours, only the elevations decrease on both sides of the ridge. Notice that the elevation numbers on the contours are as important as the curves themselves.

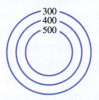

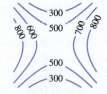

Figure 8.12: Mountain peak

Figure 8.13: Pass between two mountains

Figure 8.14: Long valley

Figure 8.15: Impossible contour lines

There are some things contours cannot do. Two contours corresponding to different elevations cannot cross each other as shown in Figure 8.15. If they did, the point of intersection of the two curves would have two different elevations, which is impossible (assuming the terrain has no overhangs).

Using Contour Diagrams

Consider the effect of different weather conditions on US corn production. What would happen if the average temperature were to increase (due to global warming, for example) or if the rainfall were to decrease (due to a drought)? One way of estimating the effect of these climatic changes is to use Figure 8.16. This map is a contour diagram giving the corn production $C = f(R, T)$ in the US as a function of the total rainfall, R, in inches, and average temperature, T, in degrees Fahrenheit, during the growing season.[4] Suppose at the present time, $R = 15$ inches and $T = 76°F$. Production is measured as a percentage of the present production; thus, the contour through $R = 15, T = 76$ is $C = 100$, that is, $C = f(15, 76) = 100$.

Example 3   Use Figure 8.16 to evaluate $f(18, 78)$ and $f(12, 76)$ and explain the answers in terms of corn production.

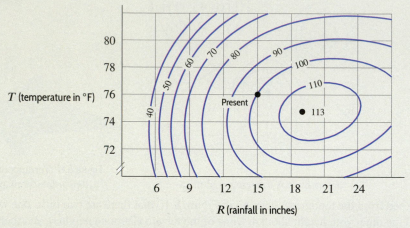

T (temperature in °F)

R (rainfall in inches)

Figure 8.16: Corn production, C, as a function of rainfall and temperature

Solution   The point with R-coordinate 18 and T-coordinate 78 is on the contour with value $C = 100$, so $f(18, 78) = 100$. This means that if the annual rainfall were 18 inches and the temperature were 78°F, the country would produce about the same amount of corn as at present, although it would be wetter and warmer than it is now. The point with R-coordinate 12 and T-coordinate 76 is about halfway between the $C = 80$ and $C = 90$ contours, so $f(12, 76) \approx 85$. This means that if the rainfall dropped to 12 inches and the temperature stayed at 76°F, then corn production would drop to about 85% of what it is now.

Example 4   Describe how corn production changes as a function of rainfall if temperature is fixed at the present value in Figure 8.16. Describe how corn production changes as a function of temperature if rainfall is held constant at the present value. Give common-sense explanations for your answers.

Solution   To see what happens to corn production if the temperature stays fixed at 76°F but the rainfall changes, look along the horizontal line $T = 76$. Starting from the present and moving left along the line $T = 76$, the values on the contours decrease. In other words, if there is a drought, corn production decreases. Conversely, as rainfall increases, that is, as we move from the present to the right along the line $T = 76$, corn production increases, reaching a maximum of more than 110% when $R = 21$, and then decreases (too much rainfall floods the fields). If, instead, rainfall remains at the present value and temperature increases, we move up the vertical line $R = 15$. Under these circumstances corn production decreases; a 2° increase causes a 10% drop in production. This makes sense since hotter temperatures lead to greater evaporation and hence drier conditions, even with rainfall constant at 15 inches. Similarly, a decrease in temperature leads to a very slight increase in production, reaching a maximum of around 102% when $T = 74$, followed by a decrease (the corn won't grow if it is too cold).

Cobb-Douglas Production Functions

Suppose you are running a small printing business, and decide to expand because you have more orders than you can handle. How should you expand? Should you start a night shift and hire more workers? Should you buy more expensive but faster computers which will enable the current staff to keep up with the work? Or should you do some combination of the two?

[4]Adapted from S. Beaty and R. Healy, "The Future of American Agriculture", *Scientific American,* Vol. 248, No. 2, February, 1983.

Obviously, the way such a decision is made in practice involves many other considerations—such as whether you could get a suitably trained night shift, or whether there are any faster computers available. Nevertheless, you might model the quantity, P, of work produced by your business as a function of two variables: your total number, N, of workers, and the total value, V, of your equipment. What might the contour diagram of the production function look like?

Example 5 Explain why the contour diagram in Figure 8.17 does not model the behavior expected of the production function, whereas the contour diagram in Figure 8.18 does.

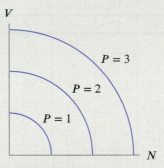

Figure 8.17: Incorrect contours for printing production

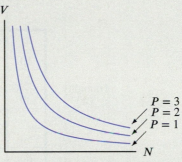

Figure 8.18: Correct contours for printing production

Solution Look at Figure 8.17. Notice that the contour $P = 1$ intersects the N- and the V- axis, suggesting that it is possible to produce work with no workers or with no equipment; this is unreasonable. However, no contours in Figure 8.18 intersect either the N- or the V-axis.

In Figure 8.18, fixing V and letting N increase corresponds to moving to the right, crossing contours less and less frequently. Production increases more and more slowly because hiring additional workers does little to boost production if the machines are already used to capacity.

Similarly, if we fix N and let V increase, Figure 8.18 shows production increasing, but at a decreasing rate. Buying machines without enough people to use them does not increase production much. Thus Figure 8.18 fits the expected behavior of the production function best.

The Cobb-Douglas Production Model

In 1928, Cobb and Douglas used a simple formula to model the production of the entire US economy in the first quarter of the 20th century. Using government estimates of P, the total yearly production between 1899 and 1922, and of K, the total capital investment over the same period, and of L, the total labor force, they found that P was well approximated by the function

$$P = 1.01 L^{0.75} K^{0.25}.$$

This function turned out to model the US economy surprisingly accurately, both for the period on which it was based and for some time afterward. The contour diagram of this function is similar to that in Figure 8.18. In general, production is often modeled by a function of the following form:

Cobb-Douglas Production Function

$$P = f(N, V) = c N^{\alpha} V^{\beta}$$

where P is the total quantity produced and c, α, and β are positive constants with $0 < \alpha < 1$ and $0 < \beta < 1$.

Contour Diagrams and Tables

Table 8.5 shows the heat index as a function of temperature and humidity. The heat index is a temperature which tells you how hot it feels as a result of the combination of the two. We can also display this function using a contour diagram. Scales for the two independent variables (temperature and humidity) go on the axes. The heat indices shown range from 64 to 151, so we will draw contours at values of 70, 80, 90, 100, 110, 120, 130, 140, and 150. How do we know where the contour for 70 goes? Table 8.5 shows that, when humidity is 0%, a heat index of 70 occurs between 75°F and 80°F, so the contour will go approximately through the point (76, 0). It also goes through the point (75, 10). Continuing in this way, we can approximate the 70 contour. See Figure 8.19. You can construct all the contours in Figure 8.20 in a similar way.

Table 8.5 *Heat index (°F)*

		70	75	80	85	90	95	100	105	110	115
	0	64	69	73	78	83	87	91	95	99	103
	10	65	70	75	80	85	90	95	100	105	111
	20	66	72	77	82	87	93	99	105	112	120
Humidity (%)	30	67	73	78	84	90	96	104	113	123	135
	40	68	74	79	86	93	101	110	123	137	151
	50	69	75	81	88	96	107	120	135	150	
	60	70	76	82	90	100	114	132	149		

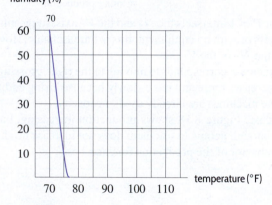

Figure 8.19: The contour for a heat index of 70

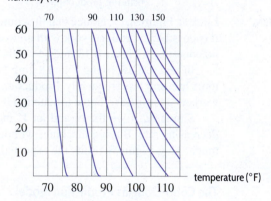

Figure 8.20: Contour diagram for the heat index

Example 6   Heat exhaustion is likely to occur where the heat index is 105 or higher. On the contour diagram in Figure 8.20, shade in the region where heat exhaustion is likely to occur.

Solution   The shaded region in Figure 8.21 shows the values of temperature and humidity at which the heat index is above 105.

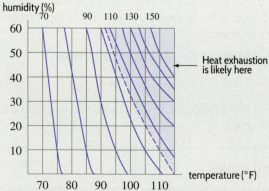

Figure 8.21: Shaded region shows conditions under which heat exhaustion is likely

Finding Contours Algebraically

Algebraic equations for the contours of a function f are easy to find if we have a formula for $f(x, y)$. A contour consists of all the points (x, y) where $f(x, y)$ has a constant value, c. Its equation is

$$f(x, y) = c.$$

Example 7   Draw a contour diagram for the airline revenue function $R = 350x + 200y$. Include contours for $R = 4000, 8000, 12000, 16000$.

Solution   The contour for $R = 4000$ is given by

$$350x + 200y = 4000.$$

This is the equation of a line with intercepts $x = 4000/350 = 11.43$ and $y = 4000/200 = 20$. (See Figure 8.22.) The contour for $R = 8000$ is given by

$$350x + 200y = 8000.$$

This is the equation of a parallel line with intercepts $x = 8000/350 = 22.86$ and $y = 8000/200 = 40$. The contours for $R = 12,000$ and $R = 16,000$ are parallel lines drawn similarly. (See Figure 8.22.)

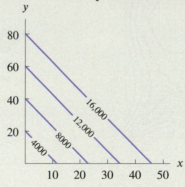

Figure 8.22: A contour diagram for $R = 350x + 200y$

Problems for Section 8.2

■ For Problems 1–3, use the contour diagram for the function $z = f(x, y)$ in Figure 8.23.

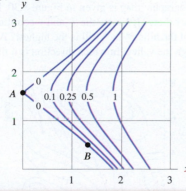

1. Approximate the coordinates of a point (x, y) with $f(x, y) = 0.5$.

2. Find the value of $f(A)$.

3. Is $f(B)$ greater than, equal to, or less than $f(A)$?

4. Figure 8.24 shows contours for the function $z = f(x, y)$. Is z an increasing or a decreasing function of x? Is z an increasing or a decreasing function of y?

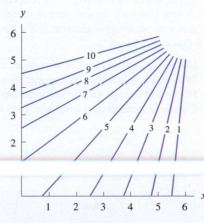

Figure 8.24

5. Figure 8.25 is a contour diagram for the sales of a product as a function of the price of the product and the amount spent on advertising. Which axis corresponds to the amount spent on advertising? Explain.

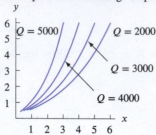

Figure 8.25

■ For Problems 6–8, use the contour diagram for the function $z = f(x, y)$ in Figure 8.26.

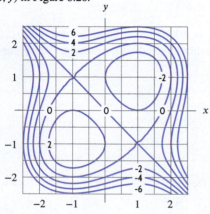

Figure 8.26

6. Find

   (a) $f(1, -1)$ **(b)** $f(2, -1)$ **(c)** $f(-1, 0)$

7. Find a value of x for which

   (a) $f(x, -1) = 2$ **(b)** $f(x, 2) = 0$

8. Find a value of y for which

   (a) $f(1, y) = -4$ **(b)** $f(0, y) = 0$

9. Figure 8.27 shows the contours of the temperature H in a room near a recently opened window. Label the three contours with reasonable values of H if the house is in the following locations.

   (a) Minnesota in winter (where winters are harsh).
   (b) San Francisco in winter (where winters are mild).
   (c) Houston in summer (where summers are hot).
   (d) Oregon in summer (where summers are mild).

Figure 8.27

10. Figure 8.28 shows the contour map of $z = f(B, t)$ which gives the balance in a bank account (in thousands of dollars) t years after an initial investment of B thousand dollars.

   (a) Label the contours with their values.
   (b) Determine the initial investment necessary so that the account has $4,000 in 15 years.

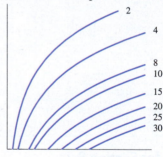

Figure 8.28

11. Figure 8.29 shows the contours of the amount, C, in mg, of medication in the blood stream, as a function of the time t since an initial dose D is administered to the patient. Which axis corresponds to the initial dosage?

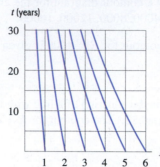

Figure 8.29

12. A topographic map is given in Figure 8.30. How many hills are there? Estimate the x- and y-coordinates of the tops of the hills. Which hill is the highest? A river runs through the valley; in which direction is it flowing?

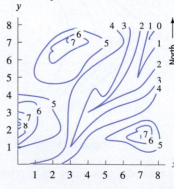

Figure 8.30

■ In Problems **13–20**, sketch a contour diagram for the function with at least four labeled contours. Describe in words the contours and how they are spaced.

13. $f(x, y) = x + y$

14. $f(x, y) = 3x + 3y$

15. $f(x, y) = x + y + 1$

16. $f(x, y) = 2x - y$

17. $f(x, y) = -x - y$

18. $f(x, y) = y - x^2$

19. $f(x, y) = x^2 + y^2$

20. $f(x, y) = xy$

21. Draw a contour diagram for $C(d, m) = 40d + 0.15m$. Include contours for $C = 50, 100, 150, 200$.

22. Maple syrup production is highest when the nights are cold and the days are warm. Make a possible contour diagram for maple syrup production as a function of the high (daytime) temperature and the low (nighttime) temperature. Label the contours with 10, 20, 30, and 40 (in liters of maple syrup).

23. Hiking on a level trail going due east, you decide to leave the trail and climb toward the mountain on your left. The farther you go along the trail before turning off, the gentler the climb. Sketch a possible topographical map showing the elevation contours.

■ Problems **24–26** refer to the map in Figure 8.8 on page 346.

24. Give the range of daily high temperatures for:

(a) Pennsylvania   (b)   North Dakota

(c) California

25. Sketch a possible graph of the predicted high temperature T on a line north-south through Topeka.

26. Sketch possible graphs of the predicted high temperature on a north-south line and an east-west line through Boise.

27. A manufacturer sells two products, one at a price of $4000 a unit and the other at a price of $13,000 a unit. A quantity q_1 of the first product and q_2 of the second product are sold at a total cost of $\$(4000 + q_1 + q_2)$ to the manufacturer.

(a) Express the manufacturer's profit, π, as a function of q_1 and q_2.

(b) Sketch contours of π for $\pi = 10,000$, $\pi = 20,000$, and $\pi = 30,000$ and the break-even curve $\pi = 0$.

28. The contour diagram in Figure 8.31 shows your happiness as a function of love and money.

(a) Describe in words your happiness as a function of:

(i) Money, with love fixed.

(ii) Love, with money fixed.

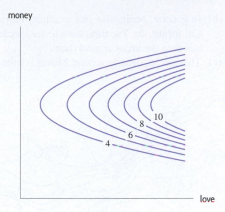

Figure 8.31

29. Sketch a contour diagram for $z = y - \sin x$. Include at least four labeled contours. Describe the contours in words and how they are spaced.

30. Each of the contour diagrams in Figure 8.32 shows population density in a certain region. Choose the contour diagram that best corresponds to each of the following situations. Many different matchings are possible. Pick any reasonable one and justify your choice.

(a) The center of the diagram is a city.

(b) The center of the diagram is a lake.

(c) The center of the diagram is a power plant.

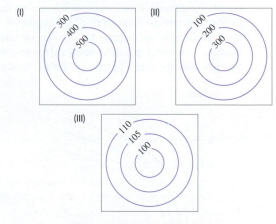

Figure 8.32

31. Figure 8.33 shows contours of the function giving the species density of breeding birds at each point in the US, Canada, and Mexico.[5] Are the following statements true or false? Explain your answers.

and two different cross-sections with money fixed.

species density increases.

[5] From the undergraduate senior thesis of Professor Robert Cook, former Director of Harvard's Arnold Arboretum.

(b) In general, peninsulas (for example, Florida, Baja California, the Yucatan) have lower species densities than the areas around them.

(c) The species density around Miami is over 100.

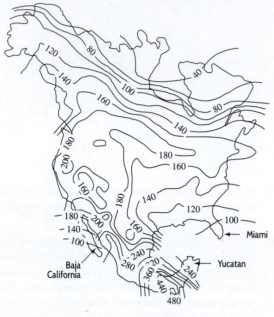

Figure 8.33

32. The concentration, C, of a drug in the blood is given by $C = f(x, t) = xte^{-t}$, where x is the amount of drug injected (in mg) and t is the number of hours since the injection. The contour diagram of $f(x, t)$ is given in Figure 8.34. Explain the diagram by varying one variable at a time: describe f as a function of x if t is held fixed, and then describe f as a function of t if x is held fixed.

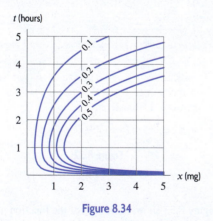

Figure 8.34

33. The wind chill tells you how cold it feels as a function of the air temperature and wind speed. Figure 8.35 is a contour diagram of wind chill (°F).

(a) If the wind speed is 15 mph, what temperature feels like −20°F?

(b) Estimate the wind chill if the temperature is 0°F and the wind speed is 10 mph.

(c) Humans are at extreme risk when the wind chill is below −50°F. If the temperature is −20°F, estimate the wind speed at which extreme risk begins.

(d) If the wind speed is 15 mph and the temperature drops by 20°F, approximately how much colder do you feel?

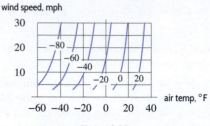

Figure 8.35

34. In a small printing business, $P = 2N^{0.6}V^{0.4}$, where N is the number of workers, V is the value of the equipment (in equipment units), and P is production, in thousands of pages per day.

(a) If this company has a labor force of 300 workers and 200 units of equipment, what is production?

(b) If the labor force is doubled (to 600 workers), how does production change?

(c) If the company purchases enough equipment to double the value of its equipment (to 400 units), how does production change?

(d) If both N and V are doubled from the values given in part (a), how does production change?

35. Figure 8.36 shows a contour map of a hill with two paths, A and B.

(a) On which path, A or B, will you have to climb more steeply?

(b) On which path, A or B, will you probably have a better view of the surrounding countryside? (Assume trees do not block your view.)

(c) Alongside which path is there more likely to be a stream?

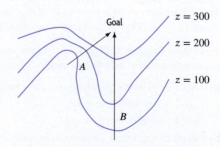

Figure 8.36

36. Figure 8.37 shows cardiac output (in liters per minute) in patients suffering from shock as a function of blood pressure in the central veins (in mm Hg) and the time in hours since the onset of shock.[6]

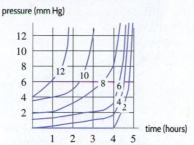

Figure 8.37

(a) In a patient with blood pressure of 4 mm Hg, what is cardiac output when the patient first goes into shock? Estimate cardiac output three hours later. How much time has passed when cardiac output is reduced to 50% of the initial value?

(b) In patients suffering from shock, is cardiac output an increasing or decreasing function of blood pressure?

(c) Is cardiac output an increasing or decreasing function of time, t, where t represents the elapsed time since the patient went into shock?

(d) If blood pressure is 3 mm Hg, explain how cardiac output changes as a function of time. In particular, does it change rapidly or slowly during the first two hours of shock? During hours 2 to 4? During the last hour of the study? Explain why this information is useful to a physician treating a patient for shock.

37. The cornea is the front surface of the eye. Corneal specialists use a TMS, or Topographical Modeling System, to produce a "map" of the curvature of the eye's surface. A computer analyzes light reflected off the eye and draws level curves joining points of constant curvature. The regions between these curves are colored different colors.

The first two pictures in Figure 8.38 are cross-sections of eyes with constant curvature, the smaller being about 38 units and the larger about 50 units. For contrast, the third eye has varying curvature.

(a) Describe in words how the TMS map of an eye of constant curvature will look.

(b) Draw the TMS map of an eye with the cross-section in Figure 8.39. Assume the eye is circular when viewed from the front, and the cross-section is the same in every direction. Put reasonable numeric labels on your level curves.

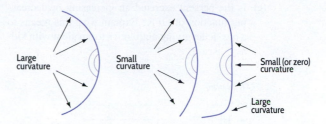

Figure 8.38: Pictures of eyes with different curvature

Figure 8.39

38. The power P produced by a windmill is proportional to the square of the diameter d of the windmill and to the cube of the speed v of the wind.[7]

(a) Write a formula for P as a function of d and v.

(b) A windmill generates 100 kW of power at a certain wind speed. If a second windmill is built having twice the diameter of the original, what fraction of the original wind speed is needed by the second windmill to produce 100 kW?

(c) Sketch a contour diagram for P.

39. Antibiotics can be toxic in large doses. If repeated doses of an antibiotic are to be given, the rate at which the medicine is excreted through the kidneys should be monitored by a physician. One measure of kidney function is the glomerular filtration rate, or GFR, which measures the amount of material crossing the outer (or glomerular) membrane of the kidney, in milliliters per minute. A normal GFR is about 125 ml/min. Figure 8.40 gives a contour diagram of the percent, P, of a dose of mezlocillin (an antibiotic) excreted, as a function of the patient's GFR and the time, t, in hours since the dose was administered.[8]

(a) In a patient with a GFR of 50, approximately how long will it take for 30% of the dose to be excreted?

(b) In a patient with a GFR of 60, approximately what percent of the dose has been excreted after 5 hours?

(c) Explain how we can tell from the graph that, for a patient with a fixed GFR, the amount excreted changes very little after 12 hours.

(d) Is the percent excreted an increasing or decreasing function of time? Explain why this makes sense.

[6] A. C. Guyton and J. E. Hall, *Textbook of Medical Physiology*, 9th ed., p. 289 (Philadelphia: W. B. Saunders, 1996).

[°]Peter G. Welling and Francis L. S. Tse, *Pharmacokinetics of Cardiovascular, Central Nervous System, and Antimicrobial Drugs*, The Royal Society of Chemistry, 1985, p. 316.

(e) Is the percent excreted an increasing or decreasing function of GFR? Explain what this means to a physician giving antibiotics to a patient with kidney disease.

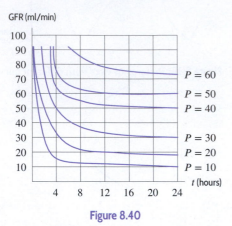

Figure 8.40

40. Each contour diagram (a)–(c) in Figure 8.41 shows satisfaction with quantities of two items X and Y combined. Match (a)–(c) with the items in (I)–(III).

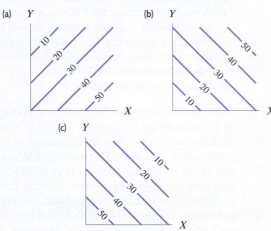

Figure 8.41

(I) X: Income; Y: Leisure time
(II) X: Income; Y: Hours worked
(III) X: Hours worked; Y: Time spent commuting

41. Figure 8.42 shows a contour plot of job satisfaction as a function of the hourly wage and the safety of the workplace (higher values mean safer). Match the jobs at points P, Q, and R with the three descriptions.

   (a) The job is so unsafe that higher pay alone would not increase my satisfaction very much.
   (b) I could trade a little less safety for a little more pay. It would not matter to me.
   (c) The job pays so little that improving safety would not make me happier.

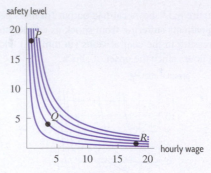

Figure 8.42

42. The total productivity $f(n, T)$ of an advertising agency (in ads per day) depends on the number n of workers and the temperature T of the office in degrees Fahrenheit. More workers create more ads, but the farther the temperature from 75°F, the slower they work. Draw a possible contour diagram for the function $f(n, T)$.

43. Figure 8.43 shows contours for a person's body mass index, BMI $= f(w, h) = 703w/h^2$, where w is weight in pounds and h is height in inches. Find the BMI contour values bounding the *underweight* and *normal* regions.

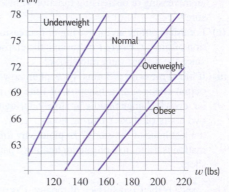

Figure 8.43

44. A shopper buys x units of item A and y units of item B, obtaining satisfaction $s(x, y)$ from the purchase. (Satisfaction is called *utility* by economists.) The contours $s(x, y) = xy = c$ are called *indifference curves* because they show pairs of purchases that give the shopper the same satisfaction.

   (a) A shopper buys 8 units of A and 2 units of B. What is the equation of the indifference curve showing the other purchases that give the shopper the same satisfaction? Sketch this curve.
   (b) After buying 4 units of item A, how many units of B must the shopper buy to obtain the same satisfaction as obtained from buying 8 units of A and 2 units of B?
   (c) The shopper reduces the purchase of item A by k, a fixed number of units, while increasing the purchase of B to maintain satisfaction. In which of the following cases is the increase in B largest?
      • Initial purchase of A is 6 units
      • Initial purchase of A is 8 units

8.3   PARTIAL DERIVATIVES

In one-variable calculus we saw how the derivative measures the rate of change of a function. We begin by reviewing this idea.

Rate of Change of Airline Revenue

In Section 8.1 we saw a two-variable function which gives an airline's revenue, R, as a function of the number of full-price tickets, x, and the number of discount tickets, y, sold:

$$R = f(x, y) = 350x + 200y.$$

If we fix the number of discount tickets at $y = 10$, we have a one-variable function

$$R = f(x, 10) = g(x) = 350x + 2000.$$

The rate of the change of revenue with respect to x is given by the one-variable derivative

$$g'(x) = 350.$$

This tells us that, if y is fixed at 10, then the revenue increases by \$350 for the next additional full-price ticket sold. We call $g'(x)$ the *partial derivative of R with respect to x* at the point $(x, 10)$. If $R = f(x, y)$, we write

$$\frac{\partial R}{\partial x} = f_x(x, 10) = g'(x) = 350.$$

Example 1   Find the rate of change of revenue, R, as y increases with x fixed at $x = 20$.

Solution   Substituting $x = 20$ into $R = 350x + 200y$ gives the one-variable function

$$R = h(y) = 350(20) + 200y = 7000 + 200y.$$

The rate of change of R as y increases with x fixed is

$$\frac{\partial R}{\partial y} = f_y(20, y) = h'(y) = 200.$$

We call $\partial R/\partial y = f_y(20, y)$ the *partial derivative of R with respect to y* at the point $(20, y)$. The fact that both partial derivatives of R are positive corresponds to the fact that the revenue is increasing as more of either type of ticket is sold.

Definition of the Partial Derivative

For any function $f(x, y)$ we study the influence of x and y separately on the value $f(x, y)$ by keeping one fixed and letting the other vary. The method of the previous example allows us to calculate the rates of change of $f(x, y)$ with respect to x and y. For all points (a, b) at which the limits exist, we make the following definitions:

Partial Derivatives of f with Respect to x and y

The *partial derivative of f with respect to x at (a, b)* is the derivative of f with y fixed at b:

$$f_x(a, b) = \begin{matrix} \text{Rate of change of } f \text{ with respect to } x \\ \text{at the point } (a, b) \end{matrix} = \lim_{h \to 0} \frac{f(a + h, b) - f(a, b)}{h}.$$

The *partial derivative of f with respect to y at (a, b)* is the derivative of f with x fixed at a:

$$f_y(a, b) = \begin{matrix} \text{Rate of change of } f \text{ with respect to } y \\ \text{at the point } (a, b) \end{matrix} = \lim_{h \to 0} \frac{f(a, b + h) - f(a, b)}{h}.$$

If we think of a and b as variables, $a = x$ and $b = y$, we have the **partial derivative functions** $f_x(x, y)$ and $f_y(x, y)$.

Just as with ordinary derivatives, there is an alternative notation:

Alternative Notation for Partial Derivatives

If $z = f(x, y)$ we can write

$$f_x(x, y) = \frac{\partial z}{\partial x} \qquad \text{and} \qquad f_y(x, y) = \frac{\partial z}{\partial y}$$

$$f_x(a, b) = \frac{\partial z}{\partial x}\bigg|_{(a,b)} \qquad \text{and} \qquad f_y(a, b) = \frac{\partial z}{\partial y}\bigg|_{(a,b)}$$

We use the symbol ∂ to distinguish partial derivatives from ordinary derivatives. In cases where the independent variables have names different from x and y, we adjust the notation accordingly. For example, the partial derivatives of $f(u, v)$ are denoted by f_u and f_v.

Estimating Partial Derivatives from a Table

Example 2 An experiment[9] done on rats to measure the toxicity of formaldehyde yielded the data shown in Table 8.6. The values in the table show the percent, P, of rats that survived an exposure with concentration c (in parts per million) after t months, so $P = f(t, c)$. Using Table 8.6, estimate $f_t(18, 6)$ and $f_c(18, 6)$. Interpret your answers in terms of formaldehyde toxicity.

Table 8.6 *Percent, P, of rat population surviving after exposure to formaldehyde vapor*

					Time t (months)								
	0	2	4	6	8	10	12	14	16	18	20	22	24
0	100	100	100	100	100	100	100	100	100	100	99	97	95
Conc. c (ppm) 2	100	100	100	100	100	100	100	100	99	98	97	95	92
6	100	100	100	99	99	98	96	96	95	93	90	86	80
15	100	100	100	99	99	99	99	96	93	82	70	58	36

Solution For $f_t(18, 6)$, we fix c at 6 ppm, and find the rate of change of percent surviving, P, with respect to t. We have

$$f_t(18, 6) \approx \frac{\Delta P}{\Delta t} = \frac{f(20, 6) - f(18, 6)}{20 - 18} = \frac{90 - 93}{20 - 18} \approx -1.5 \text{ \% per month.}$$

This is the rate of change of percent surviving, P, *in the time t direction* at the point $(18, 6)$. The

---

[9] James E. Gibson, *Formaldehyde Toxicity*, p. 125 (New York: Hemisphere Publishing Company, McGraw-Hill, 1983).

fact that it is negative means that P is decreasing as we read across the $c = 6$ row of the table in the direction of increasing t (that is, horizontally from left to right in Table 8.6). For $f_c(18, 6)$, we fix t at 18, and calculate the rate of change of P as we move in the direction of increasing c (that is, from top to bottom in Table 8.6). We have

$$f_c(18, 6) \approx \frac{\Delta P}{\Delta c} = \frac{f(18, 15) - f(18, 6)}{15 - 6} = \frac{82 - 93}{15 - 6} = -1.22\% \text{ per ppm.}$$

The rate of change of P as c increases is about -1.22% per ppm. This means that as the concentration increases by 1 ppm from 6 ppm, the percent surviving 18 months decreases by about 1.22% per unit increase ppm. The partial derivative is negative because fewer rats survive this long when the concentration of formaldehyde increases. (That is, P goes down as c goes up.)

Using Partial Derivatives to Estimate Values of the Function

Example 3   Use Table 8.6 and partial derivatives to estimate the percent of rats surviving if they are exposed to formaldehyde with a concentration of

(a)   6 ppm for 18.5 months   (b)   18 ppm for 24 months   (c)   9 ppm for 20.5 months

Solution   (a) Since $t = 18.5$ and $c = 6$, we want to evaluate $P = f(18.5, 6)$. Table 8.6 tells us that $f(18, 6) = 93\%$ and we have just calculated

$$\left.\frac{\partial P}{\partial t}\right|_{(18,6)} = f_t(18, 6) = -1.5\% \text{ per month.}$$

This partial derivative tells us that after 18 months of exposure to formaldehyde at a concentration of 6 ppm, P decreases by 1.5% for every additional month of exposure. Therefore after an additional 0.5 month, we have

$$P \approx 93 - 1.5(0.5) = 92.25\%.$$

(b) Now we wish to evaluate $f(24, 18)$. The closest entry to this in Table 8.6 is $f(24, 15) = 36$. We keep t fixed at 24 and increase c from 15 to 18. We estimate the rate of change in P as c changes; this is $\partial P / \partial c$. We see from Table 8.6 that

$$\left.\frac{\partial P}{\partial c}\right|_{(24,15)} \approx \frac{\Delta P}{\Delta c} = \frac{36 - 80}{15 - 6} = -4.89\% \text{ per ppm.}$$

The percent surviving 24 months goes down from 36% by about 4.89% for one unit increase in the formaldehyde concentration above 15 ppm. We have:

$$f(24, 18) \approx 36 - 4.89(3) = 21.33\%.$$

We estimate that only about 21% of the rats would survive for 24 months if they were exposed to formaldehyde as strong as 18 ppm. Since this figure is an extrapolation from the available data, we should use it with caution.

(c) To estimate $f(20.5, 9)$, we use the closest entry $f(20, 6) = 90$. As we move from $(20, 6)$ to $(20.5, 9)$, the percentage, P, changes both due to the change in t and due to the change in c. We estimate the two partial derivatives at $t = 20$, $c = 6$:

$$\left.\frac{\partial P}{\partial t}\right|_{(20,6)} \approx \frac{\Delta P}{\Delta t} = \frac{86 - 90}{22 - 20} = -2\% \text{ per month,}$$

$$\left.\frac{\partial P}{\partial c}\right|_{(20,6)} \approx \frac{\Delta P}{\Delta c} = \frac{70 - 90}{15 - 6} = -2.22\% \text{ per month.}$$

The change in P due to a change of $\Delta t = 0.5$ month and $\Delta c = 3$ ppm is

$$\Delta P \approx \text{Change due to } \Delta t + \text{Change due to } \Delta c$$
$$= -2(0.5) - 2.22(3)$$
$$= -7.66.$$

So for $t = 20.5, c = 9$ we have

$$f(20.5, 9) \approx f(20, 6) - 7.66 = 82.34\%.$$

In Example 3(c), we used the relationship among ΔP, Δt, and Δc. In general, the relationship between the change Δf, in function value $f(x, y)$ and the changes Δx and Δy is as follows:

> **Local Linearity**
>
> $$\begin{array}{ccccc} \text{Change} & \approx & \text{Rate of change} & \cdot \; \Delta x \; + & \text{Rate of change} & \cdot \; \Delta y \\ \text{in } f & & \text{in } x\text{-direction} & & \text{in } y\text{-direction} & \end{array}$$
>
> $$\Delta f \approx f_x \cdot \Delta x + f_y \cdot \Delta y$$

Estimating Partial Derivatives from a Contour Diagram

If we move parallel to one of the axes on a contour diagram, the partial derivative is the rate of change of the value of the function on the contours. For example, if the values on the contours are increasing in the direction of positive change, then the partial derivative must be positive.

Example 4   Figure 8.44 shows the contour diagram for the temperature $H(x, t)$ (in °F) in a room as a function of distance x (in feet) from a heater and time t (in minutes) after the heater has been turned on. What are the signs of $H_x(10, 20)$ and $H_t(10, 20)$? Estimate these partial derivatives and explain the answers in practical terms.

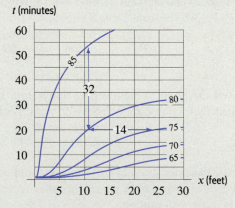

Figure 8.44: Temperature in a heated room

Solution   The point $(10, 20)$ is on the $H = 80$ contour. As x increases, we move toward the $H = 75$ contour, so H is decreasing and $H_x(10, 20)$ is negative. This makes sense because as we move farther from the heater, the temperature drops. On the other hand, as t increases, we move toward the $H = 85$ contour, so H is increasing and $H_t(10, 20)$ is positive. This also makes sense, because it says that as time passes, the room warms up.

   To estimate the partial derivatives, use a difference quotient. Looking at the contour diagram, we see there is a point on the $H = 75$ contour about 14 units to the right of $(10, 20)$. Hence, H decreases by 5 when x increases by 14, so the rate of change of H with respect to x is about $\Delta H/\Delta x = -5/14 \approx -0.36$. Thus, we find

$$H_x(10, 20) \approx -0.36°\text{F/ft}.$$

This means that near the point 10 feet from the heater, after 20 minutes the temperature drops about 1/3 of a degree for each foot we move away from the heater.

To estimate $H_t(10, 20)$, we look again at the contour diagram and notice that the $H = 85$ contour is about 32 units directly above the point $(10, 20)$. So H increases by 5 when t increases by 32. Hence,

$$H_t(10, 20) \approx \frac{\Delta H}{\Delta t} = \frac{5}{32} \approx 0.16°\text{F/min}.$$

This means that after 20 minutes the temperature is going up about 1/6 of a degree in one more minute at the point 10 ft from the heater.

Using Units to Interpret Partial Derivatives

The units of the independent and dependent variables can often be helpful in explaining the meaning of a partial derivative.

Example 5   Suppose that your weight w in pounds is a function $f(c, n)$ of the number c of calories you consume daily and the number n of minutes you exercise daily. Using the units for w, c and n, interpret in everyday terms the statements

$$\left.\frac{\partial w}{\partial c}\right|_{(2000,15)} = 0.02 \quad \text{and} \quad \left.\frac{\partial w}{\partial n}\right|_{(2000,15)} = -0.025.$$

Solution   The units of $\partial w/\partial c$ are pounds per calorie. The statement

$$\left.\frac{\partial w}{\partial c}\right|_{(2000,15)} = 0.02$$

means that if you are presently consuming 2000 calories daily and exercising 15 minutes daily, you will weigh about 0.02 pounds more for one more calorie you consume daily, or about 2 pounds for an extra 100 calories per day. The units of $\partial w/\partial n$ are pounds per minute. The statement

$$\left.\frac{\partial w}{\partial n}\right|_{(2000,15)} = -0.025$$

means that for the same calorie consumption and number of minutes of exercise, you will weigh about 0.025 pounds less for one extra minute you exercise daily, or about 1 pound less for an extra 40 minutes per day. So if you eat an extra 100 calories each day and exercise about 80 minutes more each day, your weight should remain roughly steady.

Problems for Section 8.3

■ In Problems 1–6, a point A is shown on a contour diagram of a function $f(x, y)$.
  (a) Evaluate $f(A)$.
  (b) Is $f_x(A)$ positive, negative, or zero?
  (c) Is $f_y(A)$ positive, negative, or zero?

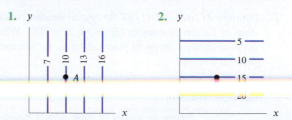

3. y

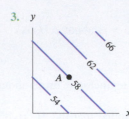

4. y

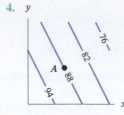

5. y

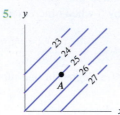

6. y

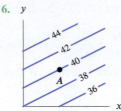

7. Using the contour diagram for $f(x, y)$ in Figure 8.45, decide whether each of these partial derivatives is positive, negative, or approximately zero.

(a) $f_x(4, 1)$ (b) $f_y(4, 1)$
(c) $f_x(5, 2)$ (d) $f_y(5, 2)$

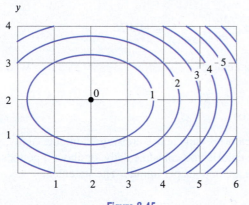

Figure 8.45

8. According to the contour diagram for $f(x, y)$ in Figure 8.45, which is larger: $f_x(3, 1)$ or $f_x(5, 2)$? Explain.

■ For Problems **9–10** refer to Table 8.5 on page 350 giving the heat index, I, in °F, as a function $f(H, T)$ of the relative humidity, H, and the temperature, T, in °F. The heat index is a temperature which tells you how hot it feels as a result of the combination of humidity and temperature.

9. Estimate $\partial I / \partial H$ and $\partial I / \partial T$ for typical weather conditions in Tucson in summer ($H = 10$, $T = 100$). What do your answers mean in practical terms for the residents of Tucson?

10. Answer the question in Problem 9 for Boston in summer ($H = 50$, $T = 80$).

---

[10]From the August 28, 1994, issue of *Parade Magazine*.

11. The demand for coffee, Q, in pounds sold per week, is a function of the price of coffee, c, in dollars per pound and the price of tea, t, in dollars per pound, so $Q = f(c, t)$.

(a) Do you expect f_c to be positive or negative? What about f_t? Explain.
(b) Interpret each of the following statements in terms of the demand for coffee:

$$f(3, 2) = 780 \quad f_c(3, 2) = -60 \quad f_t(3, 2) = 20$$

12. A drug is injected into a patient's blood vessel. The function $c = f(x, t)$ represents the concentration of the drug at a distance x mm in the direction of the blood flow measured from the point of injection and at time t seconds since the injection. What are the units of the following partial derivatives? What are their practical interpretations? What do you expect their signs to be?

(a) $\partial c / \partial x$ (b) $\partial c / \partial t$

13. The quantity Q (in pounds) of beef that a certain community buys during a week is a function $Q = f(b, c)$ of the prices of beef, b, and chicken, c, during the week. Do you expect $\partial Q / \partial b$ to be positive or negative? What about $\partial Q / \partial c$?

14. Table 8.7 gives the number of calories burned per minute, $B = f(s, w)$, for someone roller-blading,[10] as a function of the person's weight, w, and speed, s.

(a) Is f_w positive or negative? Is f_s positive or negative? What do your answers tell us about the effect of weight and speed on calories burned per minute?
(b) Estimate $f_w(160, 10)$ and $f_s(160, 10)$. Interpret your answers.

Table 8.7 *Calories burned per minute*

$w \backslash s$	8 mph	9 mph	10 mph	11 mph
120 lbs	4.2	5.8	7.4	8.9
140 lbs	5.1	6.7	8.3	9.9
160 lbs	6.1	7.7	9.2	10.8
180 lbs	7.0	8.6	10.2	11.7
200 lbs	7.9	9.5	11.1	12.6

15. Values of $f(x, y)$ are in Table 8.8. Assuming they exist, decide whether you expect the following partial derivatives to be positive or negative.

(a) $f_x(-2, -1)$ (b) $f_y(2, 1)$
(c) $f_x(2, 1)$ (d) $f_y(0, 3)$

Table 8.8

$x \backslash y$	−1	1	3	5
−2	7	3	2	1
0	8	5	3	2
2	10	7	5	4
4	13	10	8	7

16. Estimate $z_x(1,0)$ and $z_x(0,1)$ and $z_y(0,1)$ from the contour diagram for $z(x, y)$ in Figure 8.46.

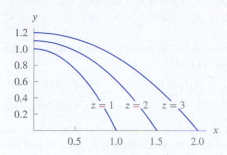

Figure 8.46

17. The monthly mortgage payment in dollars, P, for a house is a function of three variables:

$$P = f(A, r, N),$$

where A is the amount borrowed in dollars, r is the interest rate, and N is the number of years before the mortgage is paid off.

(a) $f(92000, 14, 30) = 1090.08$. What does this tell you, in financial terms?

(b) $\dfrac{\partial P}{\partial r}\bigg|_{(92000, 14, 30)} = 72.82$. What is the financial significance of the number 72.82?

(c) Would you expect $\partial P/\partial A$ to be positive or negative? Why?

(d) Would you expect $\partial P/\partial N$ to be positive or negative? Why?

18. The sales of a product, $S = f(p, a)$, are a function of the price, p, of the product (in dollars per unit) and the amount, a, spent on advertising (in thousands of dollars).

(a) Do you expect f_p to be positive or negative? Why?

(b) Explain the meaning of the statement $f_a(8, 12) = 150$ in terms of sales.

19. Figure 8.16 on page 348 gives a contour diagram of corn production as a function of rainfall, R, in inches and temperature, T, in °F. Corn production, C, is measured as a percentage of the present production, and $C = f(R, T)$. Estimate the following quantities. Give units and interpret your answers in terms of corn production:

(a) $f_R(15, 76)$ (b) $f_T(15, 76)$

20. Figure 8.47 shows a contour diagram for the monthly payment P as a function of the interest rate, $r\%$, and the amount, L, of a 5-year loan. Estimate $\partial P/\partial r$ and $\partial P/\partial L$ at the point where $r = 8$ and $L = 5000$. Give the units and the financial meaning of your answers.

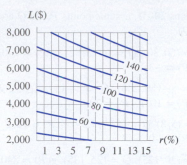

Figure 8.47

21. Use the diagram from Problem 20 in Section 8.1 to estimate $H_T(T, w)$ for $T = 10, 20, 30$ and $w = 0.1, 0.2, 0.3$. What is the practical meaning of these partial derivatives?

22. People commuting to a city can choose to go either by bus or by train. The number of people who choose either method depends in part upon the price of each. Let $f(P_1, P_2)$ be the number of people who take the bus when P_1 is the price of a bus ride and P_2 is the price of a train ride. What can you say about the signs of $\partial f/\partial P_1$ and $\partial f/\partial P_2$? Explain your answers.

23. Suppose that x is the price of one brand of gasoline and y is the price of a competing brand. Then q_1, the quantity of the first brand sold in a fixed time period, depends on both x and y, so $q_1 = f(x, y)$. Similarly, if q_2 is the quantity of the second brand sold during the same period, $q_2 = g(x, y)$. What do you expect the signs of the following quantities to be? Explain.

(a) $\partial q_1/\partial x$ and $\partial q_2/\partial y$ (b) $\partial q_1/\partial y$ and $\partial q_2/\partial x$

24. An airline's revenue, R, is a function of the number of full-price tickets, x, and the number of discount tickets, y, sold. Values of $R = f(x, y)$ are in Table 8.1 on page 340.

(a) Evaluate $f(200, 400)$, and interpret your answer.

(b) Is $f_x(200, 400)$ positive or negative? Is $f_y(200, 400)$ positive or negative? Explain.

(c) Estimate the partial derivatives in part (b). Give units and interpret your answers in terms of revenue.

25. In Problem 24 the revenue is $150,000 when 200 full-price tickets and 400 discount tickets are sold; that is, $f(200, 400) = 150,000$. Use this fact and the partial derivatives $f_x(200, 400) = 350$ and $f_y(200, 400) = 200$ to estimate the revenue when

(a) $x = 201$ and $y = 400$ (b) $x = 200$ and $y = 405$

(c) $x = 203$ and $y = 406$

26. For a function $f(x, y)$, we are given $f(100, 20) = 2750$, and $f_x(100, 20) = 4$, and $f_y(100, 20) = 7$. Estimate $f(105, 21)$.

27. For a function $f(r, s)$, we are given $f(50, 100) = 5.67$, and $f_r(50, 100) = 0.60$, and $f_s(50, 100) = -0.15$. Estimate $f(52, 108)$.

■ In Problems 28–31, assume points P and Q are close. Estimate $g(Q)$.

28. $P = (60, 80)$, $Q = (60.5, 82)$, $g(P) = 100$, $g_x(P) = 2$, $g_y(P) = -3$.

29. $P = (-150, 200)$, $Q = (-152, 203)$, $g(P) = 2500$, $g_x(P) = 10$, $g_y(P) = 20$.

30. $P = (5, 8)$, $Q = (4.97, 7.99)$, $g(P) = 12$, $g_x(P) = -0.1$, $g_y(P) = -0.2$.

31. $P = (30, 125)$, $Q = (25, 135)$, $g(P) = 840$, $g_x(P) = 4$, $g_y(P) = 1.5$.

32. Table 8.6 on page 358 gives the percent of rats surviving, P, as a function of time, t, in months and concentration of formaldehyde, c, in ppm, so $P = f(t, c)$. Use partial derivatives to estimate the percent surviving after 26 months when the concentration is 15.

33. The cardiac output, represented by c, is the volume of blood flowing through a person's heart per unit time. The systemic vascular resistance (SVR), represented by s, is the resistance to blood flowing through veins and arteries. Let p be a person's blood pressure. Then p is a function of c and s, so $p = f(c, s)$.

   (a) What does $\partial p / \partial c$ represent?

   Suppose now that $p = kcs$, where k is a constant.

   (b) Sketch the level curves of p. What do they represent? Label your axes.

   (c) For a person with a weak heart, it is desirable to have the heart pumping against less resistance, while maintaining the same blood pressure. Such a person may be given the drug nitroglycerine to decrease the SVR and the drug dopamine to increase the cardiac output. Represent this on a graph showing level curves. Put a point A on the graph representing the person's state before drugs are given and a point B for after.

   (d) Right after a heart attack, a patient's cardiac output drops, thereby causing the blood pressure to drop. A common mistake made by medical residents is to get the patient's blood pressure back to normal by using drugs to increase the SVR, rather than by increasing the cardiac output. On a graph of the level curves of p, put a point D representing the patient before the heart attack, a point E representing the patient right after the heart attack, and a third point F representing the patient after the resident has given the drugs to increase the SVR.

34. In each case, give a possible contour diagram for the function $f(x, y)$ if

   (a) $f_x > 0$ and $f_y > 0$ (b) $f_x > 0$ and $f_y < 0$

   (c) $f_x < 0$ and $f_y > 0$ (d) $f_x < 0$ and $f_y < 0$

■ In Problems 35–38, give a possible contour diagram for the function $f(x, y)$ if

35. $f_x = 0$, $f_y \neq 0$ 36. $f_y = 0$, $f_x \neq 0$

37. $f_x = 1$ 38. $f_y = -2$

39. Figure 8.48 shows contours of $f(x, y)$ with values of f on the contours omitted. If $f_x(P) > 0$, find the sign:

   (a) $f_y(P)$ (b) $f_y(Q)$ (c) $f_x(Q)$

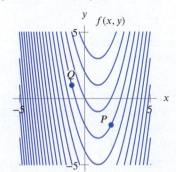

Figure 8.48

40. Figure 8.49 shows a contour diagram of Dan's happiness with snacks of different numbers of cherries and grapes.

   (a) What is the slope of the contours?

   (b) What does the slope tell you?

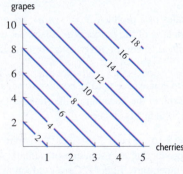

Figure 8.49

8.4 COMPUTING PARTIAL DERIVATIVES ALGEBRAICALLY

The partial derivative $f_x(x, y)$ is the ordinary derivative of the function $f(x, y)$ with respect to x with y fixed, and the partial derivative $f_y(x, y)$ is the ordinary derivative of $f(x, y)$ with respect to y with x fixed. Thus, we can use all the techniques for differentiation from single-variable calculus to find partial derivatives.

Example 1   Let $f(x, y) = x^2 + 5y^2$. Find $f_x(3, 2)$ and $f_y(3, 2)$ algebraically.

Solution   We use the fact that $f_x(3, 2)$ is the derivative of $f(x, 2)$ at $x = 3$. To find f_x, we fix y at 2:

$$f(x, 2) = x^2 + 5(2^2) = x^2 + 20.$$

Differentiating with respect to x gives

$$f_x(x, 2) = 2x \qquad \text{so} \qquad f_x(3, 2) = 2(3) = 6.$$

Similarly, $f_y(3, 2)$ is the derivative of $f(3, y)$ at $y = 2$. To find f_y, we fix x at 3:

$$f(3, y) = 3^2 + 5y^2 = 9 + 5y^2.$$

Differentiating with respect to y, we have

$$f_y(3, y) = 10y \qquad \text{so} \qquad f_y(3, 2) = 10(2) = 20.$$

Example 2   Let $f(x, y) = x^2 + 5y^2$ as in Example 1. Find f_x and f_y as functions of x and y.

Solution   To find f_x, we treat y as a constant. Thus $5y^2$ is a constant and the derivative with respect to x of this term is 0. We have

$$f_x(x, y) = 2x + 0 = 2x.$$

To find f_y, we treat x as a constant and so the derivative of x^2 with respect to y is zero. We have

$$f_y(x, y) = 0 + 10y = 10y.$$

Example 3   Find both partial derivatives of each of the following functions:
(a) $f(x, y) = 3x + e^{-5y}$ (b) $f(x, y) = x^2 y$ (c) $f(u, v) = u^2 e^{2v}$

Solution   (a)   To find f_x, we treat y as a constant, so the term e^{-5y} is a constant, and the derivative of this term is zero. Likewise, to find f_y, we treat x as a constant. We have

$$f_x(x, y) = 3 + 0 = 3 \qquad \text{and} \qquad f_y(x, y) = 0 + (-5)e^{-5y} = -5e^{-5y}.$$

(b)   To find f_x, we treat y as a constant, so the function is treated as a constant times x^2. The derivative of a constant times x^2 is the constant times $2x$, and so we have

$$f_x(x, y) = (2x)y = 2xy \qquad \text{Similarly,} \qquad f_y(x, y) = (x^2)(1) = x^2.$$

(c)   To find f_u, we treat v as a constant, and to find f_v, we treat u as a constant. We have

$$f_u(u, v) = (2u)(e^{2v}) = 2ue^{2v} \qquad \text{and} \qquad f_v(u, v) = u^2(2e^{2v}) = 2u^2 e^{2v}.$$

Example 4   The concentration C of a drug in the blood (in mg per liter) following its injection is a function of the dose x (in mg) injected and the time t (in hours) since the injection. Suppose we are told that $C = f(x, t) = xte^{-t}$. Evaluate the following quantities and explain what each one means in practical terms: (a) $f_x(1, 2)$ (b) $f_t(1, 2)$

Solution   (a)   To find f_x, we treat t as a constant and differentiate with respect to x, giving

$$f_x(x, t) = te^{-t}.$$

Substituting $x = 1, t = 2$ gives

$$f_x(1, 2) = 2e^{-2} \approx 0.27 \text{ mg/liter per milligram.}$$

To see what $f_x(1, 2)$ means, think about the function $f(x, 2)$ of which it is the derivative. The graph of $f(x, 2)$ in Figure 8.50 gives the concentration of drug in the blood as a function of dose two hours after the injection. The derivative $f_x(1, 2)$ is the slope of this graph at the point $x = 1$; it is positive because a larger dose results in a larger drug concentration. More precisely, the partial derivative $f_x(1, 2)$ gives the rate of change of the drug concentration with respect to the dose injected, namely an increase in concentration of 0.27 mg/liter per milligram of additional drug injected.

(b) To find f_t, treat x as a constant and differentiate using the product rule:

$$f_t(x, t) = x \cdot e^{-t} - xte^{-t}.$$

Substituting $x = 1, t = 2$ gives

$$f_t(1, 2) = e^{-2} - 2e^{-2} \approx -0.14 \text{ mg/liter per hour.}$$

To see what $f_t(1, 2)$ means, think about the function $f(1, t)$ of which it is the derivative. The graph of $f(1, t)$ in Figure 8.51 gives the concentration of drug in the blood at time t if the dose is 1 mg. The derivative $f_t(1, 2)$ is the slope of the graph at the point $t = 2$; it is negative because after 2 hours the drug concentration is decreasing. More precisely, the partial derivative $f_t(1, 2)$ gives the rate at which the drug concentration is changing with respect to time, namely a decrease in concentration of 0.14 mg/liter per hour.

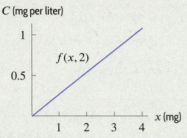

Figure 8.50: Drug concentration after 2 hours as a function of the quantity of drug injected

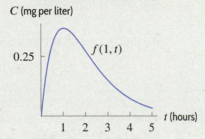

Figure 8.51: Drug concentration as a function of time if 1 mg of drug is injected

Example 5   Let's consider a small printing business where N is the number of workers, V is the value of the equipment (in units of $25,000), and P is the production, measured in thousands of pages per day. Suppose the production function for this company is given by

$$P = f(N, V) = 2N^{0.6}V^{0.4}.$$

(a) If this company has a labor force of 100 workers and 200 units' worth of equipment, what is the production output of the company?

(b) Find $f_N(100, 200)$ and $f_V(100, 200)$. Interpret your answers in terms of production.

Solution   (a) We have $N = 100$ and $V = 200$, so

$$\text{Production} = 2(100)^{0.6}(200)^{0.4} = 263.9 \text{ thousand pages per day.}$$

(b) To find f_N, we treat V as a constant and differentiate with respect to N:

$$f_N(N, V) = 2(0.6)N^{-0.4}V^{0.4}.$$

Substituting $N = 100, V = 200$ gives

$$f_N(100, 200) = 1.2(100^{-0.4})(200^{0.4}) \approx 1.583 \text{ thousand pages/worker.}$$

This tells us that if we have 200 units of equipment and increase the number of workers by 1 from 100 to 101, the production output will go up by about 1.58 units, or 1580 pages per day. Similarly, to find $f_V(100, 200)$, we treat N as a constant and differentiate with respect to V:

$$f_V(N, V) = 2(0.4)N^{0.6}V^{-0.6}.$$

Substituting $N = 100, V = 200$ gives

$$f_V(100, 200) = 0.8(100^{0.6})(200^{-0.6}) \approx 0.53 \text{ thousand pages/unit of equipment.}$$

This tells us that if we have 100 workers and increase the value of the equipment by 1 unit ($25,000) from 200 units to 201 units, the production goes up by about 0.53 units, or 530 pages per day.

Second-Order Partial Derivatives

Since the partial derivatives of a function are themselves functions, we can usually differentiate them, giving *second-order partial derivatives*. A function $z = f(x, y)$ has two first-order partial derivatives, f_x and f_y, and four second-order partial derivatives.

The Second-Order Partial Derivatives of $z = f(x, y)$

$$\frac{\partial^2 z}{\partial x^2} = f_{xx} = (f_x)_x, \qquad \frac{\partial^2 z}{\partial x \partial y} = f_{yx} = (f_y)_x,$$

$$\frac{\partial^2 z}{\partial y \partial x} = f_{xy} = (f_x)_y, \qquad \frac{\partial^2 z}{\partial y^2} = f_{yy} = (f_y)_y.$$

Second-order partial derivatives tell us how the derivatives f_x and f_y change as x and y increase. For example, if $f_{xx}(a, b) < 0$, then at (a, b), we know values of f_x decrease as x increases. In addition, it is usual to omit the parentheses, writing f_{xy} instead of $(f_x)_y$ and $\frac{\partial^2 z}{\partial y \partial x}$ instead of $\frac{\partial}{\partial y}\left(\frac{\partial z}{\partial x}\right)$.

Example 6   Use the values of the function $f(x, y)$ in Table 8.9 to estimate $f_{xy}(1, 2)$ and $f_{yx}(1, 2)$.

Table 8.9   *Values of $f(x, y)$*

		x		
		0.9	1.0	1.1
	1.8	4.72	5.83	7.06
y	2.0	6.48	8.00	9.60
	2.2	8.62	10.65	12.88

Solution   Since $f_{xy} = (f_x)_y$, we first estimate f_x:

$$f_x(1, 2) \approx \frac{f(1.1, 2) - f(1, 2)}{0.1} = \frac{9.60 - 8.00}{0.1} = 16.0,$$

$$f_x(1, 2.2) \approx \frac{f(1.1, 2.2) - f(1, 2.2)}{0.1} = \frac{12.88 - 10.65}{0.1} = 22.3.$$

Thus,

$$f_{xy}(1, 2) \approx \frac{f_x(1, 2.2) - f_x(1, 2)}{0.2} = \frac{22.3 - 16.0}{0.2} = 31.5.$$

Similarly,

$$f_{yx}(1,2) \approx \frac{f_y(1.1,2) - f_y(1,2)}{0.1} \approx \frac{1}{0.1}\left(\frac{f(1.1,2.2) - f(1.1,2)}{0.2} - \frac{f(1,2.2) - f(1,2)}{0.2}\right)$$

$$= \frac{1}{0.1}\left(\frac{12.88 - 9.60}{0.2} - \frac{10.65 - 8.00}{0.2}\right) = 31.5.$$

Observe that in this example, $f_{xy} = f_{yx}$ at the point $(1,2)$.

Example 7   Compute the four second-order partial derivatives of $f(x,y) = xy^2 + 3x^2 e^y$.

Solution   From $f_x(x,y) = y^2 + 6xe^y$ we get

$$f_{xx}(x,y) = \frac{\partial}{\partial x}(y^2 + 6xe^y) = 6e^y \quad \text{and} \quad f_{xy}(x,y) = \frac{\partial}{\partial y}(y^2 + 6xe^y) = 2y + 6xe^y.$$

From $f_y(x,y) = 2xy + 3x^2 e^y$ we get

$$f_{yx}(x,y) = \frac{\partial}{\partial x}(2xy + 3x^2 e^y) = 2y + 6xe^y \quad \text{and} \quad f_{yy}(x,y) = \frac{\partial}{\partial y}(2xy + 3x^2 e^y) = 2x + 3x^2 e^y.$$

Observe that $f_{xy} = f_{yx}$ in this example.

The Mixed Partial Derivatives Are Equal

It is not an accident that the estimates for $f_{xy}(1,2)$ and $f_{yx}(1,2)$ are equal in Example 6, because the same values of the function are used to calculate each one. The fact that $f_{xy} = f_{yx}$ in Example 7 corroborates the following general result:

> If f_{xy} and f_{yx} are continuous at (a,b), then
>
> $$f_{xy}(a,b) = f_{yx}(a,b).$$

Most of the functions we will encounter not only have f_{xy} and f_{yx} continuous, but all their higher-order partial derivatives (such as f_{xxy} or f_{xyyy}) will be continuous. We call such functions *smooth*.

Problems for Section 8.4

■ Find the partial derivatives in Problems **1–13**. The variables are restricted to a domain on which the function is defined.

1. f_x and f_y if $f(x,y) = x^2 + 2xy + y^3$

2. f_x and f_y if $f(x,y) = 2x^2 + 3y^2$

3. f_x and f_y if $f(x,y) = 100x^2 y$

4. f_u and f_v if $f(u,v) = u^2 + 5uv + v^2$

5. $\dfrac{\partial z}{\partial x}$ if $z = x^2 e^y$

6. $\dfrac{\partial Q}{\partial p}$ if $Q = 5a^2 p - 3ap^3$

7. f_t if $f(t,a) = 5a^2 t^3$

8. f_x and f_y if $f(x,y) = 5x^2 y^3 + 8xy^2 - 3x^2$

9. f_x and f_y if $f(x,y) = 10x^2 e^{3y}$

10. z_x if $z = x^2 y + 2x^5 y$

11. $\dfrac{\partial}{\partial m}\left(\dfrac{1}{2}mv^2\right)$

12. $\dfrac{\partial P}{\partial r}$ if $P = 100e^{rt}$

13. $\dfrac{\partial A}{\partial h}$ if $A = \frac{1}{2}(a+b)h$

14. If $f(x,y) = x^3 + 3y^2$, find $f(1,2)$, $f_x(1,2)$, $f_y(1,2)$.

15. If $f(u,v) = 5uv^2$, find $f(3,1)$, $f_u(3,1)$, and $f_v(3,1)$.

■ In Problems 16–19, find all points where the partial derivatives of $f(x, y)$ are both 0.

16. $f(x, y) = x^2 + y^2$

17. $f(x, y) = xe^y$

18. $f(x, y) = x^2 + 2x + y^2$

19. $f(x, y) = x^3 + 3x^2 + y^3 - 3y$

■ In Problems 20–22:

(a) Find $f_x(1, 1)$ and $f_y(1, 1)$.

(b) Use part (a) to match $f(x, y)$ with one of the contour diagrams (I)–(III), each shown centered at $(1, 1)$ with the same scale in the x and y directions.

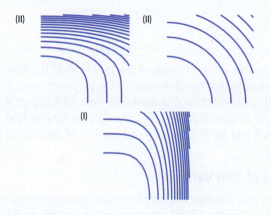

20. $f(x, y) = x^2 + y^2$

21. $f(x, y) = x^4 + y^2$

22. $f(x, y) = x^2 + y^4$

23. (a) Let $f(x, y) = x^2 + y^2$. Estimate $f_x(2, 1)$ and $f_y(2, 1)$ using the contour diagram for f in Figure 8.52.

(b) Estimate $f_x(2, 1)$ and $f_y(2, 1)$ from a table of values for f with $x = 1.9, 2, 2.1$ and $y = 0.9, 1, 1.1$.

(c) Compare your estimates in parts (a) and (b) with the exact values of $f_x(2, 1)$ and $f_y(2, 1)$ found algebraically.

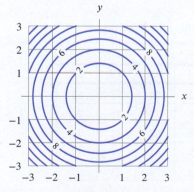

Figure 8.52

24. The amount of money, $\$B$, in a bank account earning interest at a continuous rate, r, depends on the amount deposited, $\$P$, and the time, t, it has been in the bank, where

$$B = Pe^{rt}.$$

Find $\partial B/\partial t$, $\partial B/\partial r$ and $\partial B/\partial P$ and interpret each in financial terms.

25. A company's production output, P, is given in tons, and is a function of the number of workers, N, and the value of the equipment, V, in units of \$25,000. The production function for the company is

$$P = f(N, V) = 5N^{0.75}V^{0.25}.$$

The company currently employs 80 workers, and has equipment worth \$750,000. What are N and V? Find the values of f, f_N, and f_V at these values of N and V. Give units and explain what each answer means in terms of production.

26. The cost of renting a car from a certain company is \$40 per day plus 15 cents per mile, and so we have

$$C = 40d + 0.15m.$$

Find $\partial C/\partial d$ and $\partial C/\partial m$. Give units and explain why your answers make sense.

27. Figure 8.53 is a contour diagram of $f(x, y)$. In each of the following cases, list the marked points in the diagram (there may be none or more than one) at which

(a) $f_x < 0$ (b) $f_y > 0$

(c) $f_{xx} > 0$ (d) $f_{yy} < 0$

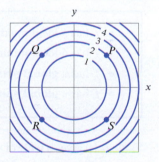

Figure 8.53

■ For Problems 28–39, calculate all four second-order partial derivatives and confirm that the mixed partials are equal.

28. $f(x, y) = x^2y$ 29. $f(x, y) = xe^y$

30. $f(x, y) = x^2 + 2xy + y^2$ 31. $f(x, y) = \dfrac{2x}{y}, \quad y \neq 0$

32. $f = 5 + x^2y^2$ 33. $f = e^{xy}$

34. $B = 5xe^{-2t}$ 35. $f(x, t) = t^3 - 4x^2t$

36. $f = 100e^{rt}$ 37. $Q = 5p_1^2p_2^{-1}, \quad p_2 \neq 0$

38. $V = \pi r^2 h$ 39. $P = 2KL^2$

40. Is there a function f which has the following partial derivatives? If so, what is it? Are there any others?

$$f_x(x, y) = 4x^3y^2 - 3y^4,$$
$$f_y(x, y) = 2x^4y - 12xy^3.$$

41. Show that the Cobb-Douglas function

$$Q = bK^\alpha L^{1-\alpha} \quad \text{where} \quad 0 < \alpha < 1$$

satisfies the equation

$$K\frac{\partial Q}{\partial K} + L\frac{\partial Q}{\partial L} = Q.$$

■ Problems **42–44** are about the money supply, M, which is the total value of all the cash and checking account balances in an economy. It is determined by the value of all the cash, B, the ratio, c, of cash to checking deposits, and the fraction, r, of checking account deposits that banks hold as cash:

$$M = \frac{c+1}{c+r}B.$$

(a) Find the partial derivative.

(b) Give its sign.

(c) Explain the significance of the sign in practical terms.

42. $\partial M / \partial B$ **43.** $\partial M / \partial r$ **44.** $\partial M / \partial c$

8.5   CRITICAL POINTS AND OPTIMIZATION

To optimize a function means to find the largest or smallest value of the function. If the function represents profit, we may want to find the conditions that maximize profit. On the other hand, if the function represents cost, we may want to find the conditions that minimize cost. In Chapter 4, we saw how to optimize a function of one variable by investigating critical points. In this section, we see how to extend the notions of critical points and local extrema to a function of more than one variable.

Local and Global Maxima and Minima for Functions of Two Variables

Functions of several variables, like functions of one variable, can have *local* and *global extrema* (that is, local and global maxima and minima). A function has a local extremum at a point where it takes on the largest or smallest value in a small region around the point. Global extrema are the largest or smallest value anywhere. For a function f defined on a domain R, we say:

- f has a **local maximum** at P_0 if $f(P_0) \geq f(P)$ for all points P near P_0
- f has a **local minimum** at P_0 if $f(P_0) \leq f(P)$ for all points P near P_0
- f has a **global maximum** at P_0 if $f(P_0) \geq f(P)$ for all points P in R
- f has a **global minimum** at P_0 if $f(P_0) \leq f(P)$ for all points P in R

Example 1   Table 8.10 gives a table of values for a function $f(x, y)$. Estimate the location and value of any global maxima or minima for $0 \leq x \leq 1$ and $0 \leq y \leq 20$.

Table 8.10   *Where are the extreme points of this function $f(x, y)$?*

					x		
		0	0.2	0.4	0.6	0.8	1.0
	0	80	84	82	76	71	65
	5	86	90	88	73	77	71
y	10	91	95	93	88	82	76
	15	87	91	89	84	78	72
	20	82	86	84	79	73	67

Solution   The global maximum value of the function appears to be 95 at the point $(0.2, 10)$. Since the table only gives certain values, we cannot be sure that this is exactly the maximum. (The function might have a larger value at, for example, $(0.3, 11)$.) The global minimum value of this function on the points given is 65 at the point $(1, 0)$.

Example 2   Figure 8.54 gives a contour diagram for a function $f(x, y)$. Estimate the location and value of any local maxima or minima. Are any of these global maxima or minima on the square shown?

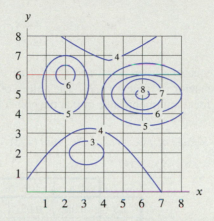

Figure 8.54: Where are the local and global extreme points of this function?

Solution   There is a local maximum of above 8 near the point $(6, 5)$, a local maximum of above 6 near the point $(2, 6)$, and a local minimum of below 3 near the point $(3, 2)$. The value above 8 is the global maximum and the value below 3 is the global minimum on the given domain.

In Example 1 and Example 2, we can estimate the location and value of extreme points, but we do not have enough information to find them exactly. This is usually true when we are given a table of values or a contour diagram. To find local or global extrema exactly, we usually need to have a formula for the function.

Finding a Local Maximum or Minimum Analytically

In one-variable calculus, the local extrema of a function occur at points where the derivative is zero or undefined. How does this generalize to the case of functions of two or more variables? Suppose that a function $f(x, y)$ has a local maximum at a point (x_0, y_0) which is not on the boundary of the domain of f. If the partial derivative $f_x(x_0, y_0)$ were defined and positive, then we could increase f by increasing x. If $f_x(x_0, y_0) < 0$, then we could increase f by decreasing x. Since f has a local maximum at (x_0, y_0), there can be no direction in which f is increasing, so we must have $f_x(x_0, y_0) = 0$. Similarly, if $f_y(x_0, y_0)$ is defined, then $f_y(x_0, y_0) = 0$. The case in which $f(x, y)$ has a local minimum is similar. Therefore, we arrive at the following conclusion:

> If a function $f(x, y)$ has a local maximum or minimum at a point (x_0, y_0) not on the boundary of the domain of f, then either
>
> $$f_x(x_0, y_0) = 0 \quad \text{and} \quad f_y(x_0, y_0) = 0$$
>
> or (at least) one partial derivative is undefined at the point (x_0, y_0). Points where each of the partial derivatives is either zero or undefined are called **critical points**.

As in the single-variable case, the fact that (x_0, y_0) is a critical point for f does not necessarily mean that f has a maximum or a minimum there.

How Do We Find Critical Points?

To find critical points of a function f, we find the points where both partial derivatives of f are zero or undefined.

Example 3   Find and analyze the critical points of $f(x, y) = x^2 - 2x + y^2 - 4y + 5$.

Solution   To find the critical points, we set both partial derivatives equal to zero:

$$f_x(x, y) = 2x - 2 = 0,$$
$$f_y(x, y) = 2y - 4 = 0.$$

Solving these equations gives $x = 1$ and $y = 2$. Hence, f has only one critical point, namely $(1, 2)$. What is the behavior of f near $(1, 2)$? The values of the function in Table 8.11 suggest that the function has a local minimum value of 0 at the point $(1, 2)$.

Table 8.11   *Values of $f(x, y)$ near the point $(1, 2)$*

				x		
		0.8	0.9	1.0	1.1	1.2
	1.8	0.08	0.05	0.04	0.05	0.08
	1.9	0.05	0.02	0.01	0.02	0.05
y	2.0	0.04	0.01	0.00	0.01	0.04
	2.1	0.05	0.02	0.01	0.02	0.05
	2.2	0.08	0.05	0.04	0.05	0.08

Example 4   A manufacturing company produces two products which are sold in two separate markets. The company's economists analyze the two markets and determine that the quantities, q_1 and q_2, demanded by consumers and the prices, p_1 and p_2 (in dollars), of each item are related by the equations

$$p_1 = 600 - 0.3q_1 \quad \text{and} \quad p_2 = 500 - 0.2q_2.$$

Thus, if the price for either item increases, the demand for it decreases. The company's total production cost is given by

$$C = 16 + 1.2q_1 + 1.5q_2 + 0.2q_1 q_2.$$

If the company wants to maximize its total profits, how much of each product should it produce? What is the maximum profit?[11]

Solution   The total revenue R is the sum of the revenues, $p_1 q_1$ and $p_2 q_2$, from each market. Substituting for p_1 and p_2, we get

$$R = p_1 q_1 + p_2 q_2$$
$$= (600 - 0.3q_1)q_1 + (500 - 0.2q_2)q_2$$
$$= 600q_1 - 0.3q_1^2 + 500q_2 - 0.2q_2^2.$$

Thus the total profit π is given by

$$\pi = R - C$$
$$= 600q_1 - 0.3q_1^2 + 500q_2 - 0.2q_2^2 - (16 + 1.2q_1 + 1.5q_2 + 0.2q_1 q_2)$$
$$= -16 + 598.8q_1 - 0.3q_1^2 + 498.5q_2 - 0.2q_2^2 - 0.2q_1 q_2.$$

---

[11] Adapted from M. Rosser and P. Lis, *Basic Mathematics for Economists,* 3rd ed., p. 351 (New York: Routledge, 2016).

To maximize π, we compute partial derivatives:

$$\frac{\partial \pi}{\partial q_1} = 598.8 - 0.6q_1 - 0.2q_2,$$
$$\frac{\partial \pi}{\partial q_2} = 498.5 - 0.4q_2 - 0.2q_1.$$

Since the partial derivatives are defined everywhere, the only critical points of π are those where the partial derivatives of π are both equal to zero. Thus, we solve the equations for q_1 and q_2,

$$598.8 - 0.6q_1 - 0.2q_2 = 0,$$
$$498.5 - 0.4q_2 - 0.2q_1 = 0,$$

giving

$$q_1 = 699.1 \approx 699 \quad \text{and} \quad q_2 = 896.7 \approx 897.$$

To see whether this is a maximum, we look at a table of values of profit π around this point. Table 8.12 suggests that profit is greatest at (699, 897). So the company should produce 699 units of the first product priced at \$390.30 per unit, and 897 units of the second product priced at \$320.60 per unit. The maximum profit is then $\pi(699, 897) = \$432{,}797$.

Table 8.12 *Does this profit function have a maximum at* (699, 897)*?*

		Quantity, q_1		
		698	699	700
Quantity, q_2	896	432,796.4	432,796.9	432,796.8
	897	432,796.7	432,797.0	432,796.7
	898	432,796.6	432,796.7	432,796.2

Is a Critical Point a Local Maximum or a Local Minimum?

We can often see whether a critical point is a local maximum or minimum or neither by looking at a table or contour diagram. The following analytic method may also be useful in distinguishing between local maxima and minima.[12] It is analogous to the Second Derivative Test in Chapter 4.

Second Derivative Test for Functions of Two Variables

Suppose (x_0, y_0) is a critical point where $f_x(x_0, y_0) = f_y(x_0, y_0) = 0$. Let

$$D = f_{xx}(x_0, y_0) f_{yy}(x_0, y_0) - f_{xy}(x_0, y_0)^2.$$

- If $D > 0$ and $f_{xx}(x_0, y_0) > 0$, then f has a local minimum at (x_0, y_0).
- If $D > 0$ and $f_{xx}(x_0, y_0) < 0$, then f has a local maximum at (x_0, y_0).
- If $D < 0$, then we say f has a *saddle point* at (x_0, y_0).
- If $D = 0$, the test is inconclusive.

If (x_0, y_0) is a saddle point of f, then it is neither a local maximum nor a local minimum of f. Figure 8.55 gives a contour diagram of a function f around a saddle point. As we move in the y-direction from the center of the diagram, the value of f decreases; but as we move in the x-direction from the center, the value of f increases.

[12] An explanation of this test can be found, for example, in W. McCallum et al., *Multivariable Calculus* (New York: John Wiley, 2017).

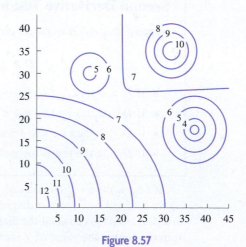

Figure 8.55: A saddle point

Example 5 Use the second derivative test to confirm that the critical point $q_1 = 699.1$, $q_2 = 896.7$ gives a local maximum of the profit function π of Example 4.

Solution To see whether or not we have found a maximum point, we compute the second-order partial derivatives:

$$\frac{\partial^2 \pi}{\partial q_1^2} = -0.6, \quad \frac{\partial^2 \pi}{\partial q_2^2} = -0.4, \quad \frac{\partial^2 \pi}{\partial q_1 \partial q_2} = -0.2.$$

Since

$$D = \frac{\partial^2 \pi}{\partial q_1^2} \frac{\partial^2 \pi}{\partial q_2^2} - \left(\frac{\partial^2 \pi}{\partial q_1 \partial q_2}\right)^2 = (-0.6)(-0.4) - (-0.2)^2 = 0.2 > 0,$$

and

$$\frac{\partial^2 \pi}{\partial q_1^2} < 0,$$

the second derivative test implies that we have found a local maximum point.

Problems for Section 8.5

1. Figure 8.56 shows contours of $f(x, y)$. List the x- and y-coordinates and the value of the function at any local maximum and local minimum points, and identify which is which. Are any of these local extrema also global extrema on the region shown? If so, which ones?

2. Figure 8.57 shows contours of $f(x, y)$. List x- and y-coordinates and the value of the function at any local maximum and local minimum points, and identify which is which. Are any of these local extrema also global extrema on the region shown? If so, which ones?

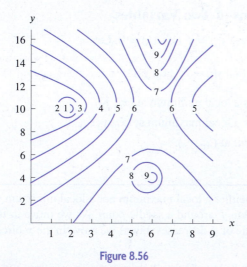

Figure 8.56

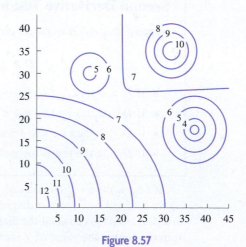

Figure 8.57

■ In Problems **3–5**, estimate the position and approximate value of the global maxima and minima on the region shown, including its boundary.

3.

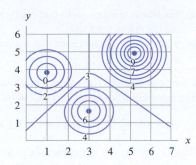

4.

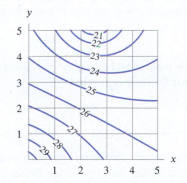

5.

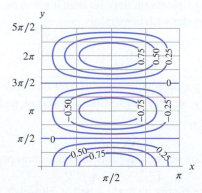

6. Values of $f(x, y)$ and its derivatives are given in Table 8.13.

(a) Identify which points in Table 8.13 are critical points of f.

(b) For each critical point, determine whether it is a local minimum, maximum, or saddle point.

Table 8.13 *Values of f and its derivatives*

(x, y)	f	f_x	f_y	f_{xx}	f_{yy}	f_{xy}
$(0, 0)$	0	0	0	−2	−3	−2
$(1, 0)$	−4	0	0	1	−1	2
$(0, 1)$	−6	0	−1	−1	−1	−2
$(−1, 0)$	−4	0	0	−1	1	2
$(0, −1)$	−6	1	0	1	−2	1

■ In Problems **7–14**, the function has a critical point at $(0, 0)$. What sort of critical point is it?

7. $f(x, y) = xy$

8. $f(x, y) = x^2 + y^2$

9. $g(x, y) = x^4 + y^3$

10. $f(x, y) = x^6 + y^6$

11. $f(x, y) = xe^y - x$

12. $f(x, y) = x - e^x - y^2$

13. $f(x, y) = x^2 - \cos y$

14. $f(x, y) = x \sin y$

■ In Problems **15–24**, find all the critical points and determine whether each is a local maximum, local minimum, a saddle point, or none of these.

15. $f(x, y) = x^2 + 4x + y^2$

16. $f(x, y) = x^2 + xy + 3y$

17. $f(x, y) = x^2 + y^2 + 6x - 10y + 8$

18. $f(x, y) = y^3 - 3xy + 6x$

19. $f(x, y) = x^3 + y^2 - 3x^2 + 10y + 6$

20. $f(x, y) = x^3 + y^3 - 6y^2 - 3x + 9$

21. $f(x, y) = x^3 + y^3 - 3x^2 - 3y + 10$

22. $f(x, y) = x^2 - 2xy + 3y^2 - 8y$

23. $f(x, y) = x^3 - 3x + y^3 - 3y$

24. $f(x, y) = 400 - 3x^2 - 4x + 2xy - 5y^2 + 48y$

25. By looking at the weather map in Figure 8.8 on page 346, find the maximum and minimum daily high temperatures in the states of Mississippi, Alabama, Pennsylvania, New York, California, Arizona, and Massachusetts.

26. A function $f(x, y)$ has partial derivatives $f_x(1, 2) = 3$, $f_y(1, 2) = 5$. Explain how you know that f does not have a minimum at $(1, 2)$.

27. For $f(x, y) = A - (x^2 + Bx + y^2 + Cy)$, what values of A, B, and C give f a local maximum value of 15 at the point $(-2, 1)$?

28. Let $f(x, y) = 3x^2 + ky^2 + 9xy$. Determine the values of k (if any) for which the critical point at $(0, 0)$ is:

(a) A saddle point
(b) A local maximum
(c) A local minimum

29. Let $f(x, y) = x^3 + ky^2 - 5xy$. Determine the values of k (if any) for which the critical point at $(0, 0)$ is:

(a) A saddle point
(b) A local maximum
(c) A local minimum

30. The quantity of a product demanded by consumers is a function of its price. The quantity of one product demanded may also depend on the price of other products. For example, if the only chocolate shop in town (a monopoly) sells milk and dark chocolates, the price it sets for each affects the demand of the other. The quantities demanded, q_1 and q_2, of two products depend on their prices, p_1 and p_2, as follows:

$$q_1 = 150 - 2p_1 - p_2$$
$$q_2 = 200 - p_1 - 3p_2.$$

(a) What does the fact that the coefficients of p_1 and p_2 are negative tell you? Give an example of two products that might be related this way.

(b) If one manufacturer sells both products, how should the prices be set to generate the maximum possible revenue? What is that maximum possible revenue?

31. Two products are manufactured in quantities q_1 and q_2 and sold at prices of p_1 and p_2, respectively. The cost of

producing them is given by

$$C = 2q_1^2 + 2q_2^2 + 10.$$

(a) Find the maximum profit that can be made, assuming the prices are fixed.

(b) Find the rate of change of that maximum profit as p_1 increases.

32. A company operates two plants which manufacture the same item and whose total cost functions are

$$C_1 = 8.5 + 0.03q_1^2 \quad \text{and} \quad C_2 = 5.2 + 0.04q_2^2,$$

where q_1 and q_2 are the quantities produced by each plant. The company is a monopoly. The total quantity demanded, $q = q_1 + q_2$, is related to the price, p, by

$$p = 60 - 0.04q.$$

How much should each plant produce in order to maximize the company's profit?[13]

8.6   CONSTRAINED OPTIMIZATION

Many real optimization problems are constrained by external circumstances. For example, a city wanting to build a public transportation system has a limited number of tax dollars available. A nation trying to maintain its balance of trade must spend less on imports than it earns on exports. In this section, we see how to find an optimum value under such constraints.

A Constrained Optimization Problem

Suppose we want to maximize the production of a company under a budget constraint. Suppose production, f, is a function of two variables, x and y, which are quantities of two raw materials, and

$$f(x, y) = x^{2/3}y^{1/3}.$$

If x and y are purchased at prices of p_1 and p_2 dollars per unit, what is the maximum production f that can be obtained with a budget of c dollars?

To increase f without regard to the budget, we simply increase x and y. However, the budget prevents us from increasing x and y beyond a certain point. Exactly how does the budget constrain us? Suppose that x and y each cost \$100 per unit, and suppose that the total budget is \$378,000. The amount spent on x and y together is given by $g(x, y) = 100x + 100y$, and since we can't spend more than the budget allows, we must have:

$$g(x, y) = 100x + 100y \leq 378{,}000.$$

The goal is to maximize the function

$$f(x, y) = x^{2/3}y^{1/3}.$$

Since we expect to exhaust the budget, we have

$$100x + 100y = 378{,}000.$$

---

[13] Adapted from M. Rosser and P. Lis, *Basic Mathematics for Economists*, 3rd ed. (New York: Routledge, 2016), p. 354.

Example 1   A company has production function $f(x, y) = x^{2/3}y^{1/3}$ and budget constraint $100x + 100y = 378,000$.

   (a) If \$100,000 is spent on x, how much can be spent on y? What is the production in this case?
   (b) If \$200,000 is spent on x, how much can be spent on y? What is the production in this case?
   (c) Which of the two options above is the better choice for the company? Do you think this is the best of all possible options?

Solution   (a) If the company spends \$100,000 on x, then it has \$278,000 left to spend on y. In this case, we have $100x = 100,000$, so $x = 1000$, and $100y = 278,000$, so $y = 2780$. Therefore,

$$\text{Production} = f(1000, 2780) = (1000)^{2/3}(2780)^{1/3} = 1406 \text{ units.}$$

   (b) If the company spends \$200,000 on x, then it has \$178,000 left to spend on y. Therefore, $x = 2000$ and $y = 1780$, and so

$$\text{Production} = f(2000, 1780) = (2000)^{2/3}(1780)^{1/3} = 1924 \text{ units.}$$

   (c) Of these two options, (b) is better since production is larger in this case. This is probably not optimal, since there are many other combinations of x and y that we have not checked.

Graphical Approach: Maximizing Production Subject to a Budget Constraint

How can we find the maximum value of production? We maximize the *objective function*

$$f(x, y) = x^{2/3}y^{1/3}$$

subject to $x \geq 0$ and $y \geq 0$ and the budget constraint

$$g(x, y) = 100x + 100y = 378,000.$$

The constraint is represented by the line in Figure 8.58. Any point on or below the line represents a pair of values of x and y that we can afford. A point on the line completely exhausts the budget, while a point below the line represents values of x and y which can be bought without using up the budget. Any point above the line represents a pair of values that we cannot afford.

Figure 8.58 also shows some contours of the production function f. Since we want to maximize f, we want to find the point which lies on the contour with the largest possible f value *and* which lies within the budget. The point we are looking for must lie on the budget constraint because we should spend all the available money. The key observation is this: The maximum occurs at a point P where the budget constraint is tangent to a production contour. (See Figure 8.58.) The reason is that if we are on the constraint line to the left of P, moving right on the constraint increases f; if we are on the line to the right of P, moving left increases f. Thus, the maximum value of f on the budget constraint line occurs at the point P.

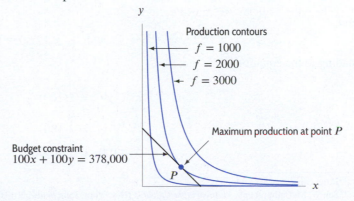

Figure 8.58: At the optimal point P the budget constraint is tangent to a production contour

In general, provided f and g are smooth, we have the following result:

> If $f(x, y)$ has a global maximum or minimum on the constraint $g(x, y) = c$, it occurs at a point where the graph of the constraint is tangent to a contour of f, or at an endpoint of the constraint.[14]

Analytical Approach: The Method of Lagrange Multipliers

Suppose we want to optimize $f(x, y)$ subject to the constraint $g(x, y) = c$. We make the following definition.

> Suppose P_0 is a point satisfying the constraint $g(x, y) = c$.
> - f has a **local maximum** at P_0 **subject to the constraint** if $f(P_0) \geq f(P)$ for all points P near P_0 satisfying the constraint.
> - f has a **global maximum** at P_0 **subject to the constraint** if $f(P_0) \geq f(P)$ for all points P satisfying the constraint.
> Local and global minima are defined similarly.

It can be shown[15] that the constraint is tangent to a contour of f at the point which satisfies the equations laid out in the following method.

> **Method of Lagrange Multipliers** To optimize $f(x, y)$ subject to the constraint $g(x, y) = c$, solve the following system of three equations:
>
> $$f_x(x, y) = \lambda g_x(x, y),$$
> $$f_y(x, y) = \lambda g_y(x, y),$$
> $$g(x, y) = c,$$
>
> for the three unknowns $x, y,$ and λ; the number λ is called the *Lagrange multiplier*. If f has a constrained global maximum or minimum, then it occurs at one of the solutions (x_0, y_0) to this system or at an endpoint of the constraint.

Example 2   Maximize $f(x, y) = x^{2/3} y^{1/3}$ subject to $100x + 100y = 378{,}000$ and $x \geq 0, y \geq 0$.

Solution   Differentiating gives

$$f_x(x, y) = \frac{2}{3} x^{-1/3} y^{1/3} \qquad \text{and} \qquad f_y(x, y) = \frac{1}{3} x^{2/3} y^{-2/3}$$

and

$$g_x(x, y) = 100 \qquad \text{and} \qquad g_y(x, y) = 100,$$

leading to the equations

$$\frac{2}{3} x^{-1/3} y^{1/3} = \lambda(100)$$

$$\frac{1}{3} x^{2/3} y^{-2/3} = \lambda(100)$$

$$100x + 100y = 378{,}000.$$

---

[14] If the constraint has endpoints.

[15] See W. McCallum et al., *Multivariable Calculus* (New York: John Wiley, 2017).

The first two equations show that we must have

$$\frac{2}{3}x^{-1/3}y^{1/3} = \frac{1}{3}x^{2/3}y^{-2/3}.$$

Using the fact that $x^{-1/3} = 1/x^{1/3}$, we can rewrite this as

$$\frac{2y^{1/3}}{3x^{1/3}} = \frac{x^{2/3}}{3y^{2/3}}.$$

Multiplying through by the denominators gives

$$2y^{1/3}(3y^{2/3}) = x^{2/3}(3x^{1/3}),$$

and simplifying using $y^{1/3} \cdot y^{2/3} = y^1$ gives

$$6y = 3x$$
$$2y = x.$$

Since we must also satisfy the constraint $100x + 100y = 378{,}000$, we substitute $x = 2y$ and get

$$100(2y) + 100y = 378{,}000$$
$$300y = 378{,}000$$
$$y = 1260.$$

Since $x = 2y$, we have $x = 2520$. The optimum value occurs at $x = 2520$ and $y = 1260$. For these values,

$$f(2520, 1260) = (2520)^{2/3}(1260)^{1/3} = 2000.1.$$

The endpoints of the constraint are the points $(3780, 0)$ and $(0, 3780)$. Since

$$f(3780, 0) = f(0, 3780) = 0,$$

we see that the maximum value of f is approximately 2000 and that it occurs at $x = 2520$ and $y = 1260$.

The Meaning of λ

In the previous example, we never found (or needed) the value of λ. However, λ does have a practical interpretation. In the production problem we maximized

$$f(x, y) = x^{2/3}y^{1/3}$$

subject to the constraint

$$g(x, y) = 100x + 100y = 378{,}000.$$

We solved the equations

$$\frac{2}{3}x^{-1/3}y^{1/3} = 100\lambda,$$

$$\frac{1}{3}x^{2/3}y^{-2/3} = 100\lambda,$$

$$100x + 100y = 378{,}000,$$

to get $x = 2520$, $y = 1260$. Continuing to find λ gives us

$$\lambda \approx 0.0053.$$

Suppose now we do another, apparently unrelated calculation. Suppose our budget is increased by $1000, from $378,000 to $379,000. The new budget constraint is

$$100x + 100y = 379{,}000.$$

The corresponding solution is at $x = 2527, y = 1263$ and the new maximum value (instead of $f = 2000.1$) is

$$f = (2527)^{2/3}(1263)^{1/3} \approx 2005.4.$$

The additional $1000 in the budget increased the production level f by 5.3 units. Notice that production increased by $5.3/1000 = 0.0053$ units per dollar, which is our value of λ. The value of λ represents the extra production achieved by increasing the budget by one dollar—in other words, the extra "bang" you get for an extra "buck" of budget.

Solving for λ in either of the equations $f_x = \lambda g_x$ or $f_y = \lambda g_y$ suggests that the Lagrange multiplier is given by the ratio of the changes:

$$\lambda \approx \frac{\Delta f}{\Delta g} = \frac{\text{Change in optimum value of } f}{\text{Change in } g}.$$

These results suggest the following interpretations of the Lagrange multiplier λ:

- The value of λ is approximately the change in the optimum value of f when the value of the constraint is increased by 1 unit.
- The value of λ represents the rate of change of the optimum value of f as the constraint increases.

Example 3   The quantity of goods produced according to the function $f(x, y) = x^{2/3}y^{1/3}$ is maximized subject to the budget constraint $100x + 100y = 378{,}000$. Suppose the budget is increased to allow a small increase in production. What price must the product sell for if it is to be worth the increased budget?

Solution   We know that $\lambda = 0.0053$. Therefore, increasing the budget by $1 increases production by about 0.0053 unit. In order to make the increase in budget profitable, the extra goods produced must sell for more than $1. If the price is p in dollars, we must have $0.0053p > 1$. Thus, we need $p > 1/0.0053 \approx$ $189.

Example 4   If x and y are the amounts of raw materials used, the quantity, Q, of a product manufactured is

$$Q = xy.$$

Assume that x costs $20 per unit, y costs $10 per unit, and the production budget is $10,000.

(a) How many units of x and y should be purchased in order to maximize production? How many units are produced at that point?

(b) Find the value of λ and interpret it.

Solution   (a) We maximize $f(x, y) = xy$ subject to the constraint $g(x, y) = 20x + 10y = 10{,}000$ and $x \geq 0$, $y \geq 0$. We have the following partial derivatives:

$$f_x = y, \qquad f_y = x, \qquad \text{and} \qquad g_x = 20, \qquad g_y = 10.$$

The method of Lagrange multipliers gives the following equations:

$$y = 20\lambda$$
$$x = 10\lambda$$
$$20x + 10y = 10{,}000.$$

Substituting values of x and y from the first two equations into the third gives

$$20(10\lambda) + 10(20\lambda) = 10{,}000$$
$$400\lambda = 10{,}000$$
$$\lambda = 25.$$

Substituting $\lambda = 25$ in the first two equations above gives $x = 250$ and $y = 500$.
The endpoints of the constraint are the point $(500, 0)$ and $(0, 1000)$. Since

$$f(500, 0) = 0 \quad \text{and} \quad f(0, 1000) = 0,$$

the maximum value of the production function is

$$f(250, 500) = 250 \cdot 500 = 125{,}000 \text{ units}.$$

The company should purchase 250 units of x and 500 units of y, giving 125,000 units of products.
(b) We have $\lambda = 25$. This tells us that if the budget is increased by \$1, we expect production to go up by about 25 units. If the budget goes up by \$1000, maximum production will increase about 25,000 to a total of roughly 150,000 units.

The Lagrangian Function

Constrained optimization problems are frequently solved using a *Lagrangian function*, $\mathcal{L}$. For example, to optimize the function $f(x, y)$ subject to the constraint $g(x, y) = c$, we use the Lagrangian function

$$\mathcal{L}(x, y, \lambda) = f(x, y) - \lambda(g(x, y) - c).$$

To see why the function $\mathcal{L}$ is useful, compute the partial derivatives of $\mathcal{L}$:

$$\frac{\partial \mathcal{L}}{\partial x} = \frac{\partial f}{\partial x} - \lambda \frac{\partial g}{\partial x},$$

$$\frac{\partial \mathcal{L}}{\partial y} = \frac{\partial f}{\partial y} - \lambda \frac{\partial g}{\partial y},$$

$$\frac{\partial \mathcal{L}}{\partial \lambda} = -(g(x, y) - c).$$

Notice that if (x_0, y_0) gives a maximum or minimum value of $f(x, y)$ subject to the constraint $g(x, y) = c$ and λ_0 is the corresponding Lagrange multiplier, then at the point (x_0, y_0, λ_0), we have

$$\frac{\partial \mathcal{L}}{\partial x} = 0 \quad \text{and} \quad \frac{\partial \mathcal{L}}{\partial y} = 0 \quad \text{and} \quad \frac{\partial \mathcal{L}}{\partial \lambda} = 0.$$

In other words, (x_0, y_0, λ_0) is a critical point for the unconstrained problem of optimization of the Lagrangian, $\mathcal{L}(x, y, \lambda)$.

We can therefore attack constrained optimization problems in two steps. First, write down the Lagrangian function $\mathcal{L}$. Second, find the critical points of $\mathcal{L}$.

Problems for Section 8.6

■ In Problems 1–10, use Lagrange multipliers to find the maximum or minimum values of $f(x, y)$ subject to the constraint.

1. $f(x, y) = x + y, \quad x^2 + y^2 = 1$

2. $f(x, y) = x^2 + 4xy, \quad x + y = 100$

3. $f(x, y) = xy, \quad 5x + 2y = 100$

4. $f(x, y) = x^2 + 3y^2 + 100, \quad 8x + 6y = 88$

5. $f(x, y) = 5xy, \quad x + 3y = 24$

6. $f(x, y) = 3x - 2y, \quad x^2 + 2y^2 = 44$

7. $f(x, y) = x^2 + y, \quad x^2 - y^2 = 1$

8. $f(x, y) = xy, \quad 4x^2 + y^2 = 8$

9. $f(x, y) = x^2 + y^2, \quad 4x - 2y = 15$

10. $f(x, y) = x^2 + y^2, \quad x^4 + y^4 = 2$

11. Figure 8.59 shows contours labeled with values of $f(x, y)$ and a constraint $g(x, y) = c$. Mark the approximate points at which:

(a) f has a maximum

(b) f has a maximum on the constraint $g = c$.

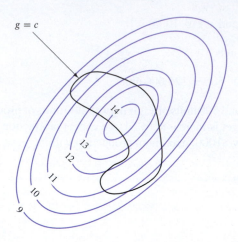

Figure 8.59

12. Figure 8.60 shows contours of $f(x, y)$ and the constraint $g(x, y) = c$. Approximately what values of x and y maximize $f(x, y)$ subject to the constraint? What is the approximate value of f at this maximum?

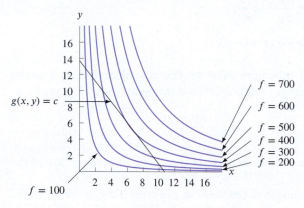

Figure 8.60

13. The quantity, Q, of a good produced depends on the quantities x_1 and x_2 of two raw materials used:

$$Q = x_1^{0.3} x_2^{0.7}.$$

A unit of x_1 costs \$10, and a unit of x_2 costs \$25. We want to maximize production with a budget of \$50 thousand for raw materials.

(a) What is the objective function?
(b) What is the constraint?

14. The quantity, Q, of a certain product manufactured depends on the quantity of labor, L, and of capital, K, used according to the function

$$Q = 900L^{1/2}K^{2/3}.$$

Labor costs \$100 per unit and capital costs \$200 per unit. What combination of labor and capital should be used to produce 36,000 units of the goods at minimum cost? What is that minimum cost?

15. The Cobb-Douglas production function for a product is

$$P = 5L^{0.8}K^{0.2},$$

where P is the quantity produced, L is the size of the labor force, and K is the amount of total equipment. Each unit of labor costs \$300, each unit of equipment costs \$100, and the total budget is \$15,000.

(a) Make a table of L and K values which exhaust the budget. Find the production level, P, for each.
(b) Use the method of Lagrange multipliers to find the optimal way to spend the budget.

16. A firm manufactures a commodity at two different factories. The total cost of manufacturing depends on the quantities, q_1 and q_2, supplied by each factory, and is expressed by the *joint cost function*,

$$C = f(q_1, q_2) = 2q_1^2 + q_1q_2 + q_2^2 + 500.$$

The company's objective is to produce 200 units, while minimizing production costs. How many units should be supplied by each factory?

17. The quantity, Q, of a product manufactured by a company is given by

$$Q = aK^{0.6}L^{0.4},$$

where a is a positive constant, K is the quantity of capital and L is the quantity of labor used. Capital costs are \$20 per unit, labor costs are \$10 per unit, and the company wants costs for capital and labor combined to be no higher than \$150. Suppose you are asked to consult for the company, and learn that 5 units each of capital and labor are being used.

(a) What do you advise? Should the company use more or less labor? More or less capital? If so, by how much?
(b) Write a one-sentence summary that could be used to sell your advice to the board of directors.

18. In Example 1, the maximum production is obtained when $2y = x$ no matter what the budget is. Assume the current values for x and y are those that maximize production.

(a) If you increase x by 3000, by how much should you increase y if production is to remain maximal?
(b) If you double x, how should you change y for production to remain maximal?
(c) If you increase x by 5%, by what percent should you change y for production to remain maximal?

19. Maximize production, $Q = 5x^{0.3}y^{0.8}$, where x and y are quantities of raw materials and the total quantity of these raw materials cannot be more than 100.

20. Maximize production for a manufacturing process with a $12 million budget and which uses x kilograms of one raw material and y kilograms of a second raw material to make $Q = 3\ln(x+1) + 2\ln(y+1)$ units of product. The first raw material costs $6 million per kilogram and the second costs $3 million per kilogram.

21. For a cost function, $f(x, y)$, the minimum cost for a production of 50 is given by $f(33, 87) = 1200$, with $\lambda = 15$. Estimate the cost if the production quota is:

    (a) Raised to 51   (b) Lowered to 49

22. A company has the production function $P(x, y)$, which gives the number of units that can be produced for given values of x and y; the cost function $C(x, y)$ gives the cost of production for given values of x and y.

    (a) If the company wishes to maximize production at a cost of $50,000, what is the objective function f? What is the constraint equation? What is the meaning of λ in this situation?

    (b) If instead the company wishes to minimize the costs at a fixed production level of 2000 units, what is the objective function f? What is the constraint equation? What is the meaning of λ in this situation?

23. You have set aside 20 hours to work on two class projects. You want to maximize your grade (measured in points), which depends on how you divide your time between the two projects.

    (a) What is the objective function for this optimization problem and what are its units?

    (b) What is the constraint?

    (c) Suppose you solve the problem by the method of Lagrange multipliers. What are the units for λ?

    (d) What is the practical meaning of the statement $\lambda = 5$?

24. A steel manufacturer can produce $P(K, L)$ tons of steel[16] using K units of capital and L units of labor, with production costs $C(K, L)$ dollars. With a budget of $800,000, the maximum production is 10,000 tons, using $550,000 of capital and $150,000 of labor. The Lagrange multiplier is $\lambda = 0.0125$.

    (a) What is the objective function?

    (b) What is the constraint?

    (c) What are the units for λ?

    (d) What is the practical meaning of the statement $\lambda = 0.0125$?

25. The quantity, q, of a product manufactured depends on the number of workers, W, and the amount of capital invested, K, and is given by the Cobb-Douglas function

$$q = 6W^{3/4}K^{1/4}.$$

In addition, labor costs are $10 per worker and capital costs are $20 per unit and the budget is $3000.

(a) What are the optimum number of workers and the optimum number of units of capital?

(b) Recompute the optimum values of W and K when the budget is increased by $1. Check that increasing the budget by $1 allows the production of λ extra units of the product, where λ is the Lagrange multiplier.

26. Each person tries to balance his or her time between leisure and work. The tradeoff is that as you work less your income falls. Therefore each person has *indifference curves* which connect the number of hours of leisure, l, and income, s. If, for example, you are indifferent between 0 hours of leisure and an income of $1125 a week on the one hand, and 10 hours of leisure and an income of $750 a week on the other hand, then the points $l = 0$, $s = 1125$, and $l = 10$, $s = 750$ both lie on the same indifference curve. Table 8.14 gives information on three indifference curves, I, II, and III.

Table 8.14

Weekly income			Weekly leisure hours		
I	II	III	I	II	III
1125	1250	1375	0	20	40
750	875	1000	10	30	50
500	625	750	20	40	60
375	500	625	30	50	70
250	375	500	50	70	90

(a) Graph the three indifference curves.

(b) You have 100 hours a week available for work and leisure combined, and you earn $10/hour. Write an equation in terms of l and s which represents this constraint.

(c) On the same axes, graph this constraint.

(d) Estimate from the graph what combination of leisure hours and income you would choose under these circumstances. Give the corresponding number of hours per week you would work.

27. If x_1 and x_2 are the number of items of two goods bought, a customer's utility is

$$U(x_1, x_2) = 2x_1x_2 + 3x_1.$$

The unit cost is $1 for the first good and $3 for the second. Use Lagrange multipliers to find the maximum value of U if the consumer's disposable income is $100. Estimate the increase in optimal utility if the consumer's disposable income increases by $6.

---

[16]www.steelonthenet.com/cost-bof.html, accessed February 3, 2017.

PROJECTS FOR CHAPTER EIGHT

1. **A Heater in a Room**

   Figure 8.61 shows the contours of the temperature along one wall of a heated room through one winter day, with time indicated as on a 24-hour clock. The room has a heater located at the leftmost corner of the wall and one window in the wall. The heater is controlled by a thermostat about 2 feet from the window.

   (a) Where is the window?   (b) When is the window open?
   (c) When is the heat on?
   (d) Draw graphs of the temperature along the wall of the room at 6 am, at 11 am, at 3 pm (15 hours) and at 5 pm (17 hours).
   (e) Draw a graph of the temperature as a function of time at the heater, at the window and midway between them.
   (f) The temperature at the window at 5 pm (17 hours) is less than at 11 am. Why do you think this might be?
   (g) To what temperature do you think the thermostat is set? How do you know?
   (h) Where is the thermostat?

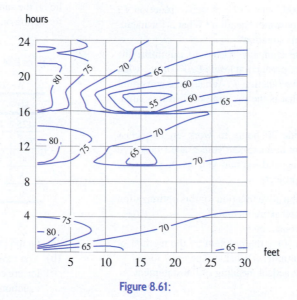

Figure 8.61:

2. **Optimizing Relative Prices for Adults and Children**

   Some items are sold at a discount to senior citizens or children. The reason is that these groups are more sensitive to price, so a discount has greater impact on their purchasing decisions. The seller faces an optimization problem: How large a discount to offer in order to maximize profits? Suppose a theater can sell q_c child tickets and q_a adult tickets at prices p_c and p_a, according to the demand functions:

   $$q_c = rp_c^{-4} \qquad \text{and} \qquad q_a = sp_a^{-2},$$

   and has operating costs proportional to the total number of tickets sold. What should be the relative price of children's and adults' tickets?

3. **Maximizing Production and Minimizing Cost: "Duality"**

   A company's production function is $P = 270x_1^{1/3}x_2^{2/3}$ for quantities x_1 and x_2 of two raw materials, costing \$4 per unit and \$27 per unit, respectively.

   (a) How much of each raw material should be used to maximize production if the budget for raw materials is \$324? What is the maximum production achieved, P_0?

(b) What is the minimum cost at which a production level of P_0 can be achieved? How much of each raw material is used at this minimum?

(c) Comment on the relationship between your answers to parts (a) and (b).

FOCUS ON THEORY

DERIVING THE FORMULA FOR A REGRESSION LINE

Suppose we want to find the "best fitting" line for some experimental data. In Appendix A, we use a computer or calculator to find the formula for this line. In this section, we derive this formula.

We decide which line fits the data best by using the following criterion. The data is plotted in the plane. The distance from a line to the data points is measured by adding the squares of the vertical distances from each point to the line. The smaller this sum of squares is, the better the line fits the data. The line with the minimum sum of square distances is called the *least-squares line*, or the *regression line*. If the data is nearly linear, the least-squares line will be a good fit; otherwise it may not be. (See Figure 8.62.)

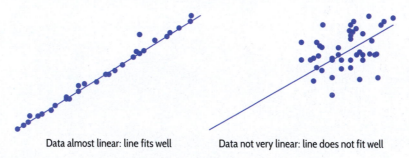

Data almost linear: line fits well   Data not very linear: line does not fit well

Figure 8.62: Fitting lines to data points

Example 1   Find a least-squares line for the following data points: $(1, 1)$, $(2, 1)$, and $(3, 3)$.

Solution   Suppose the line has equation $y = b + mx$. If we find b and m then we have found the line. So, for this problem, b and m are the two variables. We want to minimize the function $f(b, m)$ that gives the sum of the three squared vertical distances from the points to the line in Figure 8.63.

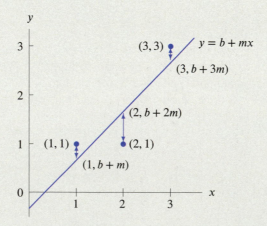

Figure 8.63: The least-squares line minimizes the sum of the squares of these vertical distances

The vertical distance from the point $(1, 1)$ to the line is the difference in the y-coordinates $1 -$

$(b + m)$; similarly for the other points. Thus, the sum of squares is

$$f(b, m) = (1 - (b + m))^2 + (1 - (b + 2m))^2 + (3 - (b + 3m))^2.$$

To minimize f we look for critical points. First we differentiate f with respect to b:

$$\begin{aligned} f_b(b, m) &= -2(1 - (b + m)) - 2(1 - (b + 2m)) - 2(3 - (b + 3m)) \\ &= -2 + 2b + 2m - 2 + 2b + 4m - 6 + 2b + 6m \\ &= -10 + 6b + 12m. \end{aligned}$$

Now we differentiate with respect to m:

$$\begin{aligned} f_m(b, m) &= 2(1 - (b + m))(-1) + 2(1 - (b + 2m))(-2) + 2(3 - (b + 3m))(-3) \\ &= -2 + 2b + 2m - 4 + 4b + 8m - 18 + 6b + 18m \\ &= -24 + 12b + 28m. \end{aligned}$$

The equations $f_b = 0$ and $f_m = 0$ give a system of two linear equations in two unknowns:

$$-10 + 6b + 12m = 0,$$
$$-24 + 12b + 28m = 0.$$

The solution to this pair of equations is the critical point $b = -1/3$ and $m = 1$. Since

$$D = f_{bb}f_{mm} - (f_{mb})^2 = (6)(28) - 12^2 = 24 \quad \text{and} \quad f_{bb} = 6 > 0,$$

we have found a local minimum. This local minimum is also the global minimum of f. Thus, the least-squares line is

$$y = x - \frac{1}{3}.$$

As a check, notice that the line $y = x$ passes through the points $(1, 1)$ and $(3, 3)$. It is reasonable that introducing the point $(2, 1)$ moves the y-intercept down from 0 to $-1/3$.

Derivation of the Formulas for the Regression Line

We use the method of Example 1 to derive the formulas for the least-squares line $y = b + mx$ generated by data points (x_1, y_1), (x_2, y_2), ..., (x_n, y_n). Notice that we are looking for the slope and y-intercept, so we think of m and b as the variables.

For each data point (x_i, y_i), the corresponding point directly above the line or below it on the line has the y-coordinate $b + mx_i$. Thus, the squares of the vertical distances from the point to the line is $(y_i - (b + mx_i))^2$. (See Figure 8.64.)

We find the sum of the n squared distances from points to the line, and think of the sum as a function of m and b:

$$f(b, m) = \sum_{i=1}^{n} (y_i - (b + mx_i))^2.$$

To minimize this function, we first find the two partial derivatives, f_b and f_m. We use the chain rule and the properties of sums.

$$f_b(b, m) = \frac{\partial}{\partial b}\left(\sum_{i=1}^{n} (y_i - (b + mx_i))^2 \right) = \sum_{i=1}^{n} \frac{\partial}{\partial b}(y_i - (b + mx_i))^2$$

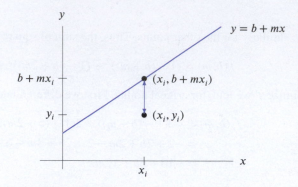

Figure 8.64: The vertical distance from a point to the line

$$= \sum_{i=1}^{n} 2(y_i - (b + mx_i)) \cdot \frac{\partial}{\partial b}(y_i - (b + mx_i))$$

$$= \sum_{i=1}^{n} 2(y_i - (b + mx_i)) \cdot (-1)$$

$$= -2 \sum_{i=1}^{n} (y_i - (b + mx_i)).$$

$$f_m(b, m) = \frac{\partial}{\partial m}\left(\sum_{i=1}^{n}(y_i - (b + mx_i))^2\right) = \sum_{i=1}^{n}\frac{\partial}{\partial m}(y_i - (b + mx_i))^2$$

$$= \sum_{i=1}^{n} 2(y_i - (b + mx_i)) \cdot \frac{\partial}{\partial m}(y_i - (b + mx_i))$$

$$= \sum_{i=1}^{n} 2(y_i - (b + mx_i)) \cdot (-x_i)$$

$$= -2 \sum_{i=1}^{n} (y_i - (b + mx_i)) \cdot x_i.$$

We now set the partial derivatives equal to zero and solve for m and b. This is easier than it looks: we simplify the appearance of the equations by temporarily substituting other symbols for the sums: write SY for $\sum y_i$, SX for $\sum x_i$, SXY for $\sum y_i x_i$ and SXX for $\sum x_i^2$. Remember, the x_i and y_i are all constants. We get a pair of simultaneous linear equations in m and b; solving for m and b gives us formulas in terms of SX, SY, SXY, and SXX. We separate $f_b(b, m)$ into three sums as shown:

$$f_b(b, m) = -2\left(\sum_{i=1}^{n} y_i - b \sum_{i=1}^{n} 1 - m \sum_{i=1}^{n} x_i\right).$$

Similarly, we can separate $f_m(b, m)$ after multiplying through by x_i:

$$f_m(b, m) = -2\left(\sum_{i=1}^{n} y_i x_i - b \sum_{i=1}^{n} x_i - m \sum_{i=1}^{n} x_i^2\right).$$

Rewriting the sums as suggested and setting $\frac{\partial f}{\partial b}$ and $\frac{\partial f}{\partial m}$ equal to zero, we have:

$$0 = SY - bn - mSX$$

$$0 = SYX - bSX - mSXX.$$

Solving this pair of simultaneous equations, we get the result:

$$b = ((SXX) \cdot (SY) - (SX) \cdot (SYX)) / (n(SXX) - (SX)^2)$$
$$m = (n(SYX) - (SX) \cdot (SY)) / (n(SXX) - (SX)^2).$$

Writing these expressions with summation notation, we arrive at the following result:

The least-squares line for data points $(x_1, y_1), (x_2, y_2), \cdots, (x_n, y_n)$ is the line $y = b + mx$ where

$$b = \left(\sum_{i=1}^{n} x_i^2 \sum_{i=1}^{n} y_i - \sum_{i=1}^{n} x_i \sum_{i=1}^{n} y_i x_i \right) / \left(n \sum_{i=1}^{n} x_i^2 - \left(\sum_{i=1}^{n} x_i \right)^2 \right)$$

$$m = \left(n \sum_{i=1}^{n} y_i x_i - \sum_{i=1}^{n} x_i \sum_{i=1}^{n} y_i \right) / \left(n \sum_{i=1}^{n} x_i^2 - \left(\sum_{i=1}^{n} x_i \right)^2 \right).$$

Example 2

Use these formulas to find the best fitting line for the data point $(1, 5), (2, 4), (4, 3)$.

Solution

We compute the sums needed in the formulas:

$$\sum_{i=1}^{3} x_i = 1 + 2 + 4 = 7$$

$$\sum_{i=1}^{3} y_i = 5 + 4 + 3 = 12$$

$$\sum_{i=1}^{3} x_i^2 = 1^2 + 2^2 + 4^2 = 1 + 4 + 16 = 21$$

$$\sum_{i=1}^{3} y_i x_i = (5)(1) + (4)(2) + (3)(4) = 5 + 8 + 12 = 25.$$

Since $n = 3$, we have:

$$b = \left(\sum_{i=1}^{3} x_i^2 \sum_{i=1}^{3} y_i - \sum_{i=1}^{3} x_i \sum_{i=1}^{3} y_i x_i \right) / \left(3 \sum_{i=1}^{3} x_i^2 - \left(\sum_{i=1}^{3} x_i \right)^2 \right)$$

$$= ((21)(12) - (7)(25)) / (3(21) - (7^2))$$

$$= 77/14 = 5.5$$

and

$$m = \left(3 \sum_{i=1}^{3} y_i x_i - \sum_{i=1}^{3} x_i \sum_{i=1}^{3} y_i \right) / \left(3 \sum_{i=1}^{3} x_i^2 - \left(\sum_{i=1}^{3} x_i \right)^2 \right)$$

$$= (3(25) - (7)(12)) / (3(21) - (7^2))$$

$$= -9/14 = -0.64.$$

The least-squares line for these three points is

$$y = 5.5 - 0.64x.$$

To check this equation, plot the line and the three points together.

Many calculators have the formulas for the least-squares line built in, so that when you enter the data, out come the values of b and m. At the same time, you get the *correlation coefficient*, which measures how close the data points actually come to fitting the least-squares line.

Problems on Deriving the Formula for Regression Lines

■ In Problems 1–2, use the method of Example 1 to find the least-squares line. Check by graphing the points with the line.

1. $(-1, 2), (0, -1), (1, 1)$ 2. $(0, 2), (1, 4), (2, 5)$

■ In Problems 3–5, use the formulas for b and m to check that you get the same result as in the problem or example specified.

3. $(-1, 2), (0, -1), (1, 1)$. See Problem 1.

4. $(0, 2), (1, 4), (2, 5)$. See Problem 2.

5. $(1, 1), (2, 1), (3, 3)$. See Example 1.

■ In Problems 6–7, we transform nonlinear data so that it looks more linear. For example, suppose the data points (x, y) fit the exponential equation

$$y = Ce^{ax},$$

where a and C are constants. Taking the natural log of both sides, we get

$$\ln y = ax + \ln C.$$

Thus, $\ln y$ is a linear function of x. To find a and C, we can use least squares for the graph of $\ln y$ against x.

6. The population of the US was about 180 million in 1960, grew to 206 million in 1970, and 226 million in 1980.

   (a) Assuming that the population was growing exponentially, use logarithms and the method of least squares to estimate the population in 1990.

(b) According to the national census, the 1990 population was 249 million. What does this say about the assumption of exponential growth?

(c) Predict the population in the year 2010.

7. A biological rule of thumb states that as the area A of an island increases tenfold, the number of animal species, N, living on it doubles. The table contains data for islands in the West Indies. Assume that N is a power function of A.

   (a) Use the biological rule of thumb to find

      (i) N as a function of A

      (ii) $\ln N$ as a function of $\ln A$

   (b) Using the data given, tabulate $\ln N$ against $\ln A$ and find the line of best fit. Does your answer agree with the biological rule of thumb?

Island	Area (sq km)	Number of species
Redonda	3	5
Saba	20	9
Montserrat	192	15
Puerto Rico	8858	75
Jamaica	10854	70
Hispaniola	75571	130
Cuba	113715	125

Chapter 9

MATHEMATICAL MODELING USING DIFFERENTIAL EQUATIONS

CONTENTS

9.1   MATHEMATICAL MODELING: SETTING UP A DIFFERENTIAL EQUATION

Sometimes we do not know a function, but we have information about its rate of change, or its derivative. Then we may be able to write a new type of equation, called a *differential equation*, from which we can get information about the original function. For example, we may use what we know about the derivative of a population function (its rate of change) to predict the population in the future.

In this section, we use a verbal description to write a differential equation.

Marine Harvesting

We begin by investigating the effect of fishing on a fish population. Suppose that, left alone, a fish population increases at a continuous rate of 20% per year. Suppose that fish are also being harvested (caught) by fishermen at a constant rate of 10 million fish per year. How does the fish population change over time?

Notice that we have been given information about the rate of change, or derivative, of the fish population. Combined with information about the initial population, we can use this to predict the population in the future. We know that

$$\begin{array}{ccc} \text{Rate of change} & = & \text{Rate of increase} & - & \text{Rate fish removed} \\ \text{of fish population} & & \text{due to breeding} & & \text{due to harvesting} \end{array}.$$

Suppose the fish population, in millions, is P and its derivative is dP/dt, where t is time in years. If left alone, the fish population increases at a continuous rate of 20% per year, so we have

$$\text{Rate of increase due to breeding} = 20\% \cdot \text{Current population}$$
$$= 0.20P \text{ million fish/year.}$$

In addition,

$$\text{Rate fish removed due to harvesting} = 10 \text{ million fish/year.}$$

Since the rate of change of the fish population is dP/dt, we have

$$\frac{dP}{dt} = 0.20P - 10.$$

This is a differential equation that models how the fish population changes. The unknown quantity in the equation is the function giving P in terms of t.

Net Worth of a Company

A company earns revenue (income) and also makes payroll payments. Assume that revenue is earned continuously, that payroll payments are made continuously, and that the only factors affecting net worth are revenue and payroll. The company's revenue is earned at a continuous annual rate of 5% times its net worth. At the same time, the company's payroll obligations are paid out at a constant rate of 200 million dollars a year.

We use this information to write a differential equation to model the net worth of the company, W, in millions of dollars, as a function of time, t, in years. We know that

$$\begin{array}{ccc} \text{Rate at which} & = & \text{Rate revenue} & - & \text{Rate payroll payments} \\ \text{net worth is changing} & & \text{is earned} & & \text{are made} \end{array}.$$

Since the company's revenue is earned at a rate of 5% of its net worth, we have

$$\text{Rate revenue is earned} = 5\% \cdot \text{Net worth} = 0.05W \text{ million dollars/year.}$$

Since payroll payments are made at a rate of 200 million dollars a year, we have

$$\text{Rate payroll payments are made} = 200 \text{ million dollars/year.}$$

Putting these two together, since the rate at which net worth is changing is dW/dt, we have

$$\frac{dW}{dt} = 0.05W - 200.$$

This is a differential equation that models how the net worth of the company changes. The unknown quantity in the equation is the function giving net worth W as a function of time t.

Pollution in a Lake

If clean water flows into a polluted lake and a stream takes water out, the level of pollution in the lake will decrease (assuming no new pollutants are added).

Example 1   The quantity of pollutant in the lake decreases at a rate proportional to the quantity present. Write a differential equation to model the quantity of pollutant in the lake. Is the constant of proportionality positive or negative? Use the differential equation to explain why the graph of the quantity of pollutant against time is decreasing and concave up, as in Figure 9.1.

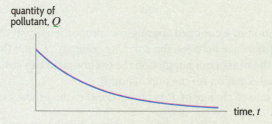

quantity of
pollutant, Q

time, t

Figure 9.1: Quantity of pollutant in a lake

Solution   Let Q denote the quantity of pollutant present in the lake at time t. The rate of change of Q is proportional to Q, so dQ/dt is proportional to Q. Thus, the differential equation is

$$\frac{dQ}{dt} = kQ.$$

Since no new pollutants are being added to the lake, the quantity Q is decreasing over time, so dQ/dt is negative. Thus, the constant of proportionality k is negative.

Why does the differential equation $dQ/dt = kQ$, with k negative, give us the graph shown in Figure 9.1? Since k is negative and Q is positive, we know kQ is negative. Thus, dQ/dt is negative, so the graph of Q against t is decreasing as in Figure 9.1. Why is it concave up? Since Q is getting smaller and k is fixed, as t increases, the product kQ is getting smaller in magnitude, and so the derivative dQ/dt is getting smaller in magnitude. Thus, the graph of Q is more horizontal as t increases. Therefore, the graph is concave up. See Figure 9.1.

The Quantity of a Drug in the Body

In the previous example, the rate at which pollutants leave a lake is proportional to the quantity of pollutants in the lake. This model works for any contaminants flowing in or out of a fluid system with complete mixing. Another example is the quantity of a drug in a person's body.

Example 2   A patient having major surgery is given the antibiotic vancomycin intravenously at a rate of 85 mg per hour. The rate at which the drug is excreted from the body is proportional to the quantity present, with proportionality constant 0.1 if time is in hours.

(a) Write a differential equation for the quantity, Q in mg, of vancomycin in the body after t hours.
(b) When $Q = 100$ mg, is Q increasing or decreasing?

Solution   (a) The quantity of vancomycin, Q, is increasing at a constant rate of 85 mg/hour and is decreasing at a rate of 0.1 times Q. The administration of 85 mg/hour makes a positive contribution to the rate of change dQ/dt. The excretion at a rate of $0.1Q$ makes a negative contribution to dQ/dt.

Putting these together, we have

$$\text{Rate of change of a quantity} = \text{Rate in} - \text{Rate out,}$$

so

$$\frac{dQ}{dt} = 85 - 0.1Q.$$

(b) When $Q = 100$ mg, the rate of change of Q is positive

$$\frac{dQ}{dt} = 85 - 0.1(100) = 75 \text{ mg/hour,}$$

so Q is increasing at that moment.

The Logistic Model

A population in a confined space grows proportionally to the product of the current population, P, and the difference between the *carrying capacity*, L, and the current population. (The carrying capacity is the maximum population the environment can sustain.) We use this information to write a differential equation for the population P.

The rate of change of P is proportional to the product of P and $L - P$, so

$$\frac{dP}{dt} = kP(L - P), \qquad \text{where } k \text{ is the positive constant of proportionality.}$$

This is called a *logistic differential equation*. What does it tell us about the graph of P? The derivative dP/dt is the product of k and P and $L - P$, so when P is small, the derivative dP/dt is small and the population grows slowly. As P increases, the derivative dP/dt increases and the population grows more rapidly. However, as P approaches the carrying capacity L, the term $L - P$ is small, and dP/dt is again small and the population grows more slowly. The *logistic growth curve* in Figure 9.2 satisfies these conditions.

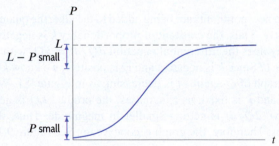

Figure 9.2: The logistic growth curve is a solution to $dP/dt = kP(L - P)$

Problems for Section 9.1

1. Match the graphs in Figure 9.3 with the following descriptions.

   (a) The temperature of a glass of ice water left on the kitchen table.
   (b) The amount of money in an interest-bearing bank account into which $50 is deposited.
   (c) The speed of a constantly decelerating car.
   (d) The temperature of a piece of steel heated in a furnace and left outside to cool.

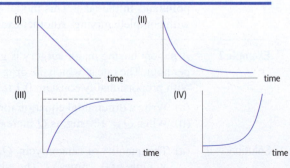

Figure 9.3

2. The graphs in Figure 9.4 represent the temperature, H(°C), of four eggs as a function of time, t, in minutes. Match three of the graphs with the descriptions (a)–(c). Write a similar description for the fourth graph, including an interpretation of any intercepts and asymptotes.

   (a) An egg is taken out of the refrigerator (just above 0°C) and put into boiling water.
   (b) Twenty minutes after the egg in part (a) is taken out of the fridge and put into boiling water, the same thing is done with another egg.
   (c) An egg is taken out of the refrigerator at the same time as the egg in part (a) and left to sit on the kitchen table.

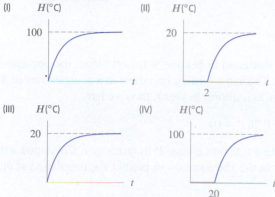

Figure 9.4

3. A population of insects grows at a rate proportional to the size of the population. Write a differential equation for the size of the population, P, as a function of time, t. Is the constant of proportionality positive or negative?

4. Money in a bank account earns interest at a continuous annual rate of 5% times the current balance. Write a differential equation for the balance, B, in the account as a function of time, t, in years.

5. Radioactive substances decay at a rate proportional to the quantity present. Write a differential equation for the quantity, Q, of a radioactive substance present at time t. Is the constant of proportionality positive or negative?

6. A bank account that initially contains $25,000 earns interest at a continuous rate of 4% per year. Withdrawals are taken out at a constant rate of $2000 per year. Write a differential equation for the balance, B, in the account as a function of the number of years, t.

7. A pollutant spilled on the ground decays at a rate of 8% a day. In addition, cleanup crews remove the pollutant at a rate of 30 gallons a day. Write a differential equation for the amount of pollutant, P, in gallons, left after t days.

8. Morphine is administered to a patient intravenously at a rate of 2.5 mg per hour. About 34.7% of the morphine is metabolized and leaves the body each hour. Write a differential equation for the amount of morphine, M, in milligrams, in the body as a function of time, t, in hours.

9. Alcohol is metabolized and excreted from the body at a rate of about one ounce of alcohol every hour. If some alcohol is consumed, write a differential equation for the amount of alcohol, A (in ounces), remaining in the body as a function of t, the number of hours since the alcohol was consumed.

10. Toxins in pesticides can get into the food chain and accumulate in the body. A person consumes 10 micrograms a day of a toxin, ingested throughout the day. The toxin leaves the body at a continuous rate of 3% every day. Write a differential equation for the amount of toxin, A, in micrograms, in the person's body as a function of the number of days, t.

11. A cup of coffee contains about 100 mg of caffeine. Caffeine is metabolized and leaves the body at a continuous rate of about 17% every hour.

    (a) Write a differential equation for the amount, A, of caffeine in the body as a function of the number of hours, t, since the coffee was consumed.
    (b) Use the differential equation to find dA/dt at the start of the first hour (right after the coffee is consumed). Use your answer to estimate the change in the amount of caffeine during the first hour.

12. A person deposits money into an account at a continuous rate of $6000 a year, and the account earns interest at a continuous rate of 7% per year.

    (a) Write a differential equation for the balance in the account, B, in dollars, as a function of years, t.
    (b) Use the differential equation to calculate dB/dt if $B = 10,000$ and if $B = 100,000$. Interpret your answers.

13. A quantity W satisfies the differential equation

$$\frac{dW}{dt} = 5W - 20.$$

    (a) Is W increasing or decreasing at $W = 10$? $W = 2$?
    (b) For what values of W is the rate of change of W equal to zero?

14. A quantity y satisfies the differential equation

$$\frac{dy}{dt} = -0.5y.$$

Under what conditions is y increasing? Decreasing?

15. An early model of the growth of Wikipedia assumed that every day a constant number, B, of articles is added by dedicated Wikipedians and that other articles are created by the general public at a rate proportional to the number of articles already there. Express this model as a differential equation for $N(t)$, the total number of Wikipedia articles t days after it started on January 15, 2001.

16. A country's infrastructure is its transportation and communication systems, power plants, and other public institutions. The Solow model asserts that the value of national infrastructure K increases due to investment and decreases due to capital depreciation. The rate of increase due to investment is proportional to national income, Y. The rate of decrease due to depreciation is proportional to the value of existing infrastructure. Write a differential equation for K.

9.2 SOLUTIONS OF DIFFERENTIAL EQUATIONS

What does it mean to "solve" a differential equation? A differential equation is an equation involving the derivative of an unknown function. The unknown is not a number but a function. A *solution* to a differential equation is any function that satisfies the differential equation.

In this section, we see how to solve a differential equation numerically and how to check whether or not a function is a solution to a differential equation. In the next section, we see how to visualize a solution.

Another Look at Marine Harvesting

Let's take another look at the fish population discussed in Section 9.1. Left alone, the population increases at a continuous rate of 20% per year. The fish are being harvested at a constant rate of 10 million fish per year. If P is the fish population, in millions, in year t, then we have

$$\frac{dP}{dt} = 0.20P - 10.$$

Solving this differential equation means finding a function giving P in terms of t. Combined with information about the initial population, we can use the equation to predict the population at any time in the future.

Solving the Differential Equation Numerically

We can approximate the solution to this differential equation by observing that the change in P is approximately $dP/dt \cdot \Delta t$ when Δt is small.

Example 1   Suppose at time $t = 0$, the fish population is 60 million. Find approximate values for $P(t)$ for $t = 1, 2, 3, 4, 5$.

Solution   We can substitute $P = 60$ into the differential equation to compute the derivative, dP/dt:

$$\text{At time } t = 0, \qquad \frac{dP}{dt} = 0.20P - 10 = 0.20(60) - 10 = 12 - 10 = 2.$$

Since at $t = 0$, the fish population is changing at a rate of 2 million fish a year, at the end of the first year, the fish population will have increased by about 2 million fish. So:

$$\text{At } t = 1, \qquad \text{we estimate} \quad P = 60 + 2 = 62.$$

We use this new value of P to estimate dP/dt during the second year:

$$\text{At time } t = 1, \qquad \frac{dP}{dt} = 0.20P - 10 = 12.4 - 10 = 2.4.$$

During the second year, the fish population increased by about 2.4 million fish, so:

$$\text{At } t = 2, \qquad \text{we estimate} \quad P = 62 + 2.4 = 64.4.$$

We use this value of P to estimate the rate of change during the third year, and so on. Continuing in this fashion, we compute the approximate values of P in Table 9.1. This table gives approximate numerical values for P at future times.

Table 9.1   *Approximate values of the fish population as a function of time*

t (years)	0	1	2	3	4	5	...
P (millions)	60	62	64.4	67.28	70.74	74.89	...

A Formula for the Solution to the Differential Equation

A function $P = f(t)$ which satisfies the differential equation

$$\frac{dP}{dt} = 0.20P - 10$$

is called a *solution* of the differential equation. Table 9.1 shows approximate numerical values of a solution. It is sometimes (but not always) possible to find a formula for the solution. In this particular case, there is a formula; it is

$$P = 50 + Ce^{0.20t}, \qquad \text{where } C \text{ is any constant.}$$

We check that this is a solution to the differential equation by substituting it into the left and right sides of the differential equation separately. We find

$$\text{Left side} = \frac{dP}{dt} = 0.20Ce^{0.20t}$$

$$\text{Right side} = 0.20P - 10 = 0.20(50 + Ce^{0.20t}) - 10$$
$$= 10 + 0.20Ce^{0.20t} - 10$$
$$= 0.20Ce^{0.20t}.$$

Since we get the same expression on both sides, we say that $P = 50 + Ce^{0.20t}$ is a solution of this differential equation. Any choice of C works, so the solutions form a family of functions with parameter C. Several members of the family of solutions are graphed in Figure 9.5.

Finding the Arbitrary Constant: Initial Conditions

To find a value for the constant C—in other words, to select a single solution from the family of solutions—we need an additional piece of information, usually the initial population. In this case, we know that $P = 60$ when $t = 0$, so substituting into

$$P = 50 + Ce^{0.20t}$$

gives

$$60 = 50 + Ce^{0.20(0)}$$
$$60 = 50 + C \cdot 1$$
$$C = 10.$$

The function $P = 50 + 10e^{0.20t}$ satisfies the differential equation *and* the initial condition that $P = 60$ when $t = 0$.

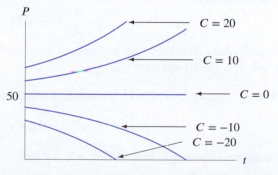

Figure 9.5: Solution curves for $dP/dt = 0.20P - 10$: Members of the family $P = 50 + Ce^{0.20t}$

General Solutions and Particular Solutions

For the differential equation $dP/dt = 0.20P - 10$, it can be shown that every solution is of the form $P = 50 + Ce^{0.20t}$ for some value of C. We say that the *general solution* of the differential equation $dP/dt = 0.20P - 10$ is the family of functions $P = 50 + Ce^{0.20t}$. The solution $P = 50 + 10e^{0.20t}$ that satisfies the differential equation together with the initial condition that $P = 60$ when $t = 0$ is called a *particular solution*. The differential equation and the initial condition together are called an *initial-value problem*.

Example 2   (a) Check that $P = Ce^{2t}$ is a solution to the differential equation

$$\frac{dP}{dt} = 2P.$$

(b) Find the particular solution satisfying the initial condition $P = 100$ when $t = 0$.

Solution   (a) Since $P = Ce^{2t}$ where C is a constant, we find expressions for each side:

$$\text{Left side} = \frac{dP}{dt} = Ce^{2t} \cdot 2 = 2Ce^{2t}$$
$$\text{Right side} = 2P = 2Ce^{2t}.$$

Since the two expressions are equal, $P = Ce^{2t}$ is a solution to the differential equation.

(b) We substitute $P = 100$ and $t = 0$ into the general solution $P = Ce^{2t}$, and solve for C:

$$100 = Ce^{2(0)}$$
$$100 = C \cdot 1$$
$$100 = C.$$

The particular solution for this initial-value problem is $P = 100e^{2t}$.

Example 3   Decide whether or not $y = e^{-2x}$ is a solution of the differential equation $y' - 2y = 0$.

Solution   Note that $y' = dy/dx$. Differentiating $y = e^{-2x}$ gives $y' = -2e^{-2x}$. Substituting, we have

$$y' - 2y = -2e^{-2x} - 2e^{-2x} = -4e^{-2x} \neq 0,$$

and so $y = e^{-2x}$ is not a solution to this differential equation.

Example 4   (a) What conditions must be imposed on the constants C and k if $y = Ce^{kt}$ is a solution to the differential equation

$$\frac{dy}{dt} = -0.5y?$$

(b) What additional conditions must be imposed on C and k if $y = Ce^{kt}$ also satisfies the initial condition that $y = 10$ when $t = 0$?

Solution   (a) If $y = Ce^{kt}$, then $dy/dt = Cke^{kt}$. Substituting into the equation $dy/dt = -0.5y$ gives

$$Cke^{kt} = -0.5(Ce^{kt}),$$

and therefore, assuming $C \neq 0$, so $Ce^{kt} \neq 0$, we have

$$k = -0.5.$$

So $y = Ce^{-0.5t}$ is a solution to the differential equation. If $C = 0$, then $Ce^{kt} = 0$ is a solution to the differential equation. No conditions are imposed on C.

(b) Since $k = -0.5$, we have $y = Ce^{-0.5t}$. Substituting $y = 10$ when $t = 0$ gives

$$10 = Ce^0$$

so

$$10 = C.$$

So $y = 10e^{-0.5t}$ is a solution to the differential equation together with the initial condition.

Problems for Section 9.2

1. Decide whether or not each of the following is a solution to the differential equation $xy' - 2y = 0$.

    (a) $y = x^2$ **(b)** $y = x^3$

2. Check that $y = t^4$ is a solution to the differential equation $t\dfrac{dy}{dt} = 4y$.

3. Find the general solution to the differential equation

$$\frac{dy}{dt} = 2t.$$

■ In Problems 4–12, use the fact that the derivative gives the slope of a curve to decide which of the graphs (A)–(F) in Figure 9.6 could represent a solution to the differential equation.

(A)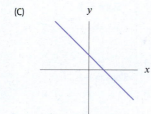

(B)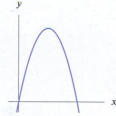

(C)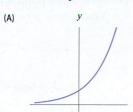

(D)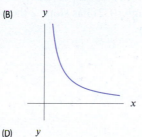

(E)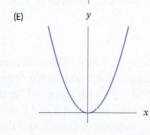

(F)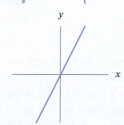

Figure 9.6

4. $\dfrac{dy}{dx} = -1$ **5.** $\dfrac{dy}{dx} = 0.1$ **6.** $\dfrac{dy}{dx} = -y^2$

7. $\dfrac{dy}{dx} = 2x$ **8.** $\dfrac{dy}{dx} = 2$ **9.** $\dfrac{dy}{dx} = y$

10. $\dfrac{dy}{dx} = -\dfrac{1}{x^2}$ **11.** $\dfrac{dy}{dx} = 1 - x$ **12.** $\dfrac{dy}{dx} = 2y$

13. Fill in the missing values in Table 9.2 given that $dy/dt = 4 - y$. Assume the rate of growth, given by dy/dt, is approximately constant over each unit time interval.

Table 9.2

t	0	1	2	3	4
y	8				

14. Fill in the missing values in Table 9.3 given that $dy/dt = 0.5t$. Assume the rate of growth, given by dy/dt, is approximately constant over each unit time interval.

Table 9.3

t	0	1	2	3	4
y	8				

15. Fill in the missing values in Table 9.4 given that $dy/dt = 0.5y$. Assume the rate of growth, given by dy/dt, is approximately constant over each unit time interval.

Table 9.4

t	0	1	2	3	4
y	8				

16. For a certain quantity y, assume that $dy/dt = \sqrt{y}$. Fill in the value of y in Table 9.5. Assume that the rate of growth, dy/dt, is approximately constant over each unit time interval.

Table 9.5

t	0	1	2	3	4
y	100				

17. If the initial population of fish is 70 million, use the differential equation $dP/dt = 0.2P - 10$ to estimate the fish population after 1, 2, 3 years.

18. Show that, for any constant P_0, the function $P = P_0 e^t$ satisfies the equation

$$\frac{dP}{dt} = P.$$

19. Suppose $Q = Ce^{kt}$ satisfies the differential equation

$$\frac{dQ}{dt} = -0.03Q.$$

What (if anything) does this tell you about the values of C and k?

20. Is there a value of n which makes $y = x^n$ a solution to the equation $13x(dy/dx) = y$? If so, what value?

21. Find the values of k for which $y = x^2 + k$ is a solution to the differential equation $2y - xy' = 10$.

22. Match solutions and differential equations. (Note: Each equation may have more than one solution, or no solution.)

(a) $\dfrac{dy}{dx} = \dfrac{y}{x}$ (I) $y = x^3$

(b) $\dfrac{dy}{dx} = 3\dfrac{y}{x}$ (II) $y = 3x$

(c) $\dfrac{dy}{dx} = 3x$ (III) $y = e^{3x}$

(d) $\dfrac{dy}{dx} = y$ (IV) $y = 3e^x$

(e) $\dfrac{dy}{dx} = 3y$ (V) $y = x$

9.3 SLOPE FIELDS

In this section, we see how to visualize a differential equation and its solutions. Let's start with the equation

$$\frac{dy}{dx} = y.$$

Any solution to this differential equation has the property that at any point in the plane, the slope of its graph is equal to its y coordinate. (That's what the equation $dy/dx = y$ is telling us!) This means that if the solution goes through the point $(0, 1)$, its slope there is 1; if it goes through a point with $y = 4$ its slope is 4. A solution going through $(0, 2)$ has slope 2 there; at the point where $y = 8$ the slope of this solution is 8. (See Figure 9.7.)

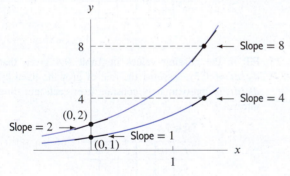

Figure 9.7: Solutions to $\frac{dy}{dx} = y$

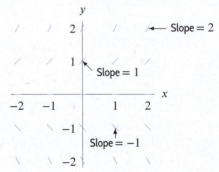

Figure 9.8: Visualizing the slope of y, if $\frac{dy}{dx} = y$

In Figure 9.8 a small line segment is drawn at the marked points showing the slope of the solution curve there. Since $dy/dx = y$, the slope at the point $(1, 2)$ is 2 (the y-coordinate), and so we draw a line segment there with slope 2. We draw a line segment at the point $(0, -1)$ with slope -1, and so on. If we draw many of these line segments, we have the *slope field* for the equation $dy/dx = y$ shown in Figure 9.9. Above the x-axis, the slopes are positive (because y is positive there), and the slopes increase as we move upward (as y increases). Below the x-axis, the slopes are negative, and get more so as we move downward. Notice that on any horizontal line (where y is constant) the slopes are constant. In the slope field you can see the ghost of the solution curve lurking. Start anywhere on the plane and move so that the slope lines are tangent to your path; you will trace out one of the solution curves. Try penciling in some solution curves on Figure 9.9, some above the x-axis and some below. The curves you draw should have the shape of exponential functions. By substituting

$y = Ce^x$ into the differential equation, you can check that each curve in the family of exponentials, $y = Ce^x$, is a solution to this differential equation.

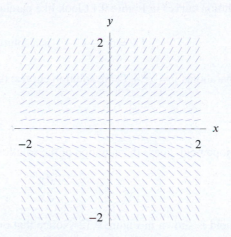

Figure 9.9: Slope field for $\frac{dy}{dx} = y$

In most problems, we are interested in getting the solution curves from the slope field. Think of the slope field as a set of signposts pointing in the direction you should go at each point. Imagine starting anywhere in the plane: look at the slope field at that point and start to move in that direction. After a small step, look at the slope field again, and alter your direction if necessary. Continue to move across the plane in the direction the slope field points, and you'll trace out a solution curve. Notice that the solution curve is not necessarily the graph of a function, and even if it is, we may not have a formula for the function. Geometrically, solving a differential equation means finding the family of solution curves.

Example 1   Figure 9.10 shows the slope field of the differential equation $\dfrac{dy}{dx} = 2x$.

(a) What do you notice about the slope field?
(b) Compare the solution curves in Figure 9.11 with the formula $y = x^2 + C$ for the solutions to this differential equation.

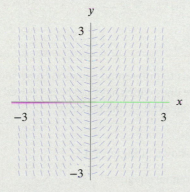

Figure 9.10: Slope field for $\frac{dy}{dx} = 2x$

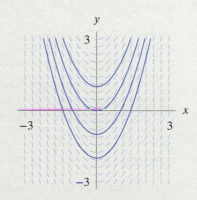

Figure 9.11: Some solutions to $\frac{dy}{dx} = 2x$

Solution   (a) In Figure 9.10, notice that on any vertical line (where x is constant) the slopes are all the same. This is because in this differential equation dy/dx depends on x only. (In the previous example, $dy/dx = y$, the slopes depended on y only.)

(b) The solution curves in Figure 9.11 look like parabolas. It is easy to check by substitution that

$$y = x^2 + C \qquad \text{is a solution to} \qquad \frac{dy}{dx} = 2x,$$

so the parabolas $y = x^2 + C$ are solution curves; they can also be obtained by using antiderivatives.

Example 2   Using the slope field, guess the equation of the solution curves of the differential equation

$$\frac{dy}{dx} = -\frac{x}{y}.$$

Solution   The slope field is shown in Figure 9.12. Notice that on the y-axis, where x is 0, the slope is 0. On the x-axis, where y is 0, the line segments are vertical and the slope is undefined. At the origin the slope is undefined and there is no line segment.

What do the solution curves of this differential equation look like? The slope field suggests they are circles centered at the origin. We guess that the general solution to this differential equation is

$$x^2 + y^2 = r^2.$$

This solution is derived in the Focus on Theory section at the end of the chapter.

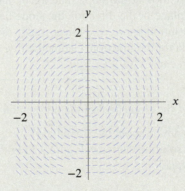

Figure 9.12: Slope field for $\frac{dy}{dx} = -\frac{x}{y}$

The previous example shows that the solutions to differential equations may sometimes be expressed as *implicit functions*. Implicit functions are ones which have not been "solved" for y; in other words, the dependent variable is not expressed as an explicit function of x.

Example 3   The slope fields for $\dfrac{dy}{dt} = 2 - y$ and $\dfrac{dy}{dt} = \dfrac{t}{y}$ are shown in Figure 9.13.

(a) Which slope field corresponds to which differential equation?

(b) Sketch solution curves on each slope field with initial conditions

(i) $y = 1$ when $t = 0$ (ii) $y = 3$ when $t = 0$ (iii) $y = 0$ when $t = 1$

(c) For each solution curve, can you say anything about the long-run behavior of y? In particular, as $t \to \infty$, what happens to the value of y?

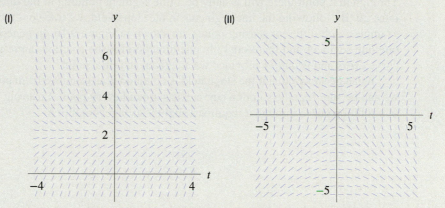

Figure 9.13: Slope fields for $\dfrac{dy}{dt} = 2 - y$ and $\dfrac{dy}{dt} = \dfrac{t}{y}$: Which is which?

Solution

(a) Consider the slopes at different points for the two differential equations. In particular, look at the line $y = 2$ in Figure 9.13. The equation $dy/dt = 2 - y$ has slope 0 all along this line, whereas the line $dy/dt = t/2$ has slope $t/2$. Since slope field (I) looks horizontal at $y = 2$, slope field (I) corresponds to $dy/dt = 2 - y$ and slope field (II) corresponds to $dy/dt = t/y$.

(b) The initial conditions (i) and (ii) give the value of y when t is 0, that is, the y-intercept. To draw the solution curve satisfying the condition (i), draw the solution curve with y-intercept 1. For (ii), draw the solution curve with y-intercept 3. For (iii), the solution goes through the point $(1, 0)$, so draw the solution curve passing through this point. See Figures 9.14 and 9.15.

(c) For $dy/dt = 2 - y$, all solution curves have $y = 2$ as a horizontal asymptote, so $y \to 2$ as $t \to \infty$. For $dy/dt = t/y$ with initial conditions $(0, 1)$ and $(0, 3)$, we see that $y \to \infty$ as $t \to \infty$. The graph has asymptotes which appear to be diagonal lines. In fact, they are $y = t$ and $y = -t$, so $y \to \pm\infty$ as $t \to \infty$.

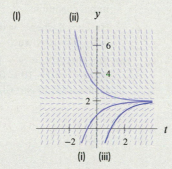

Figure 9.14: Solution curves for $\dfrac{dy}{dt} = 2 - y$

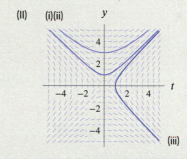

Figure 9.15: Solution curves for $\dfrac{dy}{dt} = \dfrac{t}{y}$

Existence and Uniqueness of Solutions

Since differential equations are used to model many real situations, the question of whether a solution exists and is unique can have great practical importance. If we know how the velocity of a satellite is changing, can we know its velocity for all future time? If we know the initial population of a city, and we know how the population is changing, can we predict the population in the future? Common sense says yes: if we know the initial value of some quantity and we know exactly how it is changing, we should be able to figure out the future value of the quantity.

In the language of differential equations, an initial-value problem (that is, a differential equation and an initial condition) representing a real situation almost always has a unique solution. One way to see this is by looking at the slope field. Imagine starting at the point representing the initial condition. Through that point there will usually be a line segment pointing in the direction the solution curve must go. By following the line segments in the slope field, we trace out the solution curve. Several examples with different starting points are shown in Figure 9.16. In general, at each point there is one line segment and therefore only one direction for the solution curve to go. Thus the solution curve *exists* and is *unique* provided we are given an initial point.

It can be shown that if the slope field is continuous as we move from point to point in the plane, we can be sure that the solution curve exists around every point. Ensuring that each point has only one solution curve through it requires a slightly stronger condition.

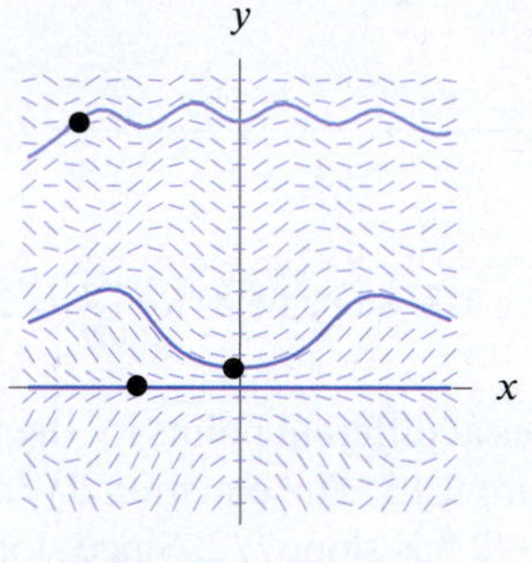

Figure 9.16: There is one and only one solution curve through each point in the plane for this slope field

Problems for Section 9.3

1. Sketch three solution curves for each of the slope fields in Figure 9.17.

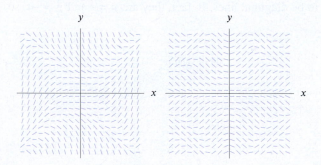

Figure 9.17

2. Sketch the slope field for $dy/dx = y^2$ at the points marked in Figure 9.18.

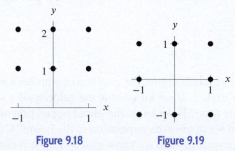

Figure 9.18   **Figure 9.19**

3. Sketch the slope field for $dy/dx = x/y$ at the points marked in Figure 9.19.

4. Figure 9.20 is a slope field for $dy/dx = y - 10$.

    (a) Draw the solution curve for each of the following initial conditions:

        (i) $y = 8$ when $x = 0$ (ii) $y = 12$ when $x = 0$
        (iii) $y = 10$ when $x = 0$

    (b) Since $dy/dx = y - 10$, when $y = 10$, we have $dy/dx = 10 - 10 = 0$. Explain why this matches your answer to part (iii).

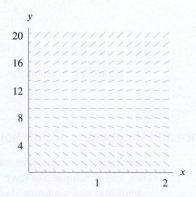

Figure 9.20: Slope field for $dy/dx = y - 10$

5. Figure 9.21 is the slope field for the equation $y' = x + y$.

   (a) Sketch the solutions that pass through the points
       (i) (0, 0)   (ii) (−3, 1)   (iii) (−1, 0)

   (b) From your sketch, guess the equation of the solution passing through (−1, 0).

   (c) Check your solution to part (b) by substituting it into the differential equation.

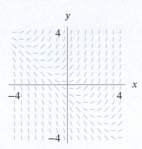

Figure 9.21: Slope field for $y' = x + y$

6. (a) For $dy/dx = x^2 - y^2$, find the slope at the following points:

   $(1, 0), \quad (0, 1), \quad (1, 1), \quad (2, 1), \quad (1, 2), \quad (2, 2)$

   (b) Sketch the slope field at these points.

7. Match the slope fields in Figure 9.22 with their differential equations:

   (a) $y' = 1 + y^2$ (b) $y' = x$ (c) $y' = \sin x$
   (d) $y' = y$ (e) $y' = x - y$ (f) $y' = 4 - y$

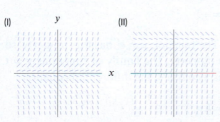

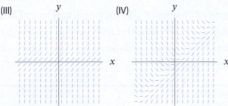

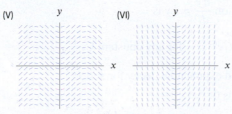

Figure 9.22: Each slope field is graphed for
$-5 \le x \le 5, -5 \le y \le 5$

8. Match the slope fields in Figure 9.23 with their differential equations. Explain your reasoning.

   (a) $y' = -y$ (b) $y' = y$ (c) $y' = x$
   (d) $y' = 1/y$ (e) $y' = y^2$

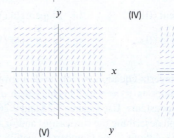

Figure 9.23: Each slope field is graphed for
$-5 \le x \le 5, -5 \le y \le 5$

9. Which one of the following differential equations best fits the slope field shown in Figure 9.24? Explain.

   I. $dP/dt = P - 1$ II. $dP/dt = P(P - 1)$

   III. $dP/dt = 3P(1 - P)$ IV. $dP/dt = 1/3 P(1 - P)$

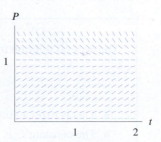

Figure 9.24

10. **(a)** Consider the slope field for $dy/dx = xy$. What is the slope of the line segment at the point $(2, 1)$? At $(0, 2)$? At $(-1, 1)$? At $(2, -2)$?

(b) Sketch part of the slope field by drawing line segments with the slopes calculated in part (a).

■ For Problems **11–16**, consider a solution curve for each of the slope fields in Problem 7. Write one or two sentences describing qualitatively the long-run behavior of y. For example, as x increases, does $y \to \infty$, or does y remain finite? You may get different limiting behavior for different starting points. In each case, your answer should discuss how the limiting behavior depends on the starting point.

11. Slope field (I)   12. Slope field (IV)

13. Slope field (III)   14. Slope field (V)

15. Slope field (II)   16. Slope field (VI)

17. The Gompertz equation, which models growth of animal tumors, is $y' = -ay\ln(y/b)$, where a and b are positive constants. Use Figures 9.25 and 9.26 to write a paragraph describing the similarities and/or differences between solutions to the Gompertz equation with $a = 1$ and $b = 2$ and solutions to the equation $y' = y(2 - y)$.

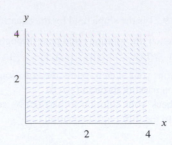

Figure 9.25: Slope field for $y' = -y\ln(y/2)$

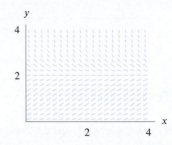

Figure 9.26: Slope field for $y' = y(2 - y)$

9.4 EXPONENTIAL GROWTH AND DECAY

What is a solution to the differential equation

$$\frac{dy}{dt} = y?$$

A solution is a function that is its own derivative. The function $y = e^t$ has this property, so $y = e^t$ is a solution. In fact, any multiple of e^t also has this property. The family of functions $y = Ce^t$ is the general solution to this differential equation. If k is a constant, the differential equation

$$\frac{dy}{dt} = ky,$$

is similar. This differential equation says that the rate of change of y is proportional to y. The constant k is the constant of proportionality. By substituting $y = Ce^{kt}$ into the differential equation, you can check that $y = Ce^{kt}$ is a solution. For another derivation of the solution, see the Focus on Theory section at the end of the chapter. We have the following result:

The general solution to the differential equation $\dfrac{dy}{dt} = ky$ is

$$y = Ce^{kt} \qquad \text{for any constant C.}$$

- This is exponential growth for $k > 0$, and exponential decay for $k < 0$.
- The constant C is the value of y when t is 0.

Graphs of solution curves for some $k > 0$ are in Figure 9.27. For $k < 0$, the graphs are reflected about the y-axis. See Figure 9.28.

Figure 9.27: Graphs of $y = Ce^{kt}$, which are solutions to $\dfrac{dy}{dt} = ky$ for some fixed $k > 0$

Figure 9.28: Graphs of $y = Ce^{kt}$, which are solutions to $\dfrac{dy}{dt} = ky$ for some fixed $k < 0$

Example 1 (a) Find the general solution to each of the following differential equations:

    (i) $\dfrac{dy}{dt} = 0.05y$ (ii) $\dfrac{dP}{dt} = -0.3P$ (iii) $\dfrac{dw}{dz} = 2z$ (iv) $\dfrac{dw}{dz} = 2w$

    (b) For differential equation (i), find the particular solution satisfying $y = 50$ when $t = 0$.

Solution (a) The differential equations given in (i), (ii), and (iv) are all examples of exponential growth or decay, since each is in the form

$$\text{Derivative} = \text{Constant} \cdot \text{Dependent variable.}$$

    Notice that differential equation (iii) is not in this form. Example 1 on page 401 showed that the solution to (iii) is $w = z^2 + C$. The general solutions are

    (i) $y = Ce^{0.05t}$ (ii) $P = Ce^{-0.3t}$ (iii) $w = z^2 + C$ (iv) $w = Ce^{2z}$.

    (b) The general solution to (i) is $y = Ce^{0.05t}$. Substituting $y = 50$ and $t = 0$ gives

$$50 = Ce^{0.05(0)}$$
$$50 = C \cdot 1.$$

    So $C = 50$, and the particular solution to this initial-value problem is $y = 50e^{0.05t}$.

Population Growth

Consider the population P of a region where there is no immigration or emigration. The rate at which the population is growing is often proportional to the size of the population. This means larger populations grow faster, as we expect since there are more people to have babies. If the population has a continuous growth rate of 2% per unit time, then we know

$$\text{Rate of growth of population} = 2\% \text{ of current population,}$$

so

$$\frac{dP}{dt} = 0.02P.$$

This equation is of the form $dP/dt = kP$ for $k = 0.02$ and has the general solution $P = Ce^{0.02t}$. If the initial population at time $t = 0$ is P_0, then $P_0 = Ce^{0.02(0)} = C$. So $C = P_0$ and we have

$$P = P_0 e^{0.02t}.$$

Continuously Compounded Interest

In Chapter 1 we introduced continuous compounding as the limiting case in which interest was added more and more often. Here we approach continuous compounding from a different point of view. We imagine interest being accrued at a rate proportional to the balance at that moment. Thus, the larger the balance, the faster interest is earned and the faster the balance grows.

Example 2   A bank account earns interest continuously at a rate of 5% of the current balance per year. Assume that the initial deposit is $1000 and that no other deposits or withdrawals are made.

(a) Write a differential equation satisfied by the balance in the account.
(b) Solve the differential equation and graph the solution.

Solution   (a) We are looking for B, the balance in the account in dollars, as a function of t, time in years. Interest is being added continuously to the account at a rate of 5% of the balance at that moment, so

Rate at which balance is increasing $= 5\%$ of current balance.

Thus, a differential equation that describes the process is

$$\frac{dB}{dt} = 0.05B.$$

Notice that it does not involve the $1000, the initial condition, because the initial deposit does not affect the process by which interest is earned.

(b) Since $B_0 = 1000$ is the initial value of B, the solution to this differential equation is

$$B = B_0 e^{0.05t} = 1000 e^{0.05t}.$$

This function is graphed in Figure 9.29.

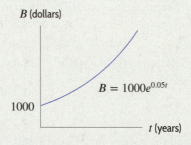

Figure 9.29: Bank balance against time

You may wonder how we can represent an amount of money by a differential equation, since money can only take on discrete values (you can't have fractions of a cent). In fact, the differential equation is only an approximation, but for large amounts of money, it is a pretty good approximation.

Pollution in the Great Lakes

In the 1960s pollution in the Great Lakes became an issue of public concern. We will set up a model for how long it would take the lakes to flush themselves clean, assuming no further pollutants were being dumped in the lake.

Let Q be the total quantity of pollutant in a lake of volume V at time t. Suppose that clean water is flowing into the lake at a constant rate r and that water flows out at the same rate. Assume that the pollutant is evenly spread throughout the lake, and that the clean water coming into the lake immediately mixes with the rest of the water.

How does Q vary with time? First, notice that since pollutants are being taken out of the lake but not added, Q decreases, and the water leaving the lake becomes less polluted, so the rate at which the pollutants leave decreases. This tells us that Q is decreasing and concave up. In addition, the pollutants will never be completely removed from the lake though the quantity remaining will become arbitrarily small. In other words, Q is asymptotic to the t-axis. (See Figure 9.30.)

Setting Up a Differential Equation for the Pollution

To understand how Q changes with time, we write a differential equation for Q. We know that

$$\begin{pmatrix} \text{Rate } Q \\ \text{changes} \end{pmatrix} = - \begin{pmatrix} \text{Rate pollutants} \\ \text{leave in outflow} \end{pmatrix}$$

where the negative sign represents the fact that Q is decreasing. At time t, the concentration of pollutants is Q/V and water containing this concentration is leaving at rate r. Thus,

$$\begin{pmatrix} \text{Rate pollutants} \\ \text{leave in outflow} \end{pmatrix} = \begin{pmatrix} \text{Rate of} \\ \text{outflow} \end{pmatrix} \times \text{Concentration} = r \cdot \frac{Q}{V}.$$

So the differential equation is

$$\frac{dQ}{dt} = -\frac{r}{V}Q$$

and its general solution is

$$Q = Q_0 e^{-rt/V}.$$

Table 9.6 contains values of r and V for four of the Great Lakes.[1] We use this data to calculate how long it would take for certain fractions of the pollution to be removed from Lake Erie.

Example 3   How long will it take for 90% of the pollution to be removed from Lake Erie? For 99% to be removed?

Solution   For Lake Erie, $r/V = 175/460 = 0.38$, so at time t we have

$$Q = Q_0 e^{-0.38t}.$$

When 90% of the pollution has been removed, 10% remains, so $Q = 0.1Q_0$. Substituting gives

$$0.1Q_0 = Q_0 e^{-0.38t}.$$

Canceling Q_0 and solving for t gives

$$t = \frac{-\ln(0.1)}{0.38} \approx 6 \text{ years.}$$

Similarly, when 99% of the pollution has been removed, $Q = 0.01Q_0$, so we solve

$$0.01Q_0 = Q_0 e^{-0.38t},$$

giving

$$t = \frac{-\ln(0.01)}{0.38} \approx 12 \text{ years.}$$

Q (quantity of pollutant)

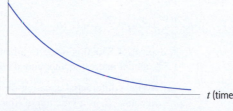

Figure 9.30: Pollutant in lake versus time

Table 9.6   *Volume and outflow in Great Lakes*

	V (km^3)	r (km^3/year)
Superior	12,200	65.2
Michigan	4900	158
Erie	460	175
Ontario	1600	209

[1] Data from William E. Boyce and Richard C. DiPrima, *Elementary Differential Equations* (New York: Wiley, 1977).

The Quantity of a Drug in the Body

As we saw in Section 9.1, the rate at which a drug leaves a patient's body is proportional to the quantity of the drug left in the body. If we let Q represent the quantity of drug left, then

$$\frac{dQ}{dt} = -kQ.$$

The negative sign indicates the quantity of drug in the body is decreasing. The solution to this differential equation is $Q = Q_0 e^{-kt}$; the quantity decreases exponentially. The constant k depends on the drug and Q_0 is the amount of drug in the body at time zero. Sometimes physicians convey information about the relative decay rate with a *half life*, which is the time it takes for Q to decrease by a factor of $1/2$.

Example 4   Valproic acid is a drug used to control epilepsy; its half-life in the human body is about 15 hours.

(a) Use the half-life to find the constant k in the differential equation $dQ/dt = -kQ$, where Q represents the quantity of drug in the body t hours after the drug is administered.
(b) At what time will 10% of the original dose remain?

Solution   (a) Since the half-life is 15 hours, we know that the quantity remaining $Q = 0.5Q_0$ when $t = 15$. We substitute into the solution to the differential equation, $Q = Q_0 e^{-kt}$, and solve for k:

$$Q = Q_0 e^{-kt}$$
$$0.5Q_0 = Q_0 e^{-k(15)}$$
$$0.5 = e^{-15k} \quad \text{(Divide through by } Q_0)$$
$$\ln 0.5 = -15k \quad \text{(Take the natural logarithm of both sides)}$$
$$k = \frac{-\ln 0.5}{15} = 0.0462. \quad \text{(Solve for } k)$$

(b) To find the time when 10% of the original dose remains in the body, we substitute $0.10Q_0$ for the quantity remaining, Q, and solve for the time, t.

$$0.10Q_0 = Q_0 e^{-0.0462t}$$
$$0.10 = e^{-0.0462t}$$
$$\ln 0.10 = -0.0462t$$
$$t = \frac{\ln 0.10}{-0.0462} = 49.84.$$

There will be 10% of the drug still in the body at $t = 49.84$, or after about 50 hours.

Problems for Section 9.4

■ Find solutions to the differential equations in Problems **1–6**, subject to the given initial condition.

1. $\dfrac{dP}{dt} = 0.02P, \quad P(0) = 20$

2. $\dfrac{dw}{dr} = 3w, \quad w = 30$ when $r = 0$

3. $\dfrac{dy}{dx} = -0.14y, \quad y = 5.6$ when $x = 0$

4. $\dfrac{dQ}{dt} = \dfrac{Q}{5}, \quad Q = 50$ when $t = 0$

5. $\dfrac{dp}{dq} = -0.1p, \quad p = 100$ when $q = 5$

6. $\dfrac{dy}{dx} + \dfrac{y}{3} = 0, \quad y(0) = 10$

7. A deposit of $5000 is made to a bank account paying 1.5% annual interest, compounded continuously.

(a) Write a differential equation for the balance in the account, B, as a function of time, t, in years.
(b) Solve the differential equation.
(c) How much money is in the account in 10 years?

8. The value of an investment grows continuously at an annual rate of r (when the interest rate is 5%, $r = 0.05$, and so on). Suppose $2000 was invested in 2010.

   (a) Write a differential equation satisfied by M, the value of the investment at time t, measured in years since 2010.
   (b) Solve the differential equation.
   (c) Sketch the solution until the year 2040 for interest rates of 5% and 10%.

9. A bank account that earns 10% interest compounded continuously has an initial balance of zero. Money is deposited into the account at a continuous rate of $1000 per year.

   (a) Write a differential equation that describes the rate of change of the balance $B = f(t)$.
   (b) Solve the differential equation.

10. The amount of ozone, Q, in the atmosphere is decreasing at a rate proportional to the amount of ozone present. If time t is measured in years, the constant of proportionality is -0.0025. Write a differential equation for Q as a function of t, and give the general solution for the differential equation. If this rate continues, approximately what percent of the ozone in the atmosphere now will decay in the next 20 years?

11. Using the model in the text and the data in Table 9.6 on page 409, find how long it would take for 90% of the pollution to be removed from Lake Michigan and from Lake Ontario, assuming no new pollutants are added. Explain how you can tell which lake will take longer to be purified just by looking at the data in the table.

12. Use the model in the text and the data in Table 9.6 on page 409 to determine which of the Great Lakes would require the longest time and which would require the shortest time for 80% of the pollution to be removed, assuming no new pollutants are being added. Find the ratio of these two times.

13. The rate at which a drug leaves the bloodstream and passes into the urine is proportional to the quantity of the drug in the blood at that time. If an initial dose of Q_0 is injected directly into the blood, 20% is left in the blood after 3 hours.

   (a) Write and solve a differential equation for the quantity, Q, of the drug in the blood after t hours.
   (b) How much of this drug is in a patient's body after 6 hours if the patient is given 100 mg initially?

14. In some chemical reactions, the rate at which the amount of a substance changes with time is proportional to the amount present. For example, this is the case as δ-glucono-lactone changes into gluconic acid.

   (a) Write a differential equation satisfied by y, the quantity of δ-glucono-lactone present at time t.
   (b) If 100 grams of δ-glucono-lactone is reduced to 54.9 grams in one hour, how many grams will remain after 10 hours?

15. Oil is pumped continuously from a well at a rate proportional to the amount of oil left in the well. Initially there were 1 million barrels of oil in the well; six years later 500,000 barrels remain.

   (a) At what rate was the amount of oil in the well decreasing when there were 600,000 barrels remaining?
   (b) When will there be 50,000 barrels remaining?

16. Hydrocodone bitartrate is used as a cough suppressant. After the drug is fully absorbed, the quantity of drug in the body decreases at a rate proportional to the amount left in the body. The half-life of hydrocodone bitartrate in the body is 3.8 hours and the dose is 10 mg.

   (a) Write a differential equation for the quantity, Q, of hydrocodone bitartrate in the body at time t, in hours since the drug was fully absorbed.
   (b) Solve the differential equation given in part (a).
   (c) Use the half-life to find the constant of proportionality, k.
   (d) How much of the 10-mg dose is still in the body after 12 hours?

17. The amount of land in use for growing crops increases as the world's population increases. Suppose $A(t)$ represents the total number of hectares of land in use in year t. (A hectare is about $2\frac{1}{2}$ acres.)

   (a) Explain why it is plausible that $A(t)$ satisfies the equation $A'(t) = kA(t)$. What assumptions are you making about the world's population and its relation to the amount of land used?
   (b) In 1966 about 4.55 billion hectares of land were in use; in 1996 the figure was 4.93 billion hectares.[2] If the total amount of land available for growing crops is thought to be 6 billion hectares, when does this model predict it will be exhausted? (Let $t = 0$ in 1966.)

9.5 APPLICATIONS AND MODELING

In the last section, we considered several situations modeled by the differential equation

$$\frac{dy}{dt} = ky.$$

[2]http://www.farmingsolutions.org/facts/factscontent_det.asp, accessed June 6, 2011.

In this section, we consider situations where the rate of change of y is a linear function of y of the form

$$\frac{dy}{dt} = k(y - A), \qquad \text{where } k \text{ and } A \text{ are constants.}$$

The Quantity of a Drug in the Body

A patient is given the drug warfarin, an anticoagulant, intravenously at the rate of 0.5 mg/hour. Warfarin is metabolized and leaves the body at the rate of about 2% per hour. A differential equation for the quantity, Q (in mg), of warfarin in the body after t hours is given by

$$\text{Rate of change} = \text{Rate in} - \text{Rate out}$$

$$\frac{dQ}{dt} = 0.5 - 0.02Q.$$

What does this tell us about the quantity of warfarin in the body for different initial values of Q?

If Q is small, then $0.02Q$ is also small and the rate the drug is excreted is less than the rate at which the drug is entering the body. Since the rate in is greater than the rate out, the rate of change is positive and the quantity of drug in the body is increasing. If Q is large enough that $0.02Q$ is greater than 0.5, then $0.5 - 0.02Q$ is negative, so dQ/dt is negative and the quantity is decreasing.

For small Q, the quantity will increase until the rate in equals the rate out. For large Q, the quantity will decrease until the rate in equals the rate out. What is the value of Q at which the rate in exactly matches the rate out? We have

$$\text{Rate in} = \text{Rate out}$$

$$0.5 = 0.02Q$$

$$Q = 25.$$

If the amount of warfarin in the body is initially 25 mg, then the amount being excreted exactly matches the amount being added. The quantity of drug Q will stay constant at 25 mg. Notice also that when $Q = 25$, the derivative dQ/dt is zero, since

$$\frac{dQ}{dt} = 0.5 - 0.02(25) = 0.5 - 0.5 = 0.$$

If the initial quantity is 25, then the solution is the horizontal line $Q = 25$. This solution is called an *equilibrium solution*.

The slope field for this differential equation is shown in Figure 9.31, with solution curves drawn for $Q_0 = 20$, $Q_0 = 25$, and $Q_0 = 30$. In each case, we see that the quantity of drug in the body is approaching the equilibrium solution of 25 mg. The solution curve with $Q_0 = 30$ should remind you of an exponential decay function. It is, in fact, an exponential decay function that has been shifted up 25 units.

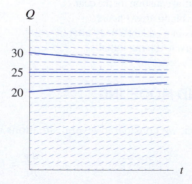

Figure 9.31: Slope field for $dQ/dt = 0.5 - 0.02Q$

Solving the Differential Equation $dy/dt = k(y - A)$

The drug concentration in the previous example satisfies a differential equation of the form

$$\frac{dy}{dt} = k(y - A).$$

Let us find the general solution to this equation. Since A is a constant, $dA/dt = 0$ so that we have

$$\frac{d}{dt}(y - A) = \frac{dy}{dt} - \frac{dA}{dt} = \frac{dy}{dt} - 0 = k(y - A).$$

Thus $y - A$ satisfies an exponential differential equation, so $y - A$ must be of the form

$$y - A = Ce^{kt}.$$

For an alternative derivation of the solution, see the Focus on Theory section at the end of the chapter.

The general solution to the differential equation

$$\frac{dy}{dt} = k(y - A)$$

is

$$y = A + Ce^{kt}, \qquad \text{for any constant } C.$$

Warning: Notice that, for differential equations of this form, the arbitrary constant C is *not* the initial value of the variable, but rather the initial value of $y - A$.

Example 1 Give the solution to each of the following differential equations:

(a) $\dfrac{dy}{dt} = 0.02(y - 50)$ (b) $\dfrac{dP}{dt} = 5(P - 10), \quad P = 8$ when $t = 0$

(c) $\dfrac{dy}{dt} = 3y - 300$ (d) $\dfrac{dW}{dt} = 500 - 0.1W$

Solution (a) The general solution is $y = 50 + Ce^{0.02t}$.
(b) The general solution is $P = 10 + Ce^{5t}$. Use the initial condition to solve for C:

$$8 = 10 + C(e^0)$$
$$8 = 10 + C.$$

So $C = -2$, and the particular solution is $P = 10 - 2e^{5t}$.
(c) First rewrite the right-hand side of the equation in the form $k(y - A)$ by factoring out a 3:

$$\frac{dy}{dt} = 3(y - 100).$$

The general solution to this differential equation is $y = 100 + Ce^{3t}$.
(d) We begin by factoring out the coefficient of W:

$$\frac{dW}{dt} = 500 - 0.1W = -0.1\left(W - \frac{500}{0.1}\right) = -0.1(W - 5000).$$

The general solution to this differential equation is $W = 5000 + Ce^{-0.1t}$.

Example 2   At the start of this section, we gave the following differential equation for the quantity of warfarin in the body:

$$\frac{dQ}{dt} = 0.5 - 0.02Q.$$

Write the general solution to this differential equation. Find particular solutions for $Q_0 = 20$, $Q_0 = 25$, and $Q_0 = 30$.

Solution   We first rewrite the differential equation in the form $dQ/dt = k(Q - A)$ by factoring out -0.02:

$$\frac{dQ}{dt} = 0.5 - 0.02Q = -0.02(Q - 25).$$

The general solution to this differential equation is

$$Q = 25 + Ce^{-0.02t}.$$

To find the particular solution when $Q_0 = 20$, we use the initial condition to solve for C:

$$20 = 25 + C(e^0)$$
$$C = -5.$$

The particular solution when $Q_0 = 20$ is $Q = 25 - 5e^{-0.02t}$.

When $Q_0 = 25$, we have $C = 0$ and the particular solution is the horizontal line $Q = 25$. When $Q_0 = 30$, we have $C = 5$ and the particular solution is $Q = 25 + 5e^{-0.02t}$. These three solutions are the three we saw earlier in Figure 9.31.

Example 3   A company's revenue is earned at a continuous annual rate of 5% of its net worth. At the same time, the company's payroll obligations are paid out at a constant rate of 200 million dollars a year. We saw in Section 9.1 that the differential equation governing the net worth, W (in millions of dollars), of this company in year t is given by

$$\text{Rate of change of } W = \text{Rate in} - \text{Rate out}$$
$$\frac{dW}{dt} = 0.05W - 200.$$

(a) Solve the differential equation, assuming an initial net worth of W_0 million dollars.
(b) Sketch the solution for $W_0 = 3000$, 4000 and 5000. For which of these values of W_0 does the company go bankrupt? In which year?

Solution   (a) Factor out 0.05 to get

$$\frac{dW}{dt} = 0.05(W - 4000).$$

The general solution is

$$W = 4000 + Ce^{0.05t}.$$

To find C we use the initial condition that $W = W_0$ when $t = 0$.

$$W_0 = 4000 + Ce^0$$
$$W_0 - 4000 = C$$

Substituting this value for C into $W = 4000 + Ce^{0.05t}$ gives

$$W = 4000 + (W_0 - 4000)e^{0.05t}.$$

(b) If $W_0 = 4000$, then $W = 4000$, the equilibrium solution.

If $W_0 = 5000$, then $W = 4000 + 1000e^{0.05t}$.

If $W_0 = 3000$, then $W = 4000 - 1000e^{0.05t}$. The graphs of these functions are in Figure 9.32. Notice that if the net worth starts with W_0 near, but not equal to, \$4000 million, then W moves further away. We see that if $W_0 = 3000$, the value of W goes to 0, and the company goes bankrupt. Solving $W = 0$ gives $t \approx 27.7$, so the company goes bankrupt in its twenty-eighth year.

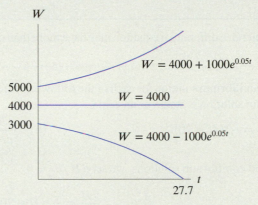

Figure 9.32: Solutions to $\frac{dW}{dt} = 0.05W - 200$

Equilibrium Solutions

Figure 9.31 shows the quantity of warfarin in the body for several different initial quantities. All these curves are solutions to the differential equation

$$\frac{dQ}{dt} = 0.5 - 0.02Q = -0.02(Q - 25),$$

and all the solutions have the form

$$Q = 25 + Ce^{-0.02t}$$

for some C. Notice that $Q \to 25$ as $t \to \infty$ for all solutions because $e^{-0.02t} \to 0$ as $t \to \infty$. In other words, in the long run, the quantity approaches the *equilibrium solution* of $Q = 25$ no matter what the initial quantity.

Notice that the equilibrium solution can be found directly from the differential equation by solving $dQ/dt = 0$:

$$\frac{dQ}{dt} = -0.02(Q - 25) = 0,$$

giving $Q = 25$. Because Q always gets closer and closer to the equilibrium value of 25 as $t \to \infty$, we call $Q = 25$ a *stable* equilibrium for Q.

A different situation is shown in Figure 9.32 with the solutions to the differential equation $dW/dt = 0.05W - 200$. We find the equilibrium by looking at the solution curves or by setting $dW/dt = 0$:

$$\frac{dW}{dt} = 0.05W - 200 = 0.05(W - 4000) = 0,$$

giving $W = 4000$ as the equilibrium solution. This equilibrium solution is called *unstable* because if W starts near, but not equal to, 4000, the net worth W moves further away from 4000 as $t \to \infty$.

- An **equilibrium solution** is constant for all values of the independent variable. The graph is a horizontal line. Equilibrium solutions can be identified by setting the derivative of the function to zero.

- An equilibrium solution is **stable** if a small change in the initial conditions gives a solution which tends toward the equilibrium as the independent variable tends to positive infinity.

- An equilibrium solution is **unstable** if a small change in the initial conditions gives a solution curve which veers away from the equilibrium as the independent variable tends to positive infinity.

In general, a differential equation may have more than one equilibrium solution or no equilibrium solution.

Example 4   Find the equilibrium solution for each of the following differential equations. Determine whether the equilibrium solution is stable or unstable.

(a) $\dfrac{dH}{dt} = -2(H - 20)$ (b) $\dfrac{dB}{dt} = 2(B - 10)$

Solution   (a) To find equilibrium solutions, we set $dH/dt = 0$:

$$\frac{dH}{dt} = -2(H - 20) = 0,$$

giving $H = 20$ as the equilibrium solution. The general solution to this differential equation is $H = 20 + Ce^{-2t}$. The solution curves for $H_0 = 10$, $H_0 = 20$, and $H_0 = 30$ are shown in Figure 9.33. We see that the equilibrium solution is stable.

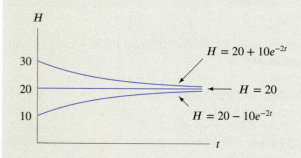

Figure 9.33: $H = 20$ is stable equilibrium

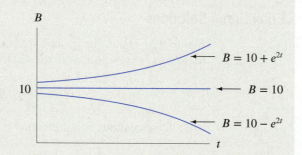

Figure 9.34: $B = 10$ is unstable equilibrium

(b) To find equilibrium solutions, we set $dB/dt = 0$:

$$\frac{dB}{dt} = 2(B - 10) = 0,$$

giving $B = 10$ as the equilibrium solution. The general solution to this differential equation is $B = 10 + Ce^{2t}$. The solution curves for $B_0 = 9$, $B_0 = 10$, and $B_0 = 11$ are shown in Figure 9.34. We see that the equilibrium solution is unstable.

Newton's Law of Heating and Cooling

Newton proposed that the temperature of a hot object decreases at a rate proportional to the difference between its temperature and that of its surroundings. Similarly, a cold object heats up at a rate proportional to the temperature difference between the object and its surroundings.

For example, a hot cup of coffee standing on a table cools at a rate proportional to the temperature difference between the coffee and the surrounding air. As the coffee cools, the rate at which it cools decreases because the temperature difference between the coffee and the air decreases. In the long run, the rate of cooling tends to zero and the temperature of the coffee approaches room temperature. Figure 9.35 shows the temperature of two cups of coffee against time, one starting at a higher temperature than the other, but both tending toward room temperature in the long run.

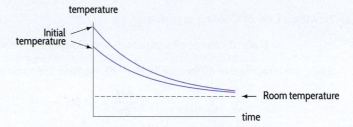

Figure 9.35: Temperature of coffee versus time

Let H be the temperature at time t of a cup of coffee in a 70°F room. Newton's Law says that the rate of change of H is proportional to the temperature difference between the coffee and the room:

$$\text{Rate of change of temperature} = \text{Constant} \cdot \text{Temperature difference}.$$

The rate of change of temperature is dH/dt. The temperature difference between the coffee and the room is $(H - 70)$, so

$$\frac{dH}{dt} = \text{Constant} \cdot (H - 70).$$

What about the sign of the constant? If the coffee starts out hotter than 70° (that is, $H - 70 > 0$), then the temperature of the coffee decreases (i.e., $dH/dt < 0$) and so the constant must be negative:

$$\frac{dH}{dt} = -k(H - 70) \qquad k > 0.$$

What can we learn from this differential equation? Suppose we take $k = 1$. The slope field for this differential equation in Figure 9.36 shows several solution curves. Notice that, as we expect, the temperature of the coffee is approaching the temperature of the room. The general solution to this differential equation is

$$H = 70 + Ce^{-t},$$

where C is an arbitrary constant.

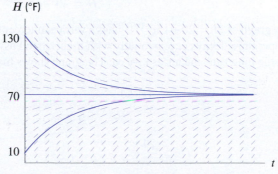

Figure 9.36: Slope field for $\dfrac{dH}{dt} = -(H - 70)$

Example 5   The body of a murder victim is found at noon in a room with a constant temperature of 20°C. At noon the temperature of the body is 35°C; two hours later the temperature of the body is 33°C.

(a) Find the temperature, H, of the body as a function of t, the time in hours since it was found.
(b) Graph H against t. What happens to the temperature in the long run?
(c) At the time of the murder, the victim's body had the normal body temperature, 37°C. When did the murder occur?

Solution   (a) Newton's Law of Cooling says that

$$\text{Rate of change of temperature} = \text{Constant} \cdot \text{Temperature difference.}$$

Since the temperature difference is $H - 20$, we have for some constant k

$$\frac{dH}{dt} = -k(H - 20).$$

The general solution is

$$H = 20 + Ce^{-kt}.$$

To determine C, we use the fact that $H = 35$ at $t = 0$:

$$35 = 20 + Ce^0$$
$$35 = 20 + C.$$

So $C = 15$ and we have

$$H = 20 + 15e^{-kt}.$$

To find k, we use the fact that $H = 33$ when $t = 2$:

$$33 = 20 + 15e^{-k(2)}.$$

We isolate the exponential and solve for k:

$$13 = 15e^{-2k}$$
$$\frac{13}{15} = e^{-2k}$$
$$\ln\left(\frac{13}{15}\right) = -2k$$
$$k = -\frac{\ln(13/15)}{2} = 0.072.$$

Therefore, the temperature, H, of the body as a function of time, t, is given by

$$H = 20 + 15e^{-0.072t}.$$

(b) The graph of $H = 20 + 15e^{-0.072t}$ has a vertical intercept of $H = 35$, the initial temperature. The temperature decays exponentially with a horizontal asymptote of $H = 20$. (See Figure 9.37.) "In the long run" means as $t \to \infty$. The graph shows that $H \to 20$ as $t \to \infty$.

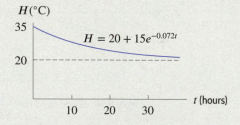

Figure 9.37: Temperature of dead body

(c) We want to know when the temperature was 37°C. We substitute $H = 37$ and solve for t:

$$37 = 20 + 15e^{-0.072t}$$

$$\frac{17}{15} = e^{-0.072t}.$$

Taking natural logs on both sides gives

$$\ln\left(\frac{17}{15}\right) = -0.072t$$

so

$$t = -\frac{\ln(17/15)}{0.072} = -1.74 \text{ hours.}$$

The murder occurred about 1.74 hours before noon, that is, about 10:15 am.

Problems for Section 9.5

■ Find particular solutions in Problems 1–8.

1. $\dfrac{dy}{dt} = 0.5(y - 200)$, $y = 50$ when $t = 0$

2. $\dfrac{dP}{dt} = P + 4$, $P = 100$ when $t = 0$

3. $\dfrac{dH}{dt} = 3(H - 75)$, $H = 0$ when $t = 0$

4. $\dfrac{dm}{dt} = 0.1m + 200$, $m(0) = 1000$

5. $\dfrac{dB}{dt} = 4B - 100$, $B = 20$ when $t = 0$

6. $\dfrac{dQ}{dt} = 0.3Q - 120$, $Q = 50$ when $t = 0$

7. $\dfrac{dB}{dt} + 2B = 50$, $B(1) = 100$

8. $\dfrac{dB}{dt} + 0.1B - 10 = 0$ $B(2) = 3$

9. Check that $y = A + Ce^{kt}$ is a solution to the differential equation

$$\frac{dy}{dt} = k(y - A).$$

10. A bank account earns 2% annual interest, compounded continuously. Money is deposited in a continuous cash flow at a rate of $1200 per year into the account.

   (a) Write a differential equation that describes the rate at which the balance $B = f(t)$ is changing.
   (b) Solve the differential equation given an initial balance $B_0 = 0$.
   (c) Find the balance after 5 years.

11. Money in an account earns interest at a continuous rate of 8% per year, and payments are made continuously out of the account at the rate of $5000 a year. The account initially contains $50,000. Write a differential equation for the amount of money in the account, B, in t years. Solve the differential equation. Does the account ever run out of money? If so, when?

12. A company earns 2% per month on its assets, paid continuously, and its expenses are paid out continuously at a rate of $80,000 per month.

   (a) Write a differential equation for the value, V, of the company as a function of time, t, in months.
   (b) What is the equilibrium solution for the differential equation? What is the significance of this value for the company?
   (c) Solve the differential equation found in part (a).
   (d) If the company has assets worth $3 million at time $t = 0$, what are its assets worth one year later?

13. A bank account earns 7% annual interest compounded continuously. You deposit $10,000 in the account, and withdraw money continuously from the account at a rate of $1000 per year.

   (a) Write a differential equation for the balance, B, in the account after t years.
   (b) What is the equilibrium solution to the differential equation? (This is the amount that must be deposited now for the balance to stay the same over the years.)
   (c) Find the solution to the differential equation.
   (d) How much is in the account after 5 years?
   (e) Graph the solution. What happens to the balance in the long run?

14. One theory on the speed an employee learns a new task claims that the more the employee already knows, the more slowly he or she learns. Suppose that the rate at which a person learns is equal to the percentage of the task not yet learned. If y is the percentage learned by time t, the percentage not yet learned by that time is

$100 - y$, so we can model this situation with the differential equation

$$\frac{dy}{dt} = 100 - y.$$

(a) Find the general solution to this differential equation.

(b) Sketch several solutions.

(c) Find the particular solution if the employee starts learning at time $t = 0$ (so $y = 0$ when $t = 0$).

15. A patient is given the drug theophylline intravenously at a rate of 43.2 mg/hour to relieve acute asthma. The rate at which the drug leaves the patient's body is proportional to the quantity there, with proportionality constant 0.082 if time, t, is in hours. The patient's body contains none of the drug initially.

   (a) Describe in words how you expect the quantity of theophylline in the patient to vary with time.

   (b) Write a differential equation satisfied by the quantity of theophylline in the body, $Q(t)$.

   (c) Solve the differential equation and graph the solution. What happens to the quantity in the long run?

16. A chain smoker smokes five cigarettes every hour. From each cigarette, 0.4 mg of nicotine is absorbed into the person's bloodstream. Nicotine leaves the body at a rate proportional to the amount present, with constant of proportionality -0.346 if t is in hours.

   (a) Write a differential equation for the level of nicotine in the body, N, in mg, as a function of time, t, in hours.

   (b) Solve the differential equation from part (a). Initially there is no nicotine in the blood.

   (c) The person wakes up at 7 am and begins smoking. How much nicotine is in the blood when the person goes to sleep at 11 pm (16 hours later)?

17. As you know, when a course ends, students start to forget the material they have learned. One model (called the Ebbinghaus model) assumes that the rate at which a student forgets material is proportional to the difference between the material currently remembered and some positive constant, a.

   (a) Let $y = f(t)$ be the fraction of the original material remembered t weeks after the course has ended. Set up a differential equation for y. Your equation will contain two constants; the constant a is less than y for all t.

   (b) Solve the differential equation.

   (c) Describe the practical meaning (in terms of the amount remembered) of the constants in the solution $y = f(t)$.

18. (a) Find the equilibrium solution of the equation

$$\frac{dy}{dt} = 0.5y - 250.$$

   (b) Find the general solution of this equation.

(c) Graph several solutions with different initial values.

(d) Is the equilibrium solution stable or unstable?

19. (a) What are the equilibrium solutions for the differential equation

$$\frac{dy}{dt} = 0.2(y - 3)(y + 2)?$$

   (b) Use a graphing calculator or computer to sketch a slope field for this differential equation. Use the slope field to determine whether each equilibrium solution is stable or unstable.

20. Figure 9.38 gives the slope field for a differential equation. Estimate all equilibrium solutions and indicate whether each is stable or unstable.

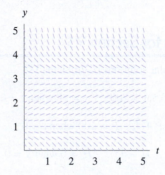

Figure 9.38

21. A yam is put in a 200°C oven and heats up according to the differential equation

$$\frac{dH}{dt} = -k(H - 200), \quad \text{for } k \text{ a positive constant.}$$

   (a) If the yam is at 20°C when it is put in the oven, solve the differential equation.

   (b) Find k using the fact that after 30 minutes the temperature of the yam is 120°C.

22. At 1:00 pm one winter afternoon, there is a power failure at your house in Wisconsin, and your heat does not work without electricity. When the power goes out, it is 68°F in your house. At 10:00 pm, it is 57°F in the house, and you notice that it is 10°F outside.

   (a) Assuming that the temperature, T, in your home obeys Newton's Law of Cooling, write the differential equation satisfied by T.

   (b) Solve the differential equation to estimate the temperature in the house when you get up at 7:00 am the next morning. Should you worry about your water pipes freezing?

   (c) What assumption did you make in part (a) about the temperature outside? Given this (probably incorrect) assumption, would you revise your estimate up or down? Why?

23. A detective finds a murder victim at 9 am. The temperature of the body is measured at 90.3°F. One hour later, the temperature of the body is 89.0°F. The temperature of the room has been maintained at a constant 68°F.

    (a) Assuming the temperature, T, of the body obeys Newton's Law of Cooling, write a differential equation for T.

    (b) Solve the differential equation to estimate the time the murder occurred.

24. A drug is administered intravenously at a constant rate of r mg/hour and is excreted at a rate proportional to the quantity present, with constant of proportionality $\alpha > 0$.

    (a) Solve a differential equation for the quantity, Q, in milligrams, of the drug in the body at time t hours. Assume there is no drug in the body initially. Your answer will contain r and α. Graph Q against t. What is Q_∞, the limiting long-run value of Q?

    (b) What effect does doubling r have on Q_∞? What effect does doubling r have on the time to reach half the limiting value, $\frac{1}{2}Q_\infty$?

    (c) What effect does doubling α have on Q_∞? On the time to reach $\frac{1}{2}Q_\infty$?

25. Some people write the solution of the initial value problem

$$\frac{dy}{dt} = k(y - A) \qquad y = y_0 \text{ at } t = 0$$

in the form

$$\frac{y - A}{y_0 - A} = e^{kt}.$$

Show that this formula gives the correct solution for y, assuming $y_0 \neq A$.

9.6 MODELING THE INTERACTION OF TWO POPULATIONS

So far we have used a differential equation to model the growth of a single quantity. We now consider the growth of two interacting populations, a situation which requires a system of two differential equations. Examples include two species competing for food, one species preying on another, or two species helping each other (symbiosis).

A Predator-Prey Model: Robins and Worms

We model a predator-prey system using what are called the Lotka-Volterra equations. Let's look at a simplified and idealized case in which robins are the predators and worms the prey.[3] Suppose there are r thousand robins and w million worms. If there were no robins, the worms would increase exponentially according to the equation

$$\frac{dw}{dt} = aw \qquad \text{where } a \text{ is a positive constant.}$$

If there were no worms, the robins would have no food and so their population would decrease according to the equation[4]

$$\frac{dr}{dt} = -br \qquad \text{where } b \text{ is a positive constant.}$$

Now imagine the effect of the two populations on one another. Clearly, the presence of the robins is bad for the worms, so

$$\frac{dw}{dt} = aw - \text{Effect of robins on worms.}$$

On the other hand, the robins do better with the worms around, so

$$\frac{dr}{dt} = -br + \text{Effect of worms on robins.}$$

How exactly do the two populations interact? Let's assume the effect of one population on the other is proportional to the number of "encounters." (An encounter is when a robin eats a worm.) The number of encounters is likely to be proportional to the product of the populations because if one population

---

[3]Based on work by Thomas A. McMahon (1943-1999), Harvard University.
[4]This assumption unrealistically predicts that the robin population will decay exponentially, rather than die out in finite time.

is held fixed, the number of encounters should be directly proportional to the other population. So we assume

$$\frac{dw}{dt} = aw - cwr \quad \text{and} \quad \frac{dr}{dt} = -br + kwr,$$

where c and k are positive constants.

To analyze this system of equations, let's look at the specific example with $a = b = c = k = 1$:

$$\frac{dw}{dt} = w - wr \quad \text{and} \quad \frac{dr}{dt} = -r + wr.$$

The Phase Plane

To see the growth of the populations, we want graphs of r and w against t. However, it is easier to obtain a graph of r against w first. If we plot a point (w, r) representing the number of worms and robins at any moment, then, as the populations change, the point moves. The wr-plane on which the point moves is called the *phase plane* and the path of the point is called the *phase trajectory*.

To find the phase trajectory, we need a differential equation relating w and r directly. We have the two differential equations

$$\frac{dw}{dt} = w - wr \quad \text{and} \quad \frac{dr}{dt} = -r + wr.$$

If we think of r as a function of w and w as a function of t, the chain rule gives

$$\frac{dr}{dt} = \frac{dr}{dw} \cdot \frac{dw}{dt}.$$

This tells us that

$$\frac{dr}{dw} = \frac{dr/dt}{dw/dt},$$

so we have

$$\frac{dr}{dw} = \frac{-r + wr}{w - wr}.$$

Figure 9.39 shows the slope field of this differential equation in the phase plane.

The Slope Field and Equilibrium Points

We can get an idea of what solutions of this equation look like from the slope field. At the point $(1, 1)$ there is no slope drawn because dr/dw is undefined there since the rates of change of both populations with respect to time are zero:

$$\frac{dw}{dt} = 1 - 1 \cdot 1 = 0, \quad \text{and} \quad \frac{dr}{dt} = -1 + 1 \cdot 1 = 0.$$

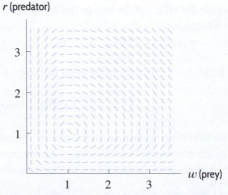

Figure 9.39: Slope field for $\dfrac{dr}{dw} = \dfrac{-r + wr}{w - wr}$

In terms of worms and robins, this means that if at some moment $w = 1$ and $r = 1$ (that is, there are 1 million worms and 1 thousand robins), then w and r remain constant forever. The point $w = 1$, $r = 1$ is therefore an equilibrium solution. The origin is also an equilibrium point, since if $w = 0$ and $r = 0$, then w and r remain constant. The slope field suggests that there are no other equilibrium points. We check this by solving

$$\frac{dw}{dt} = w - wr = 0 \qquad \text{and} \qquad \frac{dr}{dt} = -r + rw = 0,$$

which yields only $w = 0$, $r = 0$ and $w = 1$, $r = 1$ as solutions.

Trajectories in the wr-Phase Plane

Let's look at the trajectories in the phase plane. A point on a curve represents a pair of populations (w, r) existing at the same time t (though t is not shown on the graph). A short time later, the pair of populations is represented by a nearby point. As time passes, the point traces out a trajectory. It can be shown that the trajectory is a closed curve. See Figure 9.40.

In which direction does the point move on the trajectory? Look at the original pair of differential equations. They tell us how w and r change with time. Imagine, for example, that we are at the point P_0 in Figure 9.41, where $w = 2.2$ and $r = 1$; then

$$\frac{dr}{dt} = -r + wr = -1 + (2.2)(1) = 1.2 > 0.$$

Therefore, r is increasing, so the point is moving in the direction shown by the arrow in Figure 9.41.

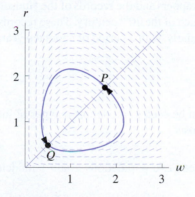

Figure 9.40: Solution curve is closed

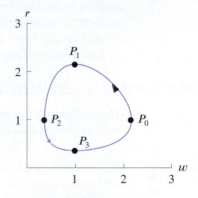

Figure 9.41: A trajectory

Example 1 Suppose that at time $t = 0$, there are 2.2 million worms and 1 thousand robins. Describe how the robin and worm populations change over time.

Solution The trajectory through the point P_0 where $w = 2.2$ and $r = 1$ is shown on the slope field in Figure 9.40 and by itself in Figure 9.41.

Initially there are lots of worms, so the robin population does well. The robin population is increasing and the worm population is decreasing until there are about 2.2 thousand robins and 1 million worms (point P_1 in Figure 9.41). At this point, there are too few worms to sustain the robin population; it begins to decrease and the worm population continues to fall as well. The robin population falls dramatically until there are about 1 thousand robins and 0.4 million worms (point P_2 in Figure 9.41). With so few robins, the worm population starts to recover, but the robin population is still decreasing. The worm population increases until there are about 0.4 thousand robins and 1 million worms (P_3 in Figure 9.41). Now there are lots of worms for the small population of robins, so both populations increase. The populations return to the starting values (since the trajectory forms a closed curve) and the cycle starts over.

Problem 17 at the end of the section shows how to calculate approximate coordinates of points on the curve.

The Populations as Functions of Time

The shape of a trajectory tells us how the populations vary with time. We use this information to graph each population against time, as in Figure 9.42. The fact that the trajectory is a closed curve means that both populations oscillate periodically. Both populations have the same period, and the worms (the prey) are at their maximum a quarter of a cycle before the robins.

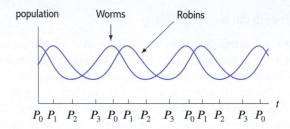

Figure 9.42: Populations of robins (in thousands) and worms (in millions) over time

Lynxes and Hares

A predator-prey system for which there are long-term data is the Canadian lynx and the hare. Both animals were of interest to fur trappers and the records of the Hudson Bay Company shed some light on their populations through much of the 20$^\text{th}$ century. These records show that both populations oscillated up and down, quite regularly, with a period of about ten years. This is the behavior predicted by Lotka-Volterra equations.

Other Forms of Species Interaction

The methods of this section can be used to model other types of interactions between two species, such as competition and symbiosis.

Example 2   Describe the interactions between two populations x and y modeled by the following systems of differential equations.

(a) $\dfrac{dx}{dt} = 0.2x - 0.5xy$

$\dfrac{dy}{dt} = 0.6y - 0.8xy$

(b) $\dfrac{dx}{dt} = -2x + 5xy$

$\dfrac{dy}{dt} = -y + 0.2xy$

(c) $\dfrac{dx}{dt} = 0.5x$

$\dfrac{dy}{dt} = -1.6y + 2xy$

(d) $\dfrac{dx}{dt} = 0.3x - 1.2xy$

$\dfrac{dy}{dt} = -0.7y + 2.5xy$

Solution   (a) If we ignore the interaction terms with xy, we have $dx/dt = 0.2x$ and $dy/dt = 0.6y$, so both populations grow exponentially. Since both interaction terms are negative, each species inhibits the other's growth, such as when deer and elk compete for food.

(b) If we ignore the interaction terms, the populations of both species decrease exponentially. However, both interaction terms are positive, meaning each species benefits from the other, so the relationship is symbiotic. An example is the pollination of plants by insects.

(c) Ignoring interaction, x grows and y decays. But the interaction term means that y benefits from x, in the way birds that build nests benefit from trees.

(d) Without the interaction, x grows and y decays. The interaction terms show that y hurts x while x benefits y. This is a predator-prey model where y is the predator and x is the prey.

Problems for Section 9.6

■ Problems **1–3** give the rates of growth of two populations, x and y, measured in thousands.

(a) Describe in words what happens to the population of each species in the absence of the other.

(b) Describe in words how the species interact with one another. Give reasons why the populations might behave as described by the equations. Suggest species that might interact in that way.

1. $\dfrac{dx}{dt} = 0.2x$

$\dfrac{dy}{dt} = 0.4xy - 0.1y$

2. $\dfrac{dx}{dt} = 0.01x - 0.05xy$

$\dfrac{dy}{dt} = -0.2y + 0.08xy$

3. $\dfrac{dx}{dt} = 0.01x - 0.05xy$

$\dfrac{dy}{dt} = 0.2y - 0.08xy$

4. The following system of differential equations represents the interaction between two populations, x and y.

$$\frac{dx}{dt} = -3x + 2xy$$
$$\frac{dy}{dt} = -y + 5xy$$

(a) Describe how the species interact. How would each species do in the absence of the other? Are they helpful or harmful to each other?

(b) If $x = 2$ and $y = 1$, does x increase or decrease? Does y increase or decrease? Justify your answers.

(c) Write a differential equation involving dy/dx.

(d) Use a computer or calculator to draw the slope field for the differential equation in part (c).

(e) Draw the trajectory starting at point $x = 2$, $y = 1$ on your slope field, and describe how the populations change as time increases.

■ Create a system of differential equations to model the situations in Problems **5–7**. You may assume that all constants of proportionality are 1.

5. The concentrations of two chemicals are denoted by x and y, respectively. Alone, each decays at a rate proportional to its concentration. Together, they interact to form a third substance. As the third substance is created, the concentrations of the initial two populations get smaller.

6. Two businesses are in competition with each other. Both businesses would do well without the other one, but each hurts the other's business. The values of the two businesses are given by x and y.

7. A population of fleas is represented by x, and a population of dogs is represented by y. The fleas need the dogs in order to survive. The dog population, however, is unaffected by the fleas.

8. Two companies, A and B, are in competition with each other. Let x represent the net worth (in millions of dollars) of Company A, and y represent the net worth (in millions of dollars) of Company B. Four trajectories are given in Figure 9.43. For each trajectory: Describe the initial conditions. Describe what happens initially: Do the companies gain or lose money early on? What happens in the long run?

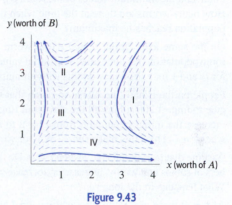

Figure 9.43

■ For Problems **9–19**, let w be the number of worms (in millions) and r the number of robins (in thousands) living on an island. Suppose w and r satisfy the following differential equations, which correspond to the slope field in Figure 9.44.

$$\frac{dw}{dt} = w - wr, \qquad \frac{dr}{dt} = -r + wr.$$

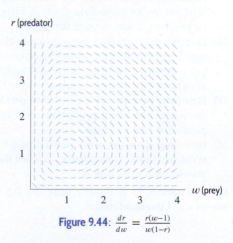

Figure 9.44: $\dfrac{dr}{dw} = \dfrac{r(w-1)}{w(1-r)}$

9. Explain why these differential equations are a reasonable model for interaction between the two populations. Why have the signs been chosen this way?

10. Solve these differential equations in the two special cases when there are no robins and when there are no worms living on the island.

11. Describe and explain the symmetry you observe in the slope field. What consequences does this symmetry have for the solution curves?

12. Assume $w = 2$ and $r = 2$ when $t = 0$. Do the numbers of robins and worms increase or decrease at first? What happens in the long run?

13. For the case discussed in Problem 12, estimate the maximum and the minimum values of the robin population. How many worms are there at the time when the robin population reaches its maximum?

14. On the same axes, graph w and r (the worm and the robin populations) against time. Use initial values of 1.5 for w and 1 for r. You may do this without units for t.

15. People on the island like robins so much that they decide to import 200 robins all the way from England, to increase the initial population from $r = 2$ to $r = 2.2$ when $t = 0$. Does this make sense? Why or why not?

16. Assume that $w = 3$ and $r = 1$ when $t = 0$. Do the numbers of robins and worms increase or decrease initially? What happens in the long run?

17. At $t = 0$ there are 2.2 million worms and 1 thousand robins.

(a) Use the differential equations to calculate the derivatives dw/dt and dr/dt at $t = 0$.
(b) Use the initial values and your answer to part (a) to estimate the number of robins and worms at $t = 0.1$.
(c) Using the method of part (a) and (b), estimate the number of robins and worms at $t = 0.2$ and 0.3.

18. (a) Assume that there are 3 million worms and 2 thousand robins. Locate the point corresponding to this situation on the slope field given in Figure 9.44. Draw the trajectory through this point.
(b) In which direction does the point move along this trajectory? Put an arrow on the trajectory and justify your answer using the differential equations for dr/dt and dw/dt given in this section.
(c) How large does the robin population get? What is the size of the worm population when the robin population is at its largest?
(d) How large does the worm population get? What is the size of the robin population when the worm population is at its largest?

19. Repeat Problem 18 if initially there are 0.5 million worms and 3 thousand robins.

20. For each system of differential equations in Example 2, determine whether x increases or decreases and whether y increases or decreases when $x = 2$ and $y = 2$.

■ For Problems 21–25, suppose x and y are the populations of two different species. Describe in words how each population changes with time.

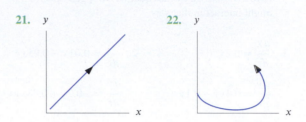

21.   22.

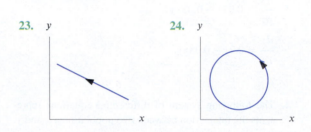

23.   24.

25. A kidney removes toxins from the blood. If a kidney does not function, the toxins can be removed by dialysis. This problem explores a model for $Q_1(t)$, the quantity of toxins in the body outside the blood, and $Q_2(t)$, the quantity of toxins in the blood, where t is the time after dialysis started.

(a) The quantity Q_1 changes for three reasons. First, toxins are created outside the blood at a constant rate, say A. Second, toxins flow into the blood at a rate proportional to the quantity outside the blood. Third, toxins flow out of the blood at a rate proportional to the quantity in the blood. Write a differential equation for Q_1.
(b) The quantity Q_2 changes for three reasons. First, dialysis removes toxins from the blood at a rate proportional to the toxins in the blood. Second and third, the same flows into and out of the blood that change Q_1 also change Q_2. Write a differential equation for Q_2.

26. For each system of equations in Example 2, write a differential equation involving dy/dx. Use a computer or calculator to draw the slope field for $x, y > 0$. Then draw the trajectory through the point $x = 3$, $y = 1$.

9.7   MODELING THE SPREAD OF A DISEASE

Differential equations can be used to predict when an outbreak of a disease becomes so severe that it is called an *epidemic*[5] and to decide what level of vaccination is necessary to prevent an epidemic. Let's consider a specific example.

Flu in a British Boarding School

In January 1978, 763 students returned to a boys' boarding school after their winter vacation. A week later, one boy developed the flu, followed immediately by two more. By the end of the month, nearly half the boys were sick. Most of the school had been affected by the time the epidemic was over in mid-February.[6]

Being able to predict how many people will get sick, and when, is an important step toward controlling an epidemic. This is one of the responsibilities of Britain's Communicable Disease Surveillance Centre and the US's Center for Disease Control and Prevention.

The S-I-R model

We apply one of the most commonly used models for an epidemic, called the S-I-R model, to the boarding school flu example. Imagine the population of the school divided into three groups:

S = the number of *susceptibles*, the people who are not yet sick
  but who could become sick

I = the number of *infecteds*, the people who are currently sick

R = the number of *recovered*, or *removed*, the people who have
  been sick and can no longer infect others or be reinfected.

In this model, the number of susceptibles decreases with time, as people become infected. We assume that the rate people become infected is proportional to the number of contacts between susceptible and infected people. We expect the number of contacts between the two groups to be proportional to both S and I. (If S doubles, we expect the number of contacts to double; similarly, if I doubles, we expect the number of contacts to double.) Thus, we assume that the number of contacts is proportional to the product, SI. In other words, we assume that for some constant $a > 0$,

$$\frac{dS}{dt} = -\left(\begin{array}{c} \text{Rate susceptibles} \\ \text{get sick} \end{array} \right) = -aSI.$$

(The negative sign is used because S is decreasing.)

The number of infecteds is changing in two ways: newly sick people are added to the infected group and others are removed. The newly sick people are exactly those people leaving the susceptible group and so accrue at a rate of aSI (with a positive sign this time). People leave the infected group either because they recover (or die), or because they are physically removed from the rest of the group and can no longer infect others. We assume that people are removed at a rate proportional to the number sick, or bI, where b is a positive constant. Thus,

$$\frac{dI}{dt} = \begin{array}{c} \text{Rate susceptibles} \\ \text{get sick} \end{array} - \begin{array}{c} \text{Rate infecteds} \\ \text{get removed} \end{array} = aSI - bI.$$

---

[5]Exactly when a disease should be called an epidemic is not always clear. The medical profession generally classifies a disease an epidemic when the frequency is higher than usually expected—leaving open the question of what is usually expected. See, for example, C. H. Hennekens and J. Buring, *Epidemiology in Medicine* (Boston: Little, Brown, 1987).

[6]Data from the Communicable Disease Surveillance Centre (UK); reported in "Influenza in a Boarding School," *British Medical Journal*, March 4, 1978, and by J. D. Murray in *Mathematical Biology*, Vol. I (New York: Springer Verlag, 3rd ed., 2002).

Assuming that those who have recovered from the disease are no longer susceptible, the recovered group increases at the rate of bI, so

$$\frac{dR}{dt} = bI.$$

We are assuming that having the flu confers immunity on a person, that is, that the person cannot get the flu again. (This is true for a given strain of flu, at least in the short run.)

We can use the fact that the total population $S + I + R$ is not changing. (The total population, the total number of boys in the school, did not change during the epidemic; see Problem 2 on page 430.) Thus, once we know S and I, we can calculate R. So we restrict our attention to the two equations

$$\frac{dS}{dt} = -aSI$$
$$\frac{dI}{dt} = aSI - bI.$$

The Constants a and b

The constant a measures how infectious the disease is—that is, how quickly it is transmitted from the infecteds to the susceptibles. In the case of the flu, we know from medical accounts that the epidemic started with one sick boy, with two more becoming sick about a day later. Thus, when $I = 1$ and $S = 762$, we have $dS/dt \approx -2$, enabling us to roughly[7] approximate a:

$$a = -\frac{dS/dt}{SI} = \frac{2}{762 \cdot 1} = 0.0026.$$

The constant b represents the rate at which infected people are removed from the infected population. In this case of the flu, boys were generally taken to the infirmary within one or two days of becoming sick. Assuming half the infected population was removed each day, we take $b \approx 0.5$. Thus, our equations are:

$$\frac{dS}{dt} = -0.0026SI$$
$$\frac{dI}{dt} = 0.0026SI - 0.5I.$$

The Phase Plane

As in Section 9.6, we look at trajectories in the phase plane. Thinking of I as a function of S, and S as a function of t, we use the chain rule to get

$$\frac{dI}{dt} = \frac{dI}{dS} \cdot \frac{dS}{dt},$$

so

$$\frac{dI}{dS} = \frac{dI/dt}{dS/dt}.$$

Substituting for dI/dt and dS/dt, we get

$$\frac{dI}{dS} = \frac{0.0026SI - 0.5I}{-0.0026SI}.$$

Assuming I is not zero, this equation simplifies to approximately

$$\frac{dI}{dS} = -1 + \frac{192}{S}.$$

---

[7] The values of a and b are close to those obtained by J. D. Murray in *Mathematical Biology,* Vol. I (New York: Springer Verlag, 3rd ed., 2002).

The slope field of this differential equation is shown in Figure 9.45. The trajectory with initial condition $S_0 = 762$, $I_0 = 1$ is shown in Figure 9.46. Time is represented by the arrow showing the direction that a point moves on the trajectory. The disease starts at the point $S_0 = 762$, $I_0 = 1$. At first, more people become infected and fewer are susceptible. In other words, S decreases and I increases. Later, I decreases as S continues to decrease.

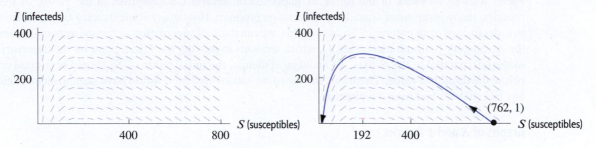

Figure 9.45: Slope field for $dI/dS = -1 + 192/S$ Figure 9.46: Trajectory for $S_0 = 762$, $I_0 = 1$

What does the SI-Phase Plane Tell Us?

To learn how the disease progresses, look at the shape of the curve in Figure 9.46. The value of I first increases, then decreases to zero. This peak value of I occurs when $S \approx 200$. We can determine exactly when the peak value occurs by solving

$$\frac{dI}{dS} = -1 + \frac{192}{S} = 0,$$

which gives

$$S = 192.$$

Notice that the peak value for I always occurs at the same value of S, namely $S = 192$. The graph shows that if a trajectory starts with $S_0 > 192$, then I first increases and then decreases to zero. On the other hand, if $S_0 < 192$, there is no peak, since I decreases right away.

For this example, the value $S = 192$ is called a *threshold population*. If S_0 is around or below 192, there is no epidemic. If S_0 is significantly greater than 192, an epidemic occurs.[8]

The phase diagram makes clear that the maximum value of I is about 300, which is the maximum number infected at any one time. In addition, the point at which the trajectory crosses the S-axis represents the time when the epidemic has passed (since $I = 0$). Thus, the S-intercept shows how many boys never get the flu and, hence, how many do get sick.

Example 1   In the boarding school model, if $S = 400$ and $I = 250$, use Figure 9.46 to decide the signs of dI/dt and dS/dt. Confirm your answers using the differential equations.

Solution   Since $S = 400$ is greater than the peak value of $S = 192$, we expect that the infected population is still increasing, so $dI/dt > 0$. The susceptible population is always decreasing (until the disease dies out), so $dS/dt < 0$. Checking the differential equations when $S = 400$ and $I = 250$, we see that $dS/dt = -0.0026SI = -260$ and $dI/dt = 0.0026SI - 0.5I = 135$.

Threshold Value

For the general SIR model, we have the following result:

$$\text{Threshold population } = \frac{b}{a}.$$

[8]Here we are using J. D. Murray's definition of an epidemic as an outbreak in which the number of infecteds increases from the initial value, I_0. See *Mathematical Biology*, Vol. I (New York: Springer Verlag, 3rd., 2002).

If S_0, the initial number of susceptibles, is above b/a, there is an epidemic; if S_0 is below b/a, there is no epidemic. See Problem 11.

How Many People Should Be Vaccinated?

Faced with an outbreak of the flu or, as happened on several US campuses in the 1980s, of the measles, many institutions consider a vaccination program. How many students must be vaccinated in order to control an outbreak? To answer this, we can think of vaccination as removing people from the S category (without increasing I), which amounts to moving the initial point on the trajectory to the left, parallel to the S-axis. To avoid an epidemic, the initial value of S_0 should be around or below the threshold value. Therefore, the boarding-school epidemic would have been avoided if all but 192 students had been vaccinated.

Graphs of S and I Against t

On the trajectory in Figure 9.46, the number of susceptible people decreases throughout the epidemic. This makes sense since people are getting sick and then well again, and thus are no longer susceptible to infection. The trajectory also shows that the number of infected people increases and then decreases. Graphs of S and I against time, t, are shown in Figure 9.47.

To get the scale on the time axis, we would need to use numerical methods. It turns out that the number of infecteds peaked after about 6 days and then dropped. The epidemic ran its course in about 20 days.

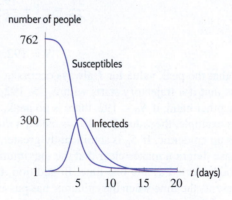

Figure 9.47: Progress of the flu over time

Problems for Section 9.7

1. Let I be the number of infected people and S be the number of susceptible people in an outbreak of a disease. Explain why it is reasonable to model the interaction between these two groups by the differential equations

$$\frac{dS}{dt} = -aSI$$

$$\frac{dI}{dt} = aSI - bI \quad \text{where } a, b \text{ are positive constants.}$$

Why have the signs been chosen this way? Why is the constant a the same in both equations?

2. Show that if S, I, and R satisfy the differential equations in Problem 1, the total population, $S + I + R$, is constant.

3. Explain how you can tell from the graph of the trajectory shown in Figure 9.46 that most people at the British boarding school eventually got sick.

4. (a) In a school of 150 students, one of the students has the flu initially. What is I_0? What is S_0?
   (b) Use these values of I_0 and S_0 and the equation

   $$\frac{dI}{dt} = 0.0026SI - 0.5I$$

   to determine whether the number of infected people initially increases or decreases. What does this tell you about the spread of the disease?

5. Repeat Problem 4 for a school with 350 students.

6. (a) On the slope field for dI/dS in Figure 9.45, draw the trajectory through the point where $I = 1$ and $S = 400$.

   (b) How many susceptible people are there when the number of infected people is at its maximum?

7. Use Figure 9.47 to estimate the maximum number of infecteds. What does this represent? When does it occur?

8. Compare the diseases modeled by each of the following differential equations with the flu model in this section. Match each set of differential equations with one of the following statements. Write a system of differential equations corresponding to each of the unmatched statements.

   (I) $\dfrac{dS}{dt} = -0.04SI$ (II) $\dfrac{dS}{dt} = -0.002SI$

   $\dfrac{dI}{dt} = 0.04SI - 0.2I$ $\dfrac{dI}{dt} = 0.002SI - 0.3I$

   (III) $\dfrac{dS}{dt} = -0.03SI$

   $\dfrac{dI}{dt} = 0.03SI$

   (a) More infectious; infecteds removed more slowly.
   (b) More infectious; infecteds removed more quickly.
   (c) Less infectious; infecteds removed more slowly.
   (d) Less infectious; infecteds removed more quickly.
   (e) Infecteds never removed.

9. For the equations (I) in Problem 8, what is the threshold value of S?

10. For the equations (II) in Problem 8, suppose $S_0 = 100$. Does the disease spread initially? What if $S_0 = 200$?

11. Let S and I satisfy the differential equations in Problem 1. Assume $I \neq 0$.

    (a) If $dI/dt = 0$, find S.
    (b) Show that I increases if S is greater than the value you found in part (a). Show that I decreases if S is less than the value you found in part (a).
    (c) Explain how you know that your answer to part (a) is the threshold value.

12. During World War I, a particularly lethal form of flu killed about 40 million people around the world.[9] The epidemic started in an army camp of 45,000 soldiers outside of Boston, where the first soldier fell sick on September 7, 1918. With time, t, in days since September 7, values of the constants in the SIR model,

    $$\dfrac{dS}{dt} = -aSI$$

    $$\dfrac{dI}{dt} = aSI - bI,$$

    are estimated to be $a = 0.000267, b = 9.865$.

    (a) What are the initial values, S_0 and I_0?
    (b) Explain how you know that this model predicts an epidemic in this case.
    (c) Find the differential equation for dI/dS. Sketch its slope field and estimate the total number of soldiers infected over the course of the disease.
    (d) Solve the differential equation for dI/dS analytically. Use the solution to solve approximately for the number of soldiers affected over the course of the disease.

PROJECTS FOR CHAPTER NINE

1. **Harvesting and Logistic Growth** In this project, we look at the effects of *harvesting* a population which is growing logistically. Harvesting could be, for example, fishing or logging. An important question is what level of harvesting leads to a *sustainable yield*. In other words, how much can be harvested without having the population depleted in the long run?

   (a) When there is no fishing, a population of fish is governed by the differential equation

   $$\dfrac{dN}{dt} = 2N - 0.01N^2,$$

   where N is the number of fish at time t in years. Sketch a graph of dN/dt against N. Mark on your graph the equilibrium values of N.

   Notice on your graph that if N is between 0 and 200, then dN/dt is positive and N increases. If N is greater than 200, then dN/dt is negative and N decreases. Check this by sketching a slope field for this differential equation. Use the slope field to sketch solutions showing N against t for various initial values. Describe what you see.

   (b) Fish are now removed by fishermen at a continuous rate of 75 fish/year. Let P be the number

---

[9]J. Taukenberger, A. Reid, T. Fanning, "Capturing a Killer Flu Virus," *Scientific American,* Vol. 292, No. 1, January 2005.

of fish at time t with harvesting. Explain why P satisfies the differential equation

$$\frac{dP}{dt} = 2P - 0.01P^2 - 75.$$

(c) Sketch dP/dt against P. Find and label the intercepts.

(d) Sketch the slope field for the differential equation for P.

(e) Recall that if dP/dt is positive for some values of P, then P increases for these values, and if dP/dt is negative for some values of P, then P decreases for these values. The value of P, however, never goes past an equilibrium value. Use this information and the graph from part (c) to answer the following questions:

  (i) What are the equilibrium values of P?

  (ii) For what initial values of P does P increase? At what value does P level off?

  (iii) For what initial values of P does P decrease?

(f) Use the slope field in part (d) to sketch graphs of P against t, with the initial values:

(i) $P(0) = 40$	**(ii)** $P(0) = 50$	**(iii)** $P(0) = 60$
(iv) $P(0) = 150$	**(v)** $P(0) = 170$	

(g) Using the graphs you drew, decide what the equilibrium values of the populations are and whether or not they are stable.

(h) We now look at the effect of different levels of fishing on a fish population. If fishing takes place at a continuous rate of H fish/year, the fish population P satisfies the differential equation

$$\frac{dP}{dt} = 2P - 0.01P^2 - H.$$

  (i) For each of the values $H = 75, 100, 200$, plot dP/dt against P.

  (ii) For which of the three values of H that you considered in part (i) is there an initial condition such that the fish population does not die out eventually?

  (iii) Looking at your answer to part (ii), decide for what values of H there is an initial value for P such that the population does not die out eventually.

  (iv) Recommend a policy to ensure long-term survival of the fish population.

2. **Population Genetics**

Population genetics is the study of hereditary traits in a population. A specific hereditary trait has two possibilities, one dominant, such as brown eyes, and one recessive, such as blue eyes.[10] Let b denote the gene responsible for the recessive trait and B denote the gene responsible for the dominant trait. Each member of the population has a pair of these genes—either BB (dominant individuals), bb (recessive individuals), or Bb (hybrid individuals). The *gene frequency* of the b gene is the total number of b genes in the population divided by the total number of all genes (b and B) controlling this trait. The gene frequency is essentially constant when there are no mutations or outside influences on the population. In this project, we consider the effect of mutations on the gene frequency.

Let q denote the gene frequency of the b gene. Then q is between 0 and 1 (since it is a fraction of a whole) and, since b and B are the only genes influencing this trait, the gene frequency of the B gene is $1 - q$. Let time t be measured in generations. Every generation, a fraction k_1 of the b genes mutate to become B genes, and a fraction k_2 of the B genes mutate to become b genes.

(a) Explain why the gene frequency q satisfies the differential equation:

$$\frac{dq}{dt} = -k_1 q + k_2(1 - q).$$

(b) If $k_1 = 0.0001$ and $k_2 = 0.0004$, simplify the differential equation for q and solve it. The initial value is q_0. Sketch the solutions with $q_0 = 0.1$ and $q_0 = 0.9$. What is the equilibrium

[10] Adapted from C. C. Li, *Population Genetics* (Chicago: University of Chicago Press, 1995).

value of q? Explain how you can tell that the gene frequency gets closer to the equilibrium value as generations pass. Explain how you can tell that the equilibrium value is completely determined by the relative mutation rates.

(c) Repeat part (b) if $k_1 = 0.00003$ and $k_2 = 0.00001$.

3. **The Spread of SARS**

In the spring of 2003, SARS (Severe Acute Respiratory Syndrome) spread rapidly in several Asian countries and Canada. Predicting the course of the disease—how many people would be infected, how long it would last—was important to officials trying to minimize the impact of the disease. This project analyzes the spread of SARS through interaction between infected and susceptible people.

The variables are S, the number of susceptibles, I, the number of infecteds who can infect others, and R, the number removed (this group includes those in quarantine and those who die, as well as those who have recovered and acquired immunity). Time, t, is in days since March 17, 2003, the date the World Health Organization (WHO) started to publish daily SARS reports. On March 17, Hong Kong reported 95 cases. In this model

$$\frac{dS}{dt} = -aSI$$

$$\frac{dI}{dt} = aSI - bI,$$

and $S + I + R = 6.8$ million, the population of Hong Kong in 2003.[11] Estimates based on WHO data give $a = 1.25 \cdot 10^{-8}$.

(a) What are S_0 and I_0, the initial values of S and I?

(b) During March 2003, the value of b was about 0.06. Using a calculator or computer, sketch the slope field for this system of differential equations and the solution trajectory corresponding to the initial conditions. (Use $0 \le S \le 7 \cdot 10^6, 0 \le I \le 0.4 \cdot 10^6$.)

(c) What does your graph tell you about the total number of people infected over the course of the disease if $b = 0.06$? What is the threshold value? What does this value tell you?

(d) During April, as public health officials worked to get the disease under control, people who had been in contact with the disease were quarantined. Explain why quarantining has the effect of raising the value of b.

(e) Using the April value, $b = 0.24$, sketch the slope field. (Use the same value of a and the same window.)

(f) What is the threshold value for $b = 0.24$? What does this tell you? Comment on the quarantine policy.

(g) Comment on the effectiveness of each of the following policies intended to prevent an epidemic and protect a city from an outbreak of SARS in a nearby region.

    I Close off the city from contact with the infected region. Shut down roads, airports, trains, and other forms of direct contact.

    II Institute a quarantine policy. Isolate anyone who has been in contact with a SARS patient or anyone who shows symptoms of SARS.

---

[11] www.census.gov, International Data Base (IDB), accessed June 8, 2004.

FOCUS ON THEORY

SEPARATION OF VARIABLES

We have seen how to sketch solution curves of a differential equation using a slope field. Now we see how to solve certain differential equations analytically, finding an equation for the solution curve.

First, we look at a familiar example, the differential equation

$$\frac{dy}{dx} = -\frac{x}{y},$$

whose solution curves are the circles

$$x^2 + y^2 = C.$$

We can check that these circles are solutions by differentiation; the question now is how they were obtained. The method of *separation of variables* works by putting all the xs on one side of the equation and all the ys on the other, giving

$$y\,dy = -x\,dx.$$

We then integrate each side separately:

$$\int y\,dy = -\int x\,dx,$$

$$\frac{y^2}{2} = -\frac{x^2}{2} + k.$$

This gives the circles we were expecting:

$$x^2 + y^2 = C \qquad \text{where } C = 2k.$$

You might worry about whether it is legitimate to separate the dx and the dy. The reason it can be done is explained at the end of this section.

The Exponential Growth and Decay Equations

We use separation of variables to derive the general solution of the equation

$$\frac{dy}{dt} = ky.$$

Separating variables, we have

$$\frac{1}{y}\,dy = k\,dt,$$

and integrating,

$$\int \frac{1}{y}\,dy = \int k\,dt,$$

gives

$$\ln|y| = kt + C \quad \text{for some constant } C.$$

Solving for $|y|$ leads to

$$|y| = e^{kt+C} = e^{kt}e^{C} = Ae^{kt}$$

where $A = e^{C}$, so A is positive. Thus,

$$y = (\pm A)e^{kt} = Be^{kt}$$

where $B = \pm A$, so B is any nonzero constant. Even though there's no C leading to $B = 0$, we can have $B = 0$ because $y = 0$ is a solution to the differential equation. We lost this solution when we divided through by y at the first step. Thus we have derived the solution used earlier in the chapter:

$$y = Be^{kt} \quad \text{for any constant } B.$$

Example 1   Find all solutions of

$$\frac{dy}{dt} = k(y - A).$$

Solution   We separate variables and integrate:

$$\int \frac{1}{y - A} \, dy = \int k \, dt.$$

This gives

$$\ln|y - A| = kt + D,$$

where D is a constant of integration. Solving for y leads to

$$|y - A| = e^{kt+D} = e^{kt}e^{D} = Be^{kt}$$

or

$$y - A = (\pm B)e^{kt} = Ce^{kt}$$
$$y = A + Ce^{kt}.$$

Also, $C = 0$ gives a solution. This is the same result we used earlier.

Example 2   Find and sketch the solution to

$$\frac{dP}{dt} = 2P - 2Pt \qquad \text{satisfying } P = 5 \text{ when } t = 0.$$

Solution   Factoring the right-hand side gives

$$\frac{dP}{dt} = P(2 - 2t).$$

Separating variables, we get

$$\int \frac{dP}{P} = \int (2 - 2t) \, dt,$$

so

$$\ln|P| = 2t - t^2 + C.$$

Solving for P leads to

$$|P| = e^{2t-t^2+C} = e^{C}e^{2t-t^2} = Ae^{2t-t^2}$$

with $A = e^{C}$, so $A > 0$. In addition, $A = 0$ gives a solution. Thus the general solution to the differential equation is

$$P = Be^{2t-t^2} \qquad \text{for any } B.$$

To find the value of B, substitute $P = 5$ and $t = 0$ into the general solution, giving

$$5 = Be^{2\cdot0-0^2} = B$$

so

$$P = 5e^{2t-t^2}.$$

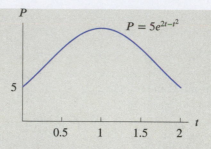

Figure 9.48: Bell-shaped solution curve

The graph of this function is in Figure 9.48. Since the solution can be rewritten as

$$P = 5e^{1-1+2t-t^2} = 5e^1 e^{-1+2t-t^2} = (5e)e^{-(t-1)^2},$$

the graph has the same shape as the graph of $y = e^{-t^2}$, the bell-shaped curve of statistics. Here the maximum, normally at $t = 0$, is shifted one unit to the right to $t = 1$.

Justification for Separation of Variables

Suppose a differential equation can be written in the form

$$\frac{dy}{dx} = g(x)f(y).$$

Provided $f(y) \neq 0$, we write $f(y) = 1/h(y)$ so the right-hand side can be thought of as a fraction,

$$\frac{dy}{dx} = \frac{g(x)}{h(y)}.$$

If we multiply through by $h(y)$ we get

$$h(y)\frac{dy}{dx} = g(x).$$

Thinking of y as a function of x, so $y = y(x)$, and $dy/dx = y'(x)$, we can rewrite the equation as

$$h(y(x)) \cdot y'(x) = g(x).$$

Now integrate both sides with respect to x:

$$\int h(y(x)) \cdot y'(x)\,dx = \int g(x)\,dx.$$

The form of the integral on the left suggests that we use the substitution $y = y(x)$. Since $dy = y'(x)\,dx$, we get

$$\int h(y)\,dy = \int g(x)\,dx.$$

If we can find antiderivatives of h and g, then this gives the equation of the solution curve.

Note that transforming the original differential equation,

$$\frac{dy}{dx} = \frac{g(x)}{h(y)},$$

into

$$\int h(y)\,dy = \int g(x)\,dx$$

looks as though we have treated dy/dx as a fraction, cross-multiplied and then integrated. Although that's not exactly what we have done, you may find this a helpful way of remembering the method. In fact, the dy/dx notation was introduced by Leibniz to allow shortcuts like this (more specifically, to make the chain rule look like cancellation).

Problems on Separation of Variables

■ In Problems 1–12, use separation of variables to find the solution to the differential equation subject to the initial condition.

1. $\dfrac{dP}{dt} = -2P, \quad P(0) = 1$

2. $\dfrac{dL}{dp} = \dfrac{L}{2}, \quad L(0) = 100$

3. $P\dfrac{dP}{dt} = 1, \quad P(0) = 1$

4. $\dfrac{dm}{ds} = m, \quad m(1) = 2$

5. $2\dfrac{du}{dt} = u^2, \quad u(0) = 1$

6. $\dfrac{dz}{dy} = zy, \quad z = 1$ when $y = 0$

7. $\dfrac{dR}{dy} + R = 1, \quad R(1) = 0.1$

8. $\dfrac{dy}{dt} = \dfrac{y}{3+t}, \quad y(0) = 1$

9. $\dfrac{dz}{dt} = te^z, \quad$ through the origin

10. $\dfrac{dy}{dx} = \dfrac{5y}{x}, \quad y = 3$ where $x = 1$

11. $\dfrac{dy}{dt} = y^2(1+t), \quad y = 2$ when $t = 1$

12. $\dfrac{dz}{dt} = z + zt^2, \quad z = 5$ when $t = 0$

13. Determine which of the following differential equations are separable. Do not solve the equations.

(a) $y' = y$ (b) $y' = x + y$

(c) $y' = xy$ (d) $y' = \sin(x + y)$

(e) $y' - xy = 0$ (f) $y' = y/x$

(g) $y' = \ln(xy)$ (h) $y' = (\sin x)(\cos y)$

(i) $y' = (\sin x)(\cos xy)$ (j) $y' = x/y$

(k) $y' = 2x$ (l) $y' = (x+y)/(x+2y)$

■ In Problems 14–19, use separation of variables to solve the differential equation. Assume a, b, and k are nonzero constants.

14. $\dfrac{dP}{dt} = P - a$

15. $\dfrac{dQ}{dt} = b - Q$

16. $\dfrac{dP}{dt} = k(P - a)$

17. $\dfrac{dR}{dt} = aR + b$

18. $\dfrac{dP}{dt} - aP = b$

19. $\dfrac{dy}{dt} = ky^2(1 + t^2)$

20. (a) Find the general solution to the differential equation modeling how a person learns:

$$\dfrac{dy}{dt} = 100 - y.$$

(b) Plot the slope field of this differential equation and sketch solutions with $y(0) = 25$ and $y(0) = 110$.

(c) For each of the initial conditions in part (b), find the particular solution and add it to your sketch.

(d) Which of these two particular solutions could represent how a person learns?

21. (a) Sketch the slope field for the differential equation $dy/dx = xy$.

(b) Sketch several solution curves.

(c) Solve the differential equation analytically.

ANSWERS TO ODD NUMBERED PROBLEMS

Section 1.1

1 (a) (IV)
  (b) (II)
  (c) (III)

3 Increasing

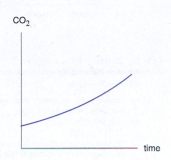

5 Decreasing

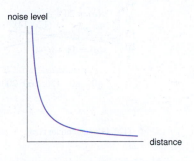

7 Argentina produced 49.2 million metric tons of wheat in 2015

9 (a) 400 ppm CO_2 in 2015
  (b) Concentration of CO_2 in 2020

11 $f(5) = 13$

13 $f(5) = 3$

15 $f(5) = 4.1$

17 $f(12) = 7.049$

19 $f(63) = f(12)$

21 (a) (III)
  (b) Potato's temperature before put in oven

23 (a) 100 species at 500 ft
  (b) k: value of N at sea level
     c: lowest elevation with no bats

25 (a) About 0.14 mg nicotine
  (b) About 4 hours
  (c) 0.4
  (d) Time nicotine level zero

27 (b) CFC consumption in 1987
  (c) Year CFC consumption is zero

29

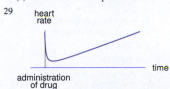

31   concentration

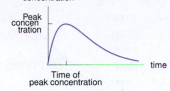

37   driving speed

39   distance from exit

Section 1.2

1 $y = (1/2)x + 2$

3 $y = (1/2)x + 2$

5 Slope: $-12/7$
  Vertical intercept: $2/7$

7 Slope: 2
  Vertical intercept: $-2/3$

9 (a) l_1
  (b) l_3
  (c) l_2
  (d) l_4

11 (a) $P = 28{,}600 + 750t$
  (b) 32,350 people
  (c) In 2023

13 (a) $C_1 = 40 + 0.15m$
     $C_2 = 50 + 0.10m$

  (b) C (cost in dollars)

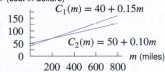

  (c) For distances less than 200 miles, C_1 is cheaper.
     For distances more than 200 miles, C_2 is cheaper.

15 (a) 300 miles
  (b) 50 mph
  (c) $D = 300 + 50t$

17 (a) $y = -2x + 27$
  (b) $s = 2t + 32$

19 (a) About 10.4 kg/hectare per mm
  (b) Additional 1 mm annual rainfall corresponds to additional 10.4 kg grass per hectare
  (c) $Q = -440 + 10.4r$

21 1939: more addl grass/addl rainfall

23 (a) $q = -(1/3)p + 8$

  (b) $p = -3q + 24$

25 (a) -2.8 days/ °C
  (b) Come out of hibernation earlier in warmer months
  (c) 17 days
  (d) $y = 171 - 2.8x$

27 (a) $f(t) = 100 + 14.222t$
  (b) D: $0 \le t \le 18$
     R: $100 \le f(t) \le 356$

29 (a) $P = 1241 + 26.9t$
  (b) 26.9 million tons per year
  (c) 1241 million tons in 1975
  (d) 2317 million tons
  (e) In the year 2021

31 (a) $V(t) = 25000 - 2000t; C(t) = 1500t$
  (b) 7.143 years
  (c) 11.750 years

33 (a) $C = 4.16 + 0.12w$
  (b) 0.12 $/gal
  (c) $4.16

35 (a) $\Delta w / \Delta h$ constant
  (b) $w = 0.68h - 34.54$; 0.68 kg/cm
  (c) $h = 1.47w + 50.79$; 1.47 cm/kg

37 (c)

39 (a) 60, 40 years
  (b) (ii)
  (c) 6.375 beats/minute more under new formula

41 No

43 (a) $13.70 + 0.0882x$ Australian dollars
  (b) $16.65 + 0.0496y$ Australian dollars
  (c) 3.92 Australian dollars
  (d) Yes; 76.425 cm (2ft 6 in)

Section 1.3

1 Concave down

3 Concave up

5 Decreasing
  Concave up

7 (a) D to E, H to I
  (b) A to B, E to F
  (c) C to D, G to H
  (d) B to C, F to G

9 -3

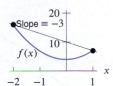

11 15 million km^2 covered by ice in Feb 2008

13 Sea ice shrunk at avg rate of 40,000 km^2/yr from 1983 to 2008

15 Sea ice grew at avg rate of 15,900 km^2/day from Feb 1 to Feb 28, 2015

17 (a) 84 million bicycles
  (b) 1.68 million bicycles per year

19 (a) 180 feet per second
  (b) 100 feet per second

21 1 meter/sec

23 $72/7 = 10.286$ cm/sec

25 (a) $36.09 billion
  (b) $3.0075 billion per year

27 (a) Approximately -7.667 mn lbs/yr

(b) Yes, between 2003 and 2004, between 2005 and 2009, between 2011 and 2012, and between 2013 and 2014

29 1490 thousand people/year
912.9 thousand people/year
1879 thousand people/year

31 (a) −$31.8 billion dollars
(b) −$2.65 billion dollars per year
(c) Yes; 2006–2010, 2014–2015

33 (a) Negative
(b) −0.087 mg/hour

35 15.468, 57.654, 135.899, 146.353, 158.549 people/min

37 (a) About −11 cm/sec
(b) About −5.5 (cm/sec)/kg

39 (a) (i) $f(2001) = 272$
(ii) $f(2014) = 525$
(b) $(f(2014) − f(2001))/(2014 − 2001) =$ 19.46 billionaires/yr
(c) $f(t) = 19.46t − 38{,}667.5$

41 (a) Concave up; no
(b) About 2.6 m/sec

43 Decreasing, concave down

45 Increases by 25%

47 Decreases by 83.3%

49 Change in 1931

51 The small class

53 (a) 100%
(b) −50%
(c) 0.1%

55 Increase by 4.3%

57 (a)

t	0	0.25	0.5	0.75
(% growth/yr)	3.8	0.8	−7.8	−4.7

(b) Yes, last two quarters

59 (a) 2007–2009
(b) 2012–2014

61 (a) 25%
(b) 12%
(c) 0.48

Section 1.4

1
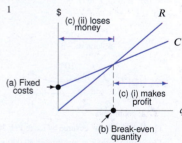
(a) Fixed costs
(b) Break-even quantity
(c) (i) makes profit
(c) (ii) loses money

3 (a) About $75; $7.50 per unit
(b) About $150

5 Fixed cost: $5.7 million
Variable cost: $2000 per unit

7 (a) Price $12, sell 60
(b) Decreasing

9 5500: Quantity demanded at price 0
100: Drop in quantity demanded if price increases $1

11 $C(q) = 500 + 6q$, $R(q) = 12q$, $\pi(q) = 6q − 500$

13 $C(q) = 5000 + 15q$, $R(q) = 60q$, $\pi(q) = 45q − 5000$

15 (a) $C(q) = 5000 + 30q$
$R(q) = 50q$
(b) $30/unit, $50/unit
(c)
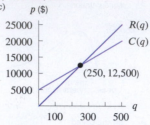
$R(q)$
$C(q)$
(250, 12,500)
(d) 250 chairs and $12,500

17 (a) First price list:
$C_1(q) = 100 + 0.03q$ dollars
Second price list:
$C_2(q) = 200 + 0.02q$ dollars
(b) First price list
(c) 10,000

19 (a) When there are more than 1000 customers
(b)
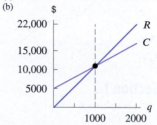
R
C

21 (a) $C(q) = 650{,}000 + 20q$
$R(q) = 70q$
$\pi(q) = 50q − 650{,}000$
(b) $20/pair, $70/pair, $50/pair
(c) More than 13,000 pairs

23 (a) $V(t) = −2000t + 50{,}000$
(b)
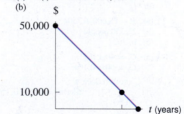
t (years)
(c) (0 years, $50,000) and (25 years, $0)

25 (a) Roughly 360 scoops
(b) Roughly 120 scoops

27 (a) First: demand curve;
Second: supply curve
(b) Roughly 14
(c) Roughly 24
(d) Lower
(e) Any price less than or equal to $143
(f) Any price greater than or equal to $110

29 (a) $C = 5q + 7000$
$R = 12q$
(b) $q = 1520$, $\pi(12) = $3640
(c) $C = 17{,}000 − 200p$
$R = 2000p − 40p^2$
$\pi(p) = −40p^2 + 2200p − 17{,}000$
(d) At $27.50 per shirt the profit is $13,250

31 (a) $q = 820 − 20p$
(b) $p = 41 − 0.05q$

33
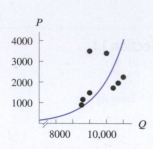
P

35 (a) $25{,}000r + 100m = 500{,}000$
(b) $m = 5000 − 250r$
(c) $r = 20 − \frac{1}{250}m$

37 (a) Higher price: customers want less but owner wants to sell more
(b) (60, 18): No
(120, 12): Yes
(c) Shaded region is all possible quantities of cakes owner is willing to sell and customers willing to buy

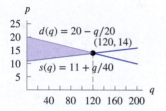
p
$d(q) = 20 − q/20$
(120, 14)
$s(q) = 11 + q/40$

(d) (120, 14): equilibrium price.

39 1.25

41 $q = 4p − 28$

43 (a) $p = 100$, $q = 500$
(b) $p = 102$, $q = 460$
(c) Consumer pays $2
Producer pays $4
(d) $2760

45 (a) Demand: $q = 100 − 2p$
Supply: $q = 2.85p − 50$
(b) New equilibrium price $p \approx $30.93
New equilibrium quantity $q \approx 38.14$ units
(c) Consumer pays $0.93
Producer pays $0.62
Total $1.55
(d) $59.12

Section 1.5

1 (a) (i), 12%
(b) (ii), 1000
(c) Yes, (iv)

3 (I) $Q = 50(1.4)^t$; (II) $Q = 50(1.2)^t$; (III) $Q = 50(0.8)^t$; (IV) $Q = 50(0.6)^t$

5 (a) a, c, p
(b) a, d, q
(c) $c = p$
(d) a and b are reciprocals
p and q are reciprocals

7 $y = 30(0.94)^t$

9 $y = 18(2/3)^x$

11 (a) $P = 1000 + 50t$
(b) $P = 1000(1.05)^t$

13 (a) $Q = 30 - 2t$

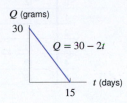

Q (grams)
30
$Q = 30 - 2t$
15
t (days)

(b) $Q = 30(0.88)^t$

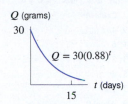

Q (grams)
30
$Q = 30(0.88)^t$
15
t (days)

15 (a) $A = 50(0.94)^t$
(b) 11.33 mg
(c) A (mg)

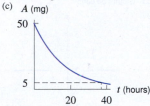

50
5
20   40   t (hours)

(d) About 37 hours

17 82.493%

19 (a)

x	0	1	2	3
e^x	1	2.72	7.39	20.09

(b)

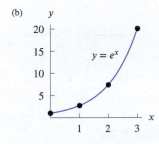

y
20
15    $y = e^x$
10
5
1   2   3   x

(c)

x	0	1	2	3
e^{-x}	1	0.37	0.14	0.05

(d)

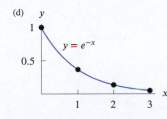

y
1
$y = e^{-x}$
0.5
1   2   3   x

21 $g(t) = 5.50(0.8)^t$

23 Table D

25 (a) $a = 1.5$ and $P_0 = 22.222$
(b) Initial quantity 22.222, growing 50% per unit time

27 (a) $a = 0.8$ and $P_0 = 976.563$
(b) Initial quantity 976.563, decaying 20% per unit time

29 About 7.33 billion; close

31 22.6% per year

33 (a) $236{,}733 + 44{,}880.3t$
(b) $W = 236{,}733(1.16194)^t = 236{,}733e^{0.1501t}$
(c)

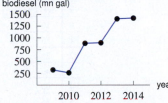

W (MW)
400,000
(3, 371,373)
300,000
236,733
1   2   3   4   t (years since 2011)

(d) Linear: 416,254 MW
Exponential: $\approx 431{,}500$ MW
Exponential model better

35 (a) $h(x) = 31 - 3x$
(b) $g(x) = 36(1.5)^x$

37 Min. wage grew 4.69% per year

39 (a) $w = 1093.965(1.1408)^t$
(b) 14.08% per year
(c) No; recent growth rate higher

41 (a) 260 million gallons, 886 million gallons
(b)

consumption of biodiesel (mn gal)
1500
1250
1000
750
500
250
2010   2012   2014   year

43 (a) 2.520 quadrillion BTUs, 3.084 quadrillion BTUs
(b)

consumption of hydro. power (quadillion BTU)
3.5
3
2.5
2010   2012   2014   year

(c) 2011, 478 trillion BTUs

45 (a) Increased by 25%: 2009, 2010, 2011; decreased: none
(b) True

Section 1.6

1 $t = (\ln 10)/(\ln 2) \approx 3.3219$

3 $t = (\ln 2)/(\ln 1.02) \approx 35.003$

5 $t = \ln 10 \approx 2.3026$

7 $t = (\ln 5)/(\ln 3) \approx 1.465$

9 $t = (\ln 100)/3 \approx 1.535$

11 $t = 30.54$

13 $t = (\ln B - \ln P)/r$

15 $t = (\ln 7 - \ln 5)/(\ln 2 - \ln 3) \approx -0.8298$

17 $t = \ln B - \ln A$

19 $t = \ln 7 - \ln 3 \approx 0.84723$

21 5; 7%

23 15; −6% (continuous)

25 $P = 15(1.2840)^t$; growth

27 $P = P_0(1.2214)^t$; growth

29 $P = 15e^{0.4055t}$

31 $P = 174e^{-0.1054t}$

33 (a) $k = 0.168$ and $P_0 = 84.575$
(b) Initial quantity 84.575, growing 16.8% per unit time

35 (a) 6%
(b) $P = 100(1.0618)^t$, 6.18%

37 8.33%

39 (a) D
(b) C
(c) B

41 (a) (i) $P = 1000(1.05)^t$
(ii) $P = 1000e^{0.05t}$
(b) (i) 1629 (ii) 1649

43 $W = 74.31e^{0.0286t}$

45 (a) $P(0.5) \approx 779$; $P(1) \approx 607$
(b) 223
(c) Approximately 4.6
(d)

Trout
1000
500    $P(t) = 1000e^{-0.5t}$
2   4   6   8   10   t

47 (a) $5 million; $3.704 million dollars
(b) 4.108 years

49 (a) $Q_0(1.0033)^x$
(b) 210.391 microgm/cubic m

51 (a) $B(t) = B_0 e^{0.067t}$
(b) $P(t) = P_0 e^{0.033t}$
(c) $t = 20.387$; in 2000

53 2054

Section 1.7

1 0.0693

3 A: continuous
B: annual
$20

5 (a) $1126.49
(b) $1127.50

7 $10,976.23

9 About 55.5 years

11 (a) 5 years
(b) 5 years
(c) Decays by 1/2 at different starting times
(d) $C = 100e^{-0.139t}$

13 (a) $W = 371e^{0.168t}$
(b) During 2016

15 347 days

17 (a) $S(t) = 219(1.05946)^t$
(b) 5.946%

19 (a) $A = 100e^{-0.17t}$
(b) $t \approx 4$ hours

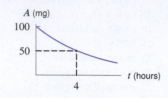

A (mg)

100

50

4   t (hours)

(c) $t = 4.077$ hours

21 (a) 47.6%
(b) 23.7%

23 8.45%

25 (a) $P(t) = (0.976)^t$
(b)

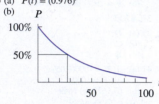

P

100%

50%

50   100 t

(c) About 29 years
(d) About 9%

27 (a) About 4 years
(b) About 4 years

29 About 173 hours

31 96.336 years

33 (a) 2024
(b) 336.49 million people

35 It is a fake

37 (a) $100{,}000e^{-0.0313t}$
(b) Larger

39 (a) 0.00664
(b) $t = 22.277$; April 11, 2032

41 $12,712.49

43 $6549.85

45 (a) Choice 1
(b) Yes. Above 25%

47 (a) 8.75 years
(b) About 9.01 years

49 (a) Option 1
(b) $2102.54, $2051.27, $2000
(c) $2000, $1951.23, $1902.46

51 (a) Option 1
(b) Option 1: $10.929 million;
Option 2: $10.530 million

53 Buy

Section 1.8

1 (a) $15x + 9$
(b) $15x - 1$
(c) $25x - 6$

3 (a) $3e^{2x}$
(b) e^{6x}
(c) $9x$

5 (a) $h^2 + 6h + 11$
(b) 11
(c) $h^2 + 6h$

7 (a) 4
(b) 2
(c) $(x+1)^2$
(d) $x^2 + 1$

(e) $t^2(t+1)$

9 (a) e
(b) e^2
(c) e^{x^2}
(d) e^{2x}
(e) $e^t t^2$

11 (a) 5
(b) 2
(c) 3
(d) 4
(e) 5
(f) 2

13 (a)

x	$f(x) + 3$
0	13
1	9
2	6
3	7
4	10
5	14

(b)

x	$f(x-2)$
2	10
3	6
4	3
5	4
6	7
7	11

(c)

x	$5g(x)$
0	10
1	15
2	25
3	40
4	60
5	75

(d)

x	$-f(x) + 2$
0	-8
1	-4
2	-1
3	-2
4	-5
5	-9

(e)

x	$g(x-3)$
3	2
4	3
5	5
6	8
7	12
8	15

(f)

x	$f(x) + g(x)$
0	12
1	9
2	8
3	12
4	19
5	26

15 (a) $y = 2^u$, $u = 3x - 1$
(b) $P = \sqrt{u}$, $u = 5t^2 + 10$
(c) $w = 2\ln u$, $u = 3r + 4$

17 $2zh + h^2$

19 $4hz$

21 6

23 3

25 4

27 1.1

29 About 0

31 $2(y-1)^3 - (y-1)^2$

33 About 18

35 Cannot be done

37

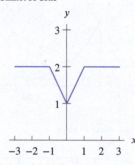

39

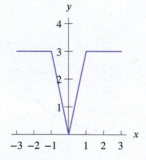

41

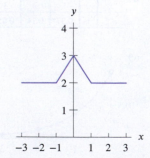

43 (a)

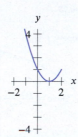

(b)

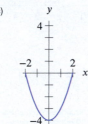

(c)

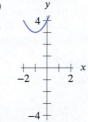

(d)

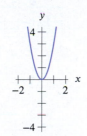

(e)

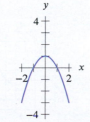

(f)

45 (a)

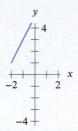

(b)

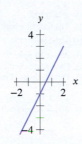

(c)

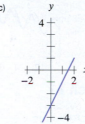

(d)

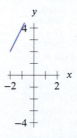

(e)

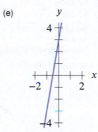

(f)

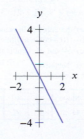

47

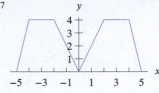

49

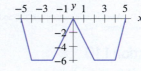

51

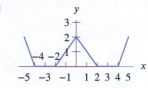

53 (a)
(b) Horizontal shift
(c) $g(t) = f(t - 4)$ for $t \geq 4$

55 $g(f(t))$ ft^3

57 $f^{-1}(30)$ min

59 (a) $c(t) = 25(0.872)^t$ or $c(t) = 25e^{-0.136966t}$
(b) 6.690 years
(c) $D(t) = 25(0.7604)^t$ or $D(t) = 25e^{-0.273932t}$
(d) $E(t) = 16.576(0.872)^t$ or $E(t) = 16.576e^{-0.136966t}$

Section 1.9

1 $y = (1/5)x$

3 $y = 8x^{-1}$

5 Not a power function.

7 $y = 9x^{10}$

9 Not a power function

11 $y = 125x^3$

13 $S = kh^2$

15 $v = d/t$

17 (a) $y = (x - 2)^3 + 1$
(b) $y = -(x + 3)^2 - 2$

19 $N = kA^{1/4}$, with $k > 0$,
Increasing, concave down

species of lizard

21 (a) $C = kW^{0.75}$
(b)
(c) Horse: 5,716 calories
Rabbit: 218 calories
(d) Mouse

23 $N = k/L^2$; small

25 (a) $T = kB^{1/4}$
(b) $k = 17.4$
(c) 50.3 seconds

27 (a) $N = kP^{0.77}$
(b) A has 5.888 times more than B
(c) Town

29 (a) $C = 115,000 - 700p$
$R = 3000p - 20p^2$

(b)

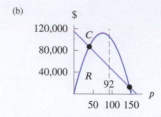

(d) When it charges between \$40 and \$145
(e) About \$92

Section 1.10

1 Amplitude = 3; Period = 2π

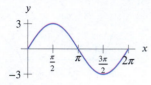

3 Amplitude = 3; Period = π

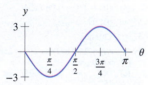

5 Amplitude = 1; Period = π

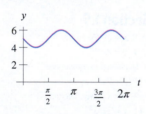

7 (b) Max: 2nd quarter; Min: 4th quarter
  (c) Period = 4 quarters or 1 year; amplitude ≈ 5 million barrels

9 (a) 5
  (b) 8
  (c) $f(x) = 5\cos((\pi/4)x)$

11 (b)

13 Estimating coefficients $H = -16\cos((\pi/6)t) + 10$

15 $0.35\cos(2\pi t/5.4) + 4$

17 $y = 7\sin(\pi t/5)$

19 $f(x) = 5\cos(x/3)$

21 $f(x) = -4\sin(2x)$

23 $f(x) = -8\cos(x/10)$

25 $f(x) = 5\sin((\pi/3)x)$

27 $f(x) = 3 + 3\sin((\pi/4)x)$

29

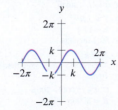

31

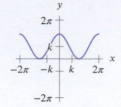

33 (a) Average depth of water
  (b) $A = 7.5$
  (c) $B = 0.507$
  (d) The time of a high tide

35 $60 - 20\cos(\pi t/12)$

37 (a) About 10 ppm
  (b) Estimating coefficients $y = (1/6) + 381$
  (c) Period: about 12 months; amplitude: about 3.5 ppm
    $y = 3.5\sin(\pi t/6)$
  (d) $h(t) = 3.5\sin(\pi t/6) + (t/6) + 381$

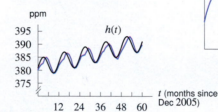

Section 2.1

1 (a) Negative
  (b) Positive
  (c) Negative
  (d) Zero
  (e) Positive
  (f) Negative
  (g) Zero
  (h) Positive

3 A and B

5 A and B

7 (a) $f(2) = 96$
  (b) $f'(2) = 16$

9 (a) Negative
  (b) $f'(1960)$

11 (a) 2.5 ft/sec
  (b) 6.5 ft/sec
  (c) 4.5 ft/sec

13 (a) Between $x = 0$ and $x = 3$
  (b) At $x = 1$
  (c) Thousands of dollars/kilogram

15 Other answers possible

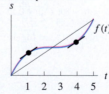

17 (a) 63 cubic millimeters
  (b) 10.5 cubic millimeters/month
  (c) About 44.4 cubic millimeters/month

19 (a) Negative
  (b) $g'(2) \approx -0.1$

21

Slope	-3	-1	0	1/2	1	2
Point	F	C	E	A	B	D

23 (a) Positive
  (b) $f'(2) \approx 0.111$, $f'(10) \approx -1.658$

25 (a) (iii)
  (b) (iii)
  (c) (i)
  (d) (ii)
  (e) (ii)

27 (a) The slopes of the two tangent lines at $x = a$ are equal for all a
  (b) A vertical shift does not change the slope

29 $(4, 25)$; $(4.2, 25.3)$; $(3.9, 24.85)$

31 0.4

33 -0.75

35

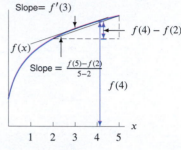

37 (a) $x = 2$ and $x = 4$
  (b) $g(4)$
  (c) $g'(4)$

39 8.34 million people/year
  8.39 million people/year

41 (a) Negative or zero;
    Positive; Positive
  (b) (i) −0.2 hr/yr, 0 hr/yr
     (ii) \$0.27/yr, \$0.47/yr
     (iii) \$8.90/yr, \$16.18/yr

Section 2.2

1 About 1.0, 0.3, −0.5, −1

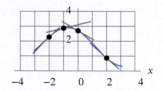

3

5

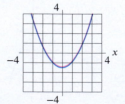

7

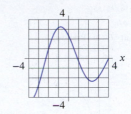

9 (a) x_3
(b) x_4
(c) x_5
(d) x_3

11 $R'(0) \approx 9.531$

13

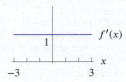

15

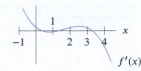

17

19 IV

21 VI

23 (V)

25 (II)

27 5.2

29

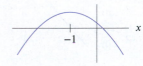

Other answers possible

31 (a) Graph II
(b) Graph I
(c) Graph III

33 $f'(2) = 4$
$f'(3) = 9$
$f'(4) = 16$
The pattern seems to be:
$f'(x) = x^2$.

35 (a) $t = 3$
(b) $t = 9$
(c) $t = 14$
(d)

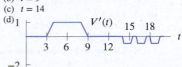

Section 2.3

1 dD/dt; feet per minute

3 dN/dD; gallons per mile

5 (a) ml; minutes
(b) ml; minutes/ml

7 (a) Mn km^2/day
Growing 73,000 km^2/day
(b) 0.365 mn km^2
Grew by 0.365 mn km^2

9 About 25 megawatts/year

11 (a) Liters per centimeter
(b) About 0.042 liters per centimeter
(c) Cannot expand much more

13 (a) meters
(b) mm (runoff depth) per meter (distance down the slope)
(c) 0.08 mm

15 f(points) = Revenue
$f'(4.3) \approx 55$ million dollars/point

17 (a) Negative
(b) °F/min

19 (a) °C/km
(b) Air temp decreases about 6.5° for a 1 km increase in altitude

21 (a) 0.0729 people/sec
(b) 14 seconds

23 (a) Consuming 1800 Calories per day results in a weight of 155 pounds; Consuming 2000 Calories per day causes neither weight gain nor loss
(b) Pounds/(Calories/day)

27 $f'(t) \approx 6$ where t is retirement age in years, $f(t)$ is age of onset in weeks

29 About 3.4, about 2.6

31 5.14

33 2.94

35 −3.98

37 (a) $\Delta V \approx 16\Delta r$
(b) 1.6
(c) 33.6

39 (a) kg/week
(b) At week 24 fetus grows about 0.096 kilograms in one more week

41 (a) Less
(b) Greater

43 (a) In 2015: Net sales 7.39 bn dollars; sales decrease by about 0.03 bn dollars in next year
(b) About 7.3 billion dollars

45 (a) 2015 Production: 23.7 bn lbs, decreasing by about 0.1 bn lbs/yr
(b) About 23.4 bn lbs

47 (a) $f(0) = 80$; $f'(0) = 0.50$
(b) $f(10) \approx 85$

49 (a) About 1.1 (liters/minute)/hour
(b) About 0.018 liter/minute
(c) $g'(2) \approx 1.1$

51 (a) Fat
(b) Protein

53 (a) About 2.0 kg/week
(b) About 0.4 kg/week
(c) About 0.3 kg/week

55 I-fat, II-protein

57 (a) $f'(a)$ is always positive
(c) $f'(100) = 2$: more
$f'(100) = 0.5$: less

59 (a) No
(b) Yes

61 (a) In 2015: 5500 thousand km^2 rain forest; decreasing by about 10.9 thousand km^2/yr
(b) Shrinking at 0.198%/yr

63 0.50

65 1.07%

67 (a) 15 billion Apps downloaded by June 2011; in June 2011 number of downloaded Apps increasing at rate of 0.93 billion Apps per month.
(b) 0.062; monthly downloaded Apps increasing at continuous rate of 6.2% per month

69 2071

Section 2.4

1 (a) Negative
(b) Negative

3 (a)

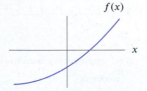

(b)

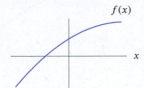

(c)

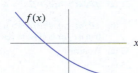

(d)

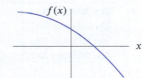

5 $f'(x) = 0$
$f''(x) = 0$

7 $f'(x) < 0$
$f''(x) > 0$

9 $f'(x) < 0$
$f''(x) < 0$

11 $f'(2) < f'(6)$

13 Derivative:
Pos. about $0 < t < 0.4$, $1.7 < t < 3.4$
Neg. about $0.4 < t < 1.7$, $3.4 < t < 4$
Second Derivative:
Pos. about $1 < t < 2.6$
Neg. about $0 < t < 1$, $2.6 < t < 4$

15 $w'(t)$: negative
$w''(t)$: positive

17

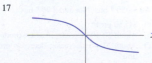

19 (a) Both negative
   (b) $f'(2) \approx -4$, $f'(8) \approx -21$

21

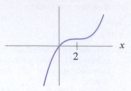

23

Point	f	f'	f''
A	$-$	0	$+$
B	$+$	0	$-$
C	$+$	$-$	$-$
D	$-$	$+$	$+$

25 (e)

29 (a) enrollment as a function of time
   (b)

   (c) $f'(t) < 0$; $f''(t) < 0$

31 (a) $dP/dt > 0$, $d^2P/dt^2 > 0$

   (b) $dP/dt < 0$, $d^2P/dt^2 > 0$
     (but dP/dt is close to zero)

33 (a) Positive

35 (a)

Time interval	Average rate (millions/month)
Sept 2014–Dec 2014	14.33
Dec 2014–Mar 2015	16.0
Mar 2015–June 2015	16.33
June 2015–Sept 2015	18.33

   (b) Positive

Section 2.5

1 About $3 per item

3 About $16.67 (answers may vary)

5 About $0.42 (answers may vary)

7 $C'(2000) \approx \$0.37$/ton
   The marginal cost is smallest on the interval
   $2500 \leq q \leq 3000$.

9 Below 4500

11 (a) About $2408
   (b) About $2192

13 (a) About $4348
   (b) $11 profit
   (c) No, company will lose money

15 Add a 50th bus; Not add a 90th bus

17 (a) $1850 profit
   (b) About $0.40 increase; increase production
   (c) About $0.45 decrease; decrease production

19

q	1	2	3	4	5
$MC(q)$	40	60	80	100	120
$MR(q)$	120	110	80	40	30

Theory: Limits, Derivatives

1

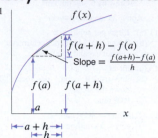

3 (a) 0/0, undefined
   (b) 6.487, 5.127, 5.013, 5.001
   (c) ≈ 5

5 (a) 0/0, undefined
   (b) 0.953, 0.995, 0.9995, 0.99995
   (c) ≈ 1

7 1.6

9 1.9

11 0

13 No; yes

15 No; yes

17 (a) Yes
   (b) Yes
   (c) No
   (d) No

19 2

21 2

23 Yes

25 Yes

27 No

29 Not continuous

31 Continuous

33 Not continuous

Section 3.1

1 3

3 $-12x^{-13}$

5 $24t^2$

7 5

9 $3q^2$

11 $18x^2 + 8x - 2$

13 $24t^2 - 8t + 12$

15 $-12x^3 - 12x^2 - 6$

17 $6.1z^{-7.1}$

19 $(1/2)x^{-1/2}$

21 $-(3/2)x^{-5/2}$

23 $2t$

25 $2z - \frac{1}{2}z^{-2}$

27 $-3t^{-2} - 8t^{-3}$

29 $(1/2)\theta^{-1/2} + \theta^{-2}$

31 $2ax + b$

33 $2at - 2b/t^3$

35 $(8\pi rb)/3$

37 a/c

39 (a) $P'(1)$: Positive;
     $P'(3)$: Zero;
     $P'(4)$: Negative

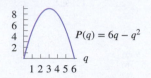

   (b) $P'(1) = 4$, $P'(3) = 0$, $P'(4) = -2$

41 (a) $2t - 4$
   (b) $f'(1) = -2$, $f'(2) = 0$

43 $f(5) =$ Height $= 625$ cm
   $f'(5) =$ Rate of change of height (erosion)
     $= -30$ cm/year

45 69.6% per year

47 -4.15

49 4800 mussels after 4 months, increasing by
   2400 mussels per month

51 $f'(t) = 6t^2 - 8t + 3$
   $f''(t) = 12t - 8$

53 (a) 2
   (b) -6
   (c) (III)

55 (a) -4
   (b) 10
   (c) (II)

57 (a) $f'(1) < f'(0) < f'(-1) < f'(4)$
   (b) $f'(1) = -1$, $f'(0) = 2$, $f'(-1) = 11$,
     $f'(4) = 26$

59 (a) $y = 7x + 1$
   (b) $y = 3x + 1$

61 (a) $y = 12x - 16$
   (b) Underestimates

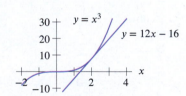

63 (a) $5159v^{-0.33}$
   (b) 11030 m^3/sec per km^3

65 (a) $C(w) = 42w^{0.75}$;
     $C'(w) = 31.5w^{-0.25}$
   (b) (i) 236.2 calories a day;
        17.7 calories/pound
     (ii) 1328 calories a day;
        10 calories/pound
     (iii) 7,469 calories a day;
         5.6 calories/pound

67 (a) $f'(m) = 4.35 \cdot m^{-0.75}$
   (b) $f(70) = 50.330$ and $f'(70) = 0.180$

69 (a) $dA/dr = 2\pi r$
   (b) Circumference of a circle

73 (a) $R(p) = 300p - 3p^2$
   (b) $240 per dollar price increase when price
     is $10
   (c) Positive for $p < 50$, negative for $p > 50$

75 (a) 770 bushels per acre

(b) 40 bushels per acre per pound of fertilizer
(c) Use more fertilizer

77 (a) $dC/dq = 0.24q^2 + 75$
(b) $C(50) = \$14,750$;
$C'(50) = \$675$ per item

79 (a) $R(q) = bq + mq^2$
(b) $R'(q) = b + 2mq$

Section 3.2

1 $9t^2 + 2e^t$

3 $3x^2 + 3^x \ln 3$

5 $5 \cdot 5^t \ln 5 + 6 \cdot 6^t \ln 6$

7 $4(\ln 10)10^x - 3x^2$

9 $5(\ln 2)(2^x) - 5$

11 $0.7e^{0.7t}$

13 $-0.2e^{-0.2t}$

15 $24e^{0.12t}$

17 $12.41(\ln 0.94)(0.94)^t$

19 Ae^t

21 $(\ln 10)10^x - 10/x^2$

23 $-1/p$

25 $2q - 2/q$

27 $Ae^t + B/t$

29 $15/(15t + 12)$

31 -7

33 $-4/t$

35 $f'(-1) \approx -0.736$
$f'(0) = -2$
$f'(1) \approx -5.437$

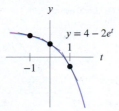

$y = 4 - 2e^t$

37 $y = 3.3x - 0.3$

39 $y = -0.899x + 8.089$

41 (a) $f'(t) = 30e^{0.03t}$
(b) $f(10) = 1349.86$ dollars;
$f'(10) = 40.50$ dollars/year

43 7.17 billion people,
0.0789 billion people per year,
8.004 billion people,
0.088 billion people per year

45 (a) $\$15,660$
(b) $V'(t) = -4.876(0.85)^t$;
thousands of dollars/year
(c) $V'(4) = -\$2545$ per year

47 (a) 9.801 million
(b) 0.0196 million/year

49 ≈ 2.41 cents/year

51 15.4 ng/ml,
-2.16 ng/ml per hour

53 (a) 0.021 micrograms/year
(b) 779.4 years old in 1988

55 (a) 1.247 million people per year
(b) 2.813 million people per year
US, since $dU/dt > dM/dt$ at $t = 0$

57 (a) $y = x - 1$
(b) 0.1; 1
(c) Yes

59 (a) $1.3 \cdot \ln 2.25 \, (2.25^t)$ thousand users/yr
(b) Increasing

Section 3.3

1 $99(x + 1)^{98}$

3 $200t(t^2 + 1)^{99}$

5 $15(5r - 6)^2$

7 $-6x + 6e^{3x}$

9 $30e^{5x} - 2xe^{-x^2}$

11 $-6te^{-3t^2}$

13 $5/(5t + 1)$

15 $2t/(t^2 + 1)$

17 $e^x/(e^x + 1)$

19 $1/(x \ln x)$

21 $3/(3t + 2)$

23 $5 + 1/(x + 2)$

25 $0.5/(x(1 + \ln x)^{0.5})$

27 $1/(4\sqrt{x}\sqrt{2 + \sqrt{x}})$

29 49.7% per year

31 1.5

33 $2/t$

35 $y = 3x - 5$

37 $\$0.20$/item

39 (a) Decrease
(b) $-5.130e^{-0.054t}$ deg C/min

41 Approx 0.8

43 Approx -0.4

45 1/2

47 -1

49 0.5

51 (a) $g'(1) = 3/4$
(b) $h'(1) = 3/2$

Section 3.4

1 $5x^4 + 10x$

3 $-2te^{-2t} + e^{-2t}$

5 $2t(3t + 1)^3 + 9t^2(3t + 1)^2$

7 $\ln x + 1$

9 $(t^3 - 4t^2 - 14t + 1)e^t$

11 $-3qe^{-q} + 3e^{-q}$

13 $1 - 3/x^2$

15 $e^{-t^2}(1 - 2t^2)$

17 $2p/(2p + 1) + \ln(2p + 1)$

19 $2we^{w^2}(5w^2 + 8)$

21 $9(te^{3t} + e^{5t})^8(e^{3t} + 3te^{3t} + 5e^{5t})$

23 $-2/(1 + t)^2$

25 $(15 + 10y + y^2)/(5 + y)^2$

27 $(ak - bc)/(cx + k)^2$

29 $ae^{-bx} - abxe^{-bx}$

31 $(1 - 2\alpha)e^{-2\alpha}e^{\alpha e^{-2\alpha}}$

33 $f'(x) = 12x + 1$ and
$f''(x) = 12$

35 $y = 5x$

37 $y = 147.781x - 236.450$

39 (a) $f'(t) = 4e^{-0.08t} - 0.32te^{-0.08t}$
(b) $f(15) = 18.072$ ng/ml;

$f'(15) = -0.241$ (ng/ml)/minute

41 (a) $f'(15) > 0$, $f'(45) < 0$

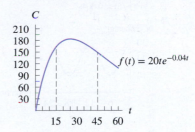

$f(t) = 20te^{-0.04t}$

(b) $f(30) \approx 181$ mg/ml, $f'(30) \approx -1.2$
mg/ml/min

43 Revenue $R(10) \approx 22,466$.
$R'(10) \approx \$449$/dollar.

45 (a) $f(140) = 15,000$:
If the cost $\$140$ per board then
15,000 skateboards are sold
$f'(140) = -100$:
Every dollar of increase from $\$140$ will
decrease the total sales by about 100
boards
(b) $dR/dp|_{p=140} = 1000$
(c) Positive
Increase by about $\$1000$

47 (a) 1.216 cell divisions per hour
(b) $0.297/(0.22 + C)^2$
(c) 0.0603 cell divisions per hour per 10^{-4}
molar
(d) $R \approx 1.216 + 0.0603(C - 2)$ cell divisions
per hour
(e) Tangent line: 1.228
Original: 1.227

49 $(f/g)'/(f/g) = (f'/f) - (g'/g)$

51 $(kc)/(k + r)^2$;
Approx change in P per unit increase in r

Section 3.5

1 $-\sin t$

3 $A \cos t$

5 $5 \cos x - 5$

7 $5 \cos(5t)$

9 $-10 \sin(5t)$

11 $AB \cos(Bt)$

13 $12 \cos(2t) - 4 \sin(4t)$

15 $2 \sin(3x) + 6x \cos(3x)$

17 $((2te^{t^2} + 1) \sin(2t) - (e^{t^2} + t)2 \cos(2t))/\sin^2(2t)$

19 $(\theta \cos \theta - \sin \theta)/\theta^2$

21 $y = -x + \pi$

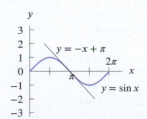

$y = -x + \pi$
$y = \sin x$

23 At $x = 0$:
$y = x$, $\sin(\pi/6) \approx 0.524$
At $x = \pi/3$:
$y = x/2 + (3\sqrt{3} - \pi)/6$,
$\sin(\pi/6) \approx 0.604$

25 (a)

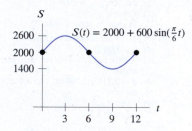

(b) $-4.712\sin(\pi t/6)$
(c) 26.8 species,
-2.36 species/month,
23.5 species,
4.08 species/month

27 CO_2 increasing at 1.999 ppm/month on Dec 1, 2008

29 CO_2 decreasing at 1.666 ppm/month on June 1, 2008

31 (a) No. Yes
(b) (iii)

33 (a) max \$2600; min \$1400; April 1

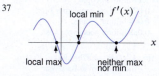

(b) $S(2) \approx 2519.62$; $S'(2) \approx 157.08$

35 (a) Falling, 0.38 m/hr
(b) Rising, 3.76 m/hr
(c) Rising, 0.75 m/hr
(d) Falling, 1.12 m/hr

37 (a) $H'(t) = (-10\pi/3)\sin((\pi/15)t)$
(b) $t = 0, 15, 30$ days; tells us when there is a full or new moon
(c) $H'(t)$ negative for $0 < t < 15$ and positive for $15 < t < 30$

Practice: Differentiation

1 $2t + 4t^3$

3 $15x^2 + 14x - 3$

5 $-2/x^3 + 5/(2\sqrt{x})$

7 $10e^{2x} - 2 \cdot 3^x(\ln 3)$

9 $2pe^{p^2} + 10p$

11 $2x\sqrt{x^2 + 1} + x^3/\sqrt{x^2 + 1}$

13 $16/(2t + 1)$

15 $2^x(\ln 2) + 2x$

17 $6(2q + 1)^2$

19 bke^{kt}

21 $2x\ln(2x + 1) + 2x^2/(2x + 1)$

23 $10\cos(2x)$

25 $15\cos(5t)$

27 $2e^x + 3\cos x$

29 $17 + 12x^{-1/2}$

31 $20x^3 - 2/x^3$

33 $1, x \neq -1$

35 $-3\cos(2 - 3x)$

37 $6/(5r + 2)^2$

39 $ae^{ax}/(e^{ax} + b)$

41 $5(w^4 - 2w)^4(4w^3 - 2)$

43 $(\cos x - \sin x)/(\sin x + \cos x)$

45 $6/w^4 + 3/(2\sqrt{w})$

47 $(2t - ct^2)e^{-ct}$

49 $(\cos\theta)e^{\sin\theta}$

51 $3x^2/a + 2ax/b - c$

53 $2r(r + 1)/(2r + 1)^2$

55 $2e^t + 2te^t + 1/(2t^{3/2})$

57 $x^2\ln x$

59 $6x(x^2 + 5)^2(3x^3 - 2)(6x^3 + 15x - 2)$

61 $(2abr - ar^4)/(b + r^3)^2$

63 $20w/(a^2 - w^2)^3$

Section 4.1

1 One

3 Four

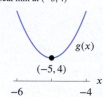

5 (a)

(b)

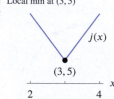

7 After 18 hours

9 (a) $x = 5$ and $x = -5$
(b) Local maximum at $x = -5$;
Local minimum at $x = 5$

11 (a) $x = 0$; $x = 2$; $x = -2$

(b) Local maximum at $x = 0$;
Local minimum at $x = 2$;
Local minimum at $x = -2$

13 (a) $x = 0$; $x = 3$; $x = -3$
(b) Neither at $x = 0$;
Local maximum at $x = -3$;
Local minimum at $x = 3$

15 Local max: $(-1.4, 6.7)$
Local min: $(1.4, -4.7)$

17 Local min: $(2.3, -13.0)$

19 Alternately incr/decr

21 Critical points: $x = -1, 1$
$x = -1$ local maximum; $x = 1$ local minimum

23 Critical points:
$x = 0$ and $x = \pm 2$
Extrema:
$f(0)$ local minimum
$f(-2)$ and $f(2)$ are not local extrema.

25 Critical points:
$x = \pm 1$
Extrema:
$f(-1)$ local minimum
$f(1)$ local maximum

27 (a) Local min at $(-5, 4)$
(b)

29 (a) Local min at $(3, 5)$
(b)

31 $x = 0$: not max/min
$x = 3/7$: local max
$x = 1$: local min

33 $x = \sqrt{B/3A}$ and $x = -\sqrt{B/3A}$

35 $x = 0$; $x = \sqrt{B/2A}$; $x = -\sqrt{B/2A}$

37

39 (a) Increasing for all x
(b) No maxima or minima

41 (a) Increasing: $-1 < x < 0$ and $x > 1$
Decreasing: $x < -1$ and $0 < x < 1$

(b) Local max: $f(0)$
Local min: $f(-1)$ and $f(1)$

43 (a) Increasing weeks 0–2 and 6–10, decreasing weeks 3–5
(b) Local max week 2–3 and week 10, local min week 0 and week 5–6

45 Increasing: $0 < t < 2, t > 4$
Decreasing: $2 < t < 4$
Local max: $t = 2, t = 5$
Local min: $t = 0, t = 4$

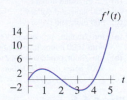

47 (a) $x = 5a/4$
(b) $a = 4.8$

49 $a = 4, b = 1$

51 $b = 2, a = 5/(2 - 2\ln 2) \approx 8.147$

53 (a) $x = 0, x = a^2/4$
(b) $a = \sqrt{20}$; Local minimum

55 (a)

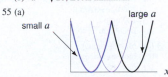

(b) Critical point moves right
(c) $x = a$

57 (a)

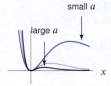

(b) Nonzero critical point moves down to the left
(c) $x = 0, 2/a$

Section 4.2

1 Two

3 Three

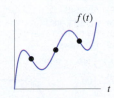

5

7

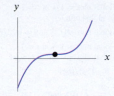

9 (a) 6 pm
(b) Noon; another between noon and 6 pm

11 $x = -1, 1/2$

13 Critical points: $x = -1$,
local max; $x = 1$, local min
Inflection point: $x = 0$

15 Critical points: $x = -2$,
local max; $x = 1$, local min
Inflection point: $x = -1/2$

17 Critical points:
$x = 0$ (neither) and $x = 1$ (local min)
Inflection points: $x = 0$ and $x = 2/3$

19 Critical points: $x = 0$ (not an extrema) and
$x = 3$ (local minimum)
Inflection points: $x = 0$ and $x = 2$

21 Critical points:
$x = -1$ (local maximum)
$x = 0$ (not an extrema)
$x = 1$ (local minimum)
Inflection points:
$x = 0$ and $x = \pm 1/\sqrt{2}$

23 (a) Critical points: $x = 0, x = \sqrt{a}$,
$x = -\sqrt{a}$
Inflection points: $x = \sqrt{a/3}$,
$x = -\sqrt{a/3}$
(b) $a = 4, b = 21$
(c) $x = \sqrt{4/3}, x = -\sqrt{4/3}$

25 (a) B, D, F
(b) B, C, E
(c) One; One

27

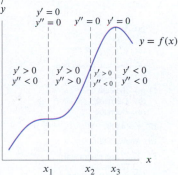

31 (a)

(b)

(c)

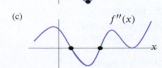

33 (a) $f'(20)$
(b) Growth more rapid at week 20 than 36

35 (a) About 1.9 cm/week
(b) About 0.75 cm/week
(c) About 1.25 cm/week

37

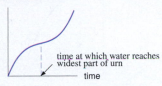

39 $y = x^3 - 3x^2 + 6$

41 $y = e^{-(x-2)^2/2}$

Section 4.3

1

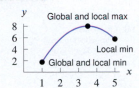

3 Global max: $x = 2$
Global min: $x = 4$

5 Global max: $x = 2$
Global min: $x = -1$

7 Global max: $x = 2$
Global min: $x = 0$

9 (a) (IV)
(b) (I)
(c) (III)
(d) (II)

11

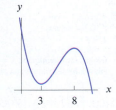

13

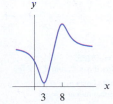

15 (a) Local minima at $x = -1$,
$x \approx 0.91, x = 4$
Local maxima at $x \approx -0.46, x \approx 3.73$
$f(0.91)$ is global minimum
$f(3.73)$ is global maximum

(b) Local minima at $x = -3$, $x \approx 0.91$
Local maxima at $x \approx -0.46$, $x = 2$
$f(-3)$ is global minimum
$f(2)$ is global maximum

17
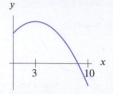

19
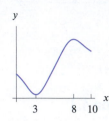

21 (a) $t \approx 50$ days
(b) Throughout interval
$t \approx 50$ days

23 (a) $f'(x) = 6x^2 - 18x + 12$,
$f''(x) = 12x - 18$.
(b) $x = 1, 2$
(c) $x = 3/2$
(d) Local minimum: $x = -0.5, 2$
Local maximum: $x = 1, 3$
Global minimum: $x = -0.5$
Global maximum: $x = 3$
(e)
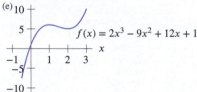
$f(x) = 2x^3 - 9x^2 + 12x + 1$

25 (a) $f'(x) = 1 + \cos x$,
$f''(x) = -\sin x$.
(b) $x = \pi$
(c) $x = 0, \pi, 2\pi$
(d) Global, local minimum: $x = 0$
Global, local maximum: $x = 2\pi$
(e)
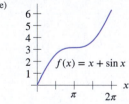
$f(x) = x + \sin x$

27 $x = -b/2a$,
Max if $a < 0$, min if $a > 0$

29 Global max at $x = -6$
Global min at $x = -5$

31 Global max at $x = 3$
Global min at $x = 1$ and $x = 5$

33 Yes, at G; No

35 2500

37 1536

39 Global max $= -1$ at $x = 2$
No global min

41 Global max $= 1/e$ at $t = 1$
No global min

43 Global min $-4/27$ at $x = \ln(2/3)$

45 Global max $= 1/2$ at $t = 1$
Global min $= -1/2$ at $t = -1$

47 $w = -(5p)/(6q)$

49 (a) g
(b) For f, $D = 300$
For g, $D = 200$

51 1250 square feet

53 (a) $x + 27/x$ meters
(b) 5.2 meters

55 (a) $(1/b, (a/b)e^{-1})$ is local max
(b) $(1/b, (a/b)e^{-1})$ is global max
Large population may have fewer
offspring than small

57 (a) 0
(b) 0.69 hours, 5mg
(c) Tends to 0

59 (b) $f(v) = v \cdot a(v)$
(c) When $a(v) = f'(v)$
(d) $a(v)$

Section 4.4

1 $5.5 < q < 12.5$ positive;
$0 < q < 5.5$ and $q > 12.5$ negative;
Maximum at $q \approx 9.5$

3 Global maximum of \$6875 at $q = 75$

5 (a)
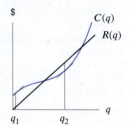

7 (a) \$9
(b) $-\$3$
(c) $C'(78) = R'(78)$

9 (a) Increase production
(b) $q = 8000$

11 (a) Increase
(b) Decrease
(c) Decrease

13 $q = 4000$

15 Above 2000

17 27.273 on equipment;
72.727 on labor;
Max prod 415.955 items

19 $R(q) = 45q - 0.01q^2$
Max revenue at $q = 2250$,
$p = \$22.50$; Revenue $= \$50,625$

21 $p = \$4.50$, $q = 3600$.

23 (a) 69.315 pounds per acre
(b) \$72,274

25 (a) $10,000 + 2q$
(b) $q = 37,820 - 5544p$
(c) $\pi = -0.00018q^2 + 4.822q - 10,000$
(d) 13,394 items, \$22,294

27 \$8378.54

29 (a) Ordering: a/q
Storage: bq
(b) $\sqrt{a/b}$

31 (a) q/r months

(b) $(ra/q) + rb$ dollars
(c) $C = (ra/q) + rb + kq/2$ dollars
(d) $q = \sqrt{2ra/k}$

33 (a) \$5 per item
(b) 25 thousand dollars
(c) 230 thousand dollars
(d) 500 thousand dollars

35 (a) tons/month, producing 400 tons/month
with 1000 hours/month labor
(b) tons/hour, produce about 2 tons more with
1001 hours than 1000 hours of labor
(d) $C = wL$, $R = pf(L)$, $\pi = R - C$

Section 4.5

1 (a) No
(b) Yes

3 (a) (i) About \$8 per unit
(ii) About \$4 per unit
(b) About 30 units

5 $MC = \$20$; $a(q) = \$25$

7 (a) 1000 m \$, 0.2q m.\$/unit, $(0.1q^2 + 1000)/q$\$m./unit
(b) 100 units
(c) Local min

9 (a) Making money
(b) Increase; increase
(c) Increase

11 (a) Decrease
(b) Can't tell

13 (a) 17.8, 16.667, 16.143, 16, 16.111, 16.4
(b) At $q = 8$
(c) Local min

17 (b) $q = [Fa/(K(1 - a))]^a$

Section 4.6

1 (a) 1.5% decrease
(b) 1.5% increase

3 4.375% decrease

7 Elastic

9 Elastic

11 $E = 2/3$; inelastic

13 (a) $E \approx 0.470$, inelastic
(c) About $P = 1.25$ and 1.50

17 \$12.91

29 1% more time gives 1.3% more sales

Section 4.7

3 Dominican Republic; 0.35 per week

5 Dominican Republic: Top graph
Guadeloupe: Lower graph
Dominica: Merged with horiz axis

9 (b) About 35%
(c) About 70%

11 (b) April 12, 2003
(d) $t = 19$; 1600 cases
(e) 1760 cases

15 $t = \frac{1}{2} \ln 6$, $y = 25$

17 (a)
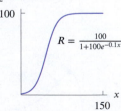
$R = \dfrac{100}{1+100e^{-0.1x}}$

(b) About 46.05

(c) Between 32.12 and 54.52 mg

21 Effective for 85%, lethal for 6%.

23 (a) $L = 1$, $k = 0.5$, $C = e^{-12.5}$
   (b) -29.4 mV, -25 mV, -20.6 mV

Section 4.8

1 (a)

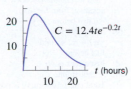

$C = 12.4te^{-0.2t}$

(b) 5 hrs, 22.8 ng/ml
   (c) 1 to 14.4 hrs
   (d) At least 20.8 hrs

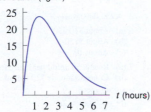

7 (a) About 33.3 minutes; about 245 ng/ml
   (b) About 191 ng/ml; about 198 ng/ml
   (c) In 3 hours

9 (b) $(\ln 2, 1/4)$, $(\ln 4, 3/16)$

11 (a) IV Fastest, P-IM slowest
   (b) IV Largest, PO smallest
   (c) IV Fastest, P-IM slowest
   (d) P-IM Longest, IV shortest
   (e) About 5 hours

Section 5.1

1 (a) 160 miles
   (b)

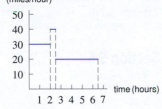

(c)

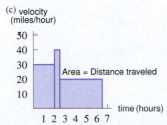

3 (a) 56 km; lower

(b) 88 km; upper

5 (a) Right sum
   (b) Lower estimate
   (c) 5
   (d) $\Delta t = 3$
   (e) Lower estimate $\approx$ 160.5

7 (a) Lower estimate = 122 ft
   Upper estimate = 298 ft
   (b)

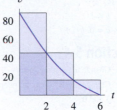

9

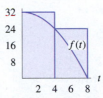

11 (a)

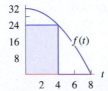

(b) 96.8 ft

13 (a) Lower estimate = 5.25 mi
   Upper estimate = 5.75 mi
   (b) Lower estimate = 11.5 mi
   Upper estimate = 14.5 mi
   (c) Every 30 seconds

15 250 meters

17 (a) $a = 44$, $b = 88$, $c = 5$
   (b) 330 feet

19 (a) About 420 kg
   (b) 336 and 504 kg

21 668.5 bn barrels

23 220 fish

25 (a) Car A
   (b) Car A
   (c) Car B

27 60 m (Other answers possible.)

29 (b) $6151

31 (a) 3400 million people
   (b) 3530 million people; differs by 130 mn

33 No

35 About 0.009 miles or 48 feet

37 (a) 663 ft
   (b) Upper estimate
   (c) 489 ft
   (d) Lower estimate
   (e) 576 ft

39 (a) 729.5 billion tons
   (b) Lower estimate
   (c) 852 billion tons
   (d) Upper estimate

41 (a) 2; 15, 17, 19, 21, 23; 10, 13, 18, 20, 30
   (b) 122; 162
   (c) 4; 15, 19, 23; 10, 18, 30
   (d) 112; 192

Section 5.2

1 (a) Right
   (b) Upper
   (c) 3
   (d) 2

3 (a) Left; larger
   (b) 2, 6, 8, 1/2

5 5.244

7 Left-hand sum = 378; Right-hand sum = 810

9 Left-hand sum = 109.5; Right-hand sum = 876

11 Left-hand sum = 77/20; Right-hand sum = 57/20

13 Left-hand sum = $1 + \sqrt{2} + \sqrt{3}$; Right-hand sum = $\sqrt{2} + \sqrt{3} + 2$

15 1096

17 About 543

19 (a) 224

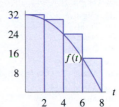

(b) 96

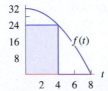

(c) About 200

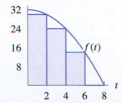

(d) About 136

21 About 350

23 About 17

25 Equal

27 1.2

29 −0.75

31 3.406

33 2.214

35 (a) 0.3

(b) 0.25

37 (a) 1.8

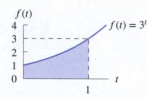

(b) 1.8205

39 IV: 93.47

41 (a) 438
 (b) 928
 (c) 592

43 (a) 80
 (b) 123
 (c) 88

Section 5.3

1 84

3 7.667

5 $\int_0^3 (3x - x^2)\,dx = 4.5$

7 (a) Negative
 (b) Positive
 (c) Negative
 (d) Positive

9 (a) About 16.5
 (b) About −3.5

11 (a) About −0.25
 (b) About 0
 (c) About 0.5

13 (a) 13
 (b) −2
 (c) 11
 (d) 15

15 II

17 III

19 (a) −2
 (b) −A/2

21 (a) 1
 (b) 2π
 (c) 2π − 1/2
 (d) π − 3/2

23 14.688

25 13.457

27 2.762

29 −0.136

31 −16

33 22/3

35 (a) 16.25
 (b) 15.75
 (c) No

37 b = 1

39 a = 0, b = 1

Section 5.4

1 Change in position; meters

3 Change in world pop; bn people

5 3.406 ft

7 (a) Removal rate 500 kg/day on day 12
 (b) Days, days, kilograms
 (c) 4000 kg removed between day 5 and day 15

9 (a) Emissions 2000–2012, m. tons CO_2 equiv
 (b) 1912.7 m. tons CO_2 equiv
    (Possible answers from 1898 to 1928)

11 (a) 2711 bn m³; 3558 bn m³
 (b) 30,960 bn m³

13 Change: 15 cm to the left
 Total distance: 15 cm

15 Change: 25 cm to the right
 Total distance: 25 cm

17 2627 acres

19 (a) $\int_0^5 (10 + 8t - t^2)\,dt$
 (b) 100
 (c) 108.33

21 1417 antibodies

23 (a) Approximately 73 minutes
 (b) Slightly less than 1800 acre-feet; about 2 mins after midnight
 (c) About 40,000 acre-feet
 (d) About 15 minutes

25 t = 1
 About 16.667 miles

27 (a) (i) $\int_{2016}^{2020} I(t)\,dt$; Total income from 2016 to 20120
    (ii) $\int_{2016}^{2020} E(t)\,dt$; Total expenditures from 2016 to 2020
    (iii) $\int_{2016}^{2020} I(t) - E(t)\,dt$; Change in value from 2016 to 2020
 (b) About 50 billion dollars

29 $\int_{2016}^{2025} I(t) - E(t)\,dt$

31 6 months: A more
 First year: B more
 Same: roughly 9 months
 A roughly 170 sales
 B roughly 250 sales

33 (a) About 430 liters
 (b) $\int_0^3 60 f(t)\,dt$
 (c) About 470 liters

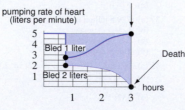

35 $23,928.74

37 Product A has a greater peak
 Product A peaks faster
 Product B provides greater overall exposure

39 About $11,600

41 (a) 346 hours
 (b) 449 millirems

43 (a) Positive: $0 \le t < 40$ and just before $t = 60$; negative: from $t = 40$ to just before $t = 60$
 (b) About 500 feet at $t = 42$ seconds
 (c) Just before $t = 60$
 (d) Just after $t = 40$
 (e) A catastrophe
 (f) Total area under curve is positive About 280 feet

Section 5.5

1 45.8°C.

3 4,250,000 riyals

5 (a) About $10,550
 (b) About $150
 (c) About 10

7 (a) Rate is 50 ng/ml per hour after 1 hour
 (b) Concentration is 730 mg/ml after 3 hours

9 (a) $18,650
 (b) $C'(400) = 28$

11 (a) $4059.24
 (b) $1940.76
 (c) $157.14

13 (a) End of the 5th week; near end of the 4th week
 (b) $P(4) < P(3) \approx P(0) < P(1) \approx P(2) < P(5)$

17 (a) Immediate-release: A
    Delayed-release: B
 (b) No; Immediate larger

19 $\int_{2t_0}^{3t_0} r(t)\,dt < \int_{t_0}^{2t_0} r(t)\,dt < \int_0^{2t_0} r(t)\,dt$

21 0.5; cost of preparing is $0.5 million

23 $\int_0^2 r(t)\,dt < \int_2^4 r(t)\,dt$

25 $\int_0^8 r(t)\,dt < 64$ million

27 (III), (II), (I)

Section 5.6

1 (a) 20
 (b) 10/3

3 2

5 (a) 0.79

7 8

9 About 17

11 (a) 527.25
 (b)

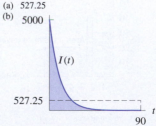

13 (a) 5423.5 barrels/day
   (b) Same because $P(t)$ linear

15 (a) \$40,000 per year
   (b) Less 35–65; more 25–35, 65–85

17 About 90 mm Hg

19 (a) 9.9 hours
   (b) 14.4 hours
   (c) 12.0 hours

21 (a) Second half

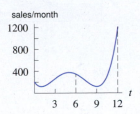

   (b) \$1531.20, \$1963.20
   (c) \$3494.40
   (d) \$291.20/month

23 $(c) < (a) < (b)$

Theory: Second Fund. Thm.

1 x^3

3 xe^x

5 (a) 0
   (b) F increases
   (c) $F(1) \approx 1.4$, $F(2) \approx 4.3$, $F(3) \approx 10.1$

9 9

11 $8c$

Section 6.1

1

b	0	0.1	0.2	0.5	1.0
$F(b)$	3	3.501	4.005	5.583	6.667

3

b	2	4	6	10	20
$f(b)$	9	10.110	10.856	11.696	12.293

5 52.545

7 13

9 $F(0) = 0$
  $F(1) = 1$
  $F(2) = 1.5$
  $F(3) = 1$
  $F(4) = 0$
  $F(5) = -1$
  $F(6) = -1.5$

11 Maximum at $(4, 3)$

13 (a) $f(x)$ increasing when $2 < x < 5$
    $f(x)$ decreasing when $x < 2$ or $x > 5$
    f has a local minimum at $x = 2$
    f has a local maximum at $x = 5$
   (b)

15 3.5, 2, 1.5, 2, 2.5

17

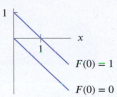

$F(0) = 1$
$F(0) = 0$

19

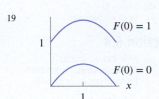

$F(0) = 1$
$F(0) = 0$

21

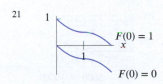

$F(0) = 1$
$F(0) = 0$

23 About 450

25 (a) (I) volume; (II) flow rate
   (b) (I) is an antiderivative of (II)

27 (a) amount of oxygen released

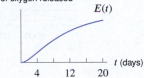

   (b) 4 days
   (c) $\int_0^{10} p(t)\, dt$ or $E(10)$
   (d) First ten

29 Critical points: $(0, 5)$, $(2, 21)$, $(4, 13)$, $(5, 15)$

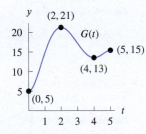

31 Min: $(1.5, -20)$, max: $(4.67, 5)$

33 (a) -16
   (b) 84

35 (a) $x = -1$, $x = 1$, $x = 3$
   (b) Local min at $x = -1$, $x = 3$;
    local max at $x = 1$
   (c)

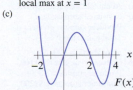

37

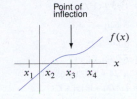

Point of inflection

39 $f(1) < f(0)$

41

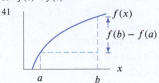

43

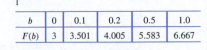

Section 6.2

1 Yes

3 No

5 No

7 (II) and (IV) are antiderivatives

9 (III), (IV) and (V) are antiderivatives

11 Number

13 Family functions

15 Number

17 $5t^2/2$

19 $t^3/3 + t^2/2$

21 $x^5/5$

23 $5q^3/3$

25 $y^3 - y^4/4$

27 $P(r) = \pi r^2 + C$

29 $2z^{3/2}/3$

31 $5x^2/2 - 2x^{3/2}/3$

33 $-1/t$

35 $F(z) = e^z + 3z + C$

37 $\ln|x| - 1/x - 1/(2x^2) + C$

39 $-e^{-3t}/3$

41 $G(t) = 5t + \sin t + C$

43 $g(x)$

45 $f(x)$

47 $F(x) = 3x$
   (only possibility)

49 $F(x) = x^2/8$
   (only possibility)

51 $F(x) = x^3/3$
   (only possibility)

53 $(5/2)x^2 + 7x + C$

55 $-20e^{-0.05t} + C$

57 $(t^{13}/13) + C$

59 $t^3/3 + 5t^2/2 + t + C$

61 $3 \ln|t| + \dfrac{2}{t} + C$

63 $2w^{3/2} + C$

65 $(e^{2t}/2) + C$

67 $x^4/4 + 5x^3/3 + 6x + C$

69 $(x^3/3) + \ln|x| + C$

71 $e^{3r}/3 + C$

73 $-625e^{-0.04q} + C$

75 $\sin\theta + C$

77 $-\cos(3x)/3 + C$

79 $2\sin(3x) + C$

81 $2e^x - 8\sin x + C$

83 $2\ln|x| + x^2/4$

85 $(2/3)x^{3/2} + 2\sqrt{x}$

87 $\cos(7x)/7 - 7\cos x$

89 $q^3 + 2q^2 + 6q + 200$

91 $F(x) = x^3/3 + x + 5$

93 $F(x) = 2e^{3x} + 3$

95 (a) I $C(t) = 1.3t + 311$
II $C(t) = 0.5t + 0.015t^2 + 311$
III $C(t) = 25e^{0.02t} + 286$
(b) I   402 ppm
II   419.5 ppm
III 387.380 ppm

Section 6.3

1 48

3 81/4

5 22

7 2

9 125

11 75/4

13 8/15

15 $609/4 - 39\pi \approx 29.728$

17 $\sin 1 - \sin(-1) = 2\sin 1$

19 $20(e^{0.15} - 1)$

21 75

23 25/2

25 1/6

27 1/2

29 $(300)^{1/3} = 6.694$

31 (a) $\int_0^5 462e^{0.019t}\,dt$
(b) About 2423 quadrillion BTUs

33 (a) First case: 19,923
Second case: 1.99 billion
(b) In both cases, 6.47 yrs
(c) 3.5 yrs

35 (a) $-(1/b) + 1$
(b) Converges to 1

37 $\int_1^\infty (1/x^2)\,dx$

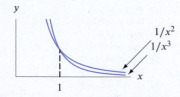

39 (a) $\int_0^\infty 1000te^{-0.5t}\,dt$

(b) r

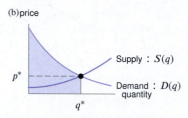

41 (a) 18, 61.2, 198
(b) $2\sqrt{b} - 2$
(c) Does not converge

Section 6.4

1 (a) $p^* = \$30,\ q^* = 6000$
(b) Consumer surplus = \$210,000;
Producer surplus $\approx$ \$70,000

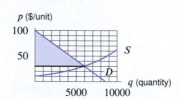

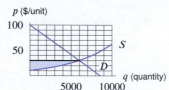
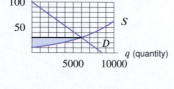

3 250

5 200

7 (a) \$400
(b) \$266.7

9 (a)

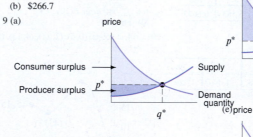

(b)

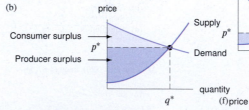

11 (a) About \$2250, \$2625, \$4875
(b) Consumer surplus: less
Producer surplus: greater
Total gains: less

13 (a) $p^* = \$6,\ q^* = 400$
(b) Consumer surplus $\approx$ \$1100,
Producer surplus $\approx$ \$800
(c) Consumer surplus $\approx$ \$1200,
Producer surplus $\approx$ \$200

15 (a) Demand: Table 6.3,
Supply: Table 6.4
(b) $p^* = 25,\ q^* = 400$
(c) Consumer surplus $\approx$ 6550;
Producer surplus $\approx$ 2850

19 (a) price

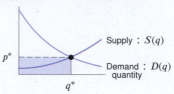

(b) price

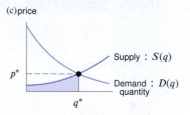

(c) price

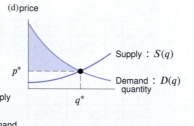

(d) price

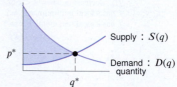

(e) price

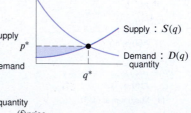

(f) price

455

Section 6.5

$/year

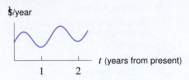

t (years from present)

3 $P = \$21,105$
   $F = \$44,680$

5 (a) $\$13,498.59$
  (b) $\$11,661.96$
  (c) Lump sum; Better to get money earlier

7 (a) $\$22,340$
  (b) $\$15,000$
  (c) $\$7,340$

9 (a) $P = \$47,216.32$
     $F = \$77,846.55$
  (b) $\$60,000; \$17,846.55$

11 Reasonable for Company A

13 2,936,142.74 euros

15 (a) $\$5820$ per year
   (b) $\$36,787.94$

17 No; present value = $\$306,279$

19 (a) $\$33.58$ billion
   (b) $\$35.66$ billion

21 (a) 80.1 and 102.1 billion dollars
   (b) 268.6 and 342.2 billion dollars

23 (a) 10.6 years
   (b) 624.9 million dollars

Section 6.6

1 (a) $2x\cos(x^2+1)$;
      $3x^2\cos(x^3+1)$
  (b) (i) $\frac{1}{2}\sin(x^2+1)+C$
      (ii) $\frac{1}{3}\sin(x^3+1)+C$
  (c) (i) $-\frac{1}{2}\cos(x^2+1)+C$
      (ii) $-\frac{1}{3}\cos(x^3+1)+C$

3 Yes

5 Yes

7 $\sqrt{x^2+4}+C$

9 $\frac{1}{3}(x^2+1)^{3/2}+C$

11 $e^{5t+2}+C$

13 $-10e^{-0.1t+4}+C$

15 $(1/33)(t^3-3)^{11}+C$

17 $(1/9)(x^2-4)^{9/2}+C$

19 $-1/(3(3x+1))+C$

21 $4\sin(x^3)+C$

23 $(1/5)x^5+2x^3+9x+C$

25 $-(1/8)(\cos\theta+5)^8+C$

27 $-(2/9)(\cos 3t)^{3/2}+C$

29 $(1/6)\ln(1+3t^2)+C$

31 $(1/4)\sin^4\alpha+C$

33 $(\sin x)^3/3+C$

35 $2\sqrt{x+e^x}+C$

37 $\frac{1}{9}(3x^2+4)^{3/2}+C$

39 $(1/3)(\ln z)^3+C$

41 $\frac{1}{10}\ln(5q^2+8)+C$

43 $2e^{\sqrt{y}}+C$

45 $(1/2)\ln(x^2+2x+19)+C$

47 $(1/35)\sin^7 5\theta+C$

49 (a) Yes; $-0.5\cos(x^2)+C$
   (b) No
   (c) No
   (d) Yes; $-1/(2(1+x^2))+C$
   (e) No
   (f) Yes; $-\ln|2+\cos x|+C$

51 $62/3$

53 $1-(1/e)$

55 2

57 $2/5$

59 $14/3$

61 $(1/2)(e^4-1)$

63 $(1/2)\ln 3$

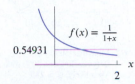

$f(x)=\frac{1}{1+x}$

0.54931

65 (a) (i) $x^3/3+5x^2+25x+C$
       (ii) $(x+5)^3/3+C$
   (b) No; differ by a constant

67 (a) $(\sin^2\theta)/2+C$
   (b) $-(\cos^2\theta)/2+C$
   (c) $-(\cos 2\theta)/4+C$
   (d) Functions differ by a constant

Section 6.7

1 $\frac{1}{5}te^{5t}-\frac{1}{25}e^{5t}+C$

3 $(1/2)y^2\ln y-(1/4)y^2+C$

5 $(1/6)q^6\ln 5q-(1/36)q^6+C$

7 $(x^4/4)\ln x-(x^4/16)+C$

9 $-2y(5-y)^{1/2}$
  $-(4/3)(5-y)^{3/2}+C$

11 $-2t(5-t)^{1/2}-(4/3)(5-t)^{3/2}$
   $-14(5-t)^{1/2}+C$

13 $-t\cos t+\sin t+C$

15 $(2/3)x^{3/2}\ln x-(4/9)x^{3/2}+C$

17 $0.386, 2\ln 2-1$

19 $5\ln 5-4\approx 4.047$

21 $(9/2)\ln 3-2\approx 2.944$

23 $t(\ln t)^2-2t\ln t+2t+C$

25 (a) Substitution
   (b) Substitution
   (c) Substitution
   (d) Substitution
   (e) Parts
   (f) Parts

27 $2\ln 2-1$

29 (a) $-(1/a)Te^{-aT}+(1/a^2)(1-e^{-aT})$
   (b) $\lim_{T\to\infty}E=1/a^2$

Chap. 6: Practice

1 $(1/18)(y^2+5)^9+C$

3 $u^5/5+5u+C$

5 $-e^{-3t}/3+C$

7 $ax^3/3+bx+C$

9 $(x^4/4)+2x^2+8x+C$

11 $q^2/2-1/(2q^2)+C$

13 $q^3/3+5q^2/2+2q+C$

15 $4\ln|x|-5/x+C$

17 $-3\cos\theta+C$

19 $p^3/3+5\ln|p|+C$

21 $-5\cos x+3\sin x+C$

23 $Aq^2/2+Bq+C$

25 $\pi hr^3/3+C$

27 $5p^3q^4+C$

29 $x^3+3e^{2x}+C$

31 $p^4/4+\ln|p|+C$

33 $\ln|y+2|+C$

35 $x^3/3+8x+e^x+C$

37 $a\ln|x|-b/x+C$

39 $\frac{1}{2}e^{2t}+5t+C$

41 $P_0e^{kt}/k+C$

43 $-(A/B)\cos(Bt)+C$

45 $\ln(2+e^x)+C$

47 $(1/2)x^2\ln x-(1/4)x^2+C$

Section 7.1

1 (a) 0.25
  (b) 0.7
  (c) 0.15

3 (a) 0.4375
  (b) 0.49
  (c) 0.2475

5 10–12: About 27%
  < 8: About 12%
  > 12: About 45%
  12–13 days

7 For small Δx around 70, fraction of families
  with incomes in that interval about $0.05\Delta x$

9 0.04

11 0.008

13 (a) 0.19
   (b) Tenth; both same
   (c) 0.02, 0.38, 0.21

15

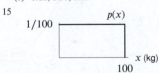

17

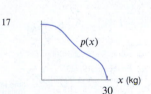

Section 7.2

1 (a)

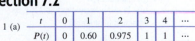

t	0	1	2	3	4	…
$P(t)$	0	0.60	0.975	1	1	…

(b) fraction of patients

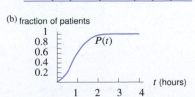

t (hours)

3 (a) Cumulative distribution increasing
(b) Vertical 0.2, horizontal 2

5 pdf; 1/4

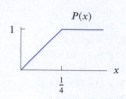

$P(x)$

7 cdf; 1

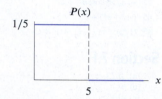

$P(x)$

1/5

5

9 cdf; 1/3

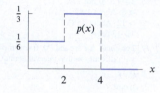

$p(x)$

2   4

11 (a)

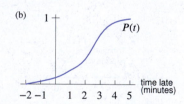

0.25

0.125

$p(t)$

-2 -1   1 2 3 4 5

time late (minutes)

(b)

1

$P(t)$

-2 -1   1 2 3 4 5

time late (minutes)

13 (a) About 2/3
(b) About 1/3
(c) Possibly many work just to pass
(d)

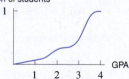

fraction of students

1

1   2   3   4

GPA

15 (a) Cumulative distribution

fraction of cost overruns

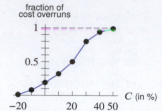

1

0.5

-20   20   40 50

C (in %)

(b)

probability density

-20   20   40

C (in %)

(c) More than 50%: 1%
Between 20% and 50%: 49%
Most likely: $C \approx 28\%$

17 (a)

fraction of bananas

1

0.75

0.5

0.25

$P(t)$

-1 1 2 3 4 5

t (weeks)

(b) About 25%

19 Fraction dead at time t

21 (a) $-e^{-2} + 1 \approx 0.865$
(b) $-(\ln 0.05)/2 \approx 1.5$ km

Section 7.3

1 About 5.4 tons

3 Mean 2/3; Median $2 - \sqrt{2} = 0.586$

5 2.48 weeks

fraction of bananas per week of age

0.4

0.3

0.2

0.1

1   2   3   4

weeks

7 (a) 0.684 : 1
(b) 1.6 hours
(c) 1.682 hours

9 (a) 0.2685
(b) 0.067

11 (a) 16.7%; 12.9%
(b) About $39,900
(c) False

Section 8.1

3 (a) $R = 100s + 5m$
(b) 125,000 dollars

5 $f(w, 60)$: 23.4, 27.3, 31.2, 35.2, 39.1

7 25

9 Increasing function

13 Decreasing function of p
Increasing function of a

15 (a) 81°F
(b) 30%

17 (a) 1596 kcal/day
(b) 1284 kcal/day
(c) Plane; weight, height, age combinations of woman whose BMR is 2000 kcal/day
(d) Lose weight

19 Incr of A and r
Decr of t

21 (a) 5200 kWh energy produced per year with 15-ft diameter, 12-mph wind
(b) 2200 kWh
(c) 2900 kWh
(d) Replace with bigger turbine

23 57.9 kg

27

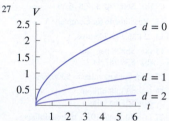

V

2.5
2
1.5
1
0.5

$d = 0$
$d = 1$
$d = 2$

1 2 3 4 5 6 t

29 (a)

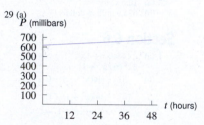

P (millibars)

700
600
500
400
300
200
100

12   24   36   48 t (hours)

(b)

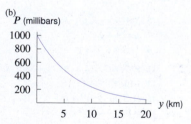

P (millibars)

1000
800
600
400
200

5   10   15   20 y (km)

Section 8.2

1 Approximately $(1.4, 2)$, other answers possible.

3 Equal.

5 x-axis: price
y-axis: advertising

7 (a) $(0, -1)$, other answers possible
(b) $(-2, 2)$, other answers are possible

9 Answers in °C:

(a)

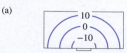

10
0
-10

(b)

(c)

(d)
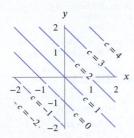

11 The horizontal axis

13 Contours evenly spaced parallel lines

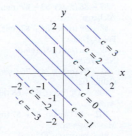

15 Contours evenly spaced parallel lines

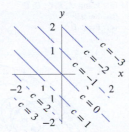

17 Contours evenly spaced parallel lines

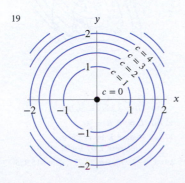

19

21
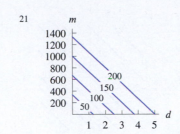

23 Elevation in meters
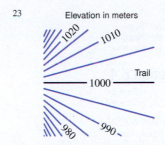

Trail
1000

25 predicted high temperature
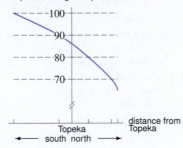

distance from Topeka

Topeka

south north

27 (a) $\pi = 3q_1 + 12q_2 - 4$
(thousand dollars)

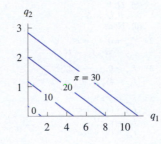

29 Contours evenly spaced

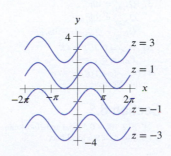

31 (a) False

(b) True
(c) False
(d) True

33 (a) About $0°F$
(b) About $-16°F$
(c) About 23 mph
(d) About $25°F$

35 (a) A
(b) B
(c) A

37

39 (a) 4 hours
(b) 40%
(c) Contours approx horizontal
(d) Increasing
(e) Increasing

41 (a) R
(b) Q
(c) P

43 Underweight: below 18.5, Normal: 18.5-25

Section 8.3

1 (a) $f(A) = 10$
(b) Positive
(c) Zero

3 (a) $f(A) = 58$
(b) Positive
(c) Positive

5 (a) $f(A) = 25$
(b) Positive
(c) Negative

7 (a) Positive
(b) Negative
(c) Positive
(d) Zero

9 $\frac{\partial I}{\partial H}\big|_{(10,100)} \approx 0.4$
$\frac{\partial I}{\partial T}\big|_{(10,100)} \approx 1$

11 (a) f_c is negative
f_t is positive

13 $\partial Q/\partial b < 0$
$\partial Q/\partial c > 0$

15 (a) Positive
(b) Negative
(c) Positive
(d) Negative

17 (c) Positive
(d) Negative

19 (a) 3.3 % / in
(b) −5% / °F

21

	w (gm/m^3)		
	0.1	0.2	0.3
10	−1.0	−6.0	−7.0
T (°C) 20	−3.0	−3.0	−4.0
30	−0.5	−1.0	−2.0

23 (a) Both negative
   (b) Both positive

25 (a) 150,350
   (b) 151,000
   (c) 152,250

27 5.67

29 2540

31 835

33 (b)

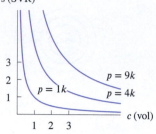

(c)

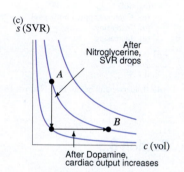

(d)

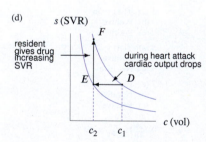

35 There are many possibilities.

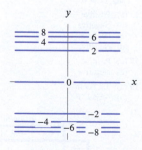

37 There are many possibilities.

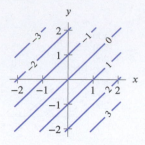

39 (a) Negative
   (b) Negative
   (c) Negative

Section 8.4

1 $f_x = 2x + 2y$
  $f_y = 2x + 3y^2$

3 $200xy$; $100x^2$

5 $2xe^y$

7 $15a^2t^2$

9 $f_x = 20xe^{3y}$,
  $f_y = 30x^2e^{3y}$.

11 $(1/2)v^2$

13 $(a + b)/2$

15 15; 5; 30

17 No such points exist

19 $(0, 1), (0, -1), (-2, 1)$ and $(-2, -1)$

21 (a) $f_x(1, 1) = 4$; $f_y(1, 1) = 2$
   (b) (III)

23 (a) 3.3, 2.5
   (b) 4.1, 2.1
   (c) 4, 2

25 80; 30; 313 tons
   2.9 tons per worker
   2.6 tons per $25,000

27 (a) Q, R
   (b) Q, P
   (c) P, Q, R, S
   (d) None

29 $f_{xx} = 0$, $f_{xy} = e^y$,
  $f_{yy} = xe^y$, $f_{yx} = e^y$

31 $f_{xx} = 0$, $f_{xy} = -2/y^2$,
  $f_{yy} = 4x/y^3$, $f_{yx} = -2/y^2$

33 $f_{xx} = y^2e^{xy}$,
  $f_{xy} = (xy + 1)e^{xy}$,
  $f_{yy} = x^2e^{xy}$,
  $f_{yx} = (xy + 1)e^{xy}$

35 $f_{xx} = -8t$, $f_{tt} = 6t$,
  $f_{xt} = f_{tx} = -8x$

37 $Q_{p_1p_1} = 10p_2^{-1}$,
  $Q_{p_2p_2} = 10p_1^2p_2^{-3}$,
  $Q_{p_1p_2} = Q_{p_2p_1} = -10p_1p_2^{-2}$

39 $P_{KK} = 0$, $P_{LL} = 4K$,
  $P_{KL} = P_{LK} = 4L$

43 (a) $-(c + 1)B/(c + r)^2$
   (b) Negative

Section 8.5

1 $f(2, 10) \approx 0.5$ local and global min
  $f(6, 4) \approx 9.5$ local max

3 Max: 11 at $(5.1, 4.9)$
  Min: -1 at $(1, 3.9)$

5 Max: 1 at $(\pi/2, 0)$; $(\pi/2, 2\pi)$
  Min: -1 at $(\pi/2, \pi)$

7 Saddle point

9 Saddle point

11 Saddle point

13 Local minimum

15 Local min: $(-2, 0)$

17 Local min: $(-3, 5)$

19 Saddle pt: $(0, -5)$
  Local min: $(2, -5)$

21 Local max: $(0, -1)$
  Saddle pts: $(0, 1), (2, -1)$
  Local min: $(2, 1)$

23 Saddle pts: $(1, -1), (-1, 1)$
  Local max: $(-1, -1)$
  Local min: $(1, 1)$

25 Mississippi:
  $87 - 88$ (max), $83 - 87$ (min)
  Alabama:
  $88 - 89$ (max), $83 - 87$ (min)
  Pennsylvania:
  $89 - 90$ (max), 80 (min)
  New York:
  $81 - 84$ (max), $74 - 76$ (min)
  California:
  $100 - 101$ (max), $65 - 68$ (min)
  Arizona:
  $102 - 107$ (max), $85 - 87$ (min)
  Massachusetts:
  $81 - 84$ (max), 70 (min)

27 $A = 10, B = 4, C = -2$

29 (a) All values of k
   (b) None
   (c) None

31 (a) $P = p_1^2/8 + p_2^2/8 - 10$
   (b) $\partial(\max P)/\partial p_1 = p_1/4$

Section 8.6

1 Min $= -\sqrt{2}$, max $= \sqrt{2}$

3 $f(10, 25) = 250$

5 $f(12, 4) = 240$

7 Min $= \frac{3}{4}$, no max

9 Min $= 11.25$; no max

11 (a) Points A, B, C, D, E
   (b) Point F
   (c) Point D

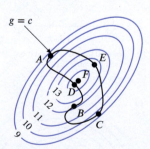

13 (a) $Q = x_1^{0.3}x_2^{0.7}$
   (b) $10x_1 + 25x_2 = 50,000$

15 (b) $L = 40, K = 30$

17 (a) Reduce K by $1/2$ unit,
     increase L by 1 unit.

19 $x = 27.273, y = 72.727, Q = 415.955$

21 (a) 1215
 (b) 1185

23 (a) Grade as function of time on each project

 (b) Total time on projects ≤ 20 hours
 (c) Points per hour
 (d) 21 hours gives 5 more points than 20 hours

25 (a) $W = 225$
  $K = 37.5$
 (b) $W = 225.075$
  $K = 37.513$
  $\lambda = 0.29$

27 1820.04; about 209

Theory: Least Squares

1 $y = 2/3 - (1/2)x$

3 $y = 2/3 - (1/2)x$

5 $y = x - 1/3$

7 (a) (i) Power function
   (ii) Linear function
 (b) $\ln N = 1.20 + 0.32 \ln A$
  Agrees with biological rule

Section 9.1

1 (a) (III)
 (b) (IV)
 (c) (I)
 (d) (II)

3 $dP/dt = kP$, $k > 0$

5 $dQ/dt = kQ$, $k < 0$

7 $dP/dt = -0.08P - 30$

9 $dA/dt = -1$

11 (a) $dA/dt = -0.17A$
 (b) -17 mg

13 (a) Increasing, decreasing
 (b) $W = 4$

15 $dN/dt = B + kN$

Section 9.2

1 (a) Yes
 (b) No

3 $y = t^2 + C$

5 F

7 E

9 A

11 D

13 4, 4, 4, 4

15 12, 18, 27, 40.5

17 74, 78.8, 84.56 million

19 $k = -0.03$ and C is any number, or $C = 0$ and k is any number

21 $k = 5$

Section 9.3

1 Other answers possible

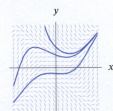

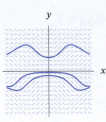

3

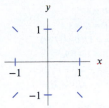

5 (a)

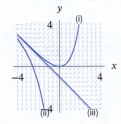

 (b) $y = -x - 1$

7 (a) III
 (b) VI
 (c) V
 (d) I
 (e) IV
 (f) II

9 III: $dP/dt = 3P(1 - P)$

11 For starting points $y > 0$:
  $y \to \infty$ as $x \to \infty$
  For starting points $y = 0$:
  $y = 0$ for all x
  For starting points $y < 0$:
  $y \to -\infty$ as $x \to \infty$

13 As $x \to \infty$, $y \to \infty$

15 $y \to 4$ as $x \to \infty$

Section 9.4

1 $P = 20e^{0.02t}$

3 $y = 5.6e^{-0.14x}$

5 $p = 164.87e^{-0.1q}$

7 (a) $dB/dt = 0.015B$
 (b) $B = 5000e^{0.015t}$
 (c) $5809.17

9 (a) $dB/dt = 0.10B + 1000$
 (b) $B = 10{,}000e^{0.1t} - 10{,}000$

11 Michigan: 72 years
  Ontario: 18 years

13 (a) $dQ/dt = -0.5365Q$
  $Q = Q_0 e^{-0.5365t}$
 (b) 4 mg

15 (a) 69,300 barrels/year
 (b) 25.9 years

17 (b) 2070

Section 9.5

1 $y = 200 - 150e^{0.5t}$

3 $H = 75 - 75e^{3t}$

5 $B = 25 - 5e^{4t}$

7 $B = 25 + 75e^{2-2t}$

11 $dB/dt = 0.08B - 5000$,
  $B = 62{,}500 - 12{,}500e^{0.08t}$,
  Yes, in 20.1 years

13 (a) $dB/dt = 0.07B - 1000$
 (b) $B \approx \$14{,}285.71$
 (c) $B = 14{,}285.71 - (4285.71)e^{0.07t}$
 (d) $B(5) \approx \$8204$
 (e) Balance is $0 in long run

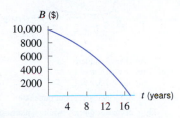

15 (b) $dQ/dt = 43.2 - 0.082Q$
 (c) $Q = 526.8 - 526.8e^{-0.082t}$
  $Q \to 526.8$ mg as $t \to \infty$

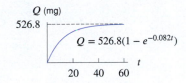

17 (a) $dy/dt = -k(y - a)$
 (b) $y = (1 - a)e^{-kt} + a$
 (c) a: fraction remembered in the long run
  k: relative rate material is forgotten

19 (a) $y = 3$ and $y = -2$
 (b) $y = 3$ is unstable
  $y = -2$ is stable

21 (a) $H = 200 - 180e^{-kt}$
 (b) $k \approx 0.027$ (if t is in minutes)

23 (a) $dT/dt = -k(T - 68)$
 (b) $T = 68 + 22.3e^{-0.06t}$;
  3:45 am.

Section 9.6

1 (a) $x \to \infty$ exponentially
  $y \to 0$ exponentially

 (b) y is helped by the presence of x

3 (a) x population grows exponentially
  y population grows exponentially
 (b) Competitor relationship

5 $dx/dt = -x - xy$,
  $dy/dt = -y - xy$

7 $dx/dt = -x + xy$,
  $dy/dt = y$

11 Symmetric about the line $r = w$;
  solutions closed curves

13 Robins:
  Max ≈ 2500
  Min ≈ 500
  When robins are at a max,
  the worm population is about 1 million

17 (a) $dw/dt = 0$
$dr/dt = 1.2$
(b) $w \approx 2.2$, $r \approx 1.1$
(c) At $t = 0.2$:
$w \approx 2.2$, $r \approx 1.3$
At $t = 0.3$
$w \approx 2.1$, $r \approx 1.4$

19 (a)

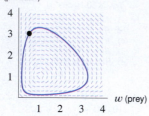

r (predator)

w (prey)

(b) Down and left
(c) $r = 3.3$, $w = 1$
(d) $w = 3.3$, $r = 1$

21 x and y increase, about same rate

23 x decreases quickly while y increases more slowly

25 (a) $dQ_1/dt = A - k_1 Q_1 + k_2 Q_2$
(b) $dQ_2/dt = -k_3 Q_2 + k_1 Q_1 - k_2 Q_2$

Section 9.7

5 (a) $I_0 = 1$, $S_0 = 349$
(b) Increases; spreads

7 About 300 boys;
$t \approx 6$ days

9 5

11 (a) b/a

Theory: Separation of Vars

1 $P = e^{-2t}$

3 $P = \sqrt{2t + 1}$

5 $u = 1/(1 - (1/2)t)$

7 $R = 1 - 0.9e^{1-y}$

9 $z = -\ln(1 - t^2/2)$

11 $y = -2/(t^2 + 2t - 4)$

13 (a) Yes   (b) No   (c) Yes
(d) No   (e) Yes   (f) Yes
(g) No   (h) Yes   (i) No
(j) Yes   (k) Yes   (l) No

15 $Q = b - Ae^{-t}$

17 $R = -(b/a) + Ae^{at}$

19 $y = -1/\left(k(t + t^3/3) + C\right)$

21 (a)

(b)

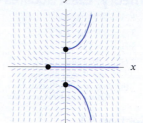

(c) $y(x) = Ae^{x^2/2}$

Cover Design: Maureen Eide/Wiley
Cover Image: © patrickzephyrphoto.com

www.wiley.com/college/hugheshallett

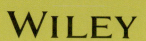